Edited by Gernot Frenking and Sason Shaik

The Chemical Bond

Related Titles

Magnasco, V.

Models for Bonding in Chemistry

2010

Print ISBN: 978-0-470-66702-6
also available in digital formats

Fleming, I.

Molecular Orbitals and Organic Chemical Reactions–Student Edition

2009

Print ISBN: 978-0-470-74659-2
also available in digital formats

Frenking, Gernot/Shaik, Sason (eds.)

The Chemical Bond

Chemical Bonding Across the Periodic Table

2014

Print ISBN: 978-3-527-33315-8
also available in digital formats

Matta, C.F., Boyd, R.J. (eds.)

The Quantum Theory of Atoms in Molecules

From Solid State to DNA and Drug Design

2007

Print ISBN: 978-3-527-30748-7
also available in digital formats

Reiher, M., Wolf, A.

Relativistic Quantum Chemistry

The Fundamental Theory of Molecular Science

2nd Edition

2014

Print ISBN: 978-3-527-33415-5
also available in digital formats

Liddle, S.T. (ed.)

Molecular Metal-Metal Bonds

Compounds, Synthesis, Properties

2014

Print ISBN: 978-3-527-33541-1
also available in digital formats

Edited by Gernot Frenking and Sason Shaik

The Chemical Bond

Fundamental Aspects of Chemical Bonding

WILEY-VCH
Verlag GmbH & Co. KGaA

The Editors

Prof. Dr. Gernot Frenking
Philipps-Universität Marburg
FB Chemie
Hans-Meerwein-Strasse
35032 Marburg
Germany

Prof. Dr. Sason Shaik
Hebrew University
Institute of Chemistry
Givat Ram Campus
91904 Jerusalem
Israel

Library of Congress Card No.: applied for

British Library Cataloguing-in-Publication Data
A catalogue record for this book is available from the British Library.

Bibliographic information published by the Deutsche Nationalbibliothek
The Deutsche Nationalbibliothek lists this publication in the Deutsche Nationalbibliografie; detailed bibliographic data are available on the Internet at <http://dnb.d-nb.de>.

Print ISBN: 978-3-527-33314-1
ePDF ISBN: 978-3-527-66472-6
ePub ISBN: 978-3-527-66471-9
mobi ISBN: 978-3-527-66470-2
oBook ISBN: 978-3-527-66469-6

Cover-Design Adam-Design, Weinheim
Typesetting Laserwords Private Limited, Chennai, India
Printing and Binding betz-druck GmbH, Darmstadt

Printed on acid-free paper

Contents

Preface

One of the fundamental territories of chemistry is the chemical bond, *the glue from which an entire chemical universe is constructed* [1]. The bond serves as a bridge between the apparent magic of chemistry (the chemical transformation) and the way this magic is conceived at present in terms of molecules changing into one another by breaking old bonds and making new ones. For a while it seemed that chemists have, by and large, abandoned their territory as if everything about bonding is known and well understood; the frontier has moved to nano and bio, leaving the original territory untended. This, however, was a wrong impression, as the interest in bonding has quickly revived to be accompanied by many interesting theoretical approaches to probe the origins of bonds, many novel bonding motifs, and even experimental studies that describe imaging of bonds being broken and remade using atomic force microscopy [2]. The bond is becoming again a central intellectual arena, and one can even find allusions to the bond as an elementary particle of chemistry, so-called "*bondon*" [3]. This "return of the bond" has prompted the two editors to edit these two volumes on bonding, and it is only fitting that their publication date is close to the centenary of the Lewis seminal paper on electron-pair bonding.

The idea of "bonding" may even have alchemical origins in the spiritual outlook of matter, where the *conjuctio* or union of the opposites is the ultimate synthesis necessary to drive the change of lower matter to gold [4]. In chemistry, the roots of the bond concept date way back to the efforts of chemists to grapple with the magic of chemistry, the formation substances and their transmutation [5]; these efforts have led to the formulation of "chemical theories" in which "affinities" between materials were considered to unite "more compounded" substances. Interestingly, the "affinity" was not defined as a static property of substances but as a dynamic one, in conformity with the "chemical magic" that a substance exists one time and then disappears making place for a new one.

In 1675, Lemery published his book *Course de Chymie* [6, 7] and used "elective affinities" as terminology that describes the selective replacement of one metal by others in the chemistry of salts [7a]. In 1718, Etienne Francois Geoffroy systematized this phenomenon in his table of *rapports*, namely, affinities. [7b] Historian Ursula Klein [8] credits Geoffroy as the first scientist to generalize the basic concept of modern chemistry – that of the "compound" with "chemical affinity" between the constituents. The tables of affinities became so important to chemists, because they documented the chemical selectivity of the known chemical bodies of the day. We learn of the importance of affinity tables from the story of Swedish chemist Torben Bergman, who assembled thousands of affinity data [9], and presented his table as a gift to the Duke of Parma. Building on Geoffroy's "theory," another Frenchman, Peirre-Joseph Macquer, published in 1749–1751 his two-volume monographs: "*Chimie Theorique*" and "*Chimie Practique,*" [10] in which he gave rules of chemical combination. As cited by the historian Siegfried, these rules sound very modern to our contemporary ear (p. 144 in Ref. [5]).

However, a dramatic change had to occur to put the "elective affinity" into the effective theoretical construct we now call "the chemical bond." This change was induced by three consecutive revolutions. The first revolution is the "compositional revolution" that started with Lavoisier and culminated in the atomic hypothesis of Dalton, which enabled him to identify the mythical term "elementary bodies" with discrete atoms; matter with "quantized weights" (p. 237 in Ref.[5]). This hypothesis has developed into quantitative compositional rules that have led eventually to the notion of the molecule with the work of Cannizzaro who demonstrated the importance of Avogadro's hypothesis (pp. 260–262 in Ref. [5]). Now, the chemical community had in its possession the elements that could bond with one another, and bonding theories started to form gradually and replace the "affinities." There were the "dualistic" electrical theories of Davy and Berzelius (p. 75 in Ref. [9]) [11], which sprung from electrochemistry. A similar theory was delineated soon after by Thomson, the discoverer of the electron, in which bonding arose from the attraction of oppositely charged ions after electron transfer between the atoms [11, 12].

Later, this theory was replaced by a more advanced one [13] where bonds were formed by Faraday "tubes" (lines of force) from one atom to another and vice versa. This has created a vogue of ionic or so-called "electromer" theory, especially amongst American scientists [1]. However, from their onset, these "ionic" type theories have been "falsified" by the fact that most organic compounds were nonpolar (nonionic), and the emergence of the concept 'valence' that united the structuralists [1], Kekulé, Couper, and Butlerov around it. This bonding dichotomy and the urge to generalize chemistry and create an Allgemeine Chemie (a unified chemistry) [14], has eventually led to the second revolution, "the electronic structure revolution," culminated in the Lewis hypothesis of the electron pair bond in 1916, which gave the clue to the nature of the atomic combination [15]. At last, the meaning of the mythical terms "elective affinity" and "elective forces" had an operational and an effective definition that allowed to construct a chemical universe. Lewis did not try to understand the "forces" between the atoms; instead, he let himself be guided by his chemical overview to hypothesize the pairing as *a quantized unit of bonding*, which gave the clue to the nature of the atomic combination [1, 16].

Starting in the late 1920s, the new quantum theory has entered chemistry and ushered the third revolution through the works of Heitler and London [17], and subsequently by Pauling [18], Slater [19], Mulliken and Hund [20], Hückel [21], and so on. Here, we have to digress for a moment and remark about the current tendency to classify chemistry as a service science and a sub-branch of physics [22], presumably so, because it falls under the sovereignty of quantum mechanics. We disagree with the description of the situation in terms of a father science and a submissive science, and have doubts it can be seriously considered by scientists who are familiar with chemistry and its creative elements [23], or even by physicists who wrote against reductionism in science [24]. A more balanced picture would be to think in terms of cross-fertilization between two independent sciences. Thus, despite the origins of quantum mechanics in physics, one must remember that chemists have been the first to employ quantum notions in their science, starting with the quantized weight of matter, going to the quantal building blocks of structure, and the quantal magic numbers, such as octet and duet rules that control the number of bonds an atom will possess. More so, the very first paper of the new quantum chemistry that dealt with the nature of the chemical bond, the paper by Heitler and London [17], was actually a dressing of the Lewis bond by the language of quantum mechanics. Pauling recognized this link and Lewis' priority in a paper [25], and in the enthusiastic letter he wrote to Lewis to let him know that his idea of a bond rests on a firm theoretical ground [26]. Chemistry was not really a submissive passive science in the story of the chemical bond; it was a partner science with a full share [1]!

During the early 1930s, Pauling has imported this emerging theory to chemistry and was referring to it as valence bond (VB) theory and thereby paying homage to the originators of the concept of valency and to his chemical heritage. In his hands, VB theory was a superposition theory of Lewis structures, which was later to be called resonance theory. Indeed, Pauling's work was a theoretical construct that cleverly and faithfully dressed the Lewis notion of a bond with a theoretical guise that could be presented in the new language of the Schrödinger equation in terms of a wave function and a Hamiltonian that extracts the energy of the molecule from this wave function [27]. The Pauling–Lewis theory constituted the localized representation of the chemical world, where the great majority of molecules would be described by a single Lewis structure having a set of localized bonds and lone pairs, whereas the minority of molecules required a few Lewis structures to be represented properly (e.g., benzene). The obvious affinity of resonance theory to the Lewis concept and to the ideas of the Ingold School of Physical Organic Chemistry [28] further entrenched the localized representation of the molecular universe in terms of local bonds and lone pairs. Most chemists still use this representation!

Quite at the same time, another theory [28], nascent from the spectroscopic studies of Hund and Mulliken [20], called molecular orbital (MO) theory was ushered into mainstream chemistry, initially by Mulliken and subsequently by Hückel [21]. Hückel devised a simple topological theory that allowed him to treat large molecules. Following Lennard-Jones' treatment of the dioxygen molecule [29], Hückel applied a clever idea of σ-π separation [21] and focused on the π-electronic structure of conjugated molecules. One of his earliest treatments was benzene and its description in terms of delocalized MOs was the archetypal example of the delocalized paradigm [30]. Here, the π-system of benzene was delocalized, the electrons were distributed all over the molecule in all the MOs, and there was no need to move electrons to and fro as in the localized picture of two resonating

Kekule structures. As discussed by Berson [23], the Hückel MO (HMO) theory was received with suspicion by chemists, who where subscribers of the Lewis–Pauling localized representation of molecular species, and who were not accustomed to the new language of secular determinants and secular equations, nor to the MOs that were spread all over the molecule. Where are my bonds? was the silent outcry of the suspicious chemists ...

While HMO was held back and awaiting its chance (that came in the 1950s), Mulliken and Pauling who have been contemporaries and influential figures were competing on charting the mental map of chemistry [28, 31]. Initially the Pauling school had the upper hand and the localized representation dominated chemical thinking. It must be recalled that, in those days, computationally both MO and VB theories were not very practical tools for investigating molecules, and with the qualitative advantages of Lewis bonding and resonance theory the localized representation seemed a superior concept. Thus, the Lewis-Pauling theory dominated chemical thought mainly because its localized representation was so much reminiscent of the primal structural concepts of chemistry [5] dating from the days of Kekulé (1829–1896), Couper (1831–1892), Frankland (1825–1899), and others through the electron-pair notion and electron-dot structures of Lewis. Pauling himself emphasized that his VB theory is a natural evolution of chemical experience, and that it emerges directly from the chemical conception of the chemical bond. This has made VB-resonance theory appear intuitive and chemically meaningful in terms of structure and valence.

By contrast, MO theory seemed alien to everything chemists had thought about the nature of the chemical bond. Even Mulliken admitted that MO theory departs from the "chemical ideology" [28a]. And to top it all, back at that period, MO theory offered no visual representation that could compete [28] with the resonance hybrid representation of VB-resonance theory, and its direct lineage to the structure of molecules, the heartland of chemistry. At the end of World War II, VB-resonance theory dominated the epistemology of chemists. However, the tide has changed in the 1950s up to the mid 1960s, when several chemical communities took on to the new theory [28]. The conceptual advantage of HMO theory was brought to light by the synthesis of new "aromatic" species, which obeyed the Hückel magic number $4n+2$, while other species with the magic number $4n$ were either unknown or known to be very reactive compared with their $4n+2$ congeners [23]. Books on HMO theory and its application were written and demonstrated the insight of the theory to practicing chemists [32]. The synthesis of molecules like ferrocene [33], for which the MO picture due to Orgel and Dunitz [33b], was so much more elegant than the resonating theoretical description of Pauling reinforced the notion that the MO representation was superior.

At the same time, semiempirical MO methods started to appear which allowed to treat large molecules and to consider σ-electrons as well [28]. Good communicators of MO theory like Coulson [34] and Dewar [35] started to offer visual representations of the delocalized picture and re-chart thereby the mental map of chemistry. The Woodward–Hoffmann Rules [36], and in retrospect also the Frontier MO (FMO) theory of Fukui [37], which patterned a great deal of data on pericyclic reactions and made many predictions that were successfully tested by chemists, highlighted the role of orbital symmetry and hence underscored the eminence of the delocalized representation. Hoffmann and his school have continued and demonstrated the unifying power of the delocalized representation in almost every area of chemistry. The connection of the delocalized MOs (canonical MOs) to observables in spectroscopy (e.g., ionization energies and the work of the late Edgar Heilbronner and Hans Bock [38] and others) has further established a credibility nexus between MO theory and experiment.

This dominance of the MO representation has further intensified with the development of ab-initio computational methods, starting some time in the 1960s but was ripen in the early 1970s when Pople and his collaborators introduced GAUSSIAN 70, which was based on MO theory and later when density functional theory (DFT) was implemented into quantum chemistry software, which enabled the calculation of molecular species and reactions, without any prior assumption (other than the used basis set and functional). The delocalized picture became the consensus and resonance and VB theories were slanted and considered to be archaic, if not simply wrong theories [27, 31].

Despite this seemingly decisive crush, still chemists have been teaching the localized picture in terms of Lewis structures, VSPER [39] (used for the 3D representation of molecules) and resonance theory and using the latter to design syntheses and/or think about reaction mechanisms. Thus, many parts of the chemical community have been functioning with split consciousness between the localized and delocalized worlds. The "conceptual war" has never really subsided, and one of the later reverberations of the struggle between localized and delocalized approaches is the classical–nonclassical controversy regarding the structure of the 2-norbornyl cation [40]. However, this "doubled consciousness" and the "conceptual wars" are arguably advantageous, because as with the Talmudic tradition, these discussions created a new culture of thought in chemistry about molecules and bonding.

The MO-based methods, augmented with electron correlation calculations, and DFT-based methods have gradually improved their computational capabilities to the point that one can now compute enzymes within their proteins and follow the reactivity of their active species [41]. A recent paper on the electronic structure and bonding of the Mn_4O_4Ca water splitting using density matrix renormalization group (DMRG) [42], is a spectacular example of the current potential capability of the MO-based computational methods.

While MO-based computational methods, including DFT, have dominated the way chemists are computing their molecules or reactions, there has been an ever-growing counter-tide of applications of the localized approach to a variety of problems in chemistry. This counter-tide is perhaps a clear message that the localized representation is more easily implemented in human thought process, even if the delocalized representation is more easily implemented by computing machines. Thus, VB theory, has continued to live very well in several communities; those of chemical dynamics [43], photochemistry [44], electron transfer chemistry, and in solid-state chemistry (e.g., conductivity), where the localized representation seemed indispensable [44].

Starting in the late 1970s, there has been an intense surge of concepts that rely on localized representation, and some examples follow: the recent interest in multiple bonding between transition metals [45] emphasized the great insight brought about using the hybridization concepts [45c] to understand unusual structural features of molecules, and the development of natural resonance theory (NRT), which allows to discuss molecular structure and bonding in terms of resonance theory [46]. The powerful "isolobal analogy" [47] uses a localized representation of bonding and moves to and fro between the representations. The reactive-bond orbital (RBO) approach finds the "identity" of the orbitals that are responsible for a given interaction between molecules, and yields good localized orbitals rather than the usual HOMO and LUMO delocalized oribitals [48]. The energy decomposition analyses [49], and the block localized wave function (BLW) method [50] enable to get VB-type information from delocalized wave functions. There is a VB approach that treats chemical reactivity and reaction mechanisms with a localized and semilocalized representations, going back and forth between the representations [51]. VB methods and models allow to derive new ideas on chemical bonding [52]. VB ideas are prominently useful in analyzing enzymatic reactivity [41, 53]. There are thriving theories of photochemical reactions based on conical intersections in the localized representation [54]. All this activity has been accompanied by a steady improvement in the computational capability of VB theory [55]. At present, a VB software such as XMVB [56] is widely available and enables to compute molecules up to the size of $(OC)_4Fe(C_2H_4)$ [55] and to explore bond breaking and bond making in chemical reactions, even in solution. Moreover, the emerging VB methodology has enabled to test many of the qualitative theories on bonding and reactivity [55]. As a result of these parallel developments, we are standing in an exciting crossroads where electronic structure theory is standing on two robust feet. The books we edited reflect this coming together of the two 'cultures.'

The advent of DFT in chemistry and the electron-density-based approaches in the 1970s contributed a powerfully effective method to compute molecules with reasonable accuracy, and at the same time it added a dimension to the MO-VB world of chemistry. On the one hand, DFT relies on the density (ρ), which is a delocalized molecular property and therefore the theory brings a measure of support to the conception of a delocalized chemical universe. On the other hand, in the eyes of many in our community, with DFT the entire simple orbital picture seemed to have collapsed altogether because, as the argument goes: ρ and the energy (E) are "the only real quantities that count." It is true that the Kohn–Sham (KS) orbitals look like MOs [57]; they are delocalized and follow the symmetry of the molecule. However, as one can, in principle, solve the DFT problem without invoking any orbitals at all, this leads to arguing that the density is self-sufficient with no need for orbitals. This, of course, is changing dramatically as chemists increasingly use the DFT method. Furthermore, as the bond concept is immensely useful, most density-only proponents devise density probes for locating bonds such as in AIM and in ELF [58]. In AIM theory, bonds are recognized by bond paths, while in

ELF theory bonds are defined by molecular basins and their population, traced by the Pauli exclusion principle. However, both these probes need to be computed and cannot be foretold without performing the calculations. In some cases, the emerging bond paths defy chemical understanding. What aggravates the situation is that there is a growing school for which the computations itself is the theory with no need for further conceptualization. If one accepts this verdict then, indeed, all our concepts seem to dissolve in the computer. However, as these two volumes show, there is ample room for bonding concepts. As the admonition attributed to Wigner says: *It is nice to know that the computer understands the problem. But I would like to understand it too.*

This two-volume project shows the immense vitality of the bond concept, and it tries encompassing the richness of the bond concept by presenting different perspectives on bonding. Volume 1 contains 11 chapters, which provide the theoretical frameworks of the various perspectives on bonding. This volume lays the foundations for the applications across the periodic table. This is done in Volume 2, which includes 18 chapters that describe applications to chemical bonding going from main group elements, through transition metals, rather earths, clusters and solids, and all the way to the weak interaction. The two volumes are not written for specialists in theoretical chemistry and, therefore, mathematical and technical details of the methods are kept to a minimum. Readers who want to learn about the background of the theoretical procedures may find this in the original literature, which is cited in the chapters.

The chapters are meant to attract the interest of the practicing chemist, teachers, and advanced students who may want to learn about the present understanding of chemical bonding. The authors were asked to present a didactic introduction into their topic and to discuss the results in a lucid style with a teaching attitude, which appeals to senior chemists and graduate students alike. We hope that the readers share our view that this goal has been achieved in the 29 chapters of "The Chemical Bond."

Jerusalem, Israel — *Sason Shaik*
Marburg, Germany — *Gernot Frenking*

References

1. Shaik, S. (2007) *J. Comput. Chem.*, **28**, 51.
2. de Oteyza, D.G., Gorman, P., Chen, Y.-C., Wickenburg, S., Riss, A., Mowbray, D.J., Etkin, G., Pendramrazi, Z., Tsai, H.-Z., Rubio, A., Crommie, M.F., and Fischer, F.R. (2013) *Science*, **340**, 1434.
3. Putz, M.V. (2010) *Int. J. Mol. Sci.*, **11**, 4227.
4. Roob, A. (1997) in *The Hermetic Museum: Alchemy and Mysticism* (ed. S. Whiteside), Taschen Verlag GmbH, London, English Translation, pp. 132–133.
5. Siegfried, R. (2002) *From Elements to Atoms: A History of Chemical Composition*, American Philosophical Society, Philadelphia, p. 16.

6. Lemerey, N.A. (1720) *Course of Chymistry* 4th English Edition translated from the French edition of 1715, London.
7. (a) Siegfried, R. (2002) *From Elements to Atoms: A History of Chemical Composition*, American Philosophical Society, Philadelphia, p. 79, 83, and 94; (b) Siegfried, R. (2002) *From Elements to Atoms: A History of Chemical Composition*, American Philosophical Society, Philadelphia, p. 76, 93–96.
8. Klein, U. (1995) *Ambix*, **42**, 79.
9. Bensaude-Vincent, B. and Stengers, I. (1996) in *A History of Chemistry* (ed. D. van Dam), Harvard University Press, Cambridge, English translation, pp. 69–70.
10. Macquer, P.-J. (1758) in *Elements of the Theory and Practice of Chemistry*, (ed. A. Reid) English translation, vol. 1, 2, London.
11. Noyes, W.A. (1917) *J. Am. Chem. Soc.*, **39**, 879.
12. Chayut, M. (1991) *Ann. Sci.*, **48**, 527.
13. Thomson, J.J. (1914) *Phil. Magaz. J. Sci.*, **27**, 757.
14. (a) Servos, J.W. (1990) *Physical Chemistry from Ostwald to Pauling*, Princeton University Press, Princeton, pp. 4–5; (b) Servos, J.W. (1984) *J. Chem. Educ.*, **61**, 5.
15. Lewis, G.N. (1916) *J. Am. Chem. Soc.*, **38**, 762.
16. Calvin, M. (1984) *J. Chem. Educ.*, **61**, 14.
17. Heitler, W. and London, F. (1927) *Zeits. für Physik.*, **44**, 455.
18. Pauling, L. (1928) *Proc. Natl. Acad. Sci. U. S. A.*, **14**, 359.
19. Slater, J.C. (1931) *Phys. Rev.*, **37**, 481.
20. (a) Mulliken, R.S. (1931) *Chem. Rev.*, **9**, 347; (b) Hund, F. (1931) *Zeits. für Physik*, **73**, 1.
21. Hückel, E. (1930) *Zeits. für Physik*, **60**, 423.
22. Knight, D. (1992) *Ideas in Chemistry: A History of Science*, Rutgers University Press, New Braunscwick, NJ.
23. Berson, J.A. (1999) *Chemical Creativity*, Wiley-VCH Verlag GmbH, Weinheim.
24. Anderson, P.W. (1972) *Science*, **177**, 393.
25. Pauling, L. (1935) *The Nature of the Chemical Bond*, Cornell University Press, Ithaca, NY 1939, p. xii and dedication page.
26. Saltzman, M.D. (1996) *Bull. Hist. Chem.*, **19**, 25.
27. Shaik, S. and Hiberty, P.C. (2008) *A Chemist's Guide to Valence Bond Theory*, John Wiley & Sons, Inc., Hoboken ch. 1.
28. (a) Brush, S.G. (1999) *Stud. Hist. Phil. Sci.*, **30**, 21; (b) Brush, S. G. (1990) *Stud. Hist. Phil. Sci.*, **30**, 263.
29. Lennard-Jones, J.E. (1929) *Trans. Faraday Soc.*, **25**, 668.
30. Hückel, E. (1931) *Zeits. für Physik*, **70**, 204.
31. Shaik, S. and Hiberty, P.C. (2004) *Rev. Comp. Chem*, **20**, 1.
32. (a) Roberts, J.D. (2004) *Acc. Chem. Res.*, **37**, 417; (b) Streitwieser, A. (2004) *Acc. Chem. Res.*, **37**, 419.
33. (a) Laszlo, P. and Hoffmann, R. (2000) *Angew. Chem. Int. Ed.*, **39**, 123; (b) Orgel, L.E. and Dunitz, J.D. (1953) *Nature*, **171**, 121.
34. Coulson, C.A. (1952) *Valence*, Oxford University Press, London.
35. Dewar, M.J.S. (1949) *Electronic Theory of Organic Chemistry*, Clarendon Press, Oxford.
36. Woodward, R.B. and Hoffmann, R. (1971) *The Conservation of Orbital Symmetry*, Verlag Chemie, Weinheim.
37. Fukui, K. (1972) *Theory of Orientation and Stereoselection*, Springer-Verlag, Heidelberg.
38. Heilbronner, E. and Bock, H. (1976) *The HMO Model and its Applications*, John Wiley & Sons, Inc., New York.
39. Gillespie, R.J. and Popelier, P.L.A. (2001) *Chemical Bonding and Molecular Geometry*, Oxford University Press, London.
40. Bunnett, J.F. (1983) *Acc. Chem. Res.*, **16**, 425.
41. Shaik, S., Cohen, Y., Wang, H., Chen, D.K., and Thiel, W. (2010) *Chem. Rev.*, **110**, 949.
42. Kurashige, Y., Chan, G.K.-L., and Yanai, T. (2013) *Nat. Chem.*, **5**, 660.
43. Truhlar, D.G. (2007) *J. Comput. Chem.*, **28**, 73.
44. Shaik, S. (2007) *New J. Chem.*, **31**, 2015.
45. (a) Roos, B.O., Borin, A.C., and Gagliardi, L. (2007) *Angew. Chem. Int. Ed.*, **46**, 1469; (b) Frenking, G. and Toner, R. (2007) *Nature*, **446**, 276;

(c) Weinhold, F. and Landis, C. (2007) *Science*, **316**, 61.

46. Weinhold, F. and Landis, C. (2005) *Valency and Bonding: A Natural Bond Orbital Donor-Acceptor Perspective*, Cambridge University Press, Cambridge.
47. Hoffmann, R. (1982) *Angew. Chem. Int. Ed. Engl.*, **21**, 711.
48. Hirao, H. (2007) *Chem. Phys. Lett.*, **443**, 141.
49. (a) Morokuma, K. (1976) *Int. J. Quantum Chem.*, **10**, 325; (b) Ziegler, T. and Rauk, A. (1979) *Inorg. Chem.*, **18**, 1558; (c) van Zeist, W.-J. and Bickelhaupt, F.M. (2010) *Org. Biomol. Chem.*, **8**, 3118–3127; (d) Von Hopffgarten, M. and Frenking, G. (2012) *Wires Comput. Mol. Sci.*, **2**, 43–62.
50. Mo, Y. and Peyerimhoff, S.D. (1998) *J. Chem. Phys.*, **109**, 1687.
51. Shaik, S. and Shurki, A. (1999) *Angew. Chem. Int. Ed.*, **38**, 586.
52. Shaik, S., Danovich, D., Silvi, B., Lauvergnat, D., and Hiberty, P.C. (2005) *Chem. Eur. J.*, **11**, 6358.
53. Braun-Sand, S., Olsson, M.H.M., and Warshel, A. (2005) *Adv. Phys. Org. Chem.*, **40**, 201.
54. Robb, M.A., Garavelli, M., Olivucci, M., and Bernardi, F. (2000) *Rev. Comput. Chem.*, **15**, 87.
55. Wu, W., Su, P., Shaik, S., and Hiberty, P.C. (2011) *Chem. Rev.*, **111**, 7557.
56. Song, L., Wu, W., Mo, Y., and Zhang, Q. (2003) *XMVB: an ab initio non-orthogonal valence bond program*, Xiamen University, China.
57. Stowaser, R. and Hoffmann, R. (1999) *J. Am. Chem. Soc.*, **121**, 3414.
58. (a) Bader, R.F.W. and Nguyen-Dang, T.T. (1981) *Adv. Quantum Chem.*, **14**, 63; (b) Silvi, B. and Savin, A. (1994) *Nature*, **371**, 683.

List of Contributors

Paul W. Ayers
McMaster University
Department of Chemistry &
Chemical Biology
1280 Main Street West
Hamilton, ON L8S 4M1
Canada

F. Matthias Bickelhaupt
VU University Amsterdam
Department of Theoretical
Chemistry and Amsterdam
Center for Multiscale Modeling
De Boelelaan 1083
1081 HV Amsterdam
The Netherlands

and

Radboud University Nijmegen
Institute for Molecules and
Materials
Heyendaalseweg 135
6525 AJ Nijmegen
The Netherlands

David Danovich
The Hebrew University
Department of Organic
Chemistry the Lise-Minerva
Center for Computational
Chemistry
91904 Jerusalem
Israel

Georg Eickerling
Universität Augsburg
Institut für Physik
Universitätsstrasse 1
86159 Augsburg
Germany

Andreas Fischer
Universität Augsburg
Institut für Physik
Universitätsstr. 1
86159 Augsburg
Germany

Gernot Frenking
Philipps-Universität Marburg
Fachbereich Chemie
Hans-Meerwein-Strasse
35043 Marburg
Germany

Paul Geerlings
Vrije Universiteit Brussel (VUB)
Eenheid Algemene Chemie
(ALGC)
Pleinlaan 2
1050 Brussels
Belgium

Yuri Grin
Max-Planck-Institut für
Chemische Physik fester Stoffe
Nöthnitzer Str. 40
01187 Dresden
Germany

Philippe C. Hiberty
Université de Paris-Sud
Laboratoire de Chimie Physique
Bat. 349 UMR CNRS 8000
Groupe de Chimie Théorique
91405 Orsay Cédex
France

Joseph Ivanic
Frederick National Laboratory for
Cancer Research
Advanced Biomedical Computing
Center
Leidos Biomedical Research, Inc.
Frederick, MD 21702
USA

Clark R. Landis
University of Wisconsin
Department of Chemistry
1101 University Avenue
Madison, WI 53706
USA

Yirong Mo
Western Michigan University
Department of Chemistry
1903 West Michigan Avenue
Kalamazoo, MI 49008
USA

Paul Lode Albert Popelier
University of Manchester
Manchester Institute of
Biotechnology (MIB)
131 Princess Street
Manchester M1 7DN
UK

and

University of Manchester
School of Chemistry
Oxford Road
Manchester M13 9PL
UK

Frank De Proft
Vrije Universiteit Brussel (VUB)
Eenheid Algemene Chemie
(ALGC)
Pleinlaan 2
1050 Brussels
Belgium

Klaus Ruedenberg
Iowa State University
Department of Chemistry and
Ames Laboratory USDOE
Ames, IA 50011
USA

Andreas Savin
Sorbonne Universités
UPMC Univ Paris 06, UMR 7616
Laboratoire de Chimie Théorique
4 Place Jussieu
75005 Paris
France

Wolfgang Scherer
Universität Augsburg
Institut für Physik
Universitätsstrasse 1
86159 Augsburg
Germany

Michael W. Schmidt
Iowa State University
Department of Chemistry and
Ames Laboratory USDOE
Ames, IA 50011
USA

Peter Schwerdtfeger
Massey University Auckland
Center for Theoretical Chemistry
and Physics (CTCP)
New Zealand Institute for
Advanced Study
Bob Tindall Building
0632 Auckland
New Zealand

and

Philipps-Universität Marburg
Fachbereich Chemie
Hans-Meerwein-Strasse
35032 Marburg
Germany

Sason Shaik
The Hebrew University
Department of Organic
Chemistry and the Lise-Minerva
Center for Computational
Chemistry
91904 Jerusalem
Israel

Bernard Silvi
CNRS, UMR 7616
Laboratoire de Chimie Théorique
4 Place Jussieu
75005 Paris
France

Frank Weinhold
University of Wisconsin
Department of Chemistry
1101 University Avenue
Madison, WI 53706
USA

Wei Wu
Xiamen University
The State Key Laboratory
of Physical Chemistry
of Solid Surfaces
Fujian Provincial Key Laboratory
of Theoretical and
Computational Chemistry
and College of Chemistry and
Chemical Engineering
Xiamen
Fujian 361005
China

1
The Physical Origin of Covalent Bonding

Michael W. Schmidt, Joseph Ivanic, and Klaus Ruedenberg

"I believe the chemical bond is not so simple as some people seem to think."
Robert S. Mulliken as quoted approvingly by Charles A. Coulson
Reviews Modern Physics ***32****, 177 (1960)*

1.1
The Quest for a Physical Model of Covalent Bonding

Up to about the seventeenth century atomists believed that there were mechanical hookups between atoms. Toward the end of that century Isaac Newton surmised that, analogous to the gravitational forces between masses, there are additional forces between atoms that are attractive at large distances and repulsive at short distances. Around 1810 Berzelius, involved in the development of electrolysis using Volta's recently discovered direct current, conjectured bonding to be due to electrostatic forces by virtue of the same permanent positive or negative charge being distributed on each atom of an element (e.g., H^+, O^-). However, in 1811 Avogadro proposed molecules like H_2 and O_2 to account for volume relationships in gas reactions, a hypothesis that implied the existence of other kinds of bonding forces between atoms. His view was confirmed by the mid-nineteenth century through the development of organic chemistry, which established what is now called *covalent bonding* [1]. In 1881 Helmholtz raised the question of how long-range electrostatic interactions could give rise to short-range bonding forces [2]. After Thomson's discovery of the electron in 1897, it was widely assumed that electrons were involved in bonding. On the basis of chemical evidence, Abegg [3] identified in 1904 what are now called positive and negative oxidation states and deduced that their maximal positive and negative values add up to 8 for each element in the second and third row of the periodic table. He inferred essentially that a special stability must be associated with an electron octet around an atom. Presuming this stability to be a driving force, in 1916 Kossel [4] rationalized ionic bonding by electron transfer whereas, also in 1916, Lewis [5] posited that covalent bonds are achieved by electron sharing.

The Chemical Bond: Fundamental Aspects of Chemical Bonding, First Edition.
Edited by Gernot Frenking, Sason Shaik.

In 1927, Heitler and London [6] showed that wave mechanics yields the covalent bond in the hydrogen molecule whereas Burreau [7] obtained covalent binding for the hydrogen molecule ion. In the subsequent years, these two systems were calculated with increasing accuracy, culminating in the work of James and Coolidge [8] who, in 1933, obtained the binding energy of H_2 within 0.6 kcal mol^{-1}. Today we know that wave mechanics does indeed yield chemical bonding in all molecules for which sufficiently accurate calculations have been made.

However, while the connection between calculation and conceptual physical reasoning is direct and immediate in classical mechanics, this is no longer the case in quantum mechanics. The question arose, therefore, how *conceptual physical reasoning* can be associated with the bond formation that is found in wave mechanical *computations*. Several intuitive answers to this question were proposed in the thirties. From a rather formalistic point of view, bonding was attributed to postulated "exchange forces." By analogy with the coupling of pendulums, "resonance" effects were posited. The recognition that bonding is associated with an accumulation of charge in the bond region led to the conjecture that the electrostatic attraction between this accumulated charge and the adjacent nuclei generates the lowering of the energy. This suggestion by Slater [9] seemed moreover to be in accordance with the virial theorem, which states that the potential energy decreases and the kinetic energy increases upon bond formation. Hellmann [10], on the other hand, advanced the view that bonding is caused by the lowering of the kinetic energy that results from electrons being able to roam over a larger area in a molecule than in an atom, similar to the effect of increasing the box length for a particle in a box. (He could not reconcile, however, the apparent inconsistency of this explanation with the virial theorem.) None of these conceptual conjectures were rigorously pursued in quantitative detail.

The present approach follows the theoretical tradition that, when a rigorous fundamental mathematical formulation with verified quantitative implications exists, then the choice of explanatory physical concepts is limited by having to reflect closely the relationships that are inherent in the theoretical framework. For the present problem, the challenge is to cast the exact energy expression, generated by accurate electronic wave functions, in a form that allows a rigorous resolution into physically interpretable parts and to find a fundamental principle that guides their interactions. The first coherent analysis of this kind was advanced and applied in the 1960s by one of the present authors and his coworkers [11]. It was based on an energy decomposition analysis of general *ab initio* wave functions of electrons in molecules and it identified the physical interactions that establish the bonds in H_2 and H_2^+. This line of reasoning was subsequently pursued by a number of authors [12]. Recent work by two of the present authors [13] as well as by Bacskay and Nordholm [14] has shed additional light on the problem.

In the present chapter, this approach is developed further. It is then used to examine the covalent bonds of H_2^+ and H_2 in detail and extended to the many-electron molecules B_2, C_2, N_2, O_2, and F_2. These analyses show that covalent bonding involves a synergism between several interactions with quite different physical attributes and quantitative characteristics. On the one hand, the lowering of the energy that establishes the bond is the result of a variational competition between the kinetic energy and potential energy. On the other hand, there occurs an intricate interplay between various intra-atomic and interatomic interactions. These basic agents have, moreover, to accommodate electron correlation. It emerges that, in all cases, the driving force of covalent bond formation is the lowering of the kinetic energy gained by the delocalization of electronic waves over more than one atom. This observation is only superficially discordant with the virial theorem which, as mentioned earlier, requires the molecule to have a higher total kinetic energy than the separated atoms. The in-depth accounting of all interconnections between the various interactions shows that the information disclosed by the *actual* total kinetic and potential energies *per se* is insufficient for drawing any inferences regarding the *origin* of covalent bonding.

1.2 Rigorous Basis for Conceptual Reasoning

The aim of the present analysis is to understand bonding features of potential energy surfaces (PESs), that is, the Born–Oppenheimer separation is assumed. Although the study is limited to ground states, the reasoning also applies to excited states when the consequences of the additional constraints of orthogonality to the lower states are accounted for.

1.2.1 Physical Origin of the Ground State

Bonding on a PES is a consequence of the *geometry dependence* of the electronic energy, specifically that this energy is lower for the molecular equilibrium geometry than for the separated atoms. Therefore, *the first objective must be to develop a conceptual understanding of the physical factors that determine the ground state energy of a system of electrons in the electrostatic field generated by fixed nuclei.*

A rigorous quantum mechanical basis for such an understanding is provided by the variation principle which states that, for all possible normalized electronic wave functions Ψ, the energy integral

$$\mathbf{E}(\Psi) = \int d\tau\, \Psi \mathcal{H} \Psi = \int d\tau\, \Psi \mathcal{T} \Psi + \int d\tau \Psi \mathcal{V} \Psi = \mathbf{T}(\Psi) + \mathbf{V}(\Psi) \qquad (1.1)$$

assumes the *lowest possible* value when Ψ is the ground state Ψ_g. Here, $\mathcal{T}$ and $\mathcal{V}$ are the kinetic and potential energy operators. Hence:

> The shape of Ψ_g is determined by the adjustments needed to minimize $\mathbf{T}(\Psi) + \mathbf{V}(\Psi)$.

The conceptual physical interpretation of $\mathbf{V}$ is self-evident: Its terms represent electron nuclear attractions, electron–electron repulsions and internuclear repulsions, all of which are classical electrostatic concepts. In systems with *bound* electrons the electron nuclear attractions dominate the potential energy integral V, which is therefore negative. Manifestly,

> The negative potential energy V is lowered by localizing Ψ in regions of low $\mathscr{V}$, in particular by contracting Ψ towards the nuclei.

It is with regard to the kinetic energy T that quantum mechanics differs fundamentally from classical mechanics and that a new type of concept has to be added to the physical reasoning. To this end, it is advantageous to transform the kinetic energy integral T into the form:

$$\mathrm{T} = \int \mathrm{d}\tau\, \Psi \mathscr{T}\, \Psi = -\frac{1}{2}\sum_k \int \mathrm{d}\tau\, \Psi (\nabla_k)^2 \Psi = +\frac{1}{2}\sum_k \int \mathrm{d}\tau\, (\nabla_k \Psi)^2 \tag{1.2}$$

where the sum $\sum_k$ goes over all electrons k. (Atomic units are used.) If Ψ is expressed in terms of its natural orbitals ψ_n and occupation numbers N_n, then T becomes

$$\mathrm{T} = -\frac{1}{2}\sum_n \mathrm{N}_n \int \mathrm{d}\tau\, (\psi_n \nabla^2 \psi_n) = \frac{1}{2}\sum_n \mathrm{N}_n \int \mathrm{d}\tau\, (\nabla \psi_n)^2 \tag{1.3}$$

The general derivations and conclusions in the subsequent analyses do not depend on whether the Laplacian or the gradient form of T is used because only the invariant integrated expectation values are involved in the inferences. In the equations, though, the gradient expression will be preferred for didactic reasons as a positive kinetic energy term always appears with a positive sign in the formulas. In some instances, the gradient expression will, moreover, prove elucidative for understanding properties of certain kinetic energy integrals by relating them to properties of the wave function. This is because, in the gradient expression of Eq. (1.2), *every* electron k makes a *positive* contribution to *every* volume element $\mathrm{d}\tau$, namely *the square of the gradient of* Ψ with respect to the individual electron coordinates. Analogously, in the gradient expression of Eq. (1.3), every orbital makes a positive contribution to every volume element. Most notably, the following fundamental property of the kinetic energy is readily deduced from these attributes of the gradient form.

Consider a single electron where $\Psi = \psi$ and

$$\mathrm{T} = \frac{1}{2}\int \mathrm{d}\tau\, (\nabla\psi)^2 = \frac{1}{2}\int \mathrm{d}x \int \mathrm{d}y \int \mathrm{d}z \left[\left(\frac{\partial\psi}{\partial x}\right)^2 + \left(\frac{\partial\psi}{\partial y}\right)^2 + \left(\frac{\partial\psi}{\partial z}\right)^2\right] \tag{1.4}$$

The following is manifest from this expression. When ψ is squeezed into a smaller space, then the normalization condition $\int \mathrm{d}\tau\, \psi^2 = 1$ requires that the maximum of ψ increases and, consequently, that the average of the gradient of ψ increases. Hence, localizing ψ will increase T. This inference is related to the uncertainty relation, which is derived in quantum mechanics texts. By virtue of the additive decomposition (Eq. (1.3)), this correlation is general and fundamental:

> Localizing an electronic wave function increases its positive kinetic energy. In the context of the variation principle, electrons therefore possess an innate drive towards delocalizing the wave function.

It is apparent that the kinetic energy and the potential energy place *opposing demands* on the wave function with regard to the aim of minimizing the total energy **E** = **T** + **V**: The kinetic energy would be minimized (viz → 0) by ultimate dilution of Ψ, and the potential energy would be minimized (viz → −∞) by complete concentration onto the nuclear centers. Hence,

> The variational energy minimum is reached by that wave function Ψ_g that achieves the optimal compromise in the variational competition between the electrostatic potential energy pull, which favors localizing contraction towards the nuclei, and the kinetic energy pressure, which drives towards delocalizing dilution.

The *variational process* can therefore be summarized as follows:

> The ground state wave function is determined by the electrostatic attractions pulling the electron cloud as close towards the nuclei as permitted by the resistance of the kinetic energy.

An additional constraint exists in the case of many electrons in that the wave function Ψ must then be antisymmetric with respect to the exchange of electrons, which entails the restriction:

> The simultaneous localization of several electrons in the same space is curbed by the exclusion principle.

It should be appreciated that the quantum mechanical variation principle for the *energy*, that is, (**T** + **V**), lends itself more easily to conceptual visualization than the variation principle of least action, that is, $\int dt(\mathbf{T} - \mathbf{V})$, does in classical mechanics. This vantage may be considered a compensation for the absence of the kind of direct simple conceptual physical interpretation of calculations that classical mechanics offers.

1.2.2
Physical Origin of Ground State Energy Differences

The prototype example for the described variational competition is the case of an electron-like particle of mass m in the field of a nucleus of charge Z with the Hamiltonian $\mathcal{H} = -(1/2m)\nabla^2 - Z/r$. Consider the normalized trial wavefunction

$$\psi = 1s_\zeta = \left(\frac{\zeta^3}{\pi}\right)^{\frac{1}{2}} \exp(-\zeta r) = \left(\frac{1}{\alpha^3\pi}\right)^{\frac{1}{2}} \exp\left(\frac{-r}{\alpha}\right) \tag{1.5}$$

Here, $\mathcal{H}$ and ψ are in atomic units. The parameter $\alpha = 1/\zeta$ is a measure of the localization of the orbital in as much as the sphere with the radius $R = 2\alpha$ includes

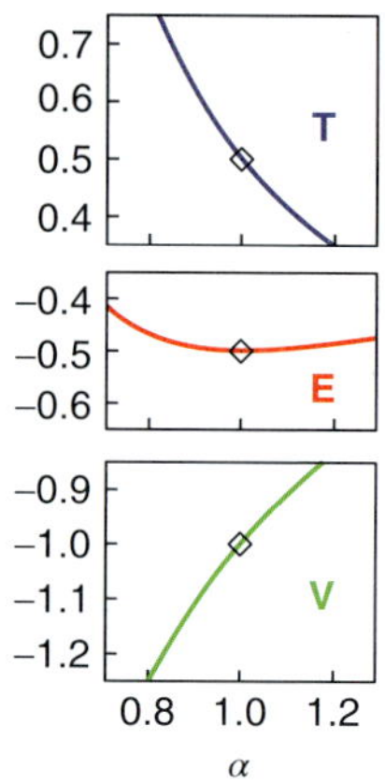

Figure 1.1 Variational competition between kinetic (blue) and potential (green) energies, and the optimal compromise for the total energy (red) that determines the ground state of the hydrogen atom. Ordinate = energy in hartree. Abscissa = orbital size as measured by the inverse orbital exponent $\alpha = 1/\zeta$ in Eq. (1.5).

94% of the orbital density ψ^2. The kinetic and potential energy integrals of ψ become

$$\mathbf{T} = \frac{\zeta^2}{2m} = \frac{1}{2}m\alpha^2 \qquad \mathbf{V} = -\zeta Z = \frac{-Z}{\alpha} \tag{1.6}$$

and they are plotted as functions of the localization parameter α in Figure 1.1. Indeed they exhibit the behavior discussed in the preceding section and the plot of the total energy $(\mathbf{T}+\mathbf{V})$ exhibits the variational competition between $\mathbf{T}$ and $\mathbf{V}$. The minimum, that is, the optimal compromise between these two functions occurs for

$$\zeta_g = mZ, \quad \mathbf{T}_g = \frac{1}{2}mZ^2, \quad \mathbf{V}_g = -mZ^2, \quad \mathbf{E}_g = -\frac{1}{2}mZ^2 \tag{1.7}$$

For the hydrogen atom, where $m = 1$ au and $Z = 1$ au, Eq. (1.7) yields the ionization potential 0.5 hartree. The physical origin of the magnitude of this fundamental quantum chemical quantity is thus the variational compromise between $\mathbf{T}$ and $\mathbf{V}$ in hydrogen.

Atomic energy units will be used in this chapter. The relation to units often used in experimental work is: 1 millihartree = 1 mh = 2.6255 kJ mol^{-1} = 0.62750 kcal mol^{-1}.

Consider now cases with a different nuclear charge Z and a different particle mass m (note, e.g., that the pi-muon has the same charge as the electron but is 180 times heavier) and let us examine *why* such systems can have *lower* ground state energies than the hydrogen atom. From the expression for $\mathbf{E}_g$ in Eq. (1.7) it is apparent that *there can be two different reasons*: (i) The nuclear charge Z is larger than the proton charge or (ii) the mass m of the particle is heavier than the electron mass.

This distinction exhibits an important aspect of the variational competition that is relevant for the elucidation of energy differences between related systems. It is illustrated in Figure 1.2, which shows the variational competition and the shift of the optimal compromise for several systems of the kind described by Eq. (1.6). To

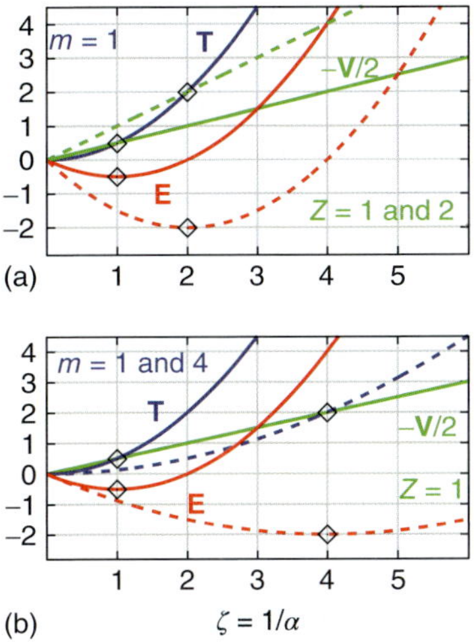

Figure 1.2 Variational competition and optimal compromise between kinetic and potential energies for hydrogen-like systems (see Eq. (1.6)). The solid curves correspond to the standard hydrogen atom, $m=1$ for **T** and $Z=1$ for **V**. The dashed curves of **T**, **V**, and **E** correspond to different choices of m and Z, viz: $Z=2$ for **V** in (a) and $m=4$ for **T** in (b). Abscissa = orbital exponent $\zeta=1/\alpha$, implying the *inverse* of the orbital size. Ordinate = energy in hartree; note that $-\mathbf{V}/2$ is plotted. Diamond markers indicate the positions of the variational compromises.

display clearly the pertinent features, the energies **E**, **T**, and $|\mathbf{V}/2|$ are plotted and the orbital exponent $\zeta=1/\alpha$ is used as abscissa. Consequently, $|\mathbf{V}/2|$ is a straight line and **T** is a quadratic. *Note that, in Eq. (1.7), the minimum of the energy always occurs at that value of* ζ *where* $\mathbf{T}=|\mathbf{V}/2|$*, that is, where the straight line* $|\mathbf{V}/2|$ *intersects the quadratic* **T**. *This is a general constraint whose rigorous origin will be discussed in the next section.*

Figure 1.2a shows the cases for $\{m=1, Z=1\}$ and $\{m=1, Z=2\}$. Changing the nuclear charge Z from 1 to 2 increases the slope of the line $|\mathbf{V}/2|$ so that it intersects the quadratic **T** at a larger value of ζ and a lower energy than in hydrogen. The variational interpretation is as follows: Increasing Z *strengthens the nuclear electrostatic pull*. In the presence of the same kinetic resistance as in hydrogen, the orbital is pulled *closer* to the nucleus than in hydrogen and the energy is thereby lowered.

Figure 1.2b shows the cases $\{m=1, Z=1\}$ and $\{m=4, Z=1\}$. According to Eq. (1.6), changing the mass m from 1 to 4 decreases the curvature of **T** so that the intersection of $|\mathbf{V}/2|$ with **T** occurs again at a larger value of ζ and a lower energy than in hydrogen. The variational interpretation is now as follows: Increasing the mass m *weakens the kinetic resistance* because the mass is in the denominator of the expression for **T**. Because the nuclear pull is the same as in hydrogen, the orbital is again pulled *closer* to the nucleus than in hydrogen and the energy is again lowered. Thus:

> The difference between the ground state energies of two systems of bound electrons can be due to a difference in the strength of the attraction towards the regions of low potential energy or it can be due to a difference in the strength of the kinetic resistance towards localization in the regions of low potential energy.

It is of course possible that both factors are simultaneously operative.

Moreover, the discussed examples exhibit another important fact. According to Eq. (1.7), the two systems $\{m=1, Z=2\}$ and $\{m=4, Z=1\}$ have the same values of $\mathbf{E}_g = -2$, $\mathbf{T}_g = 2$ and $\mathbf{V}_g = -4$ hartree. These equalities show that it is *not* possible to tell *why* either of these systems has a lower energy than hydrogen when the *only* information available are the values of $\mathbf{E}_g$, $\mathbf{T}_g$, and $\mathbf{V}_g$. Thus,

> An analysis of the physical origin of energy differences between systems requires more information than is furnished by the values of $\mathbf{E}_g$, $\mathbf{T}_g$ and $\mathbf{V}_g$.

Indeed, the preceding *variational analysis* on the basis of Figure 1.2 made use of the knowledge of *comparison values* for **E**, **T**, and **V**, in addition to those at the variational minimum.

1.2.3
Relation between Kinetic and Potential Energies

As noted in the preceding section, the optimized kinetic and potential energies of the ground states in these systems always stand in the constant ratio $\mathbf{V}_g/\mathbf{T}_g = -2$. This relationship, which is known as the virial theorem, in fact, holds rigorously for the *many* electron wave functions of *any* atom. It is a general theoretical consequence of the fact that the kinetic energy scales as (inverse length)^2 whereas the potential energy scales as (inverse length). Furthermore, it is important that this relationship is not only valid for the actual solution of the Schrödinger equation (which is optimized with respect to *all possible* variations) but also *for approximate wave functions that are merely optimized with respect to a single scale parameter for all electrons* [15]. If a many-electron wave function is expressed as a superposition of determinants of atomic orbitals (AOs), then *individual* scaling of the various AO exponents will guarantee the virial relation *a fortiori*. When *sufficiently large* basis sets with fixed exponents are employed in a calculation, then the exponential optimization of AOs will be mimicked by the linear combination of atomic orbital (LCAO) expansions of the molecular orbitals (MOs).

In a diatomic molecule, the virial relationship has the general distance-dependent form [9]

$$2\mathbf{T} + \mathbf{V} + R\left(\frac{d\mathbf{E}}{dR}\right) = 0 \tag{1.8}$$

If an approximate molecular wave function is constructed from AOs, say $\chi_{An}[\zeta_{An}(r - R_A)]$, where χ_{An} denotes orbital n on atom A at the position R_A with orbital exponent ζ_{An}, then the general virial theorem results from the following optimization [15]: Replace all orbital exponents by $s\zeta_{An}$ and optimize with respect to the one global scale parameter s. Again, optimization with respect to *all individual* orbital exponents, or a common scale parameter of groups of orbital exponents, will guarantee the virial relation *a fortiori*. These kinds of scaling manifestly represent *simultaneous shrinking or swelling with respect to all atoms* in the molecule. (Equation (1.8) can be generalized to polyatomic systems.)

It follows from Eq. (1.8) that the relation found for atoms, viz

$$2\mathbf{T} = -\mathbf{V} = |\mathbf{V}| \tag{1.9}$$

is in fact also valid *at all points where the energy gradient vanishes,* notably at equilibrium geometries and transition states. A corollary is that the relation is also valid for energy *differences such as binding energies and activation energies.* At these critical points, the satisfaction of Eq. (1.9) is a simple indicator as to whether a given wave function possesses the following property.

The virial theorem is evidence of the intrinsic balance between the intra-atomic and the interatomic electronic density distributions that is required for the optimal balance in the variational competition between the potential pull of the nuclei and the kinetic resistance of the electron wave. It is a necessary attribute of the actual wave function and thus represents a *constraint* that approximate wave functions must satisfy if successive improvements are to lead to the exact solution. Satisfaction of this constraint by an approximate wave function implies that the optimal compromise between the potential pull and the kinetic resistance has been achieved *within the formal limitation of that wave function.*

By virtue of the response of the kinetic and potential energies to orbital contraction, which was illustrated in Section 1.2.2, one readily infers that, at the critical geometries, the following assessments of an *approximate* wave function Ψ can be made:

If $2\mathbf{T}(\Psi) < |\mathbf{V}(\Psi)|$, then the actual solution Ψ_g is in some way more localized than Ψ.

If $2\mathbf{T}(\Psi) > |\mathbf{V}(\Psi)|$, then the actual solution Ψ_g is in some way less localized than Ψ.

These inequalities are helpful in understanding the changes induced by wave function optimizations.

Because the energy increases quadratically near the variational minimum, certain wave functions that do not satisfy the virial theorem can yield rough approximations to the exact *energy*, as exemplified by the wave function of Heitler and London. However, the flaws in the *wave functions*, as well as other expectation values, are considerably more serious, as exemplified by the manifestly large error in the expectation value of $1/r$.

Notwithstanding its usefulness, it should be noted that – contrary to widespread misconceptions – *the virial theorem per se generates no clues whatsoever* for answering the question *why one system is more stable than another*. This inability is for instance exhibited by the two *isoenergetic* systems $\{m=1, Z=2\}$ and $\{m=4, Z=1\}$, which were examined in Section 1.2.2. *Both* satisfy the virial theorem but, as discussed there, the two differ radically regarding their physical reasons for being more stable than hydrogen, which also satisfies the virial theorem.

1.3
Atoms in Molecules

1.3.1
Quantitative Bonding Analyses Require Quasi-Atoms in a Molecule

The chemical notion that molecules consist of atoms held together by bonds implies (i) that atoms occur not only as free entities but are also preserved in molecules, albeit possibly deformed, and (ii) that there exist interactions between them that establish the cohesion. To retrieve this model from accurate quantum mechanical wave functions presents a nontrivial challenge because, in fact, chemical binding results from electrons being *shared between several atoms* so that, for instance, very compact wave functions are typically represented in terms of MOs that are delocalized over many atoms, the simplest examples being canonical Hartree–Fock determinants. To recover and identify "atoms" in an electronic wave function Ψ requires therefore *a transformation of its representation in such a way that* Ψ *becomes constructed from subunits that exhibit atomic character*. As in previous studies, we denote such atom-like building blocks of Ψ in a molecule as *quasi-atomic*.

Important in the present context is that the expression of molecular electronic wave functions in terms of quasi-atomic components not only gratifies the chemical intuition but that *it is in fact an indispensible prerequisite for developing a rigorous quantitative analysis of the origin of covalent bonding*. This is because *the maximization of the interatomic energy lowering that creates bonding will be seen to entail a deformation of atoms into quasi-atoms.* Since, by virtue of the *atomic* variation principle, such intra-atomic deformations are necessarily antibonding, there exists a subtle competition between the intra-atomic and the bond-creating interatomic energy changes, whose elucidation cannot be bypassed if one wishes to understand the bonding mechanism.

1.3.2
Primary and Secondary Energy Contributions

For the interpretation of chemical bonding, it is helpful as well as physically justified to take into account that the influences determining atomic and molecular electron distributions can be attributed to two kinds of forces. The primary influence is the effective one-electron potential that the nuclei and the average electron distribution generate. The secondary influence is the many-electron potential that is caused by the deviations of the individual interelectronic interactions from the average, that is, the dynamic correlations. The primary potential creates a primary orbital space in which the stock [16] of the ground state wave function takes form. It can be a single-configuration (SCF) or, if several configurations compete energetically, a multi-configurational (MCSCF) function. The secondary potential generates correlation refinements of the wave function that are grafted onto the primary stock. Unfortunately, for historical formal reasons, the label "correlation," albeit "non-dynamical," "static" or "strong," is often also used for primary multi-configurational wave functions.

The primary orbital space of an *atom* is spanned by its optimized *minimal basis set orbitals*. This intuitive insight by the pioneers of the 1930s, which provided the basis for their remarkable qualitative successes, has been confirmed by the quantitative *ab initio* work of the computer-age. In the many atoms with open valence shells, that is, where the number of valence electrons is less than twice the number of minimal basis valence orbitals, the primary stock of the ground state wave function is frequently multi-configurational.

The analysis of chemical bonding is greatly clarified by separating the binding effects involving the primary stocks of the atomic wave functions from those involving the dynamic correlations. Bonds between atomic primary stocks form when at least one atom has an open valence shell because, then, the innate delocalization drive will cause at least some valence electrons to use available minimal basis set orbitals on *several* atoms, that is, the kinetic energy will be lowered by "electron sharing." In covalent bonds, this bonding in the primary orbital space typically dominates markedly over the correlation contributions. But even when dynamic correlations play a greater or even an essential part in bonding, separating the two types of contributions greatly clarifies the elucidation.

For these reasons, the bonding analyses of the present study focus primarily on MCSCF wave functions in full valence spaces. As in our earlier work, [17] we denote these *very specific* type of full space MCSCF wave functions as FORS (full optimized reaction space) wave functions. The correlation contributions to covalent bonds will then also be briefly examined.

1.3.3
Identification of Quasi-Atoms in a Molecule

In the present discussions, electronic wave functions are conceived of as being constructed from orbitals. The identification of quasi-atoms in a molecule consists then of two steps: (i) the identification of *quasi-atomic orbitals*, generically denoted as QUAOs, in terms of which the molecular electronic wave function Ψ can be expressed, and (ii) the construction of quasi-atomic *configurations* from these orbitals in terms of which the wave function Ψ can be expanded. The determination of such QUAOs and configurations requires a judicious choice of appropriate localization criteria. We consider it desirable to determine QUAOs and configurations in such a way that they *differ as little as possible from the corresponding free-atom quantities while still being capable of regenerating the molecular wave function*.

For the determination of QUAOs, the following procedure is used in the present analysis. Let ψ_k be the orbitals from which the molecular wave function is constructed, ordered by importance, for example, through a natural orbital expansion. Then the nonsymmetric, usually rectangular, overlap matrix is calculated between these MOs and the full set of orthogonalized atomic basis orbitals used on *one* atom A (such as, e.g., a cc-pVXZ basis). The singular value decomposition (SVD) of this overlap matrix determines a set of *MOs*, that is, linear combinations of the ψ_k, and an equal number of orbitals in the *AO basis space* on atom A. They form corresponding pairs with maximum possible overlap integrals, which are given by the (always positive) singular values of the SVD. This procedure is done for every atom in the molecule. Using the magnitudes of the singular values of the SVDs for the various atoms as guides, a certain number of the SVD-generated MOs is chosen from each atom as QUAOs for that atom, subject to the limitation that the total number of the quasi-atomic orbitals selected for all atoms is equal to the total number of the original MOs ψ_k. This set χ_v of QUAOs has the following properties:

- It spans the space of the MOs so that the original electronic wave function can be expressed in terms of configurations formed from these quasi-atomic orbitals.
- If a sufficient number of MOs ψ_k are used, then the QUAOs turn out to be essentially localized on the various atoms.
- The QUAOs χ_v on one atom are mutually orthogonal.
- The QUAOs χ_v from different atoms are mutually non-orthogonal. If it is expedient for some purpose, they can be symmetrically orthogonalized.

In the case of a full valence space FORS MCSCF wave function, the number of MOs is equal to the total number of conceptual minimal basis set orbitals in the molecule. For any one atom, the number of quasi-atomic orbitals is then chosen to be equal to the number of minimal basis orbitals on that atom. The quasi-atomic orbitals in fact *are* the deformed minimal basis set orbitals on that atom. The abbreviation QUAFO will be used for these quasi-atomic FORS orbitals. The corresponding linear combinations of the atomic basis orbitals on A represent the optimal *pure atomic approximate orbitals* (PAAOs) to the QUAFOs. In the molecules H_2, B_2, C_2, N_2, O_2, F_2 discussed in the following paragraphs, all overlap integrals between the QUAFOs and the corresponding PAAOs are found to be larger than 0.99, which exhibits the atomic character of the QUAFOs. Nonetheless, the PAAOs do not generate the molecular wave function and a close examination shows that the QUAFOs contain small admixtures from other atoms. (Molecular wave functions formed by configuration interaction (CI) calculations using the PAAOs yield energies a few millihartree higher than the FORS wave functions.) Details of the SVD method will become apparent from the explicit applications in the molecules discussed below in the following paragraphs. In the case of F_2, the method is also used for a wave function that includes orbitals providing some dynamic correlation.

An in-depth elaboration and discussion of the SVD approach to generate quasi-atomic orbitals in molecules will be given in a separate investigation [18]. It has been found for instance that, under certain conditions, some correlating orbitals beyond the FORS level are required to achieve satisfactory localization on atoms. On the other hand, for some wave functions, the number of available occupied MOs may be less than the total number of minimal basis set orbitals. There then exists some additional freedom that allows further atomic adaptation of the quasi-atomic orbitals. A systematic approach to this problem has been formulated previously [19] and is also discussed in Ref. [18]. In the case of polyatomic molecules, an unbiased method has furthermore been developed for forming hybrid QUAFOs that clearly exhibit directional bonding on each atom [20].

1.4 The One-Electron Basis of Covalent Binding: H_2^+

The question to be answered is: Why does the optimal compromise between the potential energy pull and the kinetic energy resistance occur for a lower total energy in H_2^+ than in H? In the end, it turns out that the kinetic energy plays the more critical role and that no *static* model can account for the physical origin of binding in H_2^+ and H_2. Furthermore, notwithstanding inner shells and other complicating factors present in atoms beyond helium, covalent bond formation appears to be generally driven by kinetic effects that are analogous to those exemplified in the prototype H_2^+ bond.

To place the analysis of this molecule on the firmest possible footing, it will be deduced from near-exact wave functions [13]. They were obtained in terms of uncontracted (14s, 6p, 3d, 2f, 1g) basis sets of 26 σ-type spherical Gaussian AOs on the two atoms, in which all orbital exponents were optimized. For the hydrogen atom, the energy found in this basis lies 0.1 μh (microhartree) above the exact value, with a virial ratio of $2T/|V| = 0.99999985$. The energy of the molecule is found to lie 0.55 μh above the exact value at the equilibrium distance, which is found to be $R = 1.99720$ Bohr. The virial ratio is $2T/|V| = 0.9999982$ at this distance [21]. The wave function can be expected to be of similar quality at intermediate distances. The quantitative results on which the discussion of the present section is based have been reported in detail in Ref. [13].

1.4.1 Molecular Wave Function as a Superposition of Quasi-Atomic Orbitals

There are two atomic minimal basis set orbitals in this system. According to the discussion in Section 1.3.2, the molecular FORS space is, therefore, also two-dimensional. However, only the bonding molecular FORS orbital is occupied whereas the second molecular FORS orbital, the antibonding orbital, is unoccupied. The latter can therefore be arbitrarily *chosen* in such a way that the quasi-atomic orbitals become as atomic-like as possible.

By collecting the basis orbitals on each atom, the wave function Ψ is cast in the form

$$\Psi = \frac{(\psi_A + \psi_B)}{\sqrt{(2 + 2S)}} \qquad S = \langle \psi_A | \psi_B \rangle = 0.588742 \tag{1.10}$$

$$\psi_x = a\mathbf{s}_x + b\mathbf{p}_x, \quad \mathbf{s} = \sum (14\text{s}), \quad \mathbf{p} = \sum (6\text{p}, 3\text{d}, 2\text{f}, 1\text{g}) \tag{1.11}$$

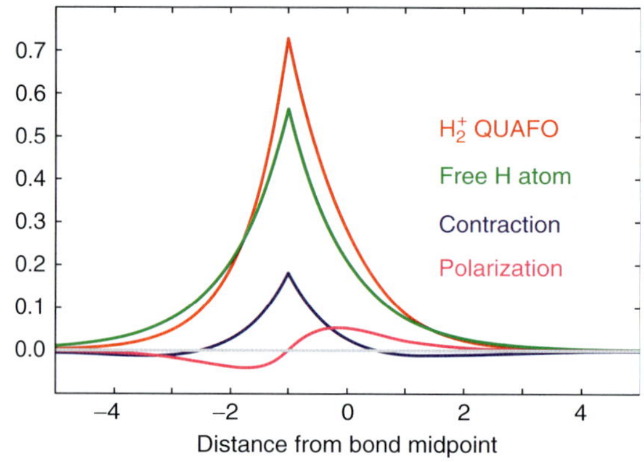

Figure 1.3 Resolution of the quasi-atomic orbital (QUAFO) on the left atom in H_2^+ as a sum of contributions from the atomic 1s orbital, the spherical deformation and the angular deformation at the equilibrium distance (see Section 1.4.1).

where ψ, **s**, and **p** are normalized. The molecular FORS space is then formed by this occupied bonding orbital Ψ and by choosing the unoccupied antibonding orbital to be

$$\Psi^* = \frac{(\psi_A - \psi_B)}{\sqrt{2-2S}} \tag{1.12}$$

Application of the SVD to the rectangular overlap matrix between these two molecular FORS orbitals and the set of all atomic basis orbitals on atom A manifestly yields the orbital ψ_A as the quasi-atomic orbital on A: the QUAFO orbital is in fact identical with the PAAO orbital. The same holds for atom B. Thus ψ_A and ψ_B are directly the quasi-atomic orbitals in this case.

Figure 1.3 exhibits the resolution (Eq. (1.11)) of the quasi-atomic orbital ψ_A on the left atom in terms of its components along the internuclear axis, for the equilibrium distance. In this figure the spherical component **s** has been further resolved in terms of the free-atom 1s orbital and the spherical deformation [s − (1s)]. The curves display the total respective contributions, *including coefficients*, so that adding the spherical deformation (blue) and the angular deformation (purple) to the free-atom 1s orbital (green) will yield the quasi-atomic orbital (red). At this distance, the spherical deformation is manifestly a contraction and the angular deformation represents a polarization.

The total relative contributions of the spherical and the angular deformations to the quasi-atomic orbital, i.e., their integrated contributions to the normalization integral, as well as their variations with the internuclear distance R are shown in Figure 1.4. The polarization deformation contributes less than 2% over the whole range, but its contribution extends to fairly large internuclear distances. The spherical deformation effectively vanishes beyond 4 Bohr, but strongly increases for shorter distance, being ~4% at the equilibrium distance.

Regarding the *normalized spherical* component **s** in Eq. (1.11), the following observation is important:

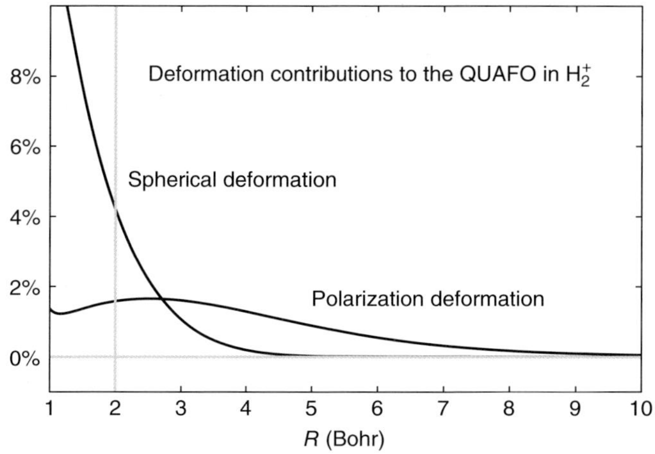

Figure 1.4 Variation of the spherical and angular deformations of the quasi-atomic orbital in H_2^+ with the internuclear distance. The gray vertical line indicates the equilibrium distance.

> The spherical component of the quasi-atomic orbital is near-identical with a scaled hydrogen-like orbital 1s* of the type formulated in Eq. (1.5).

This agreement is demonstrated in Figure 1.5 where Figure 1.5b displays the overlap integral $\langle \mathbf{s}|1s^*\rangle$ as a function of the internuclear distance R. Figure 1.5a displays the orbital exponent value ζ^* of the 1s* orbital that corresponds to the quasi-atomic orbital at each internuclear distance. This exponent was obtained by

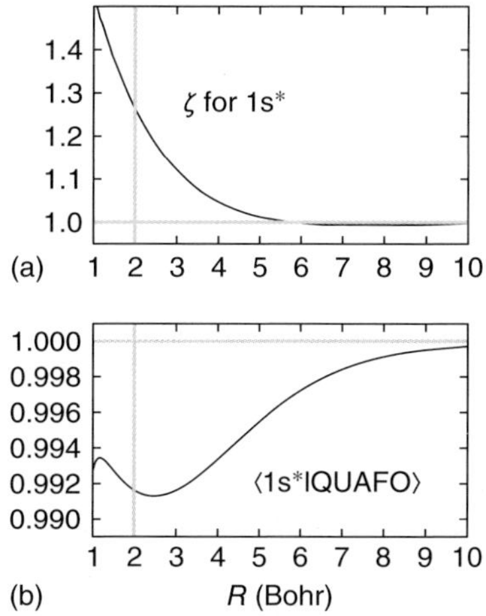

Figure 1.5 Characteristics of the exponential approximation 1s* to the spherical part of the quasi-atomic orbital (QUAFO) in H_2^+ at various internuclear distances. (a) Orbital exponent of 1s*. (b) Overlap integral between the 1s* orbital and the QUAFO. The gray vertical line indicates the equilibrium distance.

maximizing, at each internuclear distance, the overlap integral $\langle \mathbf{s}|1s^*\rangle$ with respect to the orbital exponent ζ^* in the 1s* orbital. Whereas, in the region $R > \sim 6$ Bohr, the 1s* orbital is seen to be very slightly *expanded* ($\zeta^* < 1$) with respect to the hydrogen 1s orbital ($\zeta = 1$),

> the 1s* orbital becomes considerably contracted ($\zeta^* > 1$) in the region for $R < \sim 4.5$ Bohr. At the equilibrium its orbital exponent is $\zeta^* = 1.264$.

Combining the data of Figures 1.4 and 1.5, one sees that the projection of the quasi-atomic orbital ψ_x on the scaled $1s_x$* orbital, that is, $[a \times \langle \mathbf{s}|1s^*\rangle]$, is always larger than 0.982.

1.4.2 Molecular Electron Density and Gradient Density as Sums of Intra-atomic and Interatomic Contributions

The potential energy is determined by the electron density. The kinetic energy is determined by the gradient density (see Eq. (1.4)). Therefore, both densities have to be resolved in terms of intra-atomic and interatomic contributions.

1.4.2.1 Resolution of the Molecular Density

In classical electrostatics an approximate charge distribution covering the two atoms would be a superposition of *quasi-atomic densities*, for example,

$$\rho_{qa} = \frac{1}{2}(\psi_A^2 + \psi_B^2) \tag{1.13}$$

In quantum mechanics, by contrast, the quasi-atomic *wave amplitudes* are superposed, as expressed in Eq. (1.10), so that the molecular density $\rho = \Psi^2$ differs from the *quasi-atomic* density ρ_{qa} of Eq. (1.13) as follows:

$$\Psi^2 = \rho = \rho_{qa} + \rho_{I,} \quad \rho_I = \rho - \rho_{qa} = \frac{\mathcal{I}_{AB}}{(2+2S)} \tag{1.14}$$

where

$$\mathcal{I}_{AB} = 2\psi_A\psi_B - S(\psi_A^2 + \psi_B^2) \tag{1.14a}$$

The *bond order* $(2+2S)^{-1}$ has the value 0.314714 at the equilibrium distance. In ρ, the two atomic amplitudes are summed before they are squared whereas, in ρ_{qa}, they are squared before they are summed. The difference ρ_I represents therefore the *interference density*, that is, the interference part of the total density ρ. The term $\mathcal{I}_{AB}$ represents the *interference of the quasi-atomic orbitals.*

The integrals over ρ and ρ_{qa} are both = 1. Hence, the integral over ρ_I vanishes, which implies that the interference density ρ_I represents a *charge shift* in real space. In fact, the interference density *represents the quantification of the often qualitatively invoked "accumulation of electronic charge in the bond."* This property is exhibited in panels of Figure 1.6a–c, which display the contour plots of ρ_I in a plane containing the internuclear axis *at the equilibrium distance.* With a view to the subsequent discussion, plots of ρ_I are shown not only using the exact orbitals ψ_A, ψ_B but also

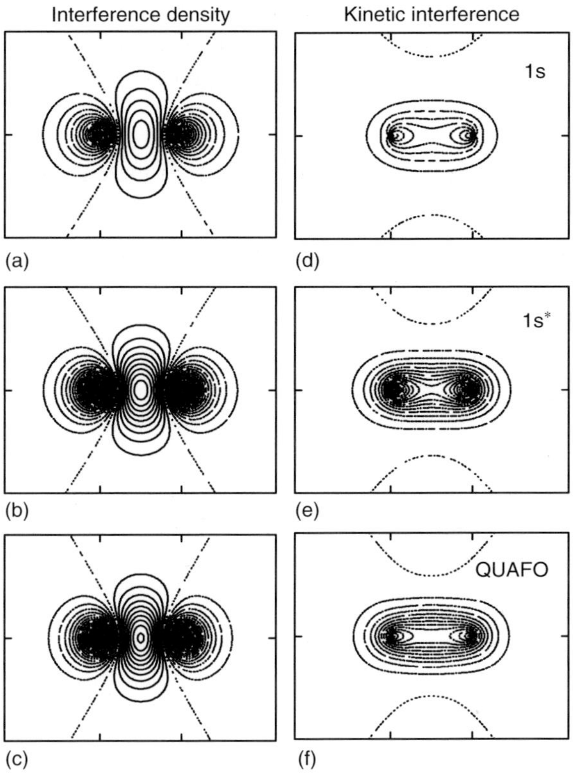

Figure 1.6 Interatomic interference in H_2^+ at the equilibrium distance. Left column of panels: interference *densities* ρ_I = charge accumulation in the bond of H_2^+. Right column of panels: kinetic interference densities = gradient attenuation. First row of panels: interference between the atomic 1s ground state orbitals. Second row of panels: interference between the scaled exponential 1s* approximations to the QUAFOs (see Figure 1.5). Third row of panels: interference between the quasi-atomic orbitals (QUAFOs) of the actual wave function of H_2^+. Contour increments: left panels: 0.002 e Bohr^{-3}; right panels: 5 millihartree Bohr^{-3}. Solid lines = positive contours. Dashed lines = negative contours. Dotted lines = zero contours.

using the hydrogen 1s orbitals as well as using the scaled 1s* orbitals discussed at the end of the preceding Section 1.4.1. Note should be made of the fact that

> for all quasi-atomic orbital choices, even when polarization is included, the charge that is accumulated in the bond is taken away from regions near the nuclei.

This aspect of the charge accumulation in the bond is often overlooked.

1.4.2.2 **Resolution of the Molecular Gradient Density**

The gradient density $(\nabla\psi)^2$ can be resolved in a similar way as the density ψ^2, that is, in terms of the average of the atomic gradient densities and the difference between this term and the molecular gradient density, that is,

$$(\nabla\Psi)^2 = \frac{[(\nabla\psi_A)^2 + (\nabla\psi_B)^2]}{2} + \left\{(\nabla\Psi)^2 - \frac{[(\nabla\psi_A)^2 + (\nabla\psi_B)^2]}{2}\right\} \tag{1.15}$$

In analogy to the interference density ρ_I, the difference term in the curly brackets on the right hand side can be considered as the *gradient interference density*. Panels in Figure 1.6d–f exhibit contours of this gradient interference densities at the equilibrium distance. *All of them are negative*, showing that the superposition of the quasi-atomic orbitals attenuates the gradient of Ψ everywhere, when compared to the average of the quasi-atomic gradient densities. Along the internuclear axis, this attenuation is manifest by comparing the plots of Ψ^2 and $(\psi_A{}^2 + \psi_B{}^2)/2$, as shown, for example, in Figure 9 of Ref. [13a].

The interference density as well as the gradient interference density result from the spread of the *orbital amplitude* from one atom to two atoms. Both are thus consequences and exhibit different aspects of the *orbital delocalization*. This inference will be confirmed by the energy analyses in Sections 1.4.5.2 (last paragraph) and 1.4.5.4 (first paragraph).

A difference in interpretation between Eqs. (1.15) and (1.14) is the following. Although both equations furnish resolutions in terms of intra-atomic contributions and interference contributions, only the sum of the intra-atomic *density* contributions in Eq. (1.14) can be considered as "quasi-classical." *This cannot be claimed for the intra-atomic gradient densities* because, as discussed in Section 1.2.1, the treatment of the kinetic energy in quantum mechanics is fundamentally different from that in classical mechanics.

1.4.3 Dependence of Delocalization and Interference on the Size of the Quasi-Atomic Orbitals

It turns out that a rather important role in the process of bonding is played by the *dependence* of the magnitude of interference on the *size of the quasi-atomic orbitals relative to the distance R between the nuclei*. This dependence is not trivially obvious.

From the definition in Eq. (1.14) one sees immediately that the interference density vanishes for $R = 0$ as well as for $R = \infty$. In between these two limiting cases, it must therefore wax and wane. On the other hand, comparing the interference contours on Figure 1.6a, obtained for the hydrogen 1s orbitals, with those on Figure 1.6b, obtained using the contracted 1s* orbitals, one notes that, at the equilibrium distance, *the accumulation of charge in the bond is stronger for the contracted 1s* orbitals than for the uncontracted 1s orbitals. Similarly, the kinetic interference density of 1s* in panel e of that figure is stronger in the bond than that of 1s in panel d. This correlation is, at first sight, surprizing since the overlap integral decreases with contraction.* The observations are put in perspective by the following explicit analysis for the case that the quasi-atomic orbitals are approximated by the scaled 1s-type orbital $1s_\zeta$ defined in Eq. (1.5).

1.4.3.1 Charge Accumulation at the Bond Midpoint

The contours in Figure 1.6a–c suggest that the maximum of the charge accumulation in the bond, which occurs at the bond midpoint, offers a rough measure of the magnitude of the interference ρ_I. By virtue of the definition in Eq. (1.14), this value is given by

$$\rho_{I,mid}(\psi_A) = \frac{\Psi^2 - (\psi_A^2 + \psi_B^2)}{2} = \psi_A^2 \times \frac{(1-S)}{(1+S)}, \quad S = \langle \psi_A | \psi_B \rangle \tag{1.16}$$

where $\psi_A{}^2$ is taken at the bond midpoint. If the scaled $1s_\zeta$ orbital of Eq. (1.5) is used as an approximation to the quasi-atomic orbital ψ_A, the Eq. (1.16) becomes

$$\rho_{I,mid}(1s_\zeta) = \left(\frac{1}{\pi R^3}\right) \left\{ \left[\frac{(1-S)}{(1+S)}\right] \sigma^3 \exp(-\sigma) \right\} \tag{1.17}$$

where

$$S = \left(1 + \sigma + \frac{\sigma^2}{3}\right) \times \exp(-\sigma), \quad \sigma = \zeta R \tag{1.18}$$

Relevant in the present context is the dependence of $\rho_{I,mid}$ on the orbital size *at any given internuclear distance.* This dependence is given by the expression in the curly brackets {} in Eq. (1.17), which is a function of $\sigma = \zeta R = R/\alpha$ = the inverse ratio of the orbital size to the internuclear distance. This dependence on σ is displayed in Figure 1.7a, which is a plot of $\rho_{I,mid}$ versus σ, where the equilbrium value has been arbitrarily chosen for the factor R^{-3} *in front of the curly bracket.* Since the maximum of the curve occurs for $\sigma = \zeta R = 4.01$, the criterion suggests that the maximal charge accumulation in the bond is obtained for a value $\zeta_m{}' > 1$ when $R < 4.01$ Bohr, but for a value $\zeta_m{}' < 1$ when $R > 4.01$ Bohr. With reference to the hydrogen 1s orbital (i.e., $\zeta = 1$), *maximal charge accumulation in the bond, according to this criterion, is therefore obtained by orbital contraction when $R < \sim 4$ Bohr, and by orbital expansion when $R > \sim 4$ Bohr.*

1.4.3.2 Total Charge Accumulation in the Bond

Alternatively, an *overall measure* of the charge accumulation in the bond can be obtained by calculating the amount of charge that interference in fact shifts from the atomic regions into the bond region, that is, from the regions with negative contours into the region with positive contours of ρ_I in Figure 1.6. These three regions are separated by the two-sheet hyperboloid on which ρ_I vanishes. For the quasi-atomic orbitals $1s_\zeta$, it is given by

$$\frac{(R_A - R_B)}{R} = \frac{\pm a(\sigma)}{\sigma}, \quad a(\sigma) = \text{arccosh}\left[\frac{1}{S(\sigma)}\right] = \ln\left\{\frac{\left[1 + \left(1 - S^2\right)^{\frac{1}{2}}\right]}{S}\right\} \tag{1.19}$$

Integration of ρ_I over the central region between the two hyperboloid sheets yields the total charge Q_I that is actually moved from the atomic regions into the bond region. It is found to be

$$Q_I(1s_\zeta) = \left[(2\sigma + \sigma^2)\left(a - \left(1 - S^2\right)^{\frac{1}{2}}\right) + a^2(1 - S^2)^{\frac{1}{2}} - \frac{a^3}{3}\right] \left[\frac{\exp(-\sigma)}{2\sigma(1+S)}\right] \tag{1.20}$$

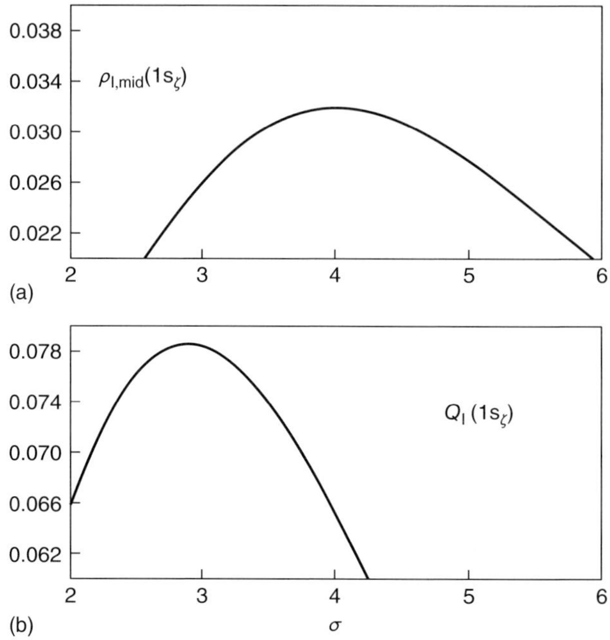

Figure 1.7 Variation of the strength of the interference density ρ_I between scaled $1s_\zeta$ orbitals (of Eq. (1.5)) with the orbital size $\alpha = 1/\zeta$ relative to the internuclear distance R. Abscissa: $\sigma = \zeta R = R/\alpha$. Ordinate of (a): value of ρ_I at the bond mid point (Eq. (1.17)). Ordinate of (b): integrated charge accumulation into the bond (Eq. (1.20)). For details, see text.

where σ, $S(\sigma)$, and $a(\sigma)$ are defined in Eqs. (1.18) and (1.19). Figure 1.7b displays the variation of the total charge accumulation $Q_I(\sigma)$ as a function of σ. Its maximal value occurs for $\sigma = 2.90$ Bohr. According to this criterion, interference is enhanced by contraction when $R < \sim 2.9$ Bohr and by expansion when $R > \sim 2.9$ Bohr.

1.4.3.3 **Origin of the Relation between Interference and Quasi-Atomic Orbital Contraction/Expansion**

The foregoing assessments show that, at shorter internuclear distances, interference is enhanced by contraction of the hydrogen AOs whereas, at larger internuclear distances, it is enhanced by expansion of these orbitals. The switchover occurs somewhere in the region around twice the equilibrium distance. A remarkable implication is that, whereas interference increases with increasing quasi-atomic orbital overlap at larger distances, *interference increases* with *decreasing overlap by contraction* at shorter distances, as noted earlier.

This consequential attribute of interference can be traced back to the following property of interfering AOs. Consider a scaled hydrogen orbital $\phi(r, \zeta) = 1s_\zeta$, as defined in Eq. (1.5). Because normalization is preserved, *the scaled orbital* $1s_\zeta = \phi(r, \zeta)$ *and the unscaled orbital* $1s = \phi(r, \zeta = 1)$ *cross over at some distance* r^* *from the* *origin*. Manifestly, when $\zeta > 1$, then $\phi(r, \zeta)$ is contracted relative to 1s and one has

$$\phi(r,\zeta) > 1\text{s} \quad \text{for } r < r^*, \quad \phi(r,\zeta) < 1\text{s} \quad \text{for } r > r^* \tag{1.21a}$$

whereas, when $\zeta < 1$, then $\phi(r, \zeta)$ is expanded relative to 1s and one has

$$\phi(r,\zeta) < 1\text{s} \quad \text{for } r < r^*, \quad \phi(r,\zeta) > 1\text{s} \quad \text{for } r > r^* \tag{1.21b}$$

Moreover, one deduces from the equality $\phi(r^*, \zeta) = 1\text{s} = \phi(r^*, 1)$ at the crossover radius r^* that the value of this radius depends on ζ according to

$$r^*(\zeta) = \left[\frac{(\log\zeta)}{(\zeta - 1)}\right] \times 1.5\,\text{bohr} \tag{1.22a}$$

and, since $r^*(\zeta)$ is a monotonic function, there exists a scaled orbital $\phi(r, \zeta)$ with a certain $\zeta = \zeta(r^*)$ for any chosen crossover point r^*. It follows from Eq. (1.22a) that

$$r^*(\zeta) < 1.5 \quad \text{for } \zeta > 1, \quad r^*(\zeta) = 1.5 \quad \text{for } \zeta = 1, \quad r^*(\zeta) > 1.5 \quad \text{for } \zeta < 1 \tag{1.22b}$$

and, conversely, that

$$\zeta(r^*) > 1 \quad \text{when } r^* < 1.5, \quad \zeta(r^*) < 1 \quad \text{when } r^* > 1.5 \tag{1.22c}$$

Consider now two scaled 1s orbitals placed at an internuclear distance $2R$ of less than 3 Bohr so that the bond midpoint is at a distance $R < 1.5$ Bohr from each nucleus. For each value of r^* in the range $R < r^* < 1.5$ Bohr, there then exists a scaled orbital $\phi(r, \zeta)$ on each nucleus that (i) has $\zeta = \zeta(r^*) > 1$, i.e., that is *contracted* relative to the respective uncontracted 1s orbital (see Eq. (1.22c), left), and (ii) is *larger* than the respective 1s orbital, that is $\phi(r, \zeta) > 1\text{s}$, for all distance $r \leq R$ from that nucleus (see Eq. (1.21a), left). This is so, in particular at the bond midpoint, where $r = R$. Therefore, according to Eq. (1.16), *the charge accumulation of the superposition of the two contracted orbitals at the bond midpoint is larger than that of the superposition of two uncontracted orbitals.* (Note that the factor $(1 - S)/(1 + S)$ in Eq. (1.16) is also larger for the contracted orbitals, because the overlap integral between them is smaller than that for the uncontracted orbitals.) Thus, *when the two orbital centers are closer than 3 Bohr, quasi-atomic orbital contraction enhances charge accumulation in the bond and, hence, interference and delocalization between these orbitals.* It is also apparent that similar inferences can be drawn for any suitably decaying quasi-atomic orbitals [14]. This aspect of interference has not been previously noted.

As shown by an analogous analysis, the enhancement increasingly diminishes as the internuclear distance becomes longer than 3 Bohr and, for sufficiently large distances, charge accumulation in the bond is enhanced by orbital expansion rather than contraction.

1.4.4
Binding Energy as a Sum of Two Intra-atomic and Three Interatomic Contributions

The well-known general shapes of the kinetic and potential components of the binding energy curve exhibit very different behaviors at larger and shorter internuclear distances. These shapes suggest that the binding energy is the result of several contributions with different distance dependencies. In fact, five contributions with distinct conceptual physical meanings can be identified. They are formulated in the present section. In the next section their quantitative properties are discussed and in the subsequent section their synergism is traced.

Substitution of the density decomposition (Eq. (1.14)) into the potential energy integral of H_2^+ yields the following resolution of the potential energy of H_2^+ in terms of three contributions:

$$\mathbf{V}(\Psi) = \int dx \left(-\frac{1}{r_A} - \frac{1}{r_B}\right)\rho + \frac{1}{R} = \mathbf{V}_a + \mathbf{V}_{qc} + \mathbf{V}_I \tag{1.23}$$

where the individual terms have the following definitions and physical meanings

$$\mathbf{V}_a = \frac{\left\{-\int dx\,\psi_A^2/r_A - \int dx\,\psi_B^2/r_B\right\}}{2}$$
$$= \text{the } \textit{intra-atomic} \text{ potential energy of the quasi-atomic orbitals} \tag{1.24}$$

$$\mathbf{V}_{qc} = \frac{\left\{-\int dx\,\psi_A^2/r_B - \int dx\,\psi_B^2/r_A\right\}}{2} + \frac{1}{R}$$
$$= \text{the } \textit{quasi-classical coulombic interatomic} \text{ potential energy} \tag{1.25}$$

$$\mathbf{V}_I = \int dx \left(-\frac{1}{r_A} - \frac{1}{r_B}\right)\rho_I$$
$$= [2(1+S)]^{-1}\int dx \left(-\frac{1}{r_A} - \frac{1}{r_B}\right)\mathcal{I}_{AB}$$
$$= \text{the } \textit{inter-atomic} \text{ potential energy due to the charge accumulation in the bond}$$
$$= \text{the potential } \textit{interference} \text{ energy. } [\mathcal{I}_{AB} \text{ was defined in Eq. (1.14a)}] \tag{1.26}$$

The potential part of the *binding energy*, that is,

$$\mathbf{V}_{\text{binding}} = \mathbf{V}(\Psi) - \mathbf{V}_H \quad \text{with } \mathbf{V}_H = -\int dx \frac{(1s_A)^2}{r_A} = -1\,\text{hartree} \tag{1.27}$$

is therefore the sum of an intra-atomic and an interatomic contribution:

$$\mathbf{V}_{\text{binding}} = \mathbf{V}_{\text{intra}} + \mathbf{V}_{\text{inter}} \tag{1.28}$$

where

$$\mathbf{V}_{\text{intra}} = \mathbf{V}_a - \mathbf{V}_H, \quad \mathbf{V}_{\text{inter}} = \mathbf{V}_{qc} + \mathbf{V}_I \tag{1.29}$$

The analogous resolution of the kinetic energy is simpler, viz,

$$\mathbf{T}(\Psi) = \frac{1}{2}\int \mathrm{d}x(\nabla\Psi)^2 = \mathbf{T}_a + \mathbf{T}_I \tag{1.30}$$

where

$$\mathbf{T}_a = \frac{\left\{\frac{1}{2}\int \mathrm{d}x\left(\nabla\psi_A\right)^2 + \frac{1}{2}\int \mathrm{d}x(\nabla\psi_B)^2\right\}}{2}$$
$$= \text{the } \textit{intra-atomic} \text{ kinetic energies of the quasi-atomic orbitals} \tag{1.31}$$

$$\mathbf{T}_I = \frac{1}{2}\int \mathrm{d}x\left\{(\nabla\Psi)^2 - \frac{[(\nabla\psi_A)^2 + (\nabla\psi_B)^2]}{2}\right\}$$
$$= \frac{1}{2}\int \mathrm{d}x\ \frac{\{2(\nabla\psi_A)\cdot(\nabla\psi_B) - [(\nabla\psi_A)^2 + (\nabla\psi_B)^2]\}}{2(1+S)}$$
$$= \text{the } \textit{interatomic} \text{ kinetic } \textit{interference} \text{ energy}$$
$$= \text{the } \textit{interatomic} \text{ kinetic energy resulting from the orbital delocalization} \tag{1.32}$$

The kinetic part of the binding energy, that is,

$$\mathbf{T}_{\text{binding}} = \mathbf{T}(\Psi) - \mathbf{T}_H \quad \text{with} \quad \mathbf{T}_H = \frac{1}{2}\int \mathrm{d}x\,\frac{(\nabla 1s_A)^2}{r_A} = 0.5\,\text{hartree} \tag{1.33}$$

is thus also the sum of an intra-atomic and an interatomic contribution:

$$\mathbf{T}_{\text{binding}} = \mathbf{T}_{\text{intra}} + \mathbf{T}_{\text{inter}} \tag{1.34}$$

with

$$\mathbf{T}_{\text{intra}} = \mathbf{T}_a - \mathbf{T}_H, \quad \mathbf{T}_{\text{inter}} = \mathbf{T}_I \tag{1.35}$$

By virtue of Eq. (1.4) each kinetic term is a sum of an x, y, and z component, where the z-direction is conventionally taken along the internuclear axis. It should again be emphasized that the definitions Eqs. (1.30)–(1.33) of the kinetic terms could all have been equally well written in terms of the Laplacian expression of the kinetic energy. As mentioned earlier, the gradient form is preferred in order that positive kinetic energy terms appear with positive signs in front of them.

Combining the preceding resolutions yields the following decomposition of the binding energy in terms of intra-atomic and interatomic contributions:

$$\mathbf{E}_{\text{Binding}} = \mathbf{E}(\Psi) - \mathbf{E}_H = \mathbf{E}_{\text{intra}} + \mathbf{E}_{\text{inter}} \tag{1.36}$$

$$\mathbf{E}_{\text{intra}} = \mathbf{E}_a - \mathbf{E}_H = \mathbf{T}_{\text{intra}} + \mathbf{V}_{\text{intra}} \tag{1.37a}$$

$$\mathbf{T}_{\text{intra}} = (\mathbf{T}_a - 0.5\ \text{hartree}), \qquad \mathbf{V}_{\text{intra}} = (\mathbf{V}_a + 1.0\ \text{hartree}) \tag{1.37b}$$

$$\mathbf{E}_{\text{inter}} = \mathbf{E}_I + \mathbf{V}_{qc} = \mathbf{T}_{\text{inter}} + \mathbf{V}_{\text{inter}}, \qquad \mathbf{E}_I = \mathbf{T}_I + \mathbf{V}_I \tag{1.38a}$$

$$\mathbf{T}_{\text{inter}} = \mathbf{T}_{\text{I}} \qquad \mathbf{V}_{\text{inter}} = \mathbf{V}_{\text{I}} + \mathbf{V}_{\text{qc}} \tag{1.38b}$$

The binding energy has thus the five basic conceptual physical components shaded in yellow in Eqs. (1.37)–(1.38), viz: the intra-atomic energy changes $\mathbf{T}_{\text{intra}}$ and $\mathbf{V}_{\text{intra}}$, which are due to the deformation from the free atom to the quasi-atom in the molecule, and the interatomic energy changes $\mathbf{V}_{\text{qc}}$, $\mathbf{V}_{\text{I}}$, and $\mathbf{T}_{\text{I}}$, which embody the quasi-classical interactions and the interference interactions that are generated by delocalization between the quasi-atomic orbitals on the two centers.

1.4.5
Quantitative Characteristics of the Five Energy Contributions

By virtue of the physical meanings of the five energy contributions, certain general features of their quantitative values can be identified that are relevant for understanding the behavior of the kinetic, potential, and total energy curves.

1.4.5.1 Intra-atomic Deformation Energy: $\mathbf{E}_{\text{intra}} = \mathbf{T}_{\text{intra}} + \mathbf{V}_{\text{intra}}$

The intra-atomic deformation energy $\mathbf{E}_{\text{intra}}$ of (Eq. (1.37)) is the energy deviation from the free-atom minimum. By virtue of the *atomic* variation principle, *it is necessarily positive,* that is, it has an antibonding effect. In particular, as discussed in detail in Section 1.2.2, upon quasi-atomic orbital *contraction*, $\mathbf{T}_{\text{intra}}$ will become more positive and $\mathbf{V}_{\text{intra}}$ will become more negative, with $\mathbf{T}_{\text{intra}}$ prevailing over $\mathbf{V}_{\text{intra}}$.

1.4.5.2 Quasi-Classical Interaction between the Atoms: $\mathbf{V}_{\text{qc}}$

By virtue of the molecular symmetry, the quasi-classical energy of Eq. (1.25) can also be written as

$$\mathbf{V}_{\text{qc}} = -\int \mathrm{d}x \frac{\psi_{\text{A}}^2}{r_{\text{B}}} + \frac{1}{R} \tag{1.39}$$

It is thus the potential energy of a neutral hydrogen atom at A, with a *fixed* density ${\psi_{\text{A}}}^2$, in the field of a proton B at a distance R from A.

If ${\psi_{\text{A}}}^2$ is spherically symmetric then, as Newton first showed [22], the first term on the right hand side of Eq. (1.39) equals $\{R^{-1}[-\int \mathrm{d}x\, {\psi_{\text{A}}}^2]\}$ where the integration goes over the sphere with radius R around A. Since this sphere encompasses less than the whole of ${\psi_{\text{A}}}^2$, the integral $\int \mathrm{d}x\, {\psi_{\text{A}}}^2$ is <1 and, hence, $\mathbf{V}_{\text{qc}}$ becomes positive, that is, *antibonding*. If the spherical quasi-atomic orbital is *contracted,* then the integral $[-\int \mathrm{d}x\, {\psi_{\text{A}}}^2]$ is closer to 1 and increases the shielding of nucleus A so that $\mathbf{V}_{\text{qc}}$ will become *less repulsive*. Expansion of the quasi-atomic orbital will have the opposite effect.

For $\mathbf{V}_{\text{qc}}$ to be attractive, the orbital ψ_{A} has to become sufficiently polarized toward nucleus B.

It may also be noted that the form of Eq. (1.39) for $\mathbf{V}_{\text{qc}}$ implies that the effect of orbital delocalization on the potential energy is entirely contained in the interference energy $\mathbf{V}_{\text{I}}$. The interference density is therefore an expression of delocalization in agreement with the discussion at the end of Section 1.4.2.2.

1.4.5.3 Potential Interference Energy: V_I

According to the analysis in Section 1.4.2.1, the interference density ρ_I shifts charge from near the nuclei toward the bond center, where the potential $(-1/r_A - 1/r_B)$ is less negative than near the nuclei. Therefore, *the potential interference energy of Eq. (1.26) is always positive, that is, antibonding.*

The rather popular hand waving conjecture that charge accumulation through overlap in the bonds of H_2 and H_2^+ *lowers* the potential energy is patently wrong.

1.4.5.4 Kinetic Interference Energy: T_I

By virtue of the molecular symmetry, the kinetic interference energy of Eq. (1.32) can be written

$$T_I = \frac{1}{2}\int dx(\nabla\Psi)^2 - \frac{1}{2}\int dx(\nabla\psi_A)^2 \tag{1.40}$$

where the wave function $\Psi \sim (\psi_A + \psi_B)$ is manifestly more delocalized than the orbital ψ_A. According to the basic insights of Section 1.2.1, the first term in this equation is therefore expected to have a lower kinetic energy than the second term so that *the kinetic interference energy will be negative, that is, bonding.* This expectation is confirmed by the contours on panels of Figure 1.6d–f in Section 1.4.2.2, which show that the integrand in Eq. (1.32) is negative everywhere. The conjunction of Eq. (1.40) and Figure 1.6 also confirms the inference at the end of Section 1.4.2.2 that the gradient interference density is an expression of delocalization.

Since superposing the AOs attenuates the derivative *along* the bond axis most strongly, the component in this direction is found to contribute 2/3 of the negative value of T_I at the equilibrium distance [13].

1.4.5.5 Interference Energies and Quasi-Atomic Orbital Contraction and Expansion

According to the discussions in Sections 1.4.2.2 and 1.4.5.2, charge accumulation in the bond goes hand in hand with orbital delocalization. The analysis in Section 1.4.3 had shown that, for $R < 3$–4 Bohr, charge accumulation in the bond as well as delocalization increase when the quasi-atomic orbitals *contract* (with respect to the free-atom 1s orbitals) whereas, for $R > 3$–4 Bohr, charge accumulation in the bond and delocalization increase when the quasi-atomic orbitals *expand*. These changes with orbital shrinking and swelling lead to the following changes in the kinetic and potential interference energies.

According to the discussion in Section 1.4.5.3, increasing the charge accumulation in the bond will make V_I more positive (antibonding). Therefore, *quasi-atomic orbital contraction of the free-atom 1s orbitals is expected to increase the antibonding effect of the potential interference energy when $R < \sim 4$ Bohr whereas the opposite holds when $R > \sim 4$ Bohr.* This inference is confirmed by explicit calculation of V_I for the case that the quasi-atomic orbitals are chosen to be of the scaled $1s_\zeta$ type defined in Eq. (1.5) with a variable exponent ζ. The potential interference energy is then found to be

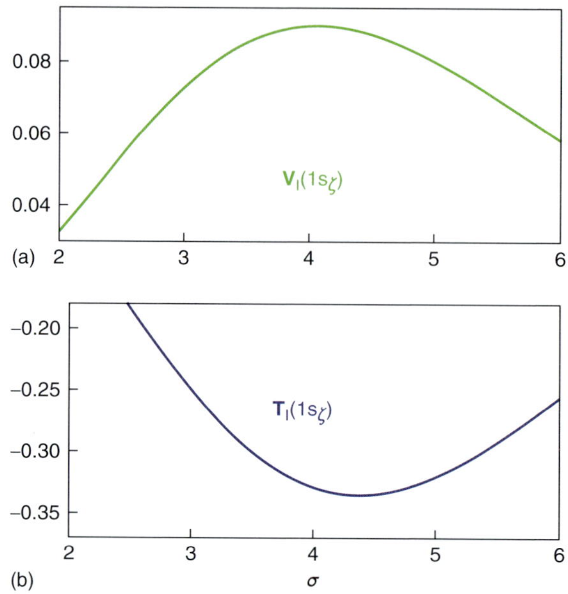

Figure 1.8 Variation of the strengths of the interference energies between scaled $1s_\zeta$ orbitals (of Eq. (1.5)) with the orbital size $\alpha = 1/\zeta$ relative to the internuclear distance R. Abscissa: $\sigma = \zeta R = R/\alpha$. (a) Potential interference energy. (b) Kinetic interference energy. For details, see text after Eq. (1.41) and (1.42).

$$\mathbf{V}_\mathrm{I}(1s_\zeta) = R^{-1}\left\{\frac{\left[S\left(1-\mathrm{e}^{-2\sigma}\right)-2\sigma\mathrm{e}^{-\sigma}\right](1+\sigma)}{(1+S)}\right\}, \quad \sigma = \zeta R \tag{1.41}$$

where $S(\sigma)$ is the overlap integral given in Eq. (1.18). *For fixed R*, the dependence on ζ is contained, through $\sigma = \zeta R$, in the expression in the large curly brackets {}. The dependence of this expression on σ is displayed in Figure 1.8a, which is a plot of $\mathbf{V}_\mathrm{I}$ versus σ where the equilibrium value has been arbitrarily chosen for the factor R^{-1} *in front of the curly bracket*. The antibonding of $\mathbf{V}_\mathrm{I}$ is seen to be maximal for $\zeta_\mathrm{m} \approx 4.07/R$. Hence, for $R < 4.07$ *Bohr, contraction (i.e., increasing* ζ_m) *will increase* the positive $\mathbf{V}_\mathrm{I}$ whereas, *for* $R > 4.07$, *expansion* will increase $\mathbf{V}_\mathrm{I}$.

According to the discussion in Section 1.4.5.4, the negative value of $\mathbf{T}_\mathrm{I}$ is due to delocalization, and therefore, increasing delocalization will make $\mathbf{T}_\mathrm{I}$ more negative (bonding). Quasi-atomic orbital contraction is therefore expected to enhance the negative kinetic interference energy for $R < \sim 4$ Bohr, whereas the opposite occurs for $R > \sim 4$ Bohr. This effect on the kinetic interference energy is confirmed through explicit calculation of $\mathbf{T}_\mathrm{I}$ by choosing as quasi-atomic orbitals the scaled 1s-type orbitals defined in Eq. (1.5) with a variable exponent ζ. The kinetic interference energy at any given internuclear distance R is then readily found to be

$$\mathbf{T}_\mathrm{I}(1s_\zeta) = R^{-2}\left\{\frac{\left(-\sigma^4\mathrm{e}^{-\sigma}\right)}{3(1+S)}\right\} \tag{1.42}$$

where σ and S are the same quantities as in Eq. (1.41). *For fixed R*, the dependence on ζ is contained, through $\sigma = \zeta R$, in the expression in the large curly brackets {}. The dependence of this expression on σ is displayed in Figure 1.8b, which is a plot of $\mathbf{T}_\mathrm{I}$ versus σ, where the equilibrium value has been arbitrarily chosen for the factor R^{-2} *in front of the curly bracket*. Its minimum occurs at $\zeta_\mathrm{m} \approx 4.38/R$, which identifies the quasi-atomic orbital with maximal interference energy lowering. *For $R < 4.38$ Bohr*, the value of ζ_m is larger than 1, which implies an *enhancement of the kinetic interference effect by quasi-atomic orbital contraction* with respect to the free-atom orbital with $\zeta = 1$. *For $R > 4.38$*, the value of ζ_m is smaller than 1, which implies an *enhancement of $\mathbf{T}_I$ with quasi-atomic orbital expansion*. It is also seen that the enhancement decreases with increasing internuclear distance as $(1/R^2)$.

Thus, the bonding character of the kinetic interference energy $\mathbf{T}_\mathrm{I}$, the antibonding character of the potential interference energy $\mathbf{V}_\mathrm{I}$, as well as the response of these quantities to quasi-atom orbital shrinking and swelling as a function of the internuclear distances, are all consequences of the delocalization of the electron from one nucleus to two nuclei.

1.4.6
Synergism of the Binding Energy Contributions along the Dissociation Curve

On the basis of the quantitative relations identified in the preceding section, the changes in the total binding energy can be understood by variational reasoning. The changes along the binding energy curve will be discussed with reference to the graphs in Figure 1.9, which exhibit the resolution of the binding energy in terms of the five components identified in Eqs. (1.36–1.38) as a function of the internuclear distance.

The three *columns* of panels exhibit the analyses that result from three different choices for the quasi-atomic orbitals ψ_A in Eq. (1.10). The last column corresponds to using the exact quasi-atomic orbitals discussed in Section 1.4.1. The second column corresponds to choosing the approximation obtained by omitting the polarization contributions in Eq. (1.10) and replacing the normalized spherically symmetric term **s** by the corresponding scaled 1s* orbital discussed at the end of Section 1.4.1. For the first column, the quasi-atomic orbitals are simply chosen to be the undeformed 1s orbital of the hydrogen atom. The *rows* of colored panels display from top to bottom: (i) the quasi-classical coulombic contributions, (ii) the interference contributions, (iii) the interatomic contributions (= the sum of the top two rows), (iv) the intra-atomic contributions, and (v) the total binding energy (= the sum of the preceding two rows).

1.4.6.1 First Column: Zeroth Order Approximation to ψ_A, ψ_B by the $1s_\mathrm{A}$, $1s_\mathrm{B}$ Hydrogen Atom Orbitals

The signs of all contributions in this column correspond exactly to the general quantitative predictions made in Section 1.4.5. The quasi-classical potential contribution (see Section 1.4.5.2) as well as the potential interference contribution

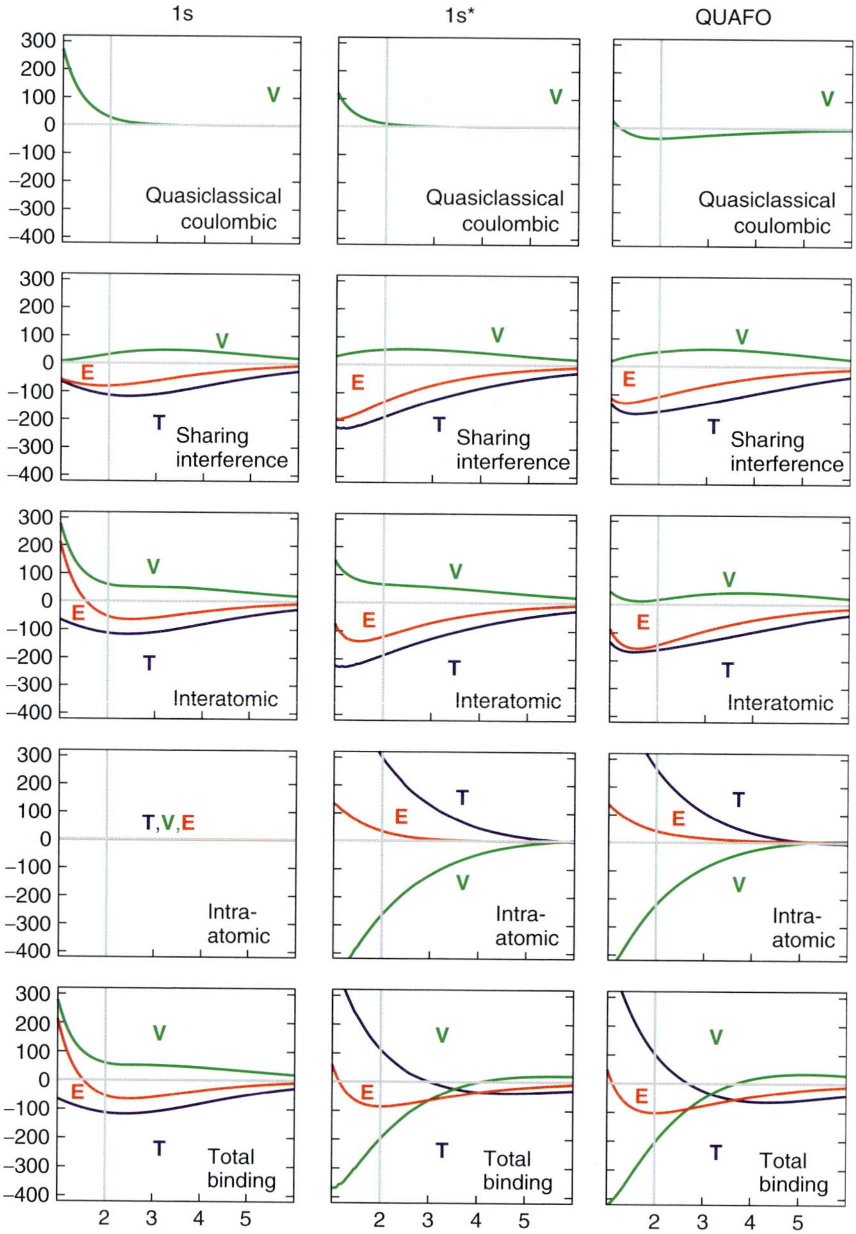

Figure 1.9 Each *column of panels* exhibits the contributions to the binding energy of H_2^+ as functions of the interatomic distance. First column: when the molecular wave function is a superposition of atomic ground state 1s orbitals. Second column: when the molecular wave function is a superposition of the scaled exponential 1s* approximations to the QUAFOs (see Figure 1.5). Third column: when the molecular wave function is a superposition of the QUAFOs, that is, for the actual wave function.

(see Section 1.4.5.3) is positive so that the *total interatomic potential energy is antibonding over the whole range. However, the negative (bonding) kinetic interference energy, which prevails at all distances, is stronger (see Section* 1.4.5.4*).* No intra-atomic deformation energy exists, of course, in this approximation. *The total molecular binding energy is entirely due to the kinetic interference energy lowering that is a result of the delocalization.*

1.4.6.2 Second Column: Optimal Spherical Approximation to ψ_A, ψ_B by the Scaled Orbitals $1s_A{}^*$, $1s_B{}^*$

As noted in the last paragraph of Section 1.4.1 (see also Figure 1.5), these quasi-atomic orbitals are very slightly expanded with respect to the hydrogen 1s orbital when $R > \sim 5$ Bohr. The reason is that the expansion yields an energy lowering beyond that of the first column because the (bonding) kinetic interference energy is enhanced by expansion in this region of R (see Section 1.4.5.5) even though the (antibonding) potential interference (Section 1.4.5.5) as well as the (antibonding) quasi-classical energy (Section 1.4.5.2) is enhanced. The slight deformation of the atom is also unfavorable (Section 1.4.5.1). However, all these effects are small at large distances so that all energy contributions differ little from those in the first column.

For $R < \sim 4$ Bohr on the other hand, the 1s* orbitals increasingly contract relative to the hydrogen 1s orbital, as discussed in Section 1.4.1. The reason is that, in this region, contraction yields a marked lowering of the total energy beyond that obtained with the uncontracted orbitals shown in the first column of Figure 1.9, mainly *because of the bonding enhancement of the kinetic interference energy by increasing orbital contraction* (see discussion in Section 1.4.5.5). In addition, the *quasi-classical* energy $\mathbf{V}_{qc}$ is less repulsive than that in the first column due to enhanced shielding of the nuclei (see Section 1.4.5.2). These effects prevail over the following two changes that oppose the contraction.

First, quasi-atomic orbital contraction enhances the antibonding potential interference energy $\mathbf{V}_I$ beyond that of the first column (see Section 1.4.5.5). This increased antibonding in fact largely offsets the decrease in the repulsive quasi-classical energy $\mathbf{V}_{qc}$ noted in the preceding paragraph. Therefore, the total interatomic potential energy contribution $\mathbf{V}_{inter}$ is about equally antibonding as in the first column, that is, for the undeformed 1s orbitals.

Secondly, the intra-atomic energy $\mathbf{E}_{intra}$ increases as a result of the orbital deformation by virtue of the *atomic* variation principle, as was discussed in Section 1.4.5.1. It is noteworthy that this, not very large change in $\mathbf{E}_{intra}$, is the sum of a strong increase in the intra-atomic kinetic energy $\mathbf{T}_{intra}$ prevailing over an only slightly less strong decrease in the intra-atomic potential energy $\mathbf{V}_{intra}$. These changes are in fact so strong that, in this region, the kinetic contribution $\mathbf{T}_{binding}$ and the potential contribution $\mathbf{V}_{binding}$ to the total binding energy have signs that are opposite to those that they have at larger distances. This feature will be commented on further in Section 1.4.7.

1.4.6.3 Third Column: Exact Quasi-Atomic Orbitals ψ_A, ψ_B

The exact quasi-atomic orbitals differ from the 1s* orbitals by the polarization terms as shown in Eq. (1.11) and Figures 1.3 and 1.4. Because these are relatively small, the energy curves in the third column of Figure 1.9 are similar to those in the second column.

The energy lowering with respect to the second column is essentially caused by the fact that the quasi-classical energy V_{qc} becomes attractive as a result of the quasi-atomic orbital polarization as mentioned in Section 1.4.5.2. This energy lowering outweighs a lesser energy increase in the interference energy and in the intra-atomic energy.

The interference energy becomes slightly less bonding because polarization slightly decreases the charge accumulation in the bond. The intra-atomic energy increases because polarization is achieved by moving some of the intra-atomic electron population from an s-type orbital to a p-type orbital of the same spatial size.

1.4.6.4 Conclusion

At all internuclear distances, it is the kinetic interference interaction between the two atoms that drives the energy lowering that establishes bonding. The response of the kinetic interference interaction to swelling and shrinking of the quasi-atomic orbitals is, moreover, responsible for the peculiar variations of the kinetic energy and the potential energy with the internuclear distance.

It should be noted that, at about half the equilibrium distance, the transition to the united atom regime begins so that, for shorter internuclear distances, the physical analysis requires modification.

1.4.7
Origin of Bonding at the Equilibrium Distance

The essential factors that emerge from the detailed preceding analysis of the bond-creating quantum physical mechanism are epitomized in Table 1.1, which lists the quantitative values of the binding energy contributions *at the equilibrium distance*. The rows and columns of this table have exactly the same meanings as those described in the second paragraph of Section 1.4.6 for Figure 1.9. In light of the assessments of the preceding section, the data of Table 1.1 lead to the following conclusions on the origin of the binding energy at the equilibrium distance.

1.4.7.1 Contributions to the Binding Energy

A full understanding of the energy lowering requires that *intra*-atomic as well as *interatomic* energy changes are accounted for.

The *intra*-atomic energy changes represent deformations of the free-atom AOs to the quasi-atomic orbitals in the molecule.

The *interatomic* interactions are of two kinds: (i) quasi-classical coulombic potential energy changes due to the electrostatic interactions between the densities of the two atoms and (ii) additional energy changes that result from the quantum

Table 1.1 Binding energy analysis of H_2^+ at R_{eq} (energies in mh).

Atomic orbitals used → Interaction type ↓		Free atoms	Contracted free atoms	Quasi-atoms in molecule
Quasi-classical	V	27.6	11.3	−36.2
Interference I	T	−113.7	−187.2	−161.8
	V	32.4	54.9	53.4
	E	−81.3	−132.3	−108.4
Interatomic total	T	−113.7	−187.2	−161.8
	V	60.1	66.2	17.2
	E	−53.6	−122.0	−144.6
Intra-atomic	T	0	298.8	264.4
	V	0	−264.0	−222.4
	E	0	34.8	42.0
Total Binding	T	−113.7	111.7	102.632
	V	60.1	−197.8	−205.266
	E	−53.6	−86.2	−102.634

mechanical delocalization of the electron orbital from one atom to both atoms ("electron sharing").

The effect of delocalization is embodied in constructive interference terms between the wave amplitudes of the AOs from the two atoms (for the antibonding orbital the interference is destructive). This constructive interference generates a shift of charge into the bond region ("charge accumulation") as well as an attenuation of the gradient in the bond region.

In the interference energy as well as in the intra-atomic deformation energy, the kinetic contribution and the potential contribution differ in their effects on bonding.

1.4.7.2 **Energy Lowering By Electron Sharing**

The energy lowering that creates the bond is driven by the *interatomic* interactions.

This is most clearly seen when the quasi-atomic orbitals are approximated by the undeformed orbitals of the hydrogen atom so that the intra-atomic energy changes vanish, as shown in the first column of Table 1.1. In this case the energy lowering at the equilibrium distance (−53.6 mh) is caused entirely by the *kinetic* interference energy (−113.7 mh) as a result of orbital delocalization. The potential interference energy and the potential quasi-classical energies are both positive, that is, antibonding (+32.4 and +27.6 mh, respectively).

In fact, beyond thrice the equilibrium distance, the exact binding energy comes about just in this way (albeit on a smaller scale), since the quasi-atomic orbitals do not deform at these distances. However, at less than twice the equilibrium distance intra-atomic orbital deformations occur and lower the energy further.

1.4.7.3 Energy Lowering by Quasi-Atomic Orbital Deformation

The deformations of the hydrogen atom orbitals into the quasi-atomic orbitals unavoidably *increase the intra-atomic* energy by virtue of the *atomic* variation principle. However, the deformations *decrease the interatomic* energy contributions even more, thus yielding a decrease of the *molecular* energy.

The deformations are of two kinds: spherical contraction and polarization. The second column of Table 1.1 contains the binding energy resolution after contraction. The third column shows the resolution after polarization is also added.

The spherical contraction of the hydrogen AO (second column) lowers the *inter-atomic* energy in two ways. Most importantly, *it increases the electron delocalization* between the quasi-atomic orbitals so that the negative kinetic interference energy is lowered by an additional −73.5 mh. The contraction also diminishes the quasi-classical electrostatic repulsion by −16.3 mh because of the increased shielding of the proton. This potential energy lowering is however smaller than the increase in the potential interference energy (+22.5 mh), which the aforementioned enhanced delocalization entails, so that the antibonding of the *total inter*atomic *potential* energy contribution increases by 6.1 mh. The total interatomic energy lowering by contraction is thus −67.4 mh and it is entirely due to the *kinetic* interference energy. This *interatomic* energy lowering is still stronger than the antibonding *intra-atomic* energy increase of 34.8 mh that occurs because AO contraction raises the intra-atomic kinetic energy more (by +298.8 mh) than lowering the intra-atomic potential energy (−264 mh), in agreement with the general discussion in Section 1.2.2. The net energy lowering due to contraction is thus −32.6 mh.

The polarization of the quasi-atomic orbitals (third column), on the other hand, lowers the *interatomic* quasi-classical electrostatic energy by −47.5 mh through shifting the *intra-atomic* charge on each nucleus slightly toward the other nucleus. However, polarization raises the negative interference energy by +23.9 mh so that the total *interatomic* energy is only lowered by −23.6 mh. Moreover, polarization increases the *intra-atomic* energy by +7.2 mh because it represents an intra-atomic charge shift that involves moving some electron population from a spherical orbital to higher angular momentum orbitals without orbital swelling. The net energy lowering due to polarization is thus only −16.4 mh.

1.4.7.4 Variational Perspective

The preceding analysis shows that, for every choice of quasi-atomic orbitals, *the interatomic kinetic energy change because of interference is the driving interaction, whereas the sum total of the interatomic potential interactions is always antibonding, as are of course the total intra-atomic energy changes.* Nonetheless, the *largest individual* energy changes are the *intra-atomic* changes in the kinetic and potential energies (+298.8 and −264 mh, respectively) and, although they nearly cancel each other, they actually invert the signs of the total contributions of T and V to the binding energy. This, somewhat puzzling, incidental result of the interplay between interatomic and intra-atomic energy changes can be understood by the following variational analysis.

In the molecule, the *interatomic* kinetic energy *lowering* is included in the variational competition between the total **T** and the total **V** and hence, the *intra-atomic* kinetic energy can *increase* beyond that of the free atom before the optimal compromise between **T** and **V** is reached. Hence, a contraction occurs that lowers the intra-atomic potential energy in a way similar to what would happen in a free atom when the kinetic energy resistance to localization is weakened by increasing the electron mass from 1 to about 1.275 (see Section 1.2.2). Thus, the *interatomic* kinetic energy lowering through delocalization can be said to open the possibility for an adjustment of the ratio between the *intra-atomic* electrostatic pull and kinetic resistance in the direction of a tighter attachment of the electron cloud to the nuclei than exists in the free atom, which leads to a corresponding energy lowering. One can also say that the weakening of the *overall* kinetic energy pressure resulting from *interatomic delocalization* allows for an increased *intra-atomic localization* through which the electron wave can exploit more effectively the attractive potential regions near the nuclei.

1.4.7.5 General Implications

One may reasonably ask whether the described interplay between interatomic and intra-atomic energy changes is a peculiar quirk of H_2^+ or whether it also occurs in other molecules. In this context, the virial theorem discussed in Section 1.2.3 is relevant. According to Eq. (1.9), the ratio $|2\mathbf{T}/\mathbf{V}|$ must be equal to 1 *at the equilibrium geometry of any molecule.* The following reasoning shows that this requirement entails the coupling of electron sharing and intra-atomic deformations, in particular, contraction.

From the first column of Table 1.1 it is seen that the *kinetic energy lowering* that results from electron sharing without quasi-atomic deformation, in conjunction with the concomitant *potential energy increase,* lowers the virial ratio $|2\mathbf{T}/\mathbf{V}|$ from the free-atom value 1 to the value 0.82 at the equilibrium distance. But it has to be 1 for the correct molecular wave function. According to the discussion in the second but last paragraph of Section 1.2.3, the exact wave function must therefore be more localized in regions of low potential energy. This localization is achieved by the intra-atomic contraction of the quasi-atomic orbitals, which entails the necessary large changes in the intra-atomic kinetic and potential energy (even though the total intra-atomic energy increase is small) that are needed to establish the virial ratio of 1.

In this regard, it is helpful to note that the contracted quasi-atomic orbital 1s* (second column of Table 1.1), which approximates the spherical component of the exact quasi-atomic orbital, is in fact very close to that contracted orbital $1s^{\#}$ that *minimizes the energy* of the approximate wave function $(1s_A{}^{\#} + 1s_B{}^{\#})/\sqrt{(2 + 2S^{\#})}$. [23] This orbital $1s^{\#}$ has the exponent $\zeta^{\#} = 1.240$, as compared to the exponent $\zeta^{*} = 1.264$ of 1s*, which was given at the end of Section 1.4.1. The corresponding molecular values of $\mathbf{E}^{\#}$, $\mathbf{T}^{\#}$, $\mathbf{V}^{\#}$ are therefore close to those for 1s*. Since energy optimization with respect to ζ guarantees the virial ratio $|2\mathbf{T}/\mathbf{V}| = 1$, this relation is indeed fulfilled by $\mathbf{T}^{\#}$ and $\mathbf{V}^{\#}$. It is therefore also justified to reason in terms of the variational competition between **T** and **V** with reference to the 1s* wave function listed in column 2 of Table 1.1. On the other hand, it is apparent that the proper compromise between electrostatic pull and kinetic resistance cannot be accomplished by a polarization deformation.

Finally, it is to be noted that the following relationships are *generally and rigorously valid*: (i) The virial ratio $|2T/V|$ is equal 1 at all equilibrium geometries. (ii) Orbital exponent optimization always establishes the virial ratio. (iii) Electron delocalization always lowers the kinetic energy. In conjunction, these observations lead to the inference that any covalent bond that is formed by electron sharing, that is, delocalization, can be expected to involve orbital deformations that will attach the electrons more firmly to the nuclei. This anticipation will be confirmed by the analyses of the other molecules examined in Sections 1.5 and 1.6.

1.5
The Effect of Electronic Interaction in the Covalent Electron Pair Bond: H_2

Since the two electrons in the ground state of H_2 have opposite spins and, thus, are allowed to occupy the same function space, they are shared in a similar manner between the two atoms. The cumulative result of the bonding generated by each electron produces the pair bond. The binding energy analysis of H_2 is therefore very similar to that of H_2^+, showing that covalent bonding is also a one-electron phenomenon within an electron pair bond. However, electron sharing enhances the mutual interpenetration of the electrons, each of which originally resides on one atom, *and this enhanced penetration increases the interelectronic repulsion*. For this reason, the binding energy of H_2 is only 85% of twice the binding energy of H_2^+. The elucidation of this interelectronic effect is the new element to be traced in the analysis of H_2.

Following the approach laid out in Section 1.3.2, the binding energy of the optimal wave function in the FORS space will be analyzed first. This two-determinant wave function recovers 87.3% of the binding energy. The remaining 12.7%, contributed by the dynamic interelectronic correlation, will then be examined by means of a full CI wave function of 3176 determinants.

All the wave functions to be discussed were calculated using an uncontracted (14s, 6p, 3d, 2f, 1g) basis of 26 σ-type spherical Gaussian AOs on each atom, which was optimized for the present problem and differs slightly from that used for H_2^+.[1)] It yields the energy −0.4999993 hartree and the virial ratio $|2T/V| = 1.0000197$ for the hydrogen atom. The quality of the present approach for the molecule was assessed by means of a full CI calculation at the internuclear distance $R = 1.4000$ Bohr, for which Kolos [24] as well as Nakatsuji [25], using the interelectronic distance in the wave function, have produced benchmark calculations of highest accuracy. The full

1) The uncontracted basis set for H_2 was optimized to minimize the FCI energy at $R = 1.4$ bohr and contains GTOs with the following orbital exponents: $s = 26110.0$, 4078.0, 952.7, 243.8, 70.77, 23.75, 8.713, 3.628, 1.904, 1.243, 0.6419, 0.2978, 0.1382, 0.06620; $p = 11.13$, 3.858, 2.261, 1.079, 0.4752, 0.1952; $d = 3.111$, 1.232, 0.4749, $f = 2.661$, 0.9583, $g = 2.271$.

CI energy obtained with the present basis was found to lie 0.1991 mh above the value of −1.1744757 hartree determined by the two mentioned authors.

The FORS calculation yields the equilibrium distance 1.425859 Bohr, the energy of 1.1522764 hartree and the virial ratio $|2\mathbf{T}/\mathbf{V}| = 1.0000023$. The full CI calculation yields the equilibrium distance 1.4012 Bohr, the energy −1.1742769 hartree and the virial ratio $|2\mathbf{T}/\mathbf{V}| = 1.0000015$. The best bond length known is 1.4011 Bohr [26]. The quantitative details on which the discussion in the present section is based will be elaborated in a separate report [27].

1.5.1 Quasi-Atomic Orbitals of the FORS Wave Function

There are two atomic minimal basis set orbitals in this system. According to the discussion of Section 1.3.2, the FORS MO space is therefore by definition two-dimensional. It can be spanned by the bonding orbital σ_g and the antibonding sigma orbital σ_u, both of which are partially occupied. At all internuclear distances, this FORS wave function was obtained by MCSCF optimization of the expression

$$\Psi(1,2) = c_g\sigma_g(1)\sigma_g(2) + c_u\sigma_u(1)\sigma_u(2) \tag{1.43}$$

which determined the coefficients c_g, c_u as well as the orbitals σ_g, σ_u in terms of the 52-dimensional AO basis mentioned above. At the equilibrium distance, the coefficients are $\{c_g, c_u\} = \{0.99370277, -0.11204818\}$. The two quasi-atomic orbitals ψ_A, ψ_B that span the same FORS orbital space, that is, the QUAFOs, were then obtained as follows.

To determine ψ_A, the (2×26) overlap matrix between the MCSCF orbital set (σ_g, σ_u) and the set of the 26 sigma basis AOs on atom A mentioned earlier was calculated. Its SVD yielded two orbitals in the FORS orbital space. At all internuclear distances, the largest SVD diagonal element is larger than 0.999780 (the value for 2.2 Bohr). The corresponding orbital in the FORS space is the QUAFO (quasi-atomic FORS orbital) ψ_A. The corresponding orbital that is spanned by the 26 basis AO's on atom A is the PAAO (Pure Atomic Approximate Orbital) to the QUAFO ψ_A. In contrast to the PAAO, the QUAFO contains a slight admixture from the basis AOs on atom B. In generalization of Eq. (1.11) for H_2^+, the QUAFO of H_2 on atom A can therefore be expressed as

$$\psi_A = a\mathbf{s}_A + b\mathbf{p}_A + \chi_B \tag{1.44}$$

$$\mathbf{s}_A = \sum(14s_A),\quad \mathbf{p}_A = \sum(6p_A, 3d_A, 2f_A, 1g_A),\quad \chi_B = \sum(\text{orbitals on atom B}) \tag{1.44a}$$

where $\mathbf{s}_A$ and $\mathbf{p}_A$ are normalized. By expressing the spherical component $\mathbf{s}_A$ as the sum of a projection on the orbital $\mathbf{h}_A = 1s(H)$ of the free hydrogen atom A and its orthogonal complement, the Eq. (1.44) can be written

$$\psi_A = a\langle\mathbf{s}_A|\mathbf{h}\rangle\mathbf{h} + a(\mathbf{s}_A - \langle\mathbf{s}_A|\mathbf{h}\rangle\mathbf{h} + b\mathbf{p}_A + \chi_B \tag{1.44b}$$

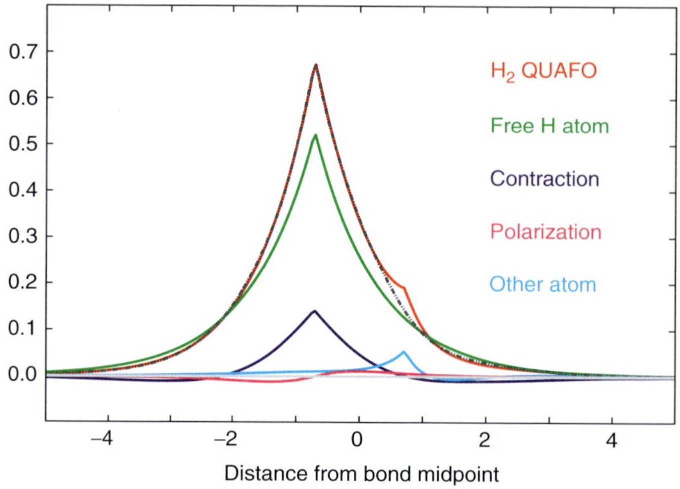

Figure 1.10 Resolution of the quasi-atomic FORS orbital (QUAFO) on the left atom in H_2 as a sum of contributions from the atomic 1s orbital, the spherical deformation, the angular deformation and the contribution from the right atom at the equilibrium distance. The dashed curve is the pure atomic approximate orbital (PAAO) to the QUAFO (see Section 1.5.1). Distances are in bohr from the bond midpoint.

The four terms of this resolution of ψ_A are displayed in Figure 1.10 for the equilibrium distance. It is apparent that, as in H_2^+, the second term represents a contraction of the spherical part.

Also shown in Figure 1.10, as a dotted curve, is the PAAO corresponding to the QUAFO ψ_A. The plot shows that the PAAO approximates the QUAFO extremely closely. The overlap integral between the two orbitals is 0.999865. A CI calculation of H_2 with the PAAOs yields an energy that is only 2 mh above that obtained with the QUAFOs, that is, the actual energy for Ψ.

The fractional contributions of contraction, polarization ($\mathbf{p}_A$), and the contribution from the other atom ($\boldsymbol{\chi}_B$) to ψ_A are exhibited in Figure 1.11 as functions of the internuclear distance. The values plotted in this Figure were obtained as follows. Since $\boldsymbol{\chi}_B$ is non-orthogonal to $\mathbf{s}_A$ and $\mathbf{p}_A$, the contribution from $\boldsymbol{\chi}_B$ that is actually different from the orbitals on A is given by $[\boldsymbol{\chi}_B - \langle\boldsymbol{\chi}_B|\mathbf{s}_A\rangle\mathbf{s}_A - \langle\boldsymbol{\chi}_B|\mathbf{p}_A\rangle\mathbf{p}_A]$ and this is the contribution shown in Figure 1.11. Accordingly, the contraction contribution shown is $[(a + \langle\boldsymbol{\chi}_B|\mathbf{s}_A\rangle)(\mathbf{s}_A - \langle\mathbf{s}_A|\mathbf{h}\rangle\mathbf{h})]$ and the polarization contribution shown is $[(b + \langle\boldsymbol{\chi}_B|\mathbf{p}_A\rangle)\ \mathbf{p}_A]$. It is apparent that, in H_2, the spherical contraction is of considerably greater relative importance than in H_2^+. The reasons will become apparent further on.

As in H_2^+, the spherical component $\mathbf{s}_A$ is very similar to a contracted hydrogen-like 1s orbital. Figure 1.12a shows the exponent ζ^* of this 1s* orbital, determined by maximizing the overlap of 1s* with $\mathbf{s}_A$ of Eq. (1.44a), as a function of the internuclear distance. The overlap between this 1s* orbital and the QUAFO ψ_A is

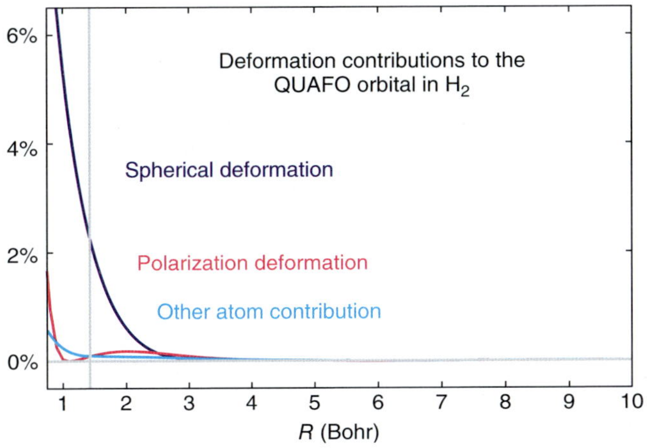

Figure 1.11 Variation with the internuclear distance for the spherical deformation, the angular deformations and the contribution from the other atom to the quasi-atomic orbital (QUAFO) in H_2. The gray vertical line indicates the equilibrium distance.

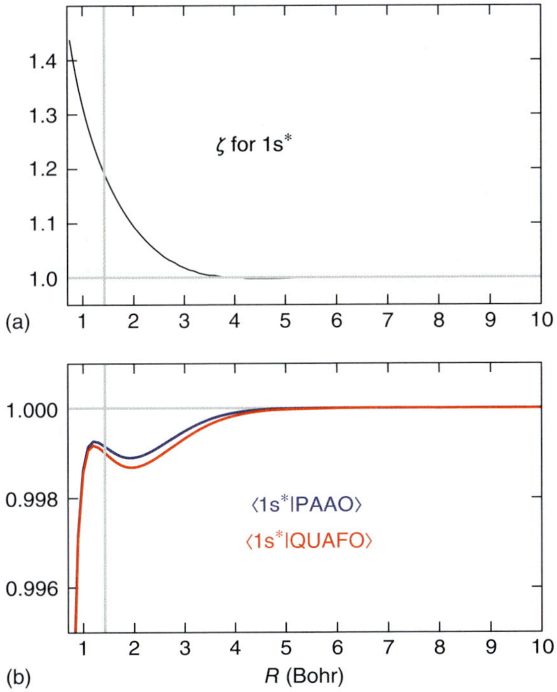

Figure 1.12 Characteristics of the scaled exponential approximation 1s* to the spherical part of the quasi-atomic orbital (QUAFO) in H_2 at various internuclear distances. (a) Orbital exponent of 1s*. (b) Overlap integral between the 1s* orbital and the quasi-atomic orbital (QUAFO) in red. Also shown: overlap integral between the 1s* orbital and the pure atomic orbital approximation (PAAO) to the QUAFO in blue.

shown in Figure 1.12b as a function of the distance. At the equilibrium distance the exponent of 1s* is $\zeta^* = 1.192$.

The QUAFO on atom B, ψ_B, was obtained in an entirely analogous way, using the 26 basis AOs on B for the SVD. The QUAFOs ψ_A and ψ_B were then chosen as the quasi-atomic basis to span the FORS space. They are related to the MCSCF bonding /antibonding MOs of Eq. (1.43) by the transformation

$$\psi_A = \sigma_g \cos\alpha + \sigma_u \sin\alpha \tag{1.45a}$$

$$\psi_B = \sigma_g \cos\alpha - \sigma_u \sin\alpha \tag{1.45b}$$

At the equilibrium distance, the value of $\cos\alpha$ is 0.920259 and the overlap integral between the non-orthogonal quasi-atomic orbitals is

$$S = \langle\psi_A|\psi_B\rangle = \cos 2\alpha = 0.693754 \tag{1.46}$$

With increasing internuclear distance, $\cos\alpha$ decreases to $1/\sqrt{2}$ and S decreases to zero.

1.5.2
FORS Wave Function and Density in Terms of Quasi-Atomic Orbitals

Using Eq. (1.45), the MCSCF optimized FORS wave function of Eq. (1.43) was then expressed in terms of the quasi-atomic orbitals, which yielded the valence-bond type expression

$$\Psi(1,2) = \left(\frac{N}{2}\right)^{\frac{1}{2}} \{\cos\gamma[\psi_A(1)\psi_B(2) + \psi_B(1)\psi_A(2)] + \sin\gamma[\psi_A(1)\psi_A(2) + \psi_B(1)\psi_B(2)]\} \tag{1.47}$$

$$N = \frac{1}{(1 + S^2 + 2S\sin 2\gamma)} \tag{1.48}$$

The coefficients $\cos\gamma$ and $\sin\gamma$ are related to the coefficients c_g, c_u, $\cos\alpha$, $\sin\alpha$ of Eqs. (1.43) and (1.45) by

$$\sin\gamma = \frac{C_1}{\sqrt{(C_1^2 + C_2^2)}} \qquad \cos\gamma = \frac{C_2}{\sqrt{(C_1^2 + C_2^2)}} \tag{1.49a}$$

$$4C_1 = \left(\frac{c_g}{\cos^2\alpha}\right) + \left(\frac{c_u}{\sin^2\alpha}\right), \quad 4C_2 = \left(\frac{c_g}{\cos^2\alpha}\right) - \left(\frac{c_u}{\sin^2\alpha}\right) \tag{1.49b}$$

At the equilibrium distance, the value of γ is 13.051068°.

The first order density matrix obtained from this wave function can be resolved into a quasi-atomic and an interference contribution between the quasi-atomic orbitals in a way that is entirely analogous to the corresponding resolution in H_2^+. One finds

$$\rho(1', 1'') = 2\int dx_2\,\Psi(1', 2)\Psi(1'', 2) = \rho_{qa}(1', 1'') + \rho_I(1', 1'') \tag{1.50}$$

where the quasi-atomic density matrix is the superposition of the atomic density matrices, that is,

$$\rho_{qa}(1', 1'') = \psi_A(1')\psi_B(1'') + \psi_B(1')\psi_A(1'') \tag{1.51}$$

and the interference density matrix is

$$\rho_I(1', 1'') = p \times \{[\psi_A(1')\psi_B(1'') + \psi_B(1')\psi_A(1'')] - S[(\psi_A(1')\psi_A(1'') + \psi_B(1')\psi_B(1'')]\} \tag{1.52}$$

with the *bond order*

$$p = \frac{(S + \sin 2\gamma)}{(1 + S^2 + 2S \sin 2\gamma)} \tag{1.53}$$

The diagonal terms, that is, the interference *density*, which is relevant for the potential energy terms, becomes therefore

$$\rho_I(1, 1) = \rho_I(1) = p \times \mathcal{I}_{AB} \tag{1.54}$$

where

$$\mathcal{I}_{AB} = 2\psi_A\psi_B - S(\psi_A^2 + \psi_B^2) \tag{1.54a}$$

is the same *orbital interference density* as the one that occurred in Eq. (1.14a) for H_2^+. The bond order p depends on the mixing ratio between the covalent and the ionic contribution to Ψ. For the Hartree–Fock wave function (i.e., $\gamma = 45°$ in Eq. (1.47)), it is $p = 1/(1 + S)$ so that the interference density expression of Eq. (1.54) becomes exactly twice that of H_2^+ given in Eq. (1.14). For the wave function of Eq. (1.47), it has the value 0.541996.

An implication of Eq. (1.54a) is that the orbital product $\psi_A\psi_B$ can be resolved into a coulombic and an interference component according to

$$2\psi_A\psi_B = S(\psi_A^2 + \psi_B^2) + \mathcal{I}_{AB} \tag{1.54b}$$

This *interference resolution of the product* $\psi_A\psi_B$ will be used to analyze the role of the electron interaction terms in the binding energy analysis.

1.5.3 Binding Energy as a Sum of Two Intra-atomic and Five Interatomic Contributions

1.5.3.1 Overall Resolution

By inserting the discussed resolution of the density into the energy expression resulting from $\langle\Psi|\mathcal{H}|\Psi\rangle$ with the Hamiltonian of the H_2 molecule, and after appropriately combining terms, one obtains the following resolution of the binding energy in terms of contributions with physical meanings.

The total binding energy is again the sum of intra-atomic and interatomic energy changes and both are again resolved into kinetic and potential contributions:

$$\mathbf{E}_{\text{Binding}} = \mathbf{E}(\Psi) - 2\mathbf{E}_H = \mathbf{E}_{\text{intra}} + \mathbf{E}_{\text{inter}} \tag{1.55}$$

$$\mathbf{E}_{\text{intra}} = \mathbf{T}_{\text{intra}} + \mathbf{V}_{\text{intra}} \tag{1.56}$$

$$\mathbf{E}_{\text{inter}} = \mathbf{T}_{\text{inter}} + \mathbf{V}_{\text{inter}} \tag{1.57}$$

The intra-atomic contributions are entirely analogous to those of H_2^+ in Eqs. (1.24), (1.31), and (1.37), except that, now, a full electron resides on each of the two atoms:

$$\mathbf{T}_{\text{intra}} = \frac{1}{2}\int dx(\nabla\psi_A)^2 + \frac{1}{2}\int dx(\nabla\psi_B)^2 - 1.0\,\text{Hartree} \tag{1.58}$$

$$\mathbf{V}_{\text{intra}} = -\int dx\frac{\psi_A^2}{r_A} - \int dx\frac{\psi_B^2}{r_B} + 2.0\,\text{Hartree} \tag{1.59}$$

The *interatomic kinetic energy* change is again the result of the interference of the quasi-atomic orbitals, that is,

$$\begin{aligned}\mathbf{T}_{\text{inter}} = \mathbf{T}_{\text{I}} &= p\times\int dx\left\{2\left(\psi_A\left(-\frac{1}{2}\nabla^2\right)\psi_B\right)\right.\\ &\quad\left.-\left[\left(\psi_A\left(-\frac{1}{2}\nabla^2\right)\psi_A\right)+\left(\psi_B\left(-\frac{1}{2}\nabla^2\right)\psi_B\right)\right]\right\}\\ &= p\times\frac{1}{2}\int dx\left\{2\left(\nabla\psi_A\right)\cdot(\nabla\psi_B)-[(\nabla\psi_A)^2+(\nabla\psi_B)^2]\right\}\end{aligned} \tag{1.60}$$

where p is the bond order given in Eq. (1.53). Equation (1.60) is entirely analogous to Eq. (1.32) for H_2^+.

The *interatomic potential interactions* contain two kinds of terms, as was the case for H_2^+, namely:

- *coulombic* terms, which are defined as electrostatic interactions between *non-zero* charges,
- *interference* terms, that is, electrostatic interactions, which involve interference charge distributions that have *zero integrated charges* and describe *charge shifts with respect to the coulombic terms.*

Thus:

$$\mathbf{V}_{\text{inter}} = \mathbf{V}_{\text{coulombic}} + \mathbf{V}_{\text{interference}} \tag{1.61}$$

The contributions of the *interelectronic repulsion terms* to $\mathbf{V}_{\text{coulombic}}$ and $\mathbf{V}_{\text{interference}}$ are obtained by *substituting the interference Eq.* (1.54b) *for the orbital products* $\psi_A\psi_B$ *in the energy expression of* H_2.

1.5.3.2 Interatomic Coulombic Contributions

By the just mentioned substitutions, the total *coulombic* part of the interelectronic interaction is found to be

$$(1-2q)\langle\langle\psi_A^2|\psi_B^2\rangle\rangle + q\langle\langle\psi_A^2|\psi_A^2\rangle\rangle + q\langle\langle\psi_B^2|\psi_B^2\rangle\rangle \tag{1.62}$$

In this as well as the following equations of this section, light gray shading identifies terms arising from interelectronic interactions and the electrostatic integral between two charge distributions $f(x, y, z)$ and $g(x, y, z)$ is denoted by

$$\langle\langle f|g\rangle\rangle = \int dx_1\int dx_2\frac{f(1)g(2)}{|r_1 - r_2|} \tag{1.63}$$

The factors q and $(1-2q)$ in Eq. (1.62) have the following origin. Since *each* electron is *shared* between both atoms, *both* electrons contribute to the *average charge of 1e on each atom* and there exists a finite probability of finding both electrons on the same atom, that is, occupying the same quasi-atomic orbital. The analysis of the second-order density shows that (with N and γ being defined in Eqs. (1.47), (1.48) and (1.49))

$$q = \frac{(1-N\cos 2\gamma)}{4} \tag{1.64}$$

is the probability of finding both electrons on the same atom and, correspondingly, that $(1-2q)$ is the probability of finding one electron on atom A and the other on atom B. The value of q is maximal {viz 0.25} for the MO wave function ($\gamma = 45°$) and minimal, but not zero {viz $(S^2/4(1+S^2) = 0.081299$} for the valence bond (VB) wave function ($\gamma = 0°$). It is 0.142673 for the wave function of Eq. (1.47).

Combining the coulombic electron interaction terms of Eq. (1.62) with the electron nuclear *coulombic* attractions and the internuclear repulsion, the total *coulombic interaction* is expressed as follows

$$\mathbf{V}_{\text{coulombic}} = \mathbf{V}_{\text{qc}} + \mathbf{V}_{\text{sc}} \tag{1.65}$$

Where

$$\mathbf{V}_{\text{qc}} = -\int dx \frac{\psi_A^2}{r_B} - \int dx \frac{\psi_B^2}{r_A} + \langle\langle \psi_A^2 | \psi_B^2 \rangle\rangle + \frac{1}{R}$$

= the quasi-classical coulombic interaction between the two atoms, each having one full electron (1.66)

$$\mathbf{V}_{\text{sc}} = q \times \left\{ \langle\langle \psi_A^2 | \psi_A^2 \rangle\rangle + \langle\langle \psi_B^2 | \psi_B^2 \rangle\rangle - 2 \langle\langle \psi_A^2 | \psi_B^2 \rangle\rangle \right\}$$

= a "correction" of the preceding term $\mathbf{V}_{\text{qc}}$ so that the sum of the interelectronic terms in $\mathbf{V}_{\text{coulombic}}$ of Eq. (1.65) becomes in fact identical with the actual terms given in Eq. (1.62) (1.67)

The quasi-classical coulombic term $\mathbf{V}_{\text{qc}}$ of Eq. (1.66) has been formulated so as to balance attractions and repulsions between the atoms in a way analogous to the quasi-classical interaction term in H_2^+ given in Eq. (1.25). The term $\mathbf{V}_{\text{sc}}$ manifestly *replaces* an appropriate amount of *interatomic* electron repulsion, included by definition in $\mathbf{V}_{\text{qc}}$, with the amount of *intra-atomic* electron repulsion required to recover the *actual* contribution given in Eq. (1.62). Since the *intra-atomic* electron repulsions are generated by electron sharing, the energy contribution $\mathbf{V}_{\text{sc}}$ is called the *coulombic sharing contribution*. It is the energetic measure of the finite probability of both electrons being on the same atom. Although rarely mentioned, it is quite strong and antibonding.

1.5.3.3 **Interatomic Interference Contributions**

The *potential interference energy* is the sum of two contributions

$$\mathbf{V}_{\text{interference}} = \mathbf{V}_{\text{I}} + \mathbf{V}_{\text{II}} \tag{1.68}$$

The first term is analogous to the potential interference energy of H_2^+ given by Eq. (1.26). Here, one obtains

$$\mathbf{V}_\mathrm{I} = \int \mathrm{d}x \left(-\frac{1}{r_\mathrm{A}} - \frac{1}{r_\mathrm{B}}\right) \rho_\mathrm{I} + \frac{1}{2} \langle\langle \psi_\mathrm{A}^2 + \psi_\mathrm{B}^2 | \rho_\mathrm{I} \rangle\rangle$$
$$= p \times \left\{ \int \mathrm{d}x \left[\left(-\frac{1}{r_\mathrm{A}} - \frac{1}{r_\mathrm{B}}\right) \mathcal{I}_\mathrm{AB} + \frac{1}{2} \langle\langle \psi_\mathrm{A}^2 + \psi_\mathrm{B}^2 | \mathcal{I}_\mathrm{AB} \rangle\rangle \right] \right\} \quad (1.69)$$

where ρ_I and $\mathcal{I}_\mathrm{AB}$ are the orbital interference terms of Eqs. (1.54) and (1.54a) and p is the bond order of Eq. (1.53). The difference between this expression for $\mathbf{V}_\mathrm{I}$ and that for H_2^+ in Eq. (1.26) is the presence of the electron repulsion terms $\langle\langle ..|..\rangle\rangle$, which have the following origin: The interference density of each electron experiences the attractions of the two *shielded* nuclei. The shielding that one electron experiences is caused by the coulombic repulsion of the other electron, which is evenly distributed over both nuclei, so that each nucleus is shielded by half an electron.

The second interference term in Eq. (1.68) is

$$\mathbf{V}_\mathrm{II} = \frac{1}{4} N \langle\langle \mathcal{I}_\mathrm{AB} | \mathcal{I}_\mathrm{AB} \rangle\rangle \quad (1.70)$$

It represents the interaction between the interference energies of the two electrons.

1.5.3.4 Binding Energy as a Sum of Two Intra-atomic and Five Interatomic Contributions

The total binding energy is then the sum of the following seven contributions with physical interpretations discussed in Sections 1.5.3.1 to 1.5.3.3:

$$\mathbf{E}_\mathrm{Binding} = \mathbf{T}_\mathrm{intra} + \mathbf{V}_\mathrm{intra} + \mathbf{T}_\mathrm{I} + \mathbf{V}_\mathrm{qc} + \mathbf{V}_\mathrm{sc} + \mathbf{V}_\mathrm{I} + \mathbf{V}_\mathrm{II} \quad (1.71)$$

This decomposition has not been arbitrarily conceived or imposed. Rather, it is the result of simply sorting out the terms in the rigorous energy expression $\langle\Psi|\mathcal{H}|\Psi\rangle$ after inserting the interference resolution of Eq. (1.54b) for $\psi_\mathrm{A}\psi_\mathrm{B}$.

1.5.4 Quantitative Synergism of the Contributions to the Binding Energy

1.5.4.1 Quantitative Characteristics

From the discussion in the preceding section it is apparent that five of the seven contributions to the binding energy Eq. (1.70), namely $\mathbf{T}_\mathrm{intra}$, $\mathbf{V}_\mathrm{intra}$, $\mathbf{V}_\mathrm{qc}$, $\mathbf{T}_\mathrm{I}$, $\mathbf{V}_\mathrm{I}$, have the same physical meanings as the corresponding contributions in H_2^+, which were discussed in Section 1.4.4.

These contributions also exhibit the same general *quantitative* characteristics as those discussed for H_2^+ in Section 1.4.5. An exception is the quasi-classical coulombic energy $\mathbf{V}_\mathrm{qc}$, which differs from twice that of H_2^+ in that one *internuclear* repulsion is replaced by the *interelectronic* repulsion, as is seen by comparing the definition of Eq. (1.66) for H_2 with that of Eq. (1.25) for H_2^+. As a consequence and in contrast to H_2^+, the term $\mathbf{V}_\mathrm{qc}$ is, therefore, *always* attractive in H_2.

More importantly, the binding energy of Eq. (1.71) also contains the *coulombic sharing contribution* V_{sc} of Eq. (1.67), which does not exist in H_2^+. It arises from both electrons spending part of the time simultaneously on the same atom. As explained after Eq. (1.67), it replaces an *interatomic* electron repulsion by *intra-atomic* repulsions. Since the latter are significantly larger than the former, this term is always positive, that is, antibonding. As a result the *sum total coulombic contribution* $V_{coulombic} = V_{qc} + V_{sc}$ *of Eq. (1.65) is in fact antibonding at all internuclear distances*, even when the quasi-atomic orbitals are polarized. It is a two-electron price to be paid for the energy lowering gained by simultaneous one-electron sharing of the two electrons in the same bond. The compromise between V_{sc} and the interference energy T_I determines what is sometimes called *left–right correlation*.

Finally, the binding energy of Eq. (1.70) contains the second interference term V_{II}. It represents the self-energy of a distribution with zero total charge and is therefore expected to have very small numerical values.

1.5.4.2 **Synergism along the Dissociation Curve**

The synergism of the various contributions to the binding energy is shown in Figure 1.13 for H_2, which corresponds to Figure 1.9 for H_2^+. The rows and columns of the 15 panels have exactly the same meaning as in Figure 1.9 and were explained in detail in the second paragraph of Section 1.4.6. The comparison of the two figures shows that, notwithstanding differences in quantitative details, all binding energy contributions exhibit essentially the same bonding and antibonding pattern over the full range of the interatomic distance.

The differences between H_2 and H_2^+, which are a result of the presence of the interelectronic repulsions, appear mainly in the panels of the first row, that is, for the coulombic contributions. Whereas only the quasi-classical coulombic term V_{qc} is present in H_2^+, the coulombic contribution of H_2 contains in addition the coulombic sharing contribution V_{sc}. In confirmation of the explanation elaborated in the preceding Section 1.5.4.1, it is seen that, *at all distances and on all three panels, i.e., for all three quasi-atomic orbital choices, the quasi-classical term* V_{qc} *is attractive, the sharing modification* V_{sc} *is repulsive and the total coulombic contribution* $V_{coulombic}$ *is repulsive*. Nonetheless, when the spherical 1s* quasi-atomic orbital approximation (second column) is replaced by the polarized exact quasi-atomic orbitals (third column), the coulombic contributions of H_2 and H_2^+ change in the same direction: In H_2^+, the contribution changes from repulsive to attractive; in H_2, it becomes less repulsive. Specifically, the quasi-classical term of H_2 becomes more attractive and the coulombic sharing contribution becomes less antibonding. Thus, in both molecules, the polarization of the quasi-atomic orbitals, which results from replacing the 1s* quasi-atomic orbital approximation (second column) by the exact quasi-atomic orbitals (third column), lowers the binding energy because it lowers the coulombic energy. This lowering outweighs the increases in the interference energy and the intra-atomic energy caused by the polarization as discussed in

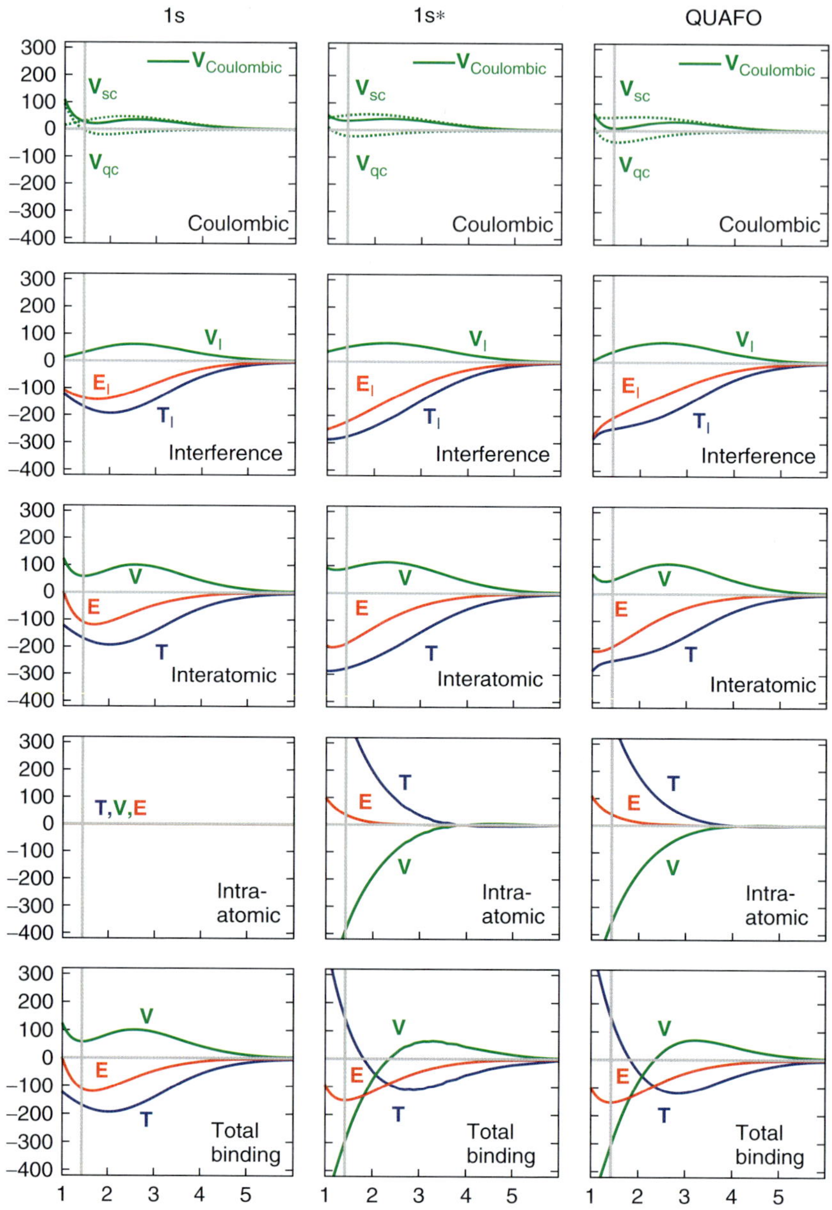

Figure 1.13 Each *column of panels* exhibits the contributions to the binding energy of H_2 as functions of the interatomic distance. First column: when the molecular wave function is a superposition of atomic ground state 1s orbitals. Second column: when the molecular wave function is a superposition of the scaled exponential 1s* approximations to the QUAFOs (see Figure 1.12). Third column: when the molecular wave function is a superposition of the QUAFOs, that is, for the actual wave function.

Section 1.4.6.3. Nonetheless, the total inter-atomic potential contribution to the binding energy remains anti-bonding over the whole range.

As noted earlier, the second interference term is expected to have very small values. In fact, these values are so small that they are indistinguishable from zero on the scale of the panels in the second row *in Figure* 1.13 and are therefore not displayed. The total potential interference energy shown in these panels is therefore essentially equal to $\mathbf{V}_I$ of Eq. (1.69).

The overall synergism of the binding energy contributions in H_2 is thus the same as that in H_2^+ and the general conclusions deduced for H_2^+ by the in-depth discussions in Sections 1.4.5 and 1.4.6 apply therefore also to H_2.

1.5.5
Origin of Bonding at the Equilibrium Distance

Detailed insights into the effect of the interelectronic interaction on the binding energy are provided by the explicit quantitative values at the equilibrium distance, which are listed in Table 1.2. The table is analogous to Table 1.1 for H_2^+ and is organized in the same manner. The rows and columns have the same meaning in the two tables and they also correspond to those in Figure 1.13, which was discussed in the preceding Section 1.5.4.2.

Table 1.2 Binding energy analysis of H_2 at R_{eq} (energies in mh).

Atomic orbitals used → **Interaction type ↓**		**Free atoms**	**Contracted free atoms**	**Quasi-atoms in molecule**
Quasi-classical coulombic	V	−4.7	−21.6	−41.0
Sharing Coulombic	V	30.8	56.2	50.4
Interference I	T	−167.8	−275.7	−245.0
	V	31.6	54.7	40.2
	E	−136.2	−221.0	−204.8
Interference II	V	1.0	1.6	1.6
Interatomic total	T	−167.8	−275.7	−245.0
	V	58.7	90.9	51.2
	E	−109.1	−184.8	−193.8
Intra-atomic	T	0	420.9	397.3
	V	0	−384.0	−355.8
	E	0	36.9	41.5
Total binding	T	−167.8	145.2	152.282
	V	58.7	−293.1	−304.559
	E	−109.1	−147.9	−152.276

The overall quantitative energetic pattern is indeed analogous to that found for H_2^+. In particular the *interatomic* energy lowering is the driver for all three quasi-atomic orbital choices (columns). There are however noteworthy differences.

1.5.5.1 The Primary Mechanism as Exhibited by Choosing the Free-Atom Orbitals as Quasi-Atomic Orbitals

As in H_2^+, the binding energy in the first column of Table 1.2 is again furnished by the *kinetic interference* energy (−167.8 mh), that is, electron delocalization. It is 54.1 mh more bonding than that of H_2^+ (−113.7 mh). But the former is less than twice the latter, presumably because the bond order p of H_2 [~0.542 see after Eq. (1.54a)] is less than twice that in H_2^+ [~0.315, see after Eq. (1.14a)].

On the other hand, the antibonding *potential interference* energy in H_2 (+31.6 + 1.0 = +32.6 mh) differs only by 0.2 mh from that of H_2^+, even though two electrons are involved. The reason is presumably that the potential acting on the interference density ρ_I of H_2 in Eq. (1.69) is weaker than the corresponding potential for H_2^+ in Eq. (1.26) because, in H_2, each nucleus is shielded by half an electron as explained after Eq. (1.69).

Although the *quasi-classical* coulombic energy of H_2 is attractive (−4.7 mh), as explained in Section 1.5.4.1, and in fact 32.3 mh lower than the repulsive quasi-classical energy in H_2^+, the *total coulombic* energy of H_2 is repulsive (+26.1 mh) and only 1.5 mh lower than that of H_2^+. This is because of the additional coulombic repulsion that is a result of the finite probability of both electrons being on the same atom, as explained in the third paragraph of Section 1.5.4.1. It gives rise to the antibonding *coulombic sharing energy* V_{SC}, of +30.8 mh!

As a result the interatomic *potential* contributions add up to the antibonding value of +58.7 mh against the bonding *kinetic* contribution of −167.8 mh, yielding the binding energy of −109.1 mh.

1.5.5.2 Effect of Quasi-Atomic Orbital Contraction

As discussed in detail in Section 1.4.3, near the equilibrium distance the contraction shown in the second column of Table 1.2 enhances interference, delocalization and charge accumulation in the bond. As in H_2^+, the *kinetic interference energy* is thus enhanced to −275.7 mh, that is, by a factor 1.64, which is near identical to the corresponding enhancement factor in H_2^+. The antibonding *potential interference energy* increases by a factor 1.73, which is also similar to that factor in H_2^+.

The contraction lowers the *quasi-classical coulombic* energy by −16.9 mh (second minus first column), that is, by almost the same amount as in H_2^+ (−16.3 mh). On the other hand however, it is manifest from the Eq. (1.67) and the subsequent discussion that quasi-atomic orbital contraction will *increase the sharing coulombic energy* V_{SC}. This increase is in fact 25.4 mh so that the antibonding of the *total coulombic energy increases* by +8.5 mh. In H_2^+ by contrast, where V_{SC} does not exist, the coulombic energy *decreases* by −16.3 mh.

As a result, contraction increases the total antibonding *interatomic potential* contributions by a factor 1.54 in H_2 as compared to a factor of only 1.1 in H_2^+. Since contraction increases the bonding *interatomic kinetic* contributions in H_2

and H_2^+ by the same factor (about 1.7, see two paragraphs earlier), contraction enhances the *total interatomic* bonding contribution of H_2 (−184.8) only by a factor 1.69 whereas that factor is 2.26 in H_2^+.

Contraction increases the intra-atomic energy by +36.9 mh, which is only 2.1 mh more than that in H_2^+ (the contraction is less than in H_2^+, but two electrons contribute).

As a result, the total binding energy of H_2 (−147.9 mh) has been enhanced by a factor of 1.36 through quasi-atomic orbital contraction. This factor is less than the factor 1.61 in H_2^+.

1.5.5.3 **Effect of Polarization**

The third column of Table 1.2 shows the effect of modifying the quasi-atomic orbitals by polarization as well as by the admixture from the other atom. For simplicity, both distortions of the exact quasi-atomic orbitals' spherical symmetry will be subsumed under the label "polarization" in the following paragraphs.

The effect of these distortions on the interference energy of H_2 is very similar to that in H_2^+: The kinetic interference contribution becomes less bonding, the potential interference contribution becomes less antibonding and the total interference energy becomes less bonding, that is, more positive. The increase in the value of the interference energy in H_2 (16.2 mh) is less than the corresponding increase in H_2^+ (23.9 mh).

As in H_2^+, polarization lowers the quasi-classical coulombic energy of H_2, that is, it makes it more bonding. But the energy lowering is less than half of that in H_2^+. It also lowers the antibonding sharing coulombic energy somewhat. The total coulombic energy lowering through polarization in H_2 (−25.2 mh) is considerably less than that in H_2^+ (−47.5 mh).

As a result the *total interatomic* bonding interactions in H_2^+ are enhanced by only −9.0 mh as compared to −23.6 mh in H_2^+.

Polarization increases the *intra-atomic* antibonding 4.6 mh, that is, 2.6 mh less than in H_2^+.

As a result, the *total binding energy* of H_2 is enhanced only by −4.4 mh through polarization whereas the enhancement in H_2^+ is −16.4 mh. The effect of the distortion of the quasi-atomic orbitals from spherical symmetry is thus much smaller in H_2 than in H_2^+.

1.5.5.4 **Binding in the Electron Pair Bond of H_2**

As in H_2^+, the driving element of binding in H_2 is the interatomic *kinetic* energy lowering through delocalization ("electron sharing"), which is embodied in the kinetic interference energy. This effect is a part of the cumulative *one-electron* energies of the two electrons. The two-electron bond is thus the cumulative result of the one-electron bonding created by each electron. The essential conclusions regarding the origin of the covalent bond in H_2^+, which were summarized in Section 1.4.7, remain therefore also valid for H_2.

These one-electron bonding effects are however diminished by the presence of the interelectronic repulsion because the bonding delocalization of *both* electrons also results in a finite probability of the two electrons being found at the same atom, and this simultaneous presence at the same site generates a strong coulombic repulsion that exists neither in the hydrogen atoms nor in the H_2^+ molecule. In consequence, the binding energy of H_2 is less than twice that of H_2^+.

In the energy analysis, the effects caused by the interelectronic repulsion appear in three places:

1) The *strong* antibonding contribution generated by the finite probability of finding both electrons near the same nucleus is embodied in the sharing coulombic energy V_{SC}, which is absent in H_2^+. Indeed, the value of V_{SC} at the equilibrium distance, viz 50.4 mh (Table 1.2) is very close to the difference of 52.0 mh between the binding energy of the FORS wave function (Table 1.2) and twice the binding energy of H_2^+ (Table 1.1).
2) Because of the electron repulsion, the delocalization of the individual electrons is moreover not as uninhibited as in H_2^+, a fact often denoted as "left–right correlation" As a result, the bond order, which weights the interference terms, is smaller than twice that in H_2^+.
3) The nuclear potentials that act on the interference density are shielded, which reduces the antibonding *potential* energy increase associated with the charge accumulation in the bond. But this enhancement of bonding is counteracted by the bond order attenuation mentioned under (ii) which decreases the bonding of the *kinetic* interference terms.

It should be noted that the large antibonding term V_{SC} is a *coulombic* interaction. Although it is generally appreciated that the *quasi-classical* coulombic energy is attractive in H_2, it seems to be much less recognized that electron sharing also generates the large repulsive coulombic term V_{SC} and that, in consequence, the *total coulombic contribution* is in fact always antibonding – in contrast to H_2^+, where it is bonding. In fact, as in H_2^+, the sum total of the inter-atomic potential contributions is always anti-bonding.

1.5.6
Electron Correlation Contribution to Bonding in H_2

As seen in the preceding parts of this section, the finite probability of both electrons being on the same atom, which is inherent in the FORS wave function, entails an electron repulsion that causes the FORS binding energy (−152.276 mh) to fall 52 mh short of twice the binding energy of H_2^+ (−102 mh). Of this difference, 22 mh are recovered by the exact wave function, in which the probability of the two electrons to find themselves in the same space element is reduced by adjustments beyond the FORS level, that are termed *dynamic correlations*. Thus, 87% of the actual binding energy (−174.476 mh) is accounted for at the FORS level and 13% is recovered by dynamic correlation.

To examine the correlating adjustments, a full configuration interaction (FCI) wave function was calculated in the configuration space spanned by the 3176 determinants that are generated by the 140 orbitals of the 14s, 6p, 3d, 2f, 1g basis mentioned in the third paragraph of Section 1.5. It yields the binding energy −174.168 at the FORS equilibrium distance (1.425859 Bohr), and −174.277 mh at the re-optimized equilibrium distance (1.4012 Bohr).

At the FORS equilibrium distance, where the FCI wave function recovers 99.8% of the binding energy, the dominant FCI natural orbitals (NOs) and their occupations are:

$1\sigma_g$	$1\sigma_u$	$2\sigma_g$	$1\pi x_u$	$1\pi y_u$	135 remaining NOs
1.96347	0.02067	0.00598	0.00428	0.00428	0.00132

The first two of these NOs closely resemble the bonding and antibonding FORS orbitals and account for 1.98414 electrons. The next three natural orbitals provide in–out and π angular correlation. The two configurations that form the FORS wave function clearly represent the dominant part of the nearly exact FCI function. According to the text after Eq. (1.43), the occupancies of the corresponding two NOs in the FORS wave function are 1.974890, and 0.025110.

To assess the effect of dynamic correlation on the bonding interactions at the equilibrium distance, the interference effects generated by the FCI wave function are compared with those of the FORS wave function in Figure 1.14. The interference densities, which show the charge accumulation in the bond, and the kinetic interference densities, which exhibit the effect of delocalization on the kinetic energy, were calculated as follows:

$$\text{Interference density} \qquad \sum_i^N n_i\phi_i^2 - \psi_A^2 - \psi_B^2$$

$$\text{Kinetic interference density} \qquad +\frac{1}{2}\sum_i^N n_i\{\nabla\phi_i\}^2 - \frac{1}{2}\{\nabla\psi_A\}^2 - \frac{1}{2}\{\nabla\psi_B\}^2$$

where the ϕ_i are the respective NOs, the n_i are their occupations and ψ_A and ψ_B are the QUAFO orbitals determined in Section 1.5.1. The upper limit N is 2 and 140 for the FORS and the FCI case respectively. These contour plots are analogous to those shown in Figure 1.6 for H_2^+.

It is apparent from the first two rows of panels in Figure 1.14 that the two-electron correlation effects modify the basic structure of the one-electron densities so little that the changes are hardly perceptible on the scale of these plots. The difference plots displayed in the third row have reduced contour increments: 4 times reduced on the left panel, 10 times reduced on the right panel. The result of electron correlation is seen to enhance very slightly the charge accumulation in the bond (lower left panel). As argued in Section 1.4.5.3 and shown by the subsequent quantitative results (e.g., Figures 1.9, 1.13 and Tables 1.1, 1.2), this enhancement is expected to slightly increase the (positive) potential energy. On the other hand, the correlation modification of the wave function slightly increases the gradient density in a very narrow region near the nuclei (lower right panel, in agreement

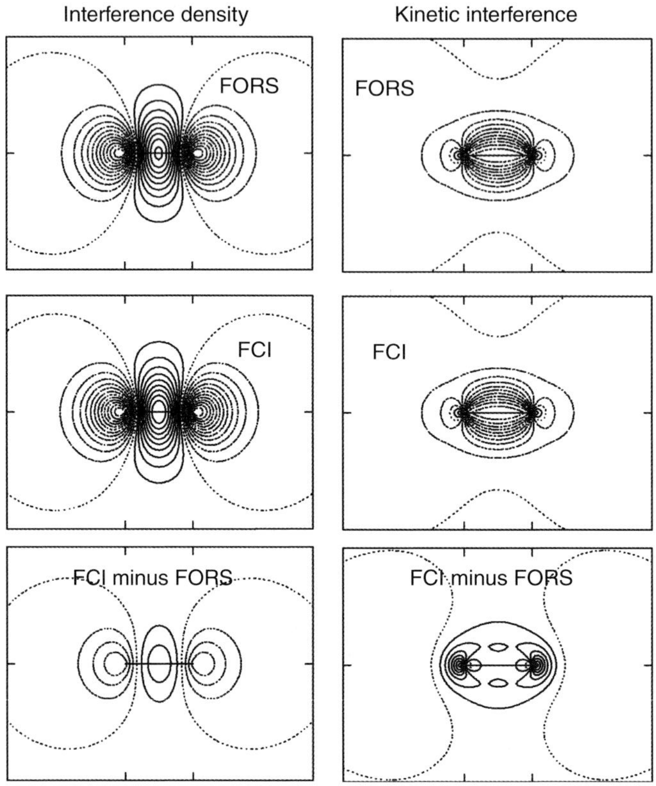

Figure 1.14 Comparison of the interference for the full CI wave function with that for the FORS wave function of H_2. Left column of panels: interference densities with contour increments = 0.004 e $Bohr^{-3}$ in the upper two panels and 0.001 e $Bohr^{-3}$ in the lowest panel. Right column of panels: kinetic interference densities with contour increments = 10 millihartree $Bohr^{-3}$ in the upper two panels and 1 millihartree $Bohr^{-3}$ in the lowest panel. Note that the increment in the upper four panel's is twice that used for the one-electron system H_2^+ in Figure 1.6.

with the lower left panel), which is expected to also increase the kinetic interference density slightly.

These inferences are confirmed by the decomposition of the binding energy in terms of its kinetic (**T**), nuclear-electronic attraction (V_{ne}), and electron–electron repulsion contributions (V_{ee}), which is documented along the internuclear distance in Figure 1.15. The two dissociation graphs at the top show that the relative roles of these three components are very similar in the FCI and the FORS wave function over the whole range.

The lower panel exhibits the modifications induced by the full CI calculations on a scale that is magnified by a factor 20. At the equilibrium distance, the one-electron energy changes are as predicted earlier from the interference plots in Figure 1.14. The increases of the one-electron energies **T** and V_{ne} are manifestly a side effect of the much larger lowering of the electron repulsion energy V_{ee} that is achieved

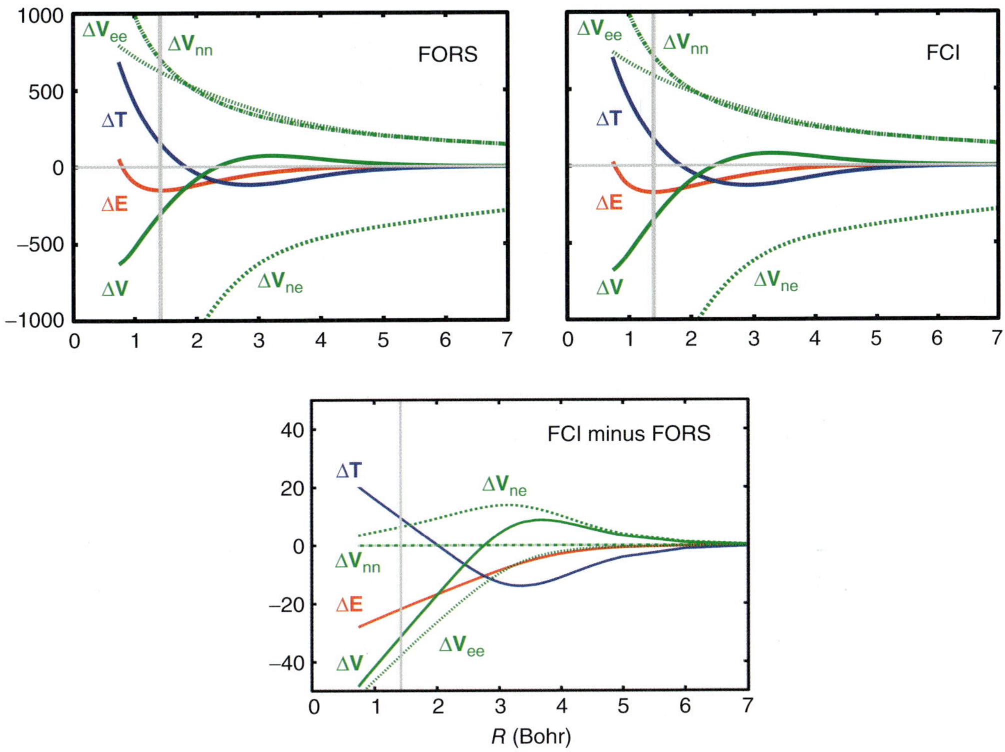

Figure 1.15 Comparison of the kinetic, potential, and total energies of the full CI wave function with those of the FORS wave function. Energy scale in millihartree.

by the correlating wave function adjustments. At large distances (>5 Bohr), on the other hand, the one-electron FCI energies still differ by about 5 micro-hartree from the corresponding FORS energies, even though the electronic interaction energy differs by less than one micro-hartree. This surprizing result may be related to the very long range of the (small) polarization effect that was noted in Figure 1.11.

The correlating wave function adjustments manifestly do not change the physical interactions that lead to bond formation at the FORS level. These interactions between the atomic minimal basis sets are left intact and remain dominant. The correlating adjustments achieve however a reduction of about 40% in the amount by which the bond energy of H_2 falls short of twice the bond energy of H_2^+.

1.6 Covalent Bonding in Molecules with More than Two Electrons: B_2, C_2, N_2, O_2, and F_2

The essential conclusion of the preceding analysis has been that covalent bond formation in H_2^+ and H_2 is the consequence of the attenuation of the kinetic energy pressure experienced by each electron due to its *interatomic* delocalization ("sharing"). This attenuation is further enhanced by intra-atomic contractions, which moreover allow the electronic wave to exploit more of the attractive potentials near the nuclei. The shrinkage toward the nuclei is signaled by large contragredient changes in the intra-atomic kinetic and potential energies. Although these changes are large, they are not the cause of the binding process inasmuch as they nearly cancel each other, leaving a small antibonding contribution. In the last paragraph of Section 1.4, it was argued that this process of covalent bonding is general. In the present section, it will be shown that, in the five diatomic molecules B_2, C_2, N_2, O_2, and F_2, covalent bonding involves, in fact, the same pattern of energy changes and thus comes about in the same way.

All molecules are treated at the full valence space MCSCF level, that is, the full FORS wave functions based on eight molecular valence orbitals is analyzed. In addition a wave function containing some additional valence correlation is considered for F_2. All calculations are performed with Dunning's quadruple-zeta cc-pVQZ basis sets [28] using the GAMESS molecular program suite [29].

1.6.1
Basis of Binding Energy Analysis

The detailed interactions in these systems are manifestly quite complex and a complete analysis is beyond the scope of this chapter. Only those aspects will be exhibited that have a bearing on the essential basic mechanism summarized in the first paragraph of this section. To this end, the following four energies are calculated for each molecule.

Energy (i): The energy of the *free atom* is calculated at the full valence space MCSCF level (FORS), the core being kept as a closed shell. This calculation is performed assuming the configurational structure that will result for the atom when the molecular calculation is done at large internuclear distances. Accordingly, the atomic calculation is performed in $C_{\infty v}$ symmetry, keeping the orbitals p_x and p_y equivalent, but the p_z orbital nonequivalent. The five minimal basis set orbitals are optimized in the quadruple-zeta AO basis (55 orbitals) of the atom.

Energy (ii): The energy of the *molecule* is calculated by a full valence space MCSCF calculation, the cores being closed shells. However, in this calculation the orbitals *are not optimized in the quadruple-zeta AO bases.* Rather, *all 10 orbitals* are determined as optimal linear combinations of the 10 optimal (core and valence) free-atom orbitals that were found in the preceding atomic calculations on the two atoms. The resulting molecular wave function represents the analogue to the wave function that was obtained for H_2 with the 1s ground state orbital of the hydrogen atom.

Energy (iii): The energy of the *molecule* is optimized by a full valence space MCSCF calculation, the cores being closed shells. In this calculation the orbitals are optimized in the full quadruple-zeta bases of both atoms. This is the FORS energy of the molecule.

Energy (iv): The energy of the *quasi-atom in the molecule* is determined, which represents the analogue to the quasi-atomic energies obtained in H_2^+ and H_2 by using the (contracted + polarized) quasi-atomic orbitals. This objective is accomplished as follows.

The first step is to determine a basis for quasi-atomic FORS orbitals (QUAFO's) that span the full FORS function space of the MCSCF MOs obtained in the preceding calculation (iii). To this end, the overlap integral matrix is calculated between, on the one hand, all 10 *molecular* FORS MCSCF orbitals (including the core orbitals) and, on the other hand, *all 55 orthogonal occupied and virtual free-atom orbitals obtained earlier in calculating the energy (i) on one of the two atoms*. Then, the SVD of this matrix is performed and those five MOs that correspond to the largest five SVD eigenvalues are taken as the optimal quasi-atomic orbital basis on that atom. By an analogous procedure, five quasi-atomic orbitals are determined for the other atom. The ten quasi-atomic orbitals obtained in this manner span the same orbital space as the molecular FORS orbitals from the molecular MCSCF calculation of step (iii). We consider these QUAFOs as the quasi-atomic orbital basis that is intrinsically embedded in the FORS wave function.

As typical examples, Figure 1.16 displays contours of the quasi-atomic MOs for the molecules N_2 and F_2. For N_2, all five QUAFOs on one atom are shown. The figure for F_2, on the other hand, omits the 1s and the equivalent p_y orbitals. It displays the three quasi-atomic FORS orbitals 2s, $2p_z$ and $2p_x$ and the two additional correlating orbitals 3s and $3p_x$, which were obtained from a wave function that included some valence correlation (see Section 1.6.4). The correlating quasi-atomic orbitals were obtained by an entirely analogous SVD algorithm involving 16 MCSCF orbitals. In view of the manifest atomic localization of these orbitals, we emphasize that all of them are *MOs* in terms of which the molecular wave functions can be expressed.

The energy of each quasi-atom in the molecule is then obtained by a full valence space MCSCF (FORS) atomic calculation with the same format as that used for the atomic calculation of Energy (i). However, the orbitals *are not optimized in the quadruple-zeta AO basis*. Rather, *all five orbitals* are determined as *optimal linear combinations of the five QUAFOs* that were obtained for that atom as described in the preceding paragraph.

1.6.2
Origin of Binding at the Equilibrium Geometry

The results obtained by these calculations for the five molecules at their theoretical equilibrium distances are contained in Table 1.3. For comparison, the corresponding values of H_2 are also included. Because the quantities of interest are the binding energies, all entries listed in this table are in fact energies *with reference to twice the free-atom energy*, i.e., $2 \times$ Energy (i) of the preceding Section 1.6.1 has been subtracted from all energies.

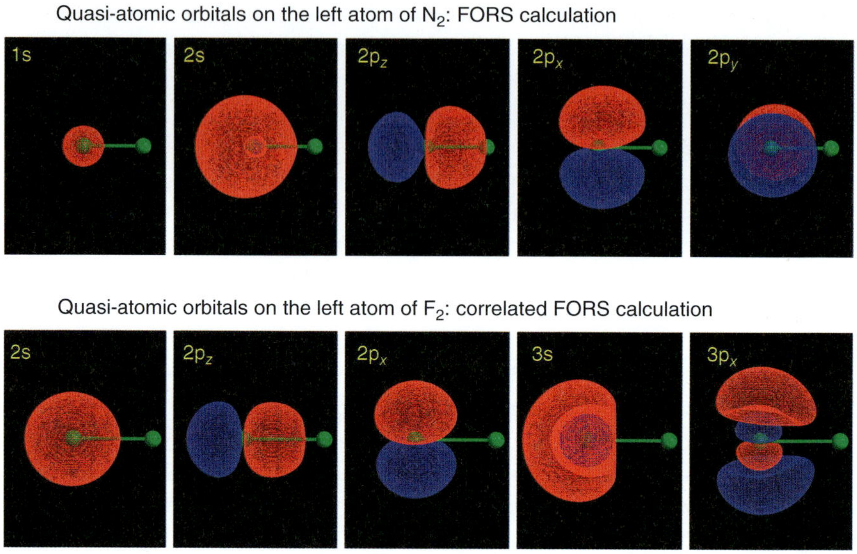

Figure 1.16 Upper panels: core and valence quasi-atomic orbitals of the FORS wave function (QUAFOs) in the N_2 molecule. Lower panels: valence quasi-atomic orbitals and two correlating quasi-atomic orbitals of a full MCSCF calculation that includes the correlating orbitals 3s, $3p_x$, $3p_y$ in the fluorine molecule (the p_y orbitals being equivalent to p_x orbitals are not shown,). For both molecules, only orbitals on one atom are shown. Note that all orbitals are *molecular* orbitals (see Section 1.3.3).

For each molecule, a panel of nine blocks of entries is displayed, each block consisting of the values for **T**, **V**, and **E**. The four corner blocks correspond to the four energies discussed in the preceding Section 1.6.1. The upper left corner corresponds to the two separated atoms. Since 2 × Energy (i) has been subtracted from all energies, this entry is always zero. The lower left corner contains the binding energy obtained using the unchanged optimal orbitals of the free atoms, i.e., it is (Energy (ii) minus 2 × Energy (i)). The upper right corner block lists the energy of the quasi-atoms in the molecule relative to the free atoms, i.e., it is 2 × (Energy (iv) minus Energy (i)). The lower right hand corner, finally, lists the actual FORS binding energy from the full MCSCF calculations, i.e., it is (Energy (iii) minus 2 × Energy (i)).

The other five blocks represent the energy *changes* between the respective adjacent blocks and their values were simply obtained by subtraction. Thus, the blocks in the *second column* show the energy differences that result from replacing the optimal free-atom orbitals by the deformed quasi-atomic orbitals of the molecule, and the blocks in the *second row of blocks* exhibit the differences between the molecular energies and the *intra-atomic* energies, i.e., the energy contributions due to the *interatomic* interactions.

Table 1.3 Energy changes from free-atom energies to molecular FORS energies, resolved in terms of intra- and interatomic contributions and in terms of free atom and deformed atomic orbitals.

Contributions to energy change		Free-atom orbitals	Orbital deformation	Quasi-atomic orbitals	Free-atom orbitals	Orbital deformation	Quasi-atomic orbitals
			H_2			B_2	
Intra-atomic contributions	T	0	397.3	397.3	0	564.6	564.6
	V	0	−355.8	−355.8	0	−507.4	−507.4
	E	0	41.5	41.5	0	57.1	57.1
Interatomic contributions	T	−167.8	−77.2	−245.0	−142.4	−325.1	−467.6
	V	58.7	−7.5	51.2	81.7	231.1	312.7
	E	−109.1	−84.7	−193.8	−60.8	−94.0	−154.8
Total binding energy	T	−167.8	320.1	152.3	−142.4	239.5	97.0
	V	58.7	−363.3	−304.6	81.7	−276.4	−194.7
	E	−109.1	−43.2	−152.3	−60.8	−36.9	−97.7
			C_2			N_2	
Intra-atomic contributions	T	0	1407.2	1407.2	0	2352	2352
	V	0	−1301.8	−1301.8	0	−2127	−2127
	E	0	105.4	105.4	0	225	225
Interatomic contributions	T	−608.0	−573.6	−1181.6	−955	−1056	−2011
	V	468.2	379.7	847.9	862	584	1446
	E	−139.8	−193.8	−333.6	−92	−472	−564
Total binding energy	T	−608.0	833.7	225.7	−955	1296	341
	V	468.2	−922.1	−453.8	862	−1543	−680
	E	−139.8	−88.4	−228.2	−92	−247	−339
			O_2			F_2	
Intra-atomic contributions	T	0	1264.9	1264.9	0	590.7	590.7
	V	0	−1119.3	−1119.3	0	−547.2	−547.2
	E	0	145.6	145.6	0	43.5	43.5
Interatomic contributions	T	−378.3	−734.8	−1113.1	−100.4	−464.1	−564.5
	V	357.2	459.3	816.4	97.2	393.3	490.5
	E	−21.2	−275.5	−296.7	−3.2	−70.8	−74.0
Total binding energy	T	−378.3	530.1	151.8	−100.4	126.7	26.3
	V	357.2	−660.0	−302.9	97.2	−153.9	−56.7
	E	−21.2	−129.9	−151.1	−3.2	−27.2	−30.4

Quadruple-zeta bases, energies in mh.

The following conclusions are manifest from these data.

1) The first column exhibits the bonding that is achieved when the free-atom orbitals are used in the molecular calculation. In all molecules, this bonding is seen to be the result of the lowering of the interatomic kinetic energy contributions that is stronger than a concomitant potential energy increase, exactly as in H_2^+ and H_2. There is no question that this binding is due to delocalization.
2) The center column in the first row of the blocks shows the *intra-atomic energy increase* when the optimal free-atom orbitals are replaced by the deformed quasi-atomic orbitals of the molecule. In all molecules, the intra-atomic *potential* energy *decreases* considerably, but the intra-atomic *kinetic* energy *increases* even more so that the total intra-atomic energy increases somewhat (as it must by the intra-atomic variation principle). These changes in the intra-atomic kinetic and potential energies show that the quasi-atomic orbital deformations are dominated by an overall contraction in all molecules, as was the case in H_2^+ and H_2.
3) The third column in the second row of the blocks shows the interatomic interactions that create the bond between the quasi-atoms calculated with the deformed quasi-atomic orbitals. These interactions are always stronger than the interactions between the undeformed atoms listed in the first column. This enhancement is indicated by blue-green highlighting of the energies in the second row of the blocks. In all molecules, as in H_2^+ and H_2, the enhancement is due to the interatomic *kinetic* contributions, which are indicated by yellow highlighting.
4) The values of the just mentioned enhancement of the interatomic interaction by quasi-atomic orbital deformation, are given in the second column of the second row of the blocks (i.e., the center block of the entire panel). It is always the result of a considerable enhancement of the interatomic *kinetic* contributions, which indicates that, the quasi-atomic orbital deformations increase the delocalization, as was discussed in detail for H_2^+ and H_2. On the other hand, the quasi-atomic orbital deformations render the interatomic *potential* contribution more antibonding in all molecules, except in H_2 where it becomes slightly less antibonding. In all systems (including H_2), the total interatomic potential energy contribution (last column, second row of the blocks) is very antibonding.
5) The lowest block in the third column shows the kinetic and potential energy decomposition of the FORS binding energy. These contributions are, respectively, positive and negative (as they must be by the virial theorem). From the data in the blocks in the first and second row of the last column, it is apparent that these signs are a consequence of the *intra-atomic orbital* contraction mentioned above, which are a side effect of the enhanced binding of the *interatomic kinetic* contributions due to orbital deformation, as has been discussed for H_2^+ and H_2.

In summary, bonding is brought about by the *kinetic energy* lowering in the *interatomic* interactions in all molecules. The kinetic energy lowering is further enhanced by a deformation of the quasi-atomic orbitals, which is an overall contraction. This enhancement prevails over the intra-atomic energy increase that is unavoidably also generated by the deformation. The intra-atomic energy increase is the result of large compensating intra-atomic kinetic and potential energy changes that cause the kinetic binding energy to be positive and the potential binding energy to be negative, but reveal no information about the origin of binding.

1.6.3
Synergism along the Dissociation Curve

The overall consistency that has been found for the essential bonding contributions at the equilibrium geometries extends to the entire dissociation curves. This similarity is exhibited in Figures 1.17–1.19 for the molecules B_2, C_2, N_2, O_2, and F_2. Each molecule is represented by one row of panels. In each row, from left to right:

- The first panel displays the kinetic, potential, and total energy curves for the *interatomic interactions obtained with the free-atom orbitals* (corresponding to item 1) in the preceding Section 1.6.2).
- The second panel displays the curves for the *interatomic interactions obtained with the deformed quasi-atomic orbitals of the molecules, that is, the QUAFOs* (corresponding to item 3) in the preceding Section 1.6.2).
- The third panel displays the curves for the *intra-atomic* energy changes caused by the deformations that change the free-atom orbitals into the QUAFOs (corresponding to item 2) in the preceding Section 1.6.2).
- The fourth panel shows the binding curves obtained by optimizing the QUAFOs at each internuclear distance (corresponding to item 5) in the preceding Section 1.6.2). The values of this FORS binding energy curve, as well as its kinetic and potential components in the fourth panel, are the sums of the corresponding values in the second and third panels.

Between the equilibrium distance, which is indicated by a gray vertical line, and infinite separation, the curves of all molecules exhibit the same overall pattern, namely:

Binding is provided by the *interatomic* interactions (first and second panel), specifically by the *kinetic* interatomic energy lowering, due to delocalization, prevailing over an interatomic potential energy increase, presumably due to charge accumulation in the bond and sharing coulombic effects. The interatomic interactions for the deformed quasi-atomic orbitals (second panel) are qualitatively similar to those for the undeformed free-atom orbitals (first panel). But the deformations enhance these contributions.

The *intra-atomic* contribution is the result of a large increase in the kinetic component prevailing slightly over a large decrease in the potential component,

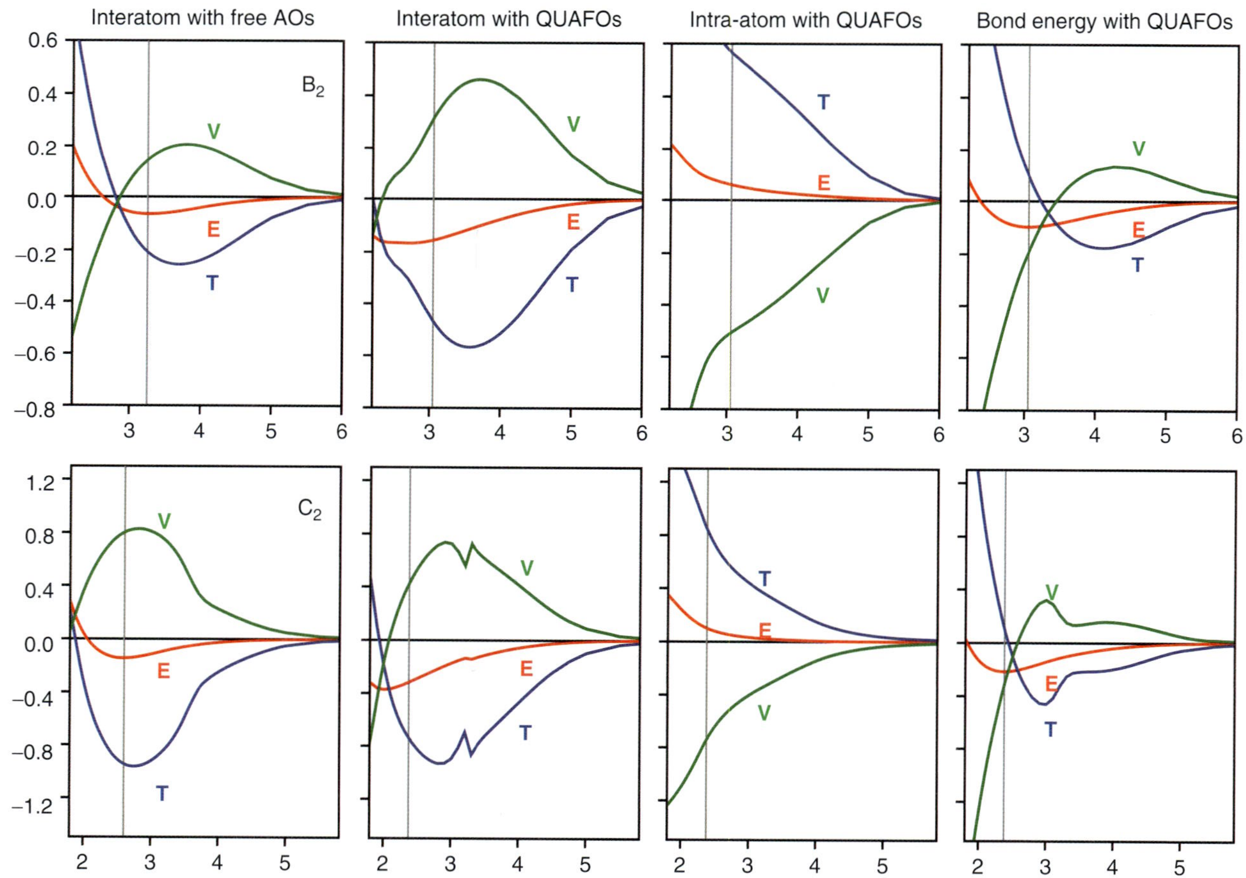

Figure 1.17 Contributions to kinetic, potential, and total binding energy curves of B_2 (first row of panels) and C_2 (second row of panels) for full valence space wave functions. First column of panels: *interatomic* contributions when the quasi-atomic orbitals are the optimized orbitals of the free atoms. Second column of panels: *interatomic* contributions when the quasi-atomic orbitals are the QUAFOs of the FORS wave function. Third column of panels: *intra-atomic* contributions due to deformation of the free-atom orbitals into QUAFOs. Fourth column of panels: FORS binding energy curves. Internuclear distance in Bohr. Energies in hartree.

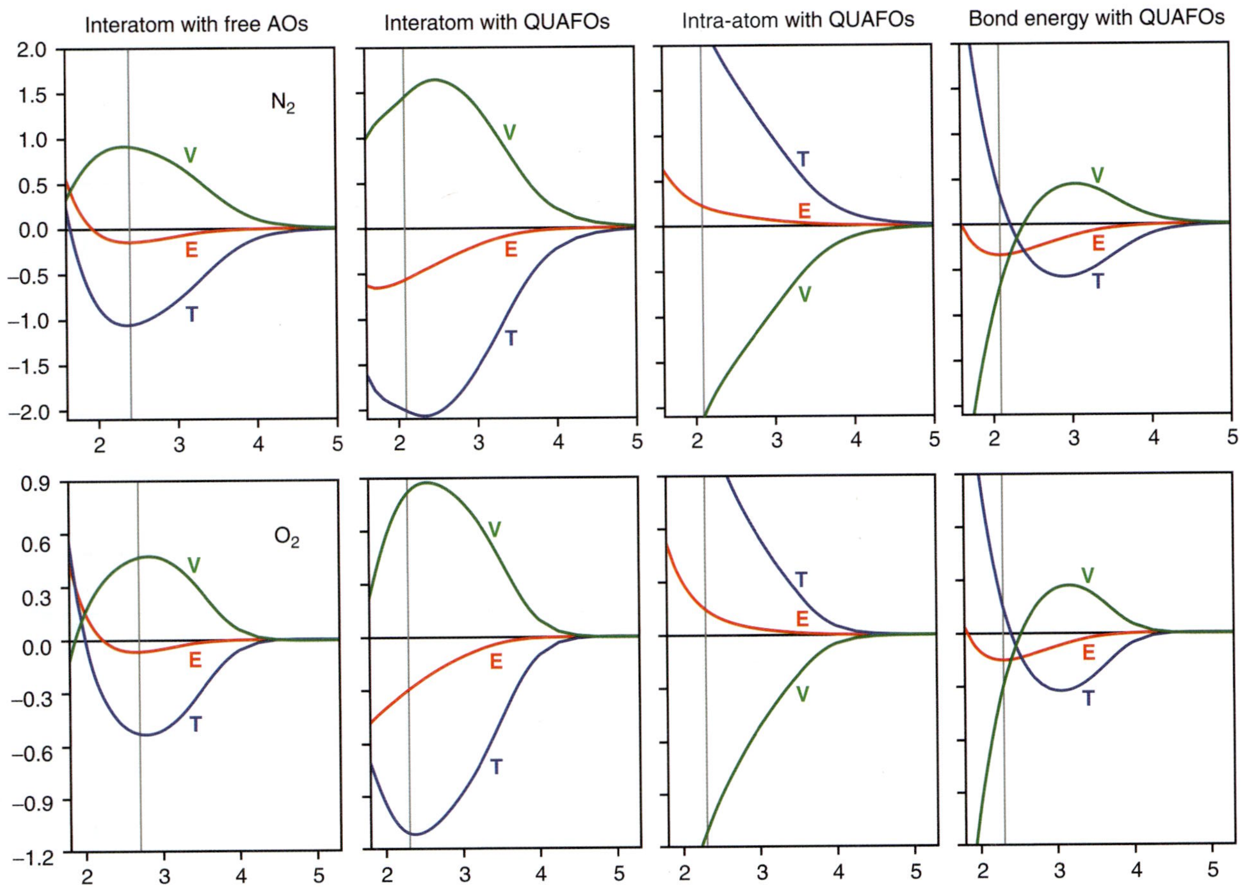

Figure 1.18 Contributions to kinetic, potential and total binding energy curves of N_2 (first row of panels) and O_2 (second row of panels) for full valence space wave functions. See caption of Figure 1.17 for a description of the column of panels.

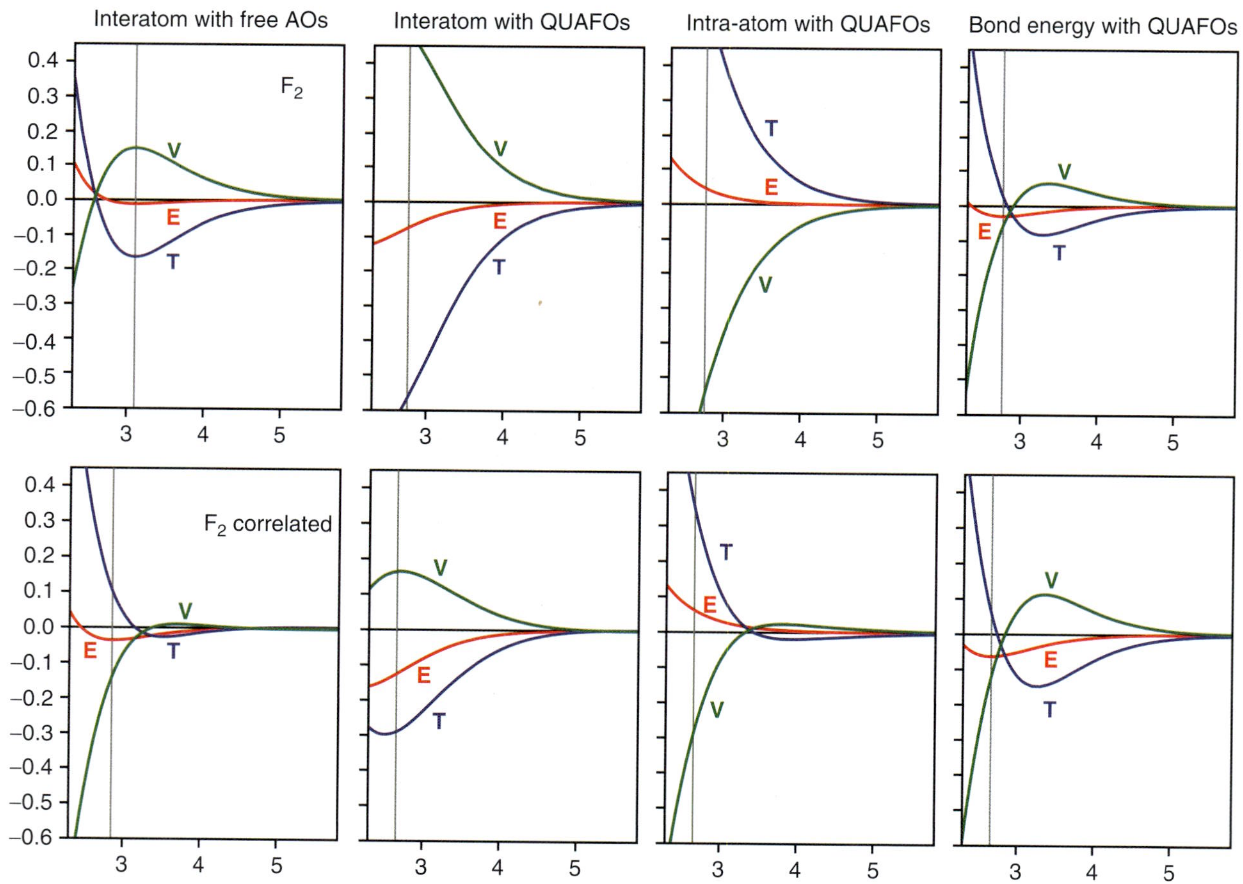

Figure 1.19 Contributions to kinetic, potential, and total binding energy curves of F_2. The first row of panels contains the data of the FORS calculations. The second row of panels contains the results of a full MCSCF calculation including, in addition, the correlating orbitals 3s, $3p_x$, $3p_y$ on both atoms (see Figure 1.16). See caption of Figure 1.17 for the description of the column of panels.

changes that imply intra-atomic contraction. Although the intra-atomic changes determine the signs of the kinetic and components of the binding energy, they yield an overall antibonding contribution to this energy.

It is evident that, in all of these molecules, the basic synergism that leads to covalent binding is entirely analogous to the one that was found for H_2^+ and H_2 and exhibited in Figures 1.9 and 1.13.

Parenthetically, it is noted that the kinks in the curves for C_2 are due to the avoided curve crossing that exists in this system. The curve shown is the lowest $^1\Sigma_g^+$ state obtained from a state-average MCSCF calculation over the three lowest states $X^1\Sigma_g^+$, $B'^1\Sigma_g^+$, $B^1\Delta_g$ (even though the Δ-state is lower at large distances). Appropriately corresponding state averaged calculations were performed for the carbon atom to obtain the dissociation limit of the lowest $^1\Sigma_g^+$ state.

1.6.4
Effect of Dynamic Correlation on Covalent Binding

The effect of dynamic electron correlation on the bond in the H_2 molecule was discussed in Section 1.5.6. In that case, it accounted for about 13% of the binding energy and it did not change the conclusions regarding the origin of covalent binding. It was also noticed that the energy lowering achieved by the correlating adjustments in the wave function are not able to compensate entirely for the adverse effect that electron repulsion has on bond formation at the FORS level through the sharing coulombic interaction.

The molecules treated in the present section differ from H_2 in that dynamic electron correlation already exists in the separate atoms. It is generally found that dynamic correlation in a molecule is larger than the sum of the dynamic correlations in the separate atoms, presumably because there is more space available for the electrons to avoid each other in the molecule than in the separate atoms. The data in the first three rows in Table 1.4 confirm this general trend for the molecules treated in the present section by furnishing a comparison of the FORS binding energies with the experimental binding energies (The experimental energy for B_2 is from Bytautas *et al.* [30] the others are from Feller and Sordo [31]). It can

Table 1.4 Dynamic correlation contributions to bond energies (mh).

	H_2	B_2	C_2	N_2	O_2	F_2
D_e FORS	−152.3	−97.7	−228.2	−339	−151.1	−30.4
D_e Experiment	−174.5	−107.9^a	−230.1^b	−364.1^b	−191.6^b	−60.9^b
% Dynamic correlation	13	9.5	0.8	7	21	50
D_e FORS+^{c}			−228.2	−360.5	−192.5	−63.9

^{a}From Bytautas *et al.*, Ref. [30]
^{b}From Feller and Sordo, Ref. [31]
^{c}See second and third but last paragraphs of Section 1.6.4.

be inferred that the bonds in B_2, C_2, N_2, O_2 are dominated by covalent bonding of the FORS wave functions, that is, by the interactions in the orbital space of the (optimized) minimal basis sets, which are the ones that were analyzed in the preceding sections.

In F_2, on the other hand, about half of the binding energy is due to a lowering of the dynamic correlation energy upon molecule formation. This does not invalidate the analysis of the FORS wave function in the preceding sections, which elucidates that part of the binding that is the result of the interactions within the orbital space of the minimal basis sets.

Remarkably, the binding energy is recovered within 3 mh when the dynamic correlation is simply accounted for by an extended MCSCF calculation that provides one correlating orbital for each of the six valence orbitals containing lone pairs. These correlating orbitals were shown in the lower panel set in Figure 1.16. The second row of Figure 1.19 exhibits the resolution of the dissociation curve obtained with this wave function in terms of contributions that are analogous to those formulated in Sections 1.6.3 and 1.6.4. It is apparent that even in this case, the basic pattern of the interatomic and intra-atomic contributions is the same as that obtained for the FORS wave function, which is shown in the first row of that figure.

In fact, analogous MCSCF calculations, based on wave functions with N valence electrons in N valence orbitals, yield good binding energies for N_2 and O_2 as well, as shown in the last row of Table 1.4, where they are denoted as FORS+ calculations.[2)] For the C_2 molecule, the minimal basis set FORS wave function provides already one orbital for each electron and this is presumably the reason why, here, the FORS calculation recovers the binding energy within 3.5 mh. All of these energies were calculated at the optimized equilibrium geometries.

A more extreme case is presented by the molecule Be_2, which would not exist without the help of the dynamic correlation interactions. However, as the detailed analysis of this bond by the present authors [32] showed, even in this case the contributions of the minimal basis set interference interactions are in fact essential for the existence of the bond.

1.7 Conclusions

As Robert Mulliken noted in the remark quoted at the beginning of this chapter, chemical bonding is more complicated than one would like it to be – a not uncommon experience in the sciences. The following précis summarizes the essential conclusions regarding covalent binding that emerge from the detailed analyses of the seven molecules examined in the preceding sections.

2) For the O_2 molecule, 14 orbitals are used, that is, the additional 3s, $3p_x$, $3p_y$ (but not $3p_z$) orbitals are added. This is because the separated atom calculation must be performed in cylindrical symmetry, which is done by a state-averaged MCSCF calculation involving the p_x as well as the p_y orbital.

1) Electronic ground states in atoms and molecules are determined by the optimal compromise in the variational competition between the electrostatic potential pulling the electrons toward the nuclei and the intrinsic kinetic delocalization drive of electron waves, which resists localization toward nuclei. Bonding occurs when the molecule offers a variational compromise with an energy lower than that available in the separated atoms.
2) The formation of a covalent bond is a consequence of the *kinetic energy lowering* that results from valence electrons *delocalizing* over several atoms. Such delocalization occurs between open-shell atoms, that is, atoms in which the number of valence electrons is less than twice the number of valence-shell *minimal basis set* orbitals, because some valence electrons can then exploit valence orbitals on *several* atoms by partial occupation.

   The delocalization is manifest when electronic wave functions are expressed in terms of delocalized bonding orbitals. When the wave functions are equivalently expressed in terms of quasi-atomic orbitals, then the kinetic energy lowering through delocalization is quantified by the negative *interference energy* between the quasi-atomic orbitals involved in the delocalization.[3]
3) The *kinetic* energy lowering through delocalization prevails over a lesser *potential energy increase* that is caused by the accumulation of charge in the bond region, which is associated with delocalization.[4]
4) At and near the equilibrium distance, delocalization has the special feature that it is *enhanced by shrinkage of the quasi-atomic wave function components* toward their respective nuclei.[5] This additional interatomic delocalization is therefore coupled with *contractions of the quasi-atomic orbitals,* whereby electronic charge is moved *from the outer regions* into the *interatomic* region as well as into the *intra-atomic* regions. The resulting energy lowering involves the following subtle interplay between *interatomic* and *intra-atomic* energy changes.

   The charge shift into the bond region enhances the *interatomic* delocalization and thereby lowers the kinetic energy further. Moreover, this *interatomic* kinetic energy lowering weakens the *overall* kinetic resistance against localization and thereby allows the nuclear attractions to pull more electronic charge toward the nuclei. As a result, the *intra-atomic potential* energy strongly decreases and the *intra-atomic kinetic* energy strongly increases until the virial ratio $2T = |V|$ is reached, which characterizes the optimal compromise in the variational competition between kinetic and potential energy at the equilibrium distance.[6]

   Even though the shift of charge into the *intra-atomic* region is thus instrumental in bringing about the compliance with the *overall* virial constraint, the *total intra-atomic energy nonetheless increases* in accordance with the *intra-atomic*

3) Interference energies are conceptually related to the resonance energies of molecular orbital and valence bond theories. However, the resolution of resonance energies into kinetic and potential components was never examined.
4) Charge accumulation in the bond *does not lower* the potential energy! See Section 1.4.5.3.
5) Even though the atomic contractions decrease the overlap integral! See Sections 1.4.3 and 1.4.5.5.
6) In accordance with the fundamental analysis in Section 1.2.3.

variation principle (the intra-atomic kinetic energy increase being stronger than the intra-atomic potential energy decrease). This *intra-atomic* energy increase is however less than the *interatomic* kinetic energy decrease through the increased delocalization that drives the contractions.

5) At larger internuclear distances the contraction discussed under item 4) does not occur and the delocalization discussed under items 2) and 3) is the only contributor to covalent bonding.[7)]

6) The detailed changes in the *potential* interactions that occur when bonds form are involved and complicated to sort out. It is found, however, that the *total contribution of all interatomic potential interactions* is positive, that is, antibonding in all molecules at all internuclear distances.

   The negative sign of the potential contribution to the total binding energy at the equilibrium geometry is the result of the intra-atomic adjustments discussed in the second paragraph of item 4) above. Since these *intra-atomic* adjustments are consequences of the *interatomic* kinetic energy attenuation, the sign of this potential contribution does not indicate what drives bond formation.[8)]

7) In as much as delocalization and kinetic energy changes are *one-electron* attributes, covalent bonding is the cumulative result of the bonding effects of the individual bonding electrons. The *interelectronic interactions* have an adverse effect on covalent bonding for the following reason.

   The delocalization of several electrons over several atoms increases the interpenetration of the respective electron clouds and, hence, strengthens the electrostatic repulsions between them, which diminishes the bond energy compared to what it would be without electron–electron interaction. This detrimental effect is mitigated, in so far as possible, by the inclusion of dynamic correlation terms in the wave function. In some molecules, dynamic correlation contributes a considerable part of the bonding energy. Nonetheless, the kinetic effects of delocalization are always essential for covalent bonding.

8) Finally, it should be noted that covalent bonding is not the only kind of chemical bonding. As Schwarz and coworkers have pointed out [12w, 33], changes in each of the terms in the electronic energy expression, viz in the kinetic terms, in the nuclear–electronic interactions, and in the electron–electron interactions can result in bonding. While covalent bonding is driven by changes in the kinetic terms, ionic bonding and long-range multipole interactions are driven by changes in the electron–nuclear interactions. Long-range dispersion forces, on the other hand, are driven by changes in the electron–electron interactions [34].

7) This is related to the less stringent form of the virial theorem in this region. See Section 1.2.3.
8) That the actual values of the kinetic and potential energies *at the variational minimum* generally do not provide sufficient information to deduce the physical origin of energy differences between systems was discussed in detail in the last paragraph of Section 1.2.2.

Acknowledgments

This work was supported in part by the US Department of Energy, Office of Basic Energy Sciences, Division of Chemical Sciences, Geosciences and Biosciences at the Ames Laboratory (for K.R.). The Ames Laboratory is operated for the US Department of Energy by Iowa State University under Contract DE-AC02-07CH11358. The work was supported in part by the National Science Foundation under Grant CHE-1147446 to Iowa State University (for M.W.S.). The work was supported in part with federal funds from the National Cancer Institute, National Institutes of Health, under Contract No. HHSN 261200800001E (for J.I.). The content of this publication does not necessarily reflect the views or policies of the Department of Health and Human Services, nor does mention of trade names, commercial products, or organizations imply endorsement by the US Government.

References

1. References to literature on the early history can be found in the article Ruedenberg, K. and Schwarz, W.H.E. (2013) Three millennia of atoms and molecules, in *Proceedings of the Symposium Pioneers in Quantum Chemistry* (eds T. Strom and A. Wilson), American Chemical Society, Washington, DC.
2. See, e.g., Helmholtz' Faraday address of 1881 at the Royal Institution in London, as reported on pp.278 to 284 in Volume II of Koenigsberger, L. (1902) *Hermann von Helmholtz*, Vieweg, Braunschweig.
3. Abegg, R. (1904) *Z. Anorg. Chem.*, **39**, 30.
4. Kossel, W. (1916) *Ann. Phys.*, **49**, 229.
5. Lewis, G.N. (1916) *J. Am. Chem. Soc.*, **38**, 762; and in the book *Valence and the Structure of Atoms and Molecules*, ACS Sponsored Monograph by the Chemical Catalog Company, 1923; republished by Dover, New York, 1966.
6. Heitler, W. and London, F. (1927) *Z. Phys.*, **44**, 455.
7. Burreau, Ø. (1927) *Det. Kgl. Danske Vid. Selskab.*, **7**, 1.
8. James, H.M. and Coolidge, A.S. (1933) *J. Chem. Phys.*, **1**, 825.
9. Slater, J.C. (1933) *J. Chem. Phys.*, **1**, 687.
10. Hellmann, H. (1933) *Z. Phys.*, **35**, 180, In his book *Quantenchemie*, Deuticke, Vienna, 1937, Hellmann seems to be more troubled about the apparent contradiction to the virial theorem.
11. (a) Ruedenberg, K. (1962) *Revs. Mod. Phys.*, **34**, 326. (b) Edmiston, C. and Ruedenberg, K. (1964) *J. Phys. Chem.*, **68**, 1628. (c) Feinberg, M.J., Ruedenberg, K., and Mehler, E. (1970) in *Advances in Quantum Chemistry*, vol. 5 (ed. P.O. Löwdin), Academic Press, p. 27. (d) Feinberg, M.J. and Ruedenberg, K. (1971) *J. Chem. Phys.*, **54**, 1495. (e) Feinberg, M.J. and Ruedenberg, K. (1971) *J. Chem. Phys.*, **55**, 5804. (f) Ruedenberg, K. (1975) in *Localization and Delocalization in Quantum Chemistry*, vol. 1 (ed. R. Daudel), Reidel Publishing Company, Dordrecht, p. 222.
12. (a) Goddard, W.A. and Wilson, C.W. (1972) *Theor. Chim. Acta*, **26**, 195 and 211. (b) Kutzelnigg, W. (1973) *Angew. Chem.*, **85**, 551 and *Angew. Chem. Int. Ed.*, **12**, 46. (c) Steiner, E. (1975) *The Determination and Interpretation of Molecular Wave Functions*, Cambridge University Press. (d) Fukui, K. (1976) *Chemical Reactions and Electronic Orbitals*, Maruzen Co, Tokyo, in Japanese only (*Kagaku Hannoh to Denshi no Kidoh*). (e) Mulliken, R.S. and Ermler, W.C. (1977) *Diatomic Molecules*, Section II.F, Academic Press, New York. (f) Harcourt, R.D., Solomon, H., Beckworth, J., and Chislett, L. (1982) *Am. J. Phys.*, **50**, 557. (g) Rozendahl, A. and Baerends, E.J. (1985) *Chem.*

Phys, **95**, 57. (h) Baird, N.C. (1986) *J. Chem. Educ.*, **63**, 660. (i) Harcourt, R.D. (1988) *Am. J. Phys.*, **56**, 660. (j) Nordholm, S. (1988) *J. Chem. Educ.*, **65**, 581. (k) Kutzelnigg, W. (1990) in *The Concept of the Chemical Bond*, vol. 2 (ed. Z.B. Maksic), Springer, Berlin, p. 1. (l) Harris, F.E. (1991) in *Encyclopedia of Physics*, 2nd edn (eds R.G. Lerner and G.L. Trig), John Wiley & Sons, Inc., New York, p. 762. (m) Melrose, M.P., Chauhan, M., and Kahn, F. (1994) *Theor. Chim. Acta*, **88**, 311. (n) Rioux, F. (1997) *Chem. Educ.*, **2** (6), 1. (o) Bacskay, G.B., Reimers, J.R., and Nordholm, S. (1997) *J. Chem. Ed.*, **74**, 1494. (p) Gordon, M.S. and Jensen, J.H. (1998) in *Encyclopedia of Computational Chemistry* (ed. P.v.R. Schleyer), John Wiley & Sons, Inc., New York, pp. 3198–3214. (q) Gordon, M.S. and Jensen, J.H. (2000) *Theor. Chem. Acc.*, **103**, 248. (r) Bickelhaupt, F.M. and Baerends, E.J. (2000) *Rev. Comput. Chem.*, **15**, 1. (s) Rioux, F. (2001) *Chem. Educ.*, **6** (5), 288. (t) Cardozo, T.M. and Nascimento, M.A.C. (2009) *J. Chem. Phys.*, **130**, 104102. (u) Cardozo, T.M. and Nascimento, M.A.C. (2009) *J. Phys. Chem. A*, **113**, 12541. (v) Cardozo, T.M., Nascimento Freitas, G., and Nascimento, M.A.C. (2010) *J. Phys. Chem. A*, **114**, 8798. (w) Bitter, T., Wang, S.G., Ruedenberg, K., and Schwarz, W.H.E. (2010) *Theor. Chem. Acc.*, **127**, 237. (x) Fantuzzi, F., Cardozo, T.M., and Nascimento, M.A.C. (2012) *Phys. Chem. Chem. Phys.*, **14**, 5479.

13. (a) Ruedenberg, K. and Schmidt, M.W. (2007) *J. Comput. Chem.*, **28**, 391. (b) Ruedenberg, K. and Schmidt, M.W. (2009) *J. Phys. Chem. A*, **113**, 1954.
14. Bacskay, G.B. and Nordholm, S. (2013) *J. Phys. Chem. A*, **117**, 7946–7958.
15. (a) Coulson, C.A. and Bell, R.P. (1945) *Trans. Faraday Soc.*, **41**, 141. (b) Löwdin, P.O. (1959) *J. Mol. Spectrosc.*, **3**, 46.
16. The word "stock" is being used with the meaning it has in horticulture, notably in arboriculture and viticulture. See, e.g., Mudge, K., Janick, J., Scofield, S., and Goldschmidt, E.E. (2009) A history of grafting. *Horticultural Rev.*, **35**, 9, *http://www.hort.purdue.edu/newcrop/janick-papers/c09.pdf* (accessed on 22 November 2013).
17. (a) Ruedenberg, K., Schmidt, M.W., Gilbert, M., and Elbert, S.T. (1982) *Chem. Phys.*, **71**, 41. (b) Ruedenberg, K., Schmidt, M.W., and Gilbert, M. (1982) *Chem. Phys.*, **71**, 51. (c) Ruedenberg, K., Schmidt, M.W., Gilbert, M., and Elbert, S.T. (1982) *Chem. Phys.*, **71**, 65.
18. West, A., Schmidt, M.W., Gordon, M.S., and Ruedenberg K. (2013) *J. Chem. Phys.*, **139**, 234107.
19. (a) Lu, W.C., Wang, C.Z., Schmidt, M.W., Bytautas, L., Ho, K.M., and Ruedenberg, K. (2004) *J. Chem. Phys.*, **120**, 2629. (b) Lu, W.C., Wang, C.Z., Schmidt, M.W., Bytautas, L., Ho, K.M., and Ruedenberg, K. (2004) *J. Chem. Phys.*, **120**, 2638. (c) Lu, W.C., Wang, C.Z., Chan, T.L., Ruedenberg, K., and Ho, K.M. (2004) *Phys. Rev. B*, **70**, 04110(R).
20. (a) Ivanic, J., Atchity, G.J., and Ruedenberg, K. (2008) *Theor. Chem. Acc.*, **120**, 281. (b) Ivanic, J. and Ruedenberg, K. (2008) *Theor. Chem. Acc.*, **120**, 295.
21. The most accurate value reported for the equilibrium distance so far is 1.9971933199699992 bohr with an energy of −0.6026346191065398 hartree by Scott, T.C., Aubert-Frecon, M., and Grotendorst, J. (2006) *Chem. Phys.*, **324**, 323.
22. Newton, I. *Philosophiae Naturalis Principia Mathematica*, London, 1687; Book I, Section XII.
23. Finkelstein, B.N. and Horowitz, G.E. (1928) *Z. Phys.*, **48**, 118.
24. Kolos, W. (1994) *J. Chem. Phys.*, **101**, 1330.
25. Nakatsuji, H. (2004) *Phys. Rev. Lett.*, **93**, 030403.
26. This is the theoretical value of Cencek, W. and Szalewicz, K. (2008) *Int. J. Quantum Chem.*, **108**, 2191. The corresponding energy is −1.174475714 hartree.
27. Schmidt, M.W., Ivanic, J., and Ruedenberg K. in preparation.
28. Dunning, T.H. (1989) *J. Chem. Phys.*, **90**, 1007. A survey of cc-type basis sets is given at the website *http://tyr0.chem.wsu.edu/~kipeters/Pages/*

cc'append.html (accessed 22 September 2013).

29. (a) Schmidt, M.W., Baldridge, K.K., Boatz, J.A., Elbert, S.T., Gordon, M.S., Jensen, J.H., Koseki, S., Matsunaga, N., Nguyen, K.A., Su, S., Windus, T.L., Dupuis, M., and Montgomery, J.A. (1993) *J. Comput. Chem.*, **14**, 1347. (b) Gordon, M.S. and Schmidt, M.W. (2005) in *Theory and Applications of Computational Chemistry, the First Forty Years* (eds C.E. Dykstra, G. Frenking, K.S. Kim, and G.E. Scuseria), Elsevier, Amsterdam, pp. 1167–1189.

30. Bytautas, L., Matsunaga, N., Scuseria, G.E., and Ruedenberg, K. (2012) *J. Phys. Chem. A*, **116**, 1717.

31. Feller, D. and Sordo, J.A. (2000) *J. Chem. Phys.*, **113**, 485.

32. Schmidt, M.W., Ivanic, J., and Ruedenberg, K. (2010) *J. Phys. Chem. A*, **114**, 8687.

33. Bitter, T., Ruedenberg, K., and Schwarz, W.H.E. (2007) *J. Comput. Chem.*, **28**, 411.

34. See e.g., (a) Buckingham, A.D., Fowler, P.W., and Hudson, J.M. (1988) *Chem. Rev.*, **88**, 963. (b) Chałasiński, G. and SzczJśniak, M.M. (2000) *Chem. Rev.*, **100**, 4227. (c) Stone, A.J. (1996) *The Theory of Intermolecular Forces*, Clarendon, Oxford.

2
Bridging Cultures

Philippe C. Hiberty and Sason Shaik

2.1
Introduction

The two general theories of chemistry, one called *valence bond (VB)* theory and the other *molecular orbital (MO)* theory, were developed at about the same time, but have quickly given rise to two different cultures that have competed, sometimes fervently, on charting the mental map and epistemology of chemistry. VB theory is close to the natural language of chemists. It views the electron pairs of a molecule as being located in bonds or lone pairs. On the contrary, MO theory views the electron pairs as completely delocalized over the entire molecule. Given these two seemingly opposing points of views, it is no surprise that many teachers and researchers have had the feeling that if one is right, the other must be wrong.

However, a deeper understanding of both theories shows that they are both correct, and mutually transformable. More so, both are exact in their most elaborate versions, and expressing the same chemical reality in different languages. This was recognized as early as 1935 by van Vleck and Sherman [1], stating that *it becomes meaningless quibbling to argue which of the two methods is the better in refined forms since they ultimately merge*. It is unfortunate that this wise statement was not more widely appreciated by the community of chemists and teachers. What could be added nowadays is that not only do the two theories ultimately converge but a number of good methods have been devised to extract some VB information from MO methods, *thus bridging the two cultures*. Our aim in this chapter is to describe these bridges. But before doing so, it is useful to briefly recall the history of the VB and MO methods and the roots of the rivalry that has still not completely subsided to these days.

2.2
A Short History of the MO/VB Rivalry

The roots of VB theory in chemistry can be traced back to the famous paper of Lewis *The Atom and The Molecule* [2], which introduced the notions of electron-pair

The Chemical Bond: Fundamental Aspects of Chemical Bonding, First Edition.
Edited by Gernot Frenking, Sason Shaik.

bonding and octet rule (so-called initially the rule of eight) [3]. This paper that predated the new quantum mechanics by 11 years constitutes the first effective formulation of bonding in terms of the covalent-ionic classification, which is still being taught, and which has formed the basis for the subsequent construction and generalization of VB theory. From then on, the notion of electron pairing as a mechanism of bonding became widespread and initiated the "electronic structure revolution" in chemistry [4]. Meanwhile, despite this great step forward, understanding the mechanism by which an electron pair could constitute a bond remained a mystery until 1927 when Heitler and London published their seminal paper, *Interaction Between Neutral Atoms and Homopolar Binding* [5], in which they showed that the bonding in H_2 originates in the quantum mechanical "resonance" interaction, which is contributed as the two electrons are allowed to exchange their positions between the two atoms.

In modern terms, the bonding in H_2 can be accounted for by the wave function, referred to as Ψ_{HLVB}, written in Eq. (2.1) (dropping normalization constant):

$$\Psi_{\text{HLVB}} = |\chi_a \overline{\chi}_b| - |\overline{\chi}_a \chi_b| \tag{2.1}$$

where χ_a and χ_b are the respective atomic spin orbitals of the two hydrogen atoms, with a bar indicating a β spin and the absence of bar meaning an α spin. The wave function in Eq. (2.1) is a superposition of two neutral situations with inverted spin distributions. Thus, the bonding in H_2 arises because of the quantum mechanical "resonance" interaction between the two patterns of spin arrangement that are required in order to form a singlet electron pair. As this "resonance energy" accounted for a good deal of the experimental bond energy of the molecule, this implied that the wave function in Eq. (2.1), which is referred to henceforth as the HL-wave function, could describe the chemical bonding in a satisfactory manner. This "resonance origin" of the bonding was a remarkable feat of the new quantum theory, because until then it was not obvious how two neutral species could be at all bonded.

The HL-wave function formed the basis for the version of VB theory that became very popular later, and which was behind some of the "failings" that were to be attributed to VB theory. In 1929, Slater presented his determinant-based method [6] and in 1931 he generalized the HL model to *n*-electrons by expressing the total wave function as a product of $n/2$ bond wave functions of the HL type [7]. In 1932, Rumer [8] showed how to write down all the possible bond pairing schemes for *n*-electrons and avoid linear dependencies between the forms, in order to obtain canonical structures. We shall refer hereafter to the kind of theory that considers only covalent structures as Heitler-London Valence Bond (HLVB).

The success of the HL model and its relation to Lewis's model, posed a wonderful opportunity for Pauling and Slater to construct a general quantum chemical theory for polyatomic molecules. In the same year, 1931, they both published a few seminal papers in which they developed the notion of hybridization, the covalent-ionic superposition, and the resonating benzene picture [7, 9–12]. The electron pair bond was described as a superposition of the covalent HL form and the two possible ionic forms of the bond, as shown in Scheme 2.1 for an A–B bond.

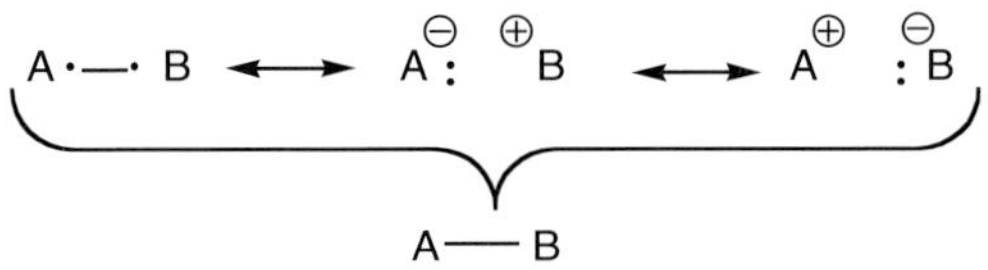

Scheme 2.1 VB representation of an A–B bond. The first resonance structure is purely covalent, while the two other ones are ionic.

The notion of orbital hybridization proved extremely useful and was used to discuss molecular geometries and bond angles in a variety of molecules, ranging from organic to transition metal compounds. These papers were followed by a stream of five papers, published during 1931–1933, and entitled *The Nature of the Chemical Bond*. This series of papers enabled the description of any bond in any molecule, and culminated in the famous monograph in which all the structural chemistry of the time was treated in terms of the covalent-ionic superposition theory, and resonance and hybridization theories [13].

At the same time that Slater and Pauling were developing their VB theory [14], Mulliken [15] and Hund [16] developed an alternative approach called *molecular orbital (MO) theory* that has a spectroscopic origin. In MO theory, the electrons in a molecule occupy delocalized orbitals made from linear combination of atomic orbitals (AOs). Eventually, it would be the work of Hückel that ushered MO theory into mainstream chemistry and gave it an impetus by forming a successful and widely applicable tool. In 1930, Hückel suggested the σ-π separation [17] and turned to solve the electronic structure of benzene using his new Hückel Molecular Orbital (HMO) approach [18]. The π-MO picture allowed Hückel to understand the special stability of benzene, owing to the closed-shell occupation of the 6-π electrons in six delocalized MOs, as shown in Scheme 2.2. Furthermore, the HMO approach enabled Hückel to make predictions on the stabilities/instabilities of $C_5H_5^-$ and $C_7H_7^+$ ions and C_4H_4 (CBD) and C_8H_8 (COT) molecules. Already in this paper

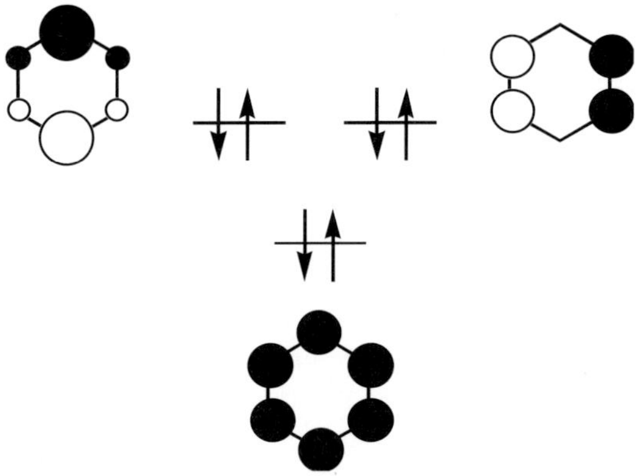

Scheme 2.2 The Hückel representation of benzene in terms of three occupied delocalized π-MOs

and in a subsequent one [19], Hückel formed the foundations for what was to become later known as the *Hückel rule*, regarding the special stability of "aromatic" molecules with $4n+2$ π-electrons [20].

On the HLVB front, Pauling and Wheland [21] described the benzene electronic structure as a combination of the five Rumer structures in Scheme 2.3, and approximated the matrix elements between the structures by retaining only close neighbor resonance interactions. This approach has enabled them to extend the treatment to naphthalene and to a variety of other species. Thus, in the HLVB approach, benzene is described as a "resonance hybrid" of the two Kekulé structures and the three Dewar structures. The pictorial representation of the wave function, the link to Kekulé's oscillation hypothesis and to Ingold's mesomerism [20, 22], which were known to chemists, made the HLVB representation very popular among practicing chemists.

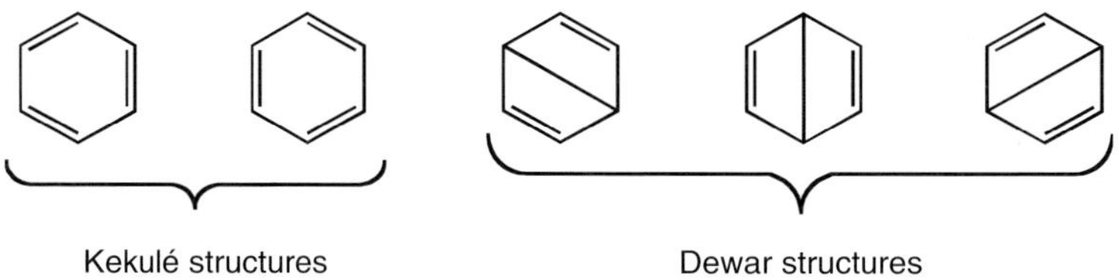

Scheme 2.3 The Pauling–Wheland representation of benzene in terms of two Kekulé structures and two Dewar structures

With these two seemingly different treatments of benzene, the chemical community was faced with two alternative descriptions of one of its molecular icons, and this began the VB–MO rivalry that seems to accompany chemistry to the twenty-first century [23].

By the end of World War II, Pauling's resonance theory was widely accepted, while most practicing chemists ignored HMO and MO theories, perhaps due to Pauling's skill as a communicator [20]. However, the most important reason for this dominance was the direct lineage of VB resonance theory to the structural concepts of chemistry dating from the days of Kekulé, Couper, and others through the electron-pair notion and electron-dot structures of Lewis. Another very important reason was the facile qualitative application of this theory to all known structural chemistry of the time, in Pauling's book [13], and to a variety of problems in organic chemistry, in Wheland's book [22]. The combination of an easily applicable general theory, and its ability to fit experiment so well, created a rare credibility nexus. By contrast, MO theory seemed alien to everything chemists had thought about the nature of the chemical bond, and offered no visual representation to compete with the resonance hybrid representation of VB resonance theory. At the end of World War II, VB resonance theory dominated the epistemology of chemists.

By the mid 1950s, the tide has started shifting slowly in favor of MO theory, gaining momentum through the mid-1960s. What had caused the shift is a combination of factors, of which the following may be decisive. One is in organic chemistry, where pericyclic reactions (Diels–Alder, sigmatropic reactions, electrocyclic reactions, and cheletropic reactions) were the main field of synthetic research at that time. Practicing chemists were desperate to understand pericyclic reactions and there were already textbooks who dubbed them "reactions without mechanism." These mysterious reactions soon became lucid, with the publication of the MO-based Woodward–Hoffmann rules [24] and the rediscovery of Fukui's frontier orbitals [25]. A second reason came from inorganic chemistry. Dewar suggested in 1951 an MO model for the bonding in Zeise's salt [26]. Chatt and Duncanson picked it up and realized the power of the model [27]. They systematically extended it to other transition metal complexes and in this way introduced MO theory into experimental inorganic chemistry. Further, there were the construction of intuitive MO theories and their wide applicability for rationalization of structures (e.g., Walsh diagrams) and spectra (electronic and ESR), the highly successful predictive application of MO theory in chemical reactivity, and the development of widely applicable MO-based computational techniques (e.g., extended Hückel and semiempirical programs). On the VB side, Pauling himself insisted that resonance theory was sufficient to deal with most problems. This oversimplified version of the HLVB theory, which simply enumerates structures without proper consideration of their interaction matrix elements, led to a wrong prediction that the resonance energy of CBD should be as large as or even larger than that of benzene. Thus, owing to these failings of HLVB, a decisive victory was won by MO theory when organic chemists were finally able to synthesize transient molecules and establish the stability patterns of $C_8H_8^{2-}$, $C_5H_5^{-,+}$, $C_3H_3^{+,-}$, and $C_7H_7^{+,-}$ during the 1950s–1960s, and C_4H_4 during the 1960s–1978 [12, 28, 29]. The results, which obeyed the Hückel rules, as well as the success of Woodward–Hoffmann's rules [24], convinced most of the organic chemists that MO theory was right, while HLVB and resonance theories were wrong. However, as would be made clear eventually, a proper VB theory that includes the ionic structures demonstrates clearly the instabilities of CBD and COT versus the stability of benzene [30]. Nevertheless, this has been ignored by the community by and large.

Another alleged "failure" of VB is associated with the dioxygen molecule, O_2. Application of the simple Pauling–Lewis recipe of hybridization and bond pairing to rationalize and predict the electronic structure of molecules fails to predict the paramagnetic nature of O_2. By contrast, using MO theory reveals this paramagneticity instantaneously [31]. Even though VB theory does not actually fail with O_2, as at all levels, including qualitative HLVB, VB describes O_2 as a triplet diradical just like MO theory [32], this "failure" of Lewis's recipe stuck to VB theory and became a fixture of the common chemical wisdom.

VB theory continued to be associated with more "failures." Thus, the development of photoelectron spectroscopy and its application to molecules in the 1970s, in the hands of Heilbronner, showed that the spectra could be easily interpreted as reflecting the energies of canonical molecular orbitals (CMOs) [33, 34]. As such, it was easy to explain the two different ionization peaks of methane, which correspond to the a_1 and t_2 MOs. Unfortunately, this has served to dismiss VB theory, because it describes electron pairs as belonging to four equivalent local bonds. Thus, on the basis of a loose reasoning, the idea spread that VB theory would predict a single ionization peak for methane. The argument is of course unfounded, as is shown later; however, with this and similar types of arguments, VB theory has eventually fallen into a state of disrepute and become known, at least in the student times of the present authors, either as a "wrong theory" or simply as a "naive theory." Alas, the warring camps seemed to enjoy the dispute more than facing the facts.

Nevertheless, despite the obituaries showered on VB theory in textbooks and in the public chemical opinion, the theory has never really died. Owing to its close affinity to classical chemistry and utmost clarity, it has remained an integral part of the thought process of many chemists, even among proponents of MO theory. Within the chemical dynamics community, the usage of the theory has never been arrested, and it lives in terms of computational methods called *LEPS, BEBO, DIM*, and so on, which were (and still are) used for generation of potential energy surfaces. Moreover, around the 1970s, but especially 1980s and onward, VB theory began to rise from its ashes, to dispel many myths about its "failures" and to offer a sound and attractive alternative to MO theory [32], while at the same time new algorithms gradually gave VB the status of a practical *ab initio* computational method [35, 36].

In brief, until the mid-1950s, VB theory had dominated chemistry, and then MO theory took over while VB theory fell into disrepute and was almost completely abandoned. The more recent period from the 1980s onward marked a comeback of VB theory, which has been ever since enjoying a renaissance both in the qualitative application of the theory and in the development of new methods for its computer implementation [37]. One of the great merits of VB theory is its pictorially intuitive wave function that is expressed as a linear combination of chemically meaningful structures. It is this feature that has made VB theory so popular in the 1930s–1950s, and it is the same feature that underlies its temporary demise and ultimate resurgence. In a complementary way, MO theory offers a great deal of insight via the symmetry of the MOs and their bonding/antibonding characters. Now that VB theory enjoys a comeback, we think it is timely to put some emphasis on the many bridges that have been built between the two theories throughout the past decades, and to encourage chemists and teachers to bury the hatchet, which has so far undermined the intellectual heritage of chemistry by eliminating one of its two thought cultures. We will thus show in the following sections that VB and MO theories are not separate or contradictory approaches to chemistry, but are the two facets of a unique description of molecular wave functions.

2.3 Mapping MO-Based Wave Functions to VB Wave Functions

What is the difference between the MO and VB descriptions of an electronic system, at the simplest level of both theories? As we shall see, in the cases of one-electron, three-electron and four-electron interactions between two centers, there is no difference between the two theories, except for the representations that look different. On the other hand, the two theories differ in their description of the two-electron bond.

Let us take once again the example of H_2, with its two AOs χ_a and χ_b, and examine first the VB description, dropping normalization factors for simplicity. At the HLVB level, only the covalent component of the two-electron bond is considered, leading to the simple wave function Ψ_{HLVB} displayed earlier in Eq. (2.1).

As early recognized [14], at the equilibrium distance the bonding is not 100% covalent, and it requires some ionic components, to be described accurately. On the other hand, at long distances the HLVB wave function is the correct state, as the ionic components necessarily drop to zero and each hydrogen atom carries one electron away through the homolytic bond breaking. The HLVB wave function *dissociates correctly*, but is quantitatively inaccurate at bonding distances. Therefore, the way to improve the HLVB description is straightforward: by simply mixing Ψ_{HLVB} with the ionic VB structures $|\chi_a\overline{\chi}_a|$ and $|\chi_b\overline{\chi}_b|$, and variationally optimizing the coefficients, by configuration interaction (CI). One then gets the wave function $\Psi_{VB\text{-}full}$, in Eq. (2.2), which contains a major covalent component and a minor ionic one.

$$\Psi_{VB-full} = \lambda(|\chi_a\overline{\chi}_b| - |\overline{\chi}_a\chi_b|) + \mu(|\chi_a\overline{\chi}_a| + |\chi_b\overline{\chi}_b|); \quad \lambda > \mu \tag{2.2}$$

Let us now turn to the MO description. Bringing together two hydrogen atoms leads to the formation of two MOs, σ and σ^*, bonding and antibonding, respectively, (Eq. (2.3), dropping normalization constants):

$$\sigma = \chi_a + \chi_b; \quad \sigma^* = \chi_a - \chi_b \tag{2.3}$$

At the simple MO level, the ground state of H_2 is described by the configuration σ^2, in which the bonding σ MO is doubly occupied. Expansion of this MO determinant into its AO determinant constituents leads to Eq. (2.4) (again dropping normalization constants):

$$\sigma^2 = |\sigma\overline{\sigma}| = |\chi_a\overline{\chi}_b| - |\overline{\chi}_a\chi_b| + |\chi_a\overline{\chi}_a| + |\chi_b\overline{\chi}_b| \tag{2.4}$$

It is apparent from Eq. (2.4) that the first half of the expansion is nothing but the HL function Ψ_{HLVB} (Eq. (2.1)), while the remaining part is ionic. It follows that the MO description of the homonuclear 2-e bond will always be half-covalent and half-ionic, irrespective of the bonding distance. Qualitatively, it is already clear that in the MO wave function, the ionic weight is excessive at bonding distances, and becomes strictly wrong at long distances, where the weight of the ionic structures should drop to zero in accord with the homolytic cleavage. The simple MO description *does not dissociate correctly* and this is the reason why it is inappropriate for the description of stretched bonds, as, for example, those found in transition states. The remedy for this poor description is CI, specifically the mixing of the ground configuration, σ^2, with the diexcited one, σ^{*2}. The reason why this mixing resizes the covalent versus ionic weights is the following: if one expands the diexcited configuration, σ^{*2}, into its VB constituents, one finds the same covalent and ionic components as in Eq. (2.4), but coupled with a negative sign as in Eq. (2.5):

$$\sigma^{*2} = |\sigma^{*}\overline{\sigma^{*}}| = -(|\chi_a\overline{\chi}_b| - |\overline{\chi}_a\chi_b|) + (|\chi_a\overline{\chi}_a| + |\chi_b\overline{\chi}_b|) \tag{2.5}$$

It follows that mixing the two configurations σ^2 and σ^{*2} with different coefficients as in Eq. (2.6) will lead to a wave function Ψ_{MOCI} in which the covalent and ionic components

$$\Psi_{\text{MOCI}} = c_1|\sigma\overline{\sigma}| - c_2|\sigma^{*}\overline{\sigma^{*}}|; \quad c_1, c_2 > 0 \tag{2.6}$$

have unequal weights, as shown by an expansion of Ψ_{MOCI} into AO determinants in Eq. (2.7):

$$\Psi_{\text{MOCI}} = (c_1 + c_2)(|\chi_a\overline{\chi}_b| - |\overline{\chi}_a\chi_b|) + (c_1 - c_2)(|\chi_a\overline{\chi}_a| + |\chi_b\overline{\chi}_b|) \tag{2.7a}$$

$$(c_1 + c_2) = \lambda; \quad c_1 - c_2 = \mu \tag{2.7b}$$

Because c_1 and c_2 are variationally optimized, expansion of Ψ_{MOCI} should lead to exactly the same VB function as $\Psi_{\text{VB-full}}$ in Eq. (2.2), leading to the equalities expressed in Eq. (2.7b) and to the equivalence of Ψ_{MOCI} and $\Psi_{\text{VB-full}}$.

To summarize, the simple MO level describes the bond as being too ionic, while the simple HLVB level defines it as being purely covalent. Both theories converge to the right description when CI is introduced. It follows that *the accurate description of two-electron bonding is half-way in-between the simple MO and simple HLVB levels; elaborated MO and VB levels become equivalent and converge to the same description, in which the bond is mostly covalent but has a substantial contribution from ionic structures.* This equivalence clearly projects that the MO–VB rivalry is unfortunate and senseless.

Up to this point, we restricted ourselves to the simple case of determinants involving no more than two orbitals. However, the MO–VB correspondence is general, and, in fact, any MO or MO–CI wave function can be exactly transformed into a VB wave function, provided it is a spin-eigenfunction (i.e., not a spin-unrestricted wave function). While this is a trivial matter for small determinants, larger ones require a bit of algebra and a systematic method is discussed in a recent book [32, 38]. The general procedure consists of projecting the MO wave function onto a basis of VB structures, and involves the following steps: (i) determine a complete and linearly independent basis set of VB structures for the electronic system at hand; (ii) expand the MO determinants of the MO wave function as linear combinations of AO-based determinants; at this stage, the MO wave function can already be interpreted in terms of neutral or ionic structures [39]. Finally, to get the final wave function, (iii) organize the expression obtained in step (ii) in terms of the basis set of VB structures generated in step (i). The transformation is exact and straightforward in minimal basis set. For wave functions calculated in large basis sets, other procedures have been devised, which are described in Section 7.

Incidentally, it is interesting to apply the MO–VB projection to MO wave functions before and after CI, to visualize the effect of CI on the VB weights.

This has been done, among others, on the ozone molecule as calculated at the Hartree–Fock (MO–HF) and MO–CI levels [38]. According to Rumer's method, a complete and linearly independent basis of VB structures involves structures **1–6** for ozone (see Scheme 2.4). *A priori,* one expects structures **4–6** to be marginal, as in these cases one oxygen atom has a sextet structure. On the other hand, for **1**, **2**, and **3** it is difficult to predict the relative weights on a qualitative basis.

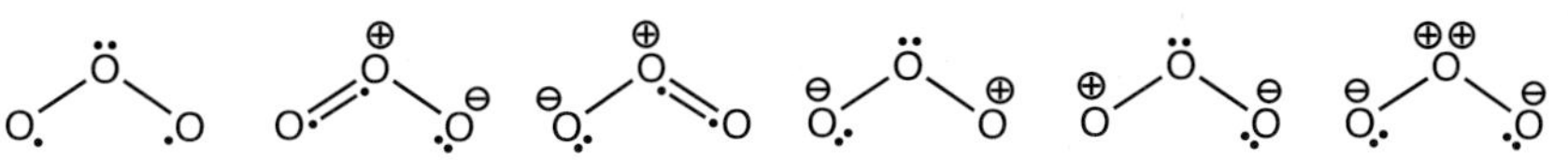

1 2 3 4 5 6

Scheme 2.4 The six VB structures of ozone. Electrons of the π system are represented by dots, while σ electrons are omitted.

The results of the MO–VB projection are displayed in Table 2.1. It can be seen that, at the MO–HF level, structures **4–6** are minor but not negligible, having a cumulative weight of 33.4%. On the other hand, the diradical structure **1** is more important with a weight of 0.21, but less than each of the ionic structures **2** and **3**. The picture changes drastically when CI is applied. At the MO–CI level (full CI among the π space of orbitals), the diradical structure **1** becomes largely dominant with a weight of 0.59, to the detriment of structures **4–6** which become completely negligible. This shows that, as in the H_2 case, the role of electron correlation is to diminish the weight of the ionic structures, thus reducing the repulsion between the valence electrons.

The same type of MO–VB projection has been carried out for other 1,3-dipoles, benzene, pyridine, furane, and other aromatics [38, 40–42]. In all cases, it is seen that CI greatly diminishes the ionic contributions to the benefit of the covalent ones. Interestingly, in the benzene case [40], a comparison could be made between the VB weights obtained by projection of an MO–CI wave function, including complete CI in the π space, with those arising from a direct VB calculation by Tantardini *et al.* [43]; the weights are *exactly* the same, showing that elaborated MO and VB levels become equivalent and converge to the right description, yielding two wave functions that show two different representations of reality but are transformable and mathematically equivalent.

Table 2.1 Weights of VB structures **1–6** in the ground state of ozone, at the Hartree–Fock and π-CI levels in the MO framework, in minimal basis set (Ref. [38]).

Level of theory	1	2	3	4	5	6
Hartree–Fock	0.213	0.226	0.226	0.107	0.107	0.120
π-CI	0.593	0.184	0.184	0.008	0.008	0.023

2.4 Localized Bond Orbitals – A Pictorial Bridge between MO and VB Wave Functions

The standard MO wave function involves CMOs, which are permitted to delocalize over the entire molecule. However, it is well known [44, 45] that an MO wave function based on CMOs can be transformed to an MO wave function that is based on localized MOs, known also as *localized bond orbitals (LBOs)* [34]. This transformation is called *unitary transformation*, and, as such, it changes the representation of the orbitals without affecting the total energy or the wave function. This equivalence is expressed in Eq. (2.8):

$$|\ldots \varphi_i^{\mathrm{cmo}} .. \varphi_j^{\mathrm{cmo}} ..| = |\ldots \varphi_i^{\mathrm{lbo}} .. \varphi_j^{\mathrm{lbo}} ..| \tag{2.8}$$

where φ_i^{cmo} corresponds to a CMO while φ_i^{lbo} is an LBO.

A unitary transformation involves simple subtractions and additions of orbitals within the complete set of the occupied CMOs. To illustrate such a transformation we choose a simple molecule, BeH_2, for which the procedure may be done in a pictorial manner without resort to equations. Figure 2.1 shows the valence-occupied CMOs of BeH_2, the lowest of the two is made from the bonding combination of the 2s AO of Be and the positive combination of the 1s AOs of the two hydrogen atoms, while the higher one is the bonding orbital between the $2p_z$(Be) orbital and the negative combination of the 1s(H) AOs. We can now make two linear combinations of these orbitals, one negative and one positive, as in Eq. (2.9), dropping normalization constants:

$$\sigma_{\mathrm{R}} = \varphi_1^{\mathrm{cmo}} + \varphi_2^{\mathrm{cmo}}; \quad \sigma_{\mathrm{L}} = \varphi_1^{\mathrm{cmo}} - \varphi_2^{\mathrm{cmo}} \tag{2.9}$$

These linear combinations, shown on the right-hand side of Figure 2.1, are seen to generate two LBOs made from sp hybrids on the Be and the 1s AOs of the hydrogens. One of these LBOs, σ_{R}, is a two-center bonding orbital localized on the right-hand side of the molecule, while the other, σ_{L}, is equivalent to the former but localized on the left-hand side. Of course, as the coefficients of the hydrogens in $\varphi_1^{\mathrm{cmo}}$are not exactly equal in absolute value to those in φ_2^{cmo}, the localization is not perfect, and each LBO contains a small component out of the bonding region, called *delocalization tail*, which is however very small. The wave function based on these localized orbitals possesses two doubly occupied LBOs and is completely equivalent to the starting wave function based on CMOs, as expressed in Eq. (2.10):

$$\Psi(\mathrm{BeH_2}) = |\varphi_1^{\mathrm{cmo}}\overline{\varphi_1^{\mathrm{cmo}}}\varphi_2^{\mathrm{cmo}}\overline{\varphi_2^{\mathrm{cmo}}}| = |\sigma_{\mathrm{R}}\overline{\sigma_{\mathrm{R}}}\sigma_{\mathrm{L}}\overline{\sigma_{\mathrm{L}}}| \tag{2.10}$$

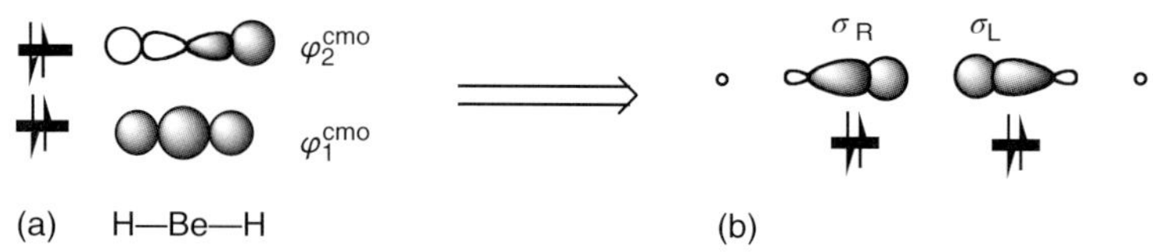

Figure 2.1 MOs of BeH_2 at the Hartree-Fock level. (a) canonical MOs; (b) localized MOs.

This LBO-based wave function is not a VB wave function. Nevertheless, it represents a Lewis structure, and hence also a pictorial analog of a perfect-pairing (PP) VB wave function. The difference between the LBO and VB wave functions is that the latter involves electron correlation, whereas the former does not. As such, in the LBOs of Eq. (2.10), each localized Be–H bond involves a 50 : 50 covalent-ionic combination as in the Hartree–Fock wave function of any two-electron bond, Eq. (2.4), and therefore exaggerates the bond ionicity.

For molecules involving many bonds, the localizing unitary transformations are more complicated than in the BeH_2 case, and are usually done by means of a computer program, which is available in all current *ab initio* codes. As there is an infinity of unitary transformations of orbitals that leave the Slater determinant unchanged, the localizing transformations are determined so as to best satisfy some specific criteria, for example, by requiring that the total spread of the localized orbitals be minimal, as in the Foster–Boys method [45]. On the other hand, it is impossible to find a set of well-localized orbitals for molecules whose electronic system is intrinsically delocalized, such as benzene or, to a much lesser extent, butadiene. Moving between localized and delocalized orbitals is extremely useful as shown masterfully by R. Hoffmann [46] in his derivation of, for example, the isolobal analogy for understanding bonding in organometallic complexes and building bridges to organic molecules.

2.5
Block-Localized Wave Function Method

The block-localized wave function (BLW) method is another simple bridge between MO and VB methods that provides VB-type information (e.g., resonance energies) [47]. It is described in detail in Chapter 6 of this book. The basic principle of BLW consists of partitioning the full basis set of orbitals into subsets, each centered on a given fragment. The MOs are then optimized in a Hartree–Fock way, with the restriction that each orbital is expanded only on its own fragment. The MOs of a given fragment are orthogonal among themselves, but the orbitals of different fragments have finite overlaps.

The applications of the BLW method are designed primarily to evaluate the electronic delocalization and charge transfer effects between fragments/molecules. Thus, the BLW represents a reference for evaluating delocalization energies relative to the fully localized reference wave function.

A typical application of the BLW method is the energy calculation of a specific resonance structure in the context of resonance theory. As a resonance structure is composed, by definition, of local bonds plus core and lone pairs, a bond between atoms A and B will be represented as a bonding MO strictly localized on the A and B centers; a lone pair will be an AO localized on a single center, and so on. With these restrictions on orbital extension, the final BLW is optimized at the constrained Hartree–Fock level and is expressed by a single Slater determinant. Consequently, the energy difference between the Hartree–Fock wave function, where all electrons are free to delocalize in the whole system, and the BLW function, where electrons are confined to specific zones of the system, is defined as the *electron delocalization energy*. Recently, the BLW method has been extended to density functional theory (DFT) [48], by replacing the Hartree–Fock exchange potential by a DFT exchange-correlation (XC) potential in the Roothaan SCF procedure. This improved BLW method, referred to as BLW-DFT, has the advantages over the original method by implicit inclusion of electron correlation to both the ground-state and the individual structures.

2.6 Generalized Valence Bond Theory: a Simple Bridge from VB to MOs

A great step forward in the direction of accuracy and compactness of the VB wave function was made by Coulson and Fischer [49], who proposed to describe the two-electron bond as a formally covalent singlet coupling between two orbitals φ_a and φ_b, the latter being optimized with all freedom to delocalize over the two centers, as exemplified in Eq. (2.11) for H_2:

$$\Psi_{CF} = |\varphi_a\overline{\varphi}_b| - |\overline{\varphi}_a\varphi_b| \tag{2.11a}$$

$$\varphi_a = \chi_a + \varepsilon\chi_b \tag{2.11b}$$

$$\varphi_b = \chi_b + \varepsilon\chi_a \tag{2.11c}$$

where χ_a and χ_b are purely localized AOs. In fact, experience shows that the Coulson–Fischer orbitals φ_a and φ_b that result from the energy minimization are generally not much delocalized ($\varepsilon < 1$), so that they can be viewed as "distorted" orbitals that remain atomic-like in nature. However, as minor as it may look, the slight delocalization tail makes the Coulson–Fischer wave function equivalent to the three-structure classical VB wave function $\Psi_{\text{VB-full}}$ of Eq. (2.2), as evidenced in Eq. (2.12) where the Coulson–Fischer wave function is expanded in classical AO determinants.

$$\Psi_{CF} = (1+\varepsilon^2)(|\chi_a\overline{\chi}_b| - |\overline{\chi}_a\chi_b|) + 2\varepsilon(|\chi_a\overline{\chi}_a| + |\chi_b\overline{\chi}_b|) \tag{2.12}$$

Thus, the Coulson–Fischer representation has the advantage of keeping the PP simple picture while adequately treating the electron correlation of the bonding electrons, by implicitly including the effect of the ionic components of the bond in a variational way.

The Coulson–Fischer proposal was later generalized to polyatomic molecules and developed as the generalized valence bond (GVB) method [50] by Goddard, and as the spin-coupled (SC) method [51] by Gerratt and his coworkers. In the latter two methods, the valence electrons are described by a single configuration of singly occupied orbitals, and the various spin-coupled structures are allowed to mix. Thus, each bond in a polyatomic VB structure is viewed as a pair of singlet-coupled orbitals, which are quasi-atomic and display a strong mutual overlap.

The GVB method is generally used in the form referred to as GVB-PP, which introduces two simplifications. The first one is the PP approximation, by which only one VB structure is generated in the calculation. The wave function may then be expressed as in Eq. (2.13), where each term in parentheses is a so-called geminal two-electron function, which takes the form of a singlet-coupled GVB pair (φ_{ia}, φ_{ib}) and is associated to one particular bond or lone pair.

$$\Psi_{\text{GVB}} = |(\varphi_{1a}\overline{\varphi}_{1b} - \overline{\varphi}_{1a}\varphi_{1b})(\varphi_{2a}\overline{\varphi}_{2b} - \overline{\varphi}_{2a}\varphi_{2b}) \ldots (\varphi_{na}\overline{\varphi}_{nb} - \overline{\varphi}_{na}\varphi_{nb})| \quad (2.13)$$

The second simplification is the *strong orthogonality* constraint, by which all the orbitals in Eq. (2.13) are required to be orthogonal to each other unless they are singlet paired, that is,

$$\langle \varphi_{ia} | \varphi_{ib} \rangle \neq 0 \qquad (\text{a} - \text{b paired}) \quad (2.14\text{a})$$

$$\langle \varphi_i | \varphi_j \rangle = 0 \quad \text{otherwise.} \quad (2.14\text{b})$$

This strong orthogonality constraint is not a drastic simplification, since it applies to orbitals that are not expected to overlap significantly. On the other hand, the orbitals that are coupled together in the same GVB pair display, of course, a strong overlap.

The GVB-PP method forms a bridge to the MO–based multiconfigurational self-consistent field (MCSCF) method (see Textbox 2.1). Indeed, each geminal in Eq. (2.13) can be rewritten, by simple orbital transformation, as an expansion in terms of natural orbitals (NOs), in a manner analogous to the transformation displayed in Eqs. (2.6) and (7) for H_2.

$$|\varphi_{ia}\overline{\varphi}_{ib} - \overline{\varphi}_{ia}\varphi_{ib}| = |C_i\phi_i\overline{\phi}_i + C_i^*\phi_i^*\overline{\phi}_i^*| \quad (2.15)$$

This alternative form of the geminal contains two terms of closed-shell form. The NOs ϕ_i and ϕ_i*, in Eq. (2.15), have the shapes of localized MOs, respectively bonding and antibonding, which are mutually orthogonal. They are connected to the GVB pairs by the simple transformation here:

$$\varphi_{ia} = \frac{\phi_i + \lambda\phi_i^*}{\sqrt{1+\lambda^2}} \quad (2.16\text{a})$$

$$\varphi_{ib} = \frac{\phi_i - \lambda\phi_i^*}{\sqrt{1+\lambda^2}} \quad (2.16\text{b})$$

$$\lambda^2 = -\frac{C_i^*}{C_i} \quad (2.16\text{c})$$

It follows that the GVB wave function of Eq. (2.13) can be readily rewritten in MO terms as in Eq. (2.17):

$$\Psi_{\text{GVB}} = |(C_1\phi_1\overline{\phi}_1 + C_1^*\phi_1^*\overline{\phi}_1^*)(C_2\phi_2\overline{\phi}_2 + C_2^*\phi_2^*\overline{\phi}_2^*) \ldots (C_n\phi_n\overline{\phi}_n + C_n^*\phi_n^*\overline{\phi}_n^*)| \quad (2.17)$$

As can be seen, all the "bonds" are described now by bonding/antibonding MOs, much like in Eq. (2.6) for H_2. The transformation, which is very simple and can be done by hand, is exact and requires no approximation. It creates an immediate bridge between the "two cultures": the VB culture in which a polyatomic molecule is represented as a single configuration displaying 2 by 2 singlet-coupled overlapping AOs, which can be hybridized, and the MO culture in which the same wave function takes a multiconfigurational form involving delocalized bonding and antibonding MOs.

Textbox 2.1

The MCSCF (multiconfigurational self-consistent field) method generates a wave function involving several configurations, for the sake of improving the Hartree–Fock single determinant by bringing electron correlation. This method differs from the simple MO–CI method in that both the coefficients of the configurations and the MOs are optimized simultaneously so as to minimize the total energy. The configurations that are involved constitute the "active space." When all possible configurations that can be formed from a given set of occupied and virtual MOs are involved in the calculation, the method is called *complete active space self-consistent field (CASSCF)*. A CASSCF calculation captures all correlation, nondynamical and dynamical, within the active space and thus it is equivalent to full CI within this space. However, the contribution of the dynamical correlation, which originates from the vacant orbitals above the active space, is very important for a quantitative account of the total correlation even in cases when the active space covers all valence orbitals.

2.7
VB Reading of CASSCF Wave Functions

While the MO–CI to VB transformation is simple and exact in the minimal basis set, it is less simple in basis sets bigger than minimal, because in such a case the AOs that compose the various MOs are slightly different from one MO to the other. Although the direct and exact transformation as described can, in principle, still be performed, it may lead to a very large number of VB structures, some of which are far from chemical intuition [52]. For these reasons, alternative bridging methods have been developed, based on the fact that CASSCF wave functions are invariant under unitary transformations of the orbitals within the active space (see Textbox 2.1).

Thus, just as was done in Section 3 with the Hartree–Fock single determinant, we can perform unitary transformations of the MOs to localize them as much as possible. Now, however, unlike the Hartree–Fock case, the CASSCF wave function does not have the constraint that the MOs must be doubly occupied. It follows that a CASSCF wave function with localized orbitals resembles very much a VB wave function, with pairs of singly occupied quasi-AOs that are singlet coupled into a bond. To maximize this resemblance, two different procedures have been devised, both sharing the acronym "CASVB" [53, 54]. These methods are briefly described in the Appendix.

2.8
Natural Bonding Orbitals and Natural Resonance Theory – a Direct Bridge between MO and VB

Natural bond orbital (NBO) theory and natural resonance theory (NRT) are described in another chapter of this book, and in two recently published monographs [55, 56]. Therefore, these theories are not described in detail in the present chapter, in which we only want to show how they constitute a good bridge between VB and MO methods.

2.8.1
Natural Bonding Orbitals

Natural orbitals (NOs) are obtained by diagonalizing the one-electron reduced density matrix that is provided by any MO- or DFT-based computational methods. In so doing, one obtains a set of NOs with maximum occupancies. These NOs form a complete set of maximum occupancy orbitals that are strictly orthogonal, unique, and intrinsic to the wave function.

By restricting the search for maximal occupancy orbitals to a set of diatomic A–B bonding regions, one can obtain a set of NBOs with occupancies n_i ($0 < n_i < 2$). The NBOs divide naturally into a leading high-occupancy set and a residual low-occupancy set, which can usually be neglected for chemical purposes. For an N-electron system, the $N/2$ "Lewis type" NBOs of highest occupancies (n_i close to 2), can be directly associated with the localized electron pairs of a given Lewis structure. This set of NBOs typically includes two-center bond orbitals, while NOs restricted to a single atom are one-center core orbitals or one-center valence lone pairs. Thus, one first result of an NBO analysis is the possibility of visualizing the shapes of the lone pairs and bond orbitals, estimating the hybridization of their atomic components, and in brief examining the bonding from a Lewis-like perspective.

Of course, the NBOs bear resemblance to the Boys-localized orbitals presented in the previous section, and arising from unitary transformations of the CMOs after a Hartree–Fock calculation. However, an important difference between Boys-localized orbitals and NBOs is that the former are necessarily obtained from a single-determinant wave function (Hartree–Fock level), while NBOs are deduced from a density matrix that can be calculated at any level of sophistication in the MO–CI framework, or by means of DFT.

The localized wave function corresponding to a Lewis structure i is made of the antisymmetrized product of the $N/2$ Lewis-type corresponding NBOs, which in turn yields a localized density matrix D_i^{L}.

2.8.2
Natural Resonance Theory

The approximation of a single Lewis ("resonance") structure is often inadequate when the density matrix belongs to a delocalized species, such as benzene. In the VB framework, this means that the electronic state in question is best represented as a combination of Lewis structures rather than a single one. In such a case, the true density matrix, D^{true}, can be approximated as a linear combination D^{w} of localized density matrices, as in Eq. (2.18)

$$D^{\mathrm{w}} = \sum_i w_i D_i^{\mathrm{L}} \tag{2.18}$$

where each D_i^{L} is the localized density matrix associated with the Lewis structure i, and w_{i} is the corresponding weight, the value of which is, of course, of primary interest.

Practically, the NRT method begins by calculating the idealized D_i^{L} for each candidate resonance structure i by a directed NBO search, then it optimizes the weights of each localized density matrix in Eq. (2.18) so as to minimize the root-mean square deviation of D^{w} from the true density matrix D^{true}. It is clear that the respective optimized weights of the localized density matrices can be interpreted as the weights of the Lewis structures that are implicitly contained in the electronic state in question.

As such, NRT constitutes a direct bridge between VB and non-VB methods as it allows one to interpret an electronic state of a molecule, calculated by any method, in terms of Lewis structures that can be given a quantified weight. Note that NRT can be used in a spin-unrestricted form, that is, just as UHF calculations gives different orbitals and density matrices for different spins, α- and β-densities, which can be analyzed separately and give different Lewis structures for different spins, much as in Linnett's version of the VB theory [57].

The agreement of NRT with direct VB calculations is generally good, as can be seen from Table 2.2, where the weights of Lewis structures for the X_3^- linear

Table 2.2 Weights of Lewis structures as extracted from an NRT analysis of a DFT calculation (C. Landis, personal communication to PCH).

	X—X :X⁻	X⁻: X—X	•X X:⁻ X•
X = F	0.43 (0.37)	0.43 (0.37)	0.13 (0.26)
X = Cl	0.44 (0.42)	0.44 (0.42)	0.12 (0.15)
X = Br	0.43 (0.41)	0.43 (0.41)	0.14 (0.15)
X = I	0.43 (0.41)	0.43 (0.41)	0.14 (0.15)

By comparison, weights arising from direct *ab initio* VB calculations (Ref. [58]) are indicated in parentheses.

trihalogen anion (X = F, Cl, Br, I), as calculated from an NRT analysis of DFT-calculated densities, (C. Landis, personal communication to PCH) are compared with weights arising from true VB calculations [58]. As can be seen, the agreement is fair for F_3^-, and excellent for Cl_3^-, Br_3^- and I_3^-.

2.9
The Mythical Conflict of Hybrid Orbitals with Photoelectron Spectroscopy

A good feature of the Hartree–Fock model is that the energies of the CMOs are good approximations of the ionization potentials (IPs) of the molecule, by virtue of Koopmans' theorem [59]. This is because the CMOs are eigenfunctions of an effective monoelectronic Hamiltonian (the Fock operator), which represents the energy of an electron in the average electric field of the others. Now it is well known (see Section 2.4) that the CMOs can be transformed into localized orbitals by unitary transformations without changing the polyelectronic wave function, a single Slater determinant in the Hartree–Fock model. Doing this transformation for methane, and choosing the transformation that minimizes the repulsions between the localized electron pairs (e.g., by Boys method [45]), yields a set of four equivalent Boys-localized MOs (φ_a, φ_b, φ_c, φ_d in Scheme 2.5), which are each made of a bonding combination of a 1s orbital of a hydrogen atom and an sp^3 hybridized atomic orbital (HAO) of the central carbon atom.

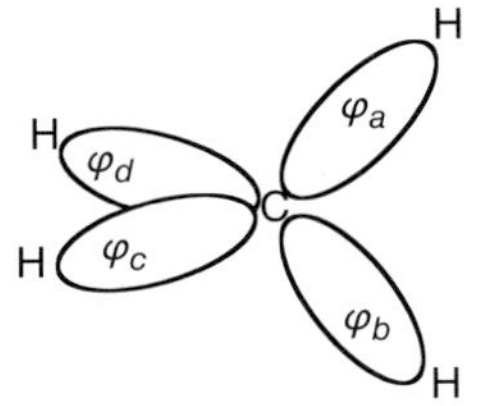

Scheme 2.5 A schematic representation of the four valence LBOs of methane.

As the Slater determinant with localized MOs is exactly equivalent to the original one with CMOs, both wave functions must give exactly the same densities, total energies, net charges, and all measurable properties for a given molecule. Therefore, its is clear that either CMOs or Boys-localized MOs with hybrid orbitals are equivalent descriptions of the same reality, and that assertions such as "hybrid orbitals do not exist and do not appropriately describe molecular bonding", or, "... are inappropriate models for the description of electronic energies and electron density within a molecule" and so on [60], are unfounded. Unfortunately, such dismissals of the concept of hybridization still appear here and there in some oral chemistry courses, and even recently in the primary literature [60]. They all use a popular and deceptively convincing argument based on the photoelectron spectrum of methane, which displays two different ionization peaks, while, allegedly, the HAO model with the four identical C–H bonds would display a single ionization peak. Actually, this argument is not based on any proper theory or experiment but arises from a misuse of Koopmans' theorem. To examine the argument in detail, let us take the example of methane and its Hartree–Fock wave function in terms of Boys-localized MOs (Eq. (2.19)):

$$\Psi(\mathrm{CH_4}) = |\varphi_a\overline{\varphi}_a\varphi_b\overline{\varphi}_b\varphi_c\overline{\varphi}_c\varphi_d\overline{\varphi}_d| \tag{2.19}$$

where the core orbital of carbon is omitted and each φ_i is a hybridized localized MO, as in Scheme 2.5.

The reasoning of those who dismiss the localized MO model goes as follows: "since the HAO model puts four electron pairs into four equivalent localized orbitals, then extracting an electron from anyone of these four orbitals should always cost the same energy, leading to a single unique IP." The error in this reasoning is straightforward: it completely ignores the quantum mechanical requirement that any wave function must match the symmetry of the molecule. Indeed, starting from $\Psi(\mathrm{CH_4})$ in Eq. (2.19) and simply extracting an electron from a localized orbital, say φ_d, would lead to a single determinant, which is the first of the series of four in Eq. (2.20). However, this *determinant alone can in no way represent an ionized state of CH_4*, because it lacks the T_d symmetry. The electron can be ejected from any one of the four localized orbitals and the chemically correct wave function of $\mathrm{CH_4}^+$ will be precisely a symmetry-adapted combination of the four corresponding determinants, as shown in Eq. (2.20).

$$\begin{aligned}\Psi(CH_4)^+ = {} & |\varphi_a\overline{\varphi}_a\varphi_b\overline{\varphi}_b\varphi_c\overline{\varphi}_c\varphi_d| \leftrightarrow |\varphi_a\overline{\varphi}_a\varphi_b\overline{\varphi}_b\varphi_c\varphi_d\overline{\varphi}_d| \\ & \leftrightarrow |\varphi_a\overline{\varphi}_a\varphi_b\varphi_c\overline{\varphi}_c\varphi_d\overline{\varphi}_d| \leftrightarrow |\varphi_a\varphi_b\overline{\varphi}_b\varphi_c\overline{\varphi}_c\varphi_d\overline{\varphi}_d|\end{aligned} \tag{2.20}$$

Using elementary group theory, we can find a triply degenerate $^2T_2$ state and one single $^2A_1$ state as shown in Figure 2.2. Thus, the figure shows clearly two IPs in conformity to experiment and to the CMO model. This is a simple textbook exercise that can be found in Ref. [32] (pp. 104–106). An even simpler demonstration shows that the HAO model for water, with the popular "rabbit-ear" shapes for the lone pairs, also yields two distinct IPs upon ionization [32]. Thus, the mythical claim that HAOs are in conflict with experimental IP measurements has no real foundation. Both canonical and localized MOs match the photoelectron spectrum. The two cultures MO and VB are brought together again.

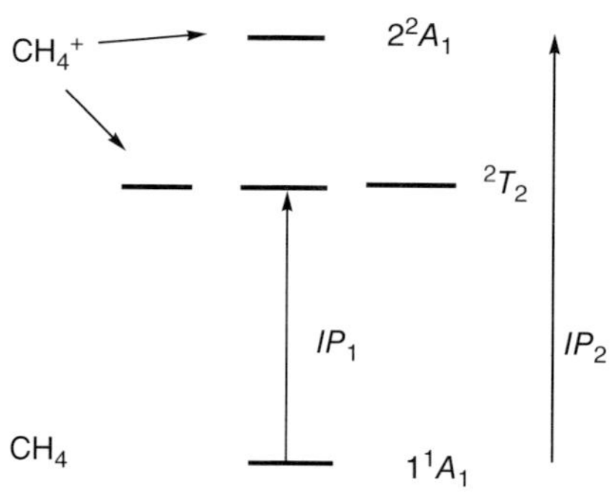

Figure 2.2 The ionized states of CH_4 within the HAO framework

2.10 Conclusion

The two cultures of chemical theory are just one culture in two disguises. Even if the struggle between the two camps of the MO and VB theories may have been important in the past, what was understandable in the fifties is now completely obsolete and counterproductive. Currently, a wide majority of theoreticians and experimental chemists agree that the MO and VB approaches are complementary, rather than exclusive of each other. They tackle the chemical questions from different angles but lead in the end to the same result, if applied at a sufficiently high level.

Of course, each representation has its specific advantages. MO theory is more portable in those problems where the delocalized description of electrons is essential, as in organometallic complexes, pericyclic reactions, photoelectron spectra, and so on. Also, MO methods, including DFT, are still more computer efficient than VB ones, although the latter are currently progressing at a fast pace [36, 61]. On the other hand, VB methods bring some complementary insight of a different kind, with a compact description of an electronic state, involving static and dynamic correlation, in terms of a few resonance structures that can be given a quantitative weight or can be computed separately.

The simplest way to see the MO–VB equivalence in polyatomic molecules is to start from the MO Hartree–Fock single determinant and to localize the orbitals. This does not change the total wave function and gives a picture of the molecule that is very close to the VB one, in terms of local bonds and lone pairs. The equivalence carries over to the electron-correlated level, with the GVB method that can be couched in VB or MO forms, by a simple back-of-an envelope transformation. More generally, any MO-based wave function can be written as a linear combination of VB structures and the reverse is, of course, true also. Methods for mapping MO-based wave functions to VB wave functions exploit this possibility, and so do the CASVB methods for VB-reading CASSCF MO wave functions. Besides, the calculation of resonance energies and energies of diabatic states can be done in the VB framework by means of the BLW method. Finally, the NBO and NRT methods are able to extract VB information from electron densities provided by any MO–CI or DFT methods, not even needing an explicit wave function. Moreover, highly correlated methods based on coupled cluster theory are known to converge more efficiently by usage of localized orbitals. The recent CCSD method, which is based on pair natural orbitals (PNOs) [62] is virtually as efficient as DFT, its formulation is akin to VB theory based on fragment orbitals [32].

All these methods have been developed because many researchers felt that the delocalized view of the electronic structure in the original MO theory needed to be complemented by a local view, characteristic of VB theory, because in the end, most chemical problems are local. One can only hope that this dual single-culture will gain ground among theoreticians and experimentalists. Chemistry has much to gain from this unification.

Appendix

In the CASVB method of Thorsteinsson [53], one transforms the canonical CASSCF orbitals so that the wave function involves a dominant component of VB type, Ψ_{VB}, which is chosen in advance and may be single- or multiconfigurational, as in Eq. (2.21):

$$\Psi_{CAS} = S_{VB}\Psi_{VB} + (1 - S_{VB}^2)\Psi_{VB}^{\perp} \tag{2.21}$$

Here, $\Psi^{\perp}$ is the orthogonal complement of Ψ_{VB} to the CASSCF wave function, and S_{VB} is the overlap between Ψ_{CAS} and Ψ_{VB}. To ensure that the obtained VB function is as close as possible to the starting CASSCF one, an obvious procedure is to transform the orbitals so as to maximize the overlap S_{VB}. An alternative is to minimize the energy of the VB function Ψ_{VB}. This latter procedure is however more expensive than the first one. As both methods generally yield similar sets of orbitals, the method of S_{VB} maximization tends to be preferred [53c].

The CASVB method of Hirao *et al.* [54] differs from the previous one in the requirement that after the transformation of the CASSCF canonical orbitals, the CASVB wave function remains strictly equivalent to the starting CASSCF wave function. The price of this strict equivalence is that the orbitals that are used to construct the VB structures remain more or less of the delocalized MO type; however, these VB-type MOs are localized as much as possible following Ruedenberg's localization procedure [63] that yields quasi-atomic CASSCF MOs that have maximum overlap with the AOs of the free atoms.

References

1. van Vleck, J.H. and Sherman, A. (1935) *Rev. Mod. Phys.*, **7**, 167.
2. Lewis, G.N. (1916) *J. Am. Chem. Soc.*, **38**, 762.
3. Servos, J.W. (1990) *Physical Chemistry from Ostwald to Pauling*, Princeton University Press, Princeton, NJ.
4. Shaik, S. (2007) *J. Comput. Chem.*, **28**, 51.
5. Heitler, W. and London, F. (1927) *Z. Phys.*, **44**, 455.
6. Slater, J.C. (1929) *Phys. Rev.*, **34**, 1293.
7. Slater, J.C. (1931) *Phys. Rev.*, **38**, 1109.
8. Rumer, G. (1932) *Göttinger Nachr.*, 337.
9. Slater, J.C. (1931) *Phys. Rev.*, **37**, 481.
10. Slater, J.C. (1931) *Phys. Rev.*, **41**, 255.
11. Pauling, L. (1931) *J. Am. Chem. Soc.*, **53**, 1367.
12. Pauling, L. (1931) *J. Am. Chem. Soc.*, **53**, 3225.
13. Pauling, L. (1939) *The Nature of the Chemical Bond*, 3rd edn, 1960, Cornell University Press, Ithaca, NY.
14. Gallup, G.A. (2002) in *Valence Bond Theory* (ed. D.L. Cooper), Elsevier, Amsterdam, pp. 1–40. A Short History of VB Theory.
15. (a) Mullliken, R.S. (1928) *Phys. Rev.*, **32**, 186; (b) Mullliken, R.S. (1928) *Phys. Rev.*, **32**, 761; (c) Mullliken, R.S. (1929) *Phys. Rev.*, **33**, 730; (d) Mullliken, R.S. (1932) *Phys. Rev.*, **41**, 49.
16. (a) Hund, F. (1931) *Z. Phys.*, **73**, 1. (b) Hund, F. (1928) *Z. Phys.*, **51**, 759.
17. Hückel, E. (1930) *Z. Phys.*, **60**, 423.
18. Hückel, E. (1931) *Z. Phys.*, **70**, 204.
19. Hückel, E. (1932) *Z. Phys.*, **76**, 628.
20. Brush, S.G. (1999) *Stud. Hist. Philos. Sci.*, **30**, 21.
21. L. Pauling, G. W. Wheland, *J. Chem. Phys.* 1933, **1**, 362. The Nature of the

Chemical Bond. V. The Quantum-Mechanical Calculation of the Resonance Energy of Benzene and Naphthalene and the Hydrocarbon Free Radicals.

22. Wheland, G.W. (1955) *Resonance in Organic Chemistry*, John Wiley & Sons, Inc, New York, p. 4, 39, 148.
23. Hoffmann, R., Shaik, S., and Hiberty, P.C. (2003) *Acc. Chem. Res.*, **36**, 750.
24. Woodward, R.B. and Hoffmann, R. (1971) *The Conservation of Orbital Symmetry*, Verlag Chemie, Weinheim.
25. Fukui, K., Yonezawa, T., and Shingu, H. (1952) *J. Chem. Phys.*, **20**, 722.
26. Dewar, M.J.S. (1951) *Bull. Soc. Chim. Fr.*, **18**, C79.
27. (a) Chatt, J. and Duncanson, L.A. (1953) *J. Chem. Soc.*, 2939. (b) Chatt, J., Duncanson, L.A., and Venanzi, L.M. (1955) *J. Chem. Soc.*, 4456.
28. Brush, S.G. (1999) *Stud. Hist. Philos. Sci.*, **30**, 263.
29. Berson, J.A. (1999) *Chemical Creativity. Ideas from the Work of Woodward, Hückel, Meerwein, and Others*, Wiley-VCH Verlag GmbH, NewYork.
30. For this issue and its earlier discussions, see: Shurki, A., Hiberty, P.C., Dijkstra, F., and Shaik, S. (2003) *J. Phys. Org. Chem.*, **16**, 731.
31. Lennard-Jones, J.E. (1929) *Trans. Faraday Soc.*, **25**, 668.
32. Shaik, S. and Hiberty, P.C. (2008) *A Chemist's Guide to Valence Bond Theory*, John Wiley & Sons, Inc., Hoboken, NJ.
33. Heilbronner, E. and Bock, H. (1976) *The HMO Model and its Applications*, John Wiley & Sons, Inc., New York.
34. Honegger, E. and Heilbronner, E. (1991) in *Theoretical Models of Chemical Bonding*, vol. 3 (ed. Z.B. Maksic), Springer-Verlag, Berlin, Heidelberg, pp. 100–151.
35. (a) Verbeek, J., Langenberg, J.H., Byrman, C.P., Dijkstra, F., and van Lenthe, J.H. TURTLE: An Ab Initio VB/VBSCF Program (1998-2000) (b) van Lenthe, J.H., Dijkstra, F., and Havenith, W.A. (2002) in *Valence Bond Theory* (ed. D.L. Cooper), Elsevier, Amsterdam, pp. 79–116. TURTLE — A Gradient VBSCF Program Theory and Studies of Aromaticity.
36. (a) Song, L., Mo, Y., Zhang, Q., and Wu, W. (2005) *J. Comput. Chem.*, **26**, 514; (b) Song, L., Song, J., Mo, Y., and Wu, W. (2009) *J. Comput. Chem.*, **30**, 399.
37. Cooper, D.L. (ed.) (2002) *Valence Bond Theory*, Elsevier, Amsterdam.
38. Hiberty, P.C. and Leforestier, C. (1978) *J. Am. Chem. Soc.*, **100**, 2012.
39. (a) Karafiloglou, P. and Malrieu, J.-P. (1986) *Chem. Phys.*, **104**, 383. (b) Karafiloglou, P. and Ohanessian, G. (1991) *J. Chem. Educ.*, **68**, 583. (c) Karafiloglou, P. (2008) *J. Phys. Chem. A*, **112**, 8839.
40. Hiberty, P.C. and Ohanessian, G. (1985) *Int. J. Quantum Chem.*, **27**, 259.
41. Hiberty, P.C. and Ohanessian, G. (1984) *J. Am. Chem. Soc.*, **106**, 6963.
42. Hiberty, P.C., Ohanessian, G., and Delbecq, F. (1985) *J. Am. Chem. Soc.*, **107**, 3095.
43. Tantardini, G.F., Raimondi, M., and Simonetta, M. (1977) *J. Am. Chem. Soc.*, **99**, 2913.
44. Edmiston, C. and Ruendenberg, K. (1963) *Rev. Mod. Phys.*, **35**, 457.
45. Boys, S.F. (1968) in *Quantum Theory of Atoms, Molecules, and the Solid State* (ed. P.-O. Löwdin), Academic Press, New York, p. 253.
46. Hoffmann, R. (1982) *Angew. Chem., Int. Ed. Engl.*, **21**, 711.
47. (a) Mo, Y., Song, L., Wu, W., and Zhang, Q. (2004) *J. Am. Chem. Soc.*, **126**, 3974–3982; (b) Mo, Y. (2003) *J. Chem. Phys.*, **119**, 1300–1306; (c) Mo, Y. and Gao, J. (2000) *J. Phys. Chem. A*, **104**, 3012–3020; (d) Mo, Y., Zhang, Y., and Gao, J. (1999) *J. Am. Chem. Soc.*, **121**, 5737–5742; (e) Mo, Y. and Peyerimhoff, S.D. (1998) *J. Chem. Phys.*, **109**, 1687–1697.
48. Mo, Y., Song, L., and Lin, Y. (2007) *J. Phys. Chem. A*, **111**, 8291–8301.
49. Coulson, C.A. and Fischer, I. (1949) *Philos. Mag.*, **40**, 386.
50. (a) Hunt, W.J., Hay, P.J., and Goddard, W.A. (1972) *J. Chem. Phys.*, **57**, 738; (b) Goddard, W.A. and Harding, L.B. (1978) *Annu. Rev. Phys. Chem.*, **29**, 363; (c) Bobrowicz, F.B. and Goddard, W.A.

(1977) in *Methods of Electronic Structure Theory* (ed. H.F. Schaefer), Plenum Press, New York, pp. 79–127.

51. (a) Cooper, D.L., Gerratt, J., and Raimondi, M. (1987) *Adv. Chem. Phys.*, **69**, 319; (b) Cooper, D.L., Gerratt, J., and Raimondi, M. (1988) *Int. Rev. Phys. Chem.*, **7**, 59; (c) Cooper, D.L., Gerratt, J., and Raimondi, M. (1990) in *Valence Bond Theory and Chemical Structure* (eds D.J. Klein and N. Trinajstic), Elsevier, p. 287; (d) Cooper, D.L., Gerratt, J., and Raimondi, M. (1990) in *Advances in the Theory of Benzenoid Hydrocarbons*, Topics in Current Chemistry, vol. 153 (eds I. Gutman and S.J. Cyvin), p. 41.

52. Hiberty, P.C. and Ohanessian, G. (1982) *J. Am. Chem. Soc.*, **104**, 66.

53. (a) Thorsteinsson, T., Cooper, D.L., Gerratt, J., Karadakov, P.B., and Raimondi, R. (1996) *Theor. Chim. Acta*, **93**, 343; (b) Cooper, D.L., Thorsteinsson, T., and J., Gerratt (1998) *Adv. Quantum Chem.*, **32**, 51; (c) Cooper D. L., Karadakov P. B., Thorsteinsson T., In *Valence Bond Theory*, Cooper D. L., Ed.; Elsevier: Amsterdam, 2002, pp. 41-53.

54. (a) Hirao, K., Nakano, H., Nakayama, K., and Dupuis, M. (1996) *J. Chem. Phys.*, **105**, 9227; (b) Hirao, K., Nakano, H., and Nakayama, K. (1998) *Int. J. Quantum Chem.*, **66**, 157.

55. Weinhold, F. and Landis, C. (2005) *Valency and Bonding*, Cambridge University Press, Cambridge.

56. Weinhold, F. and Landis, C. (2012) *Discovering Chemistry with Natural Bond Orbitals*, John Wiley & Sons, Inc., Hoboken, NJ.

57. Linnett, J.W. (1961) *J. Am. Chem. Soc.*, **83**, 2643.

58. Braida, B. and Hiberty, P.C. (2008) *J. Phys. Chem. A*, **112**, 13045.

59. Koopmans, T. (1934) *Physica*, **1**, 104.

60. Grushow, A. (2011) *J. Chem. Educ.*, **88**, 860.

61. Hiberty, P.C. and Shaik, S. (2007) *J. Comput. Chem.*, **28**, 137.

62. See discussion and references in:Riplinger, C. and Neese, F. (2013) *J. Chem. Phys.*, **138**, 034106.

63. (a) Ruedenberg, K., Schmidt, M.W., Gilbert, M.M., and Elbert, S.T. (1982) *Chem. Phys.*, **71**, 41; (b) Ruedenberg, K., Schmidt, M.W., and Gilbert, M.M. (1982) *Chem. Phys.*, **71**, 51.

3
The NBO View of Chemical Bonding

Clark R. Landis and Frank Weinhold

3.1
Introduction

> "The more accurate the computations become, the more the concepts tend to vanish into thin air"
>
> *R. S. Mulliken, J. Chem. Phys.* ***43****, S2 (1965).*

> "It is at least arguable that, from the point of view of quantum chemistry as usually practiced, the supercomputer has dissolved the bond"
>
> *B. T. Sutcliffe, Int. J. Quantum Chem.* ***58****, 645 (1996).*

Before the advent of quantum mechanics, empirical data drove development of chemical structural theory and the subsequent Lewis model [1]. Qualitative concepts such as the chemical bond, valency, and Lewis dot structures provided the dominant framework for understanding the compositions and physical properties of chemical substances. Even after the first applications of quantum mechanics to chemical bonding, *semiempirical* concepts such as electronegativity, hybridization, resonance, and bond order formed the basic vocabulary for describing interatomic attractions, because accurate solution of the Schrödinger equation was intractably difficult for all but the simplest cases. Thus, in the third edition of the *Nature of the Chemical Bond*, Pauling noted that "only a few accurate nonempirical quantum-mechanical calculations of the properties of substances in which the chemist is interested have been made," [2] thereby justifying the application of more qualitative, semiempirical calculations based on valence bond (VB) concepts.

Skip to the present day. Accurate *ab initio* wavefunctions and analysis schemes now exist for many molecules and molecular properties of interest to chemists. Most commonly, these wavefunctions are expressed in molecular orbital (MO) approximation and retain the highly delocalized, symmetry-adapted forms of canonical molecular orbitals (CMOs). As reflected by the introductory comments of Mulliken and Sutcliffe, localized chemical structure concepts such as bonds, Lewis structures, and hybridization simply disappear in such a basis. In the modern

The Chemical Bond: Fundamental Aspects of Chemical Bonding, First Edition.
Edited by Gernot Frenking, Sason Shaik.

quantum chemical era, we may well ask whether the classical chemical concepts of valency and bonding are still meaningful, or merely quaint cultural relics.

An ultimate goal of *natural bond orbital* (NBO) theory[1)] is to connect the numerical content of the modern wavefunction with easily understood concepts of bonding theory. The starting point is a high-quality wavefunction, or at least a wavefunction that is sufficient to capture the important physical properties of the molecule. As described later, NBO algorithms serve to express the wavefunction as accurately as possible in localized Lewis-like form (employing numerically optimized hybrids and bonding pattern), with an associated measure of non-Lewis (NL) "error." NBO analyses of molecules spanning the periodic table demonstrate the broad consistency of modern *ab initio* wavefunctions with classical Lewis-like concepts (dot diagram of shared electron-pair bonds and lone pairs) and related semiempirical constructs (hybridization, electronegativity, resonance, etc.). These concepts constitute a model of chemical bonding that is robust, Lewis-like, and understandable to all chemists, enhanced by the accuracy of modern wavefunction technology.

Not all molecules are described well by simple localized bonding. The NBO method identifies instances for which delocalization effects are strong, such as aromatic compounds, peptides, and hypervalent species. The nature and importance of such resonance delocalizations are captured by a simple donor–acceptor paradigm, of broad applicability to both older and newer forms of "unusual bonding." [9] Lewis-like concepts broadened with such resonance-type donor–acceptor corrections constitute a robust bonding framework that enlightens our understanding of chemical bonding phenomena while remaining fully consistent with computed electron density distributions of the most accurate available modern wavefunctions.

In this chapter, we sketch some heuristic aspects of how NBO theory "works" to express a computed electron density in Lewis-like form, and how it "works out" in illustrative molecular species chosen for pedagogical simplicity. Deeper introductions to NBO theory [3] and practice [4], chemical applications across the periodic table [5], and relationships to other analysis methods [6] are presented elsewhere, including online tutorials and bibliography on the NBO website [10].

3.2 Natural Bond Orbital Methods

Bonding descriptions that emerge from MO and VB calculations differ because of different initial assumptions about the form of the wavefunction. NBO analysis makes *no* assumptions about the form of the *N*-electron wavefunction, but begins instead with the *reduced* form of the wavefunction as given by the one-electron density operator $\boldsymbol{\Gamma}$ [11],

1) In addition to review and historical sources cited later (e.g., Refs [3–8]), see the tutorials (e.g., "MO vs NBO Analysis: What's the Difference?," <*http://nbo6.chem.wisc.edu/tut˙cmo.htm*>) and other introductory materials (e.g., "What are NBOs (and other 'natural'-type orbitals)?", <*http://nbo6.chem.wisc.edu/webnbo˙css.htm*>) on the NBO6 website Ref. [9].

$$\boldsymbol{\Gamma} = N \int |\Psi(1, 2, \ldots, N)|^2 d\tau_2 d\tau_3 \ldots d\tau_N \tag{3.1}$$

which can be obtained for variational, perturbative, or density functional theory (DFT) descriptions of many types, up to and including the exact wavefunction[2)]. On the basis of this input, the NBO bonding picture is derived from (rather than imposed upon) the chosen wavefunction.

A visual impression of the successive "natural" transformations of the density matrix representation of $\boldsymbol{\Gamma}$ from starting atomic orbital (AO) basis functions to localized natural atomic orbitals (NAOs), NBOs, natural localized molecular orbitals (NLMOs), and CMOs is depicted in the panels of Figure 3.1. The figure visually depicts what happens to the density matrix in a simple MO-type calculation for the HF molecule, starting in a 3-21G AO basis [12] (upper left corner) and ending with the five final CMOs of double occupancy (diagonal black blocks, with all other elements zero). The shadings of blocks in each panel suggest proximity to the limiting value 2.00 of double occupancy, as achieved *exactly* in either CMO or NLMO basis, and nearly so in NBO basis. Labels along the left edge identify the AO, NAO, or NBO orbital type in the corresponding density matrix. We can heuristically describe the NBO localization algorithms that lead to VB-like bonds and lone pairs of the Lewis dot diagram by following the successive steps shown in Figure 3.1 (with neglect of many technical details in the present pedagogical context).

The molecular density matrix can be considered to derive from atomic block contributions from individual atoms A, B, …

$$\boldsymbol{\Gamma} = \boldsymbol{\Gamma}_{\mathrm{A}} + \boldsymbol{\Gamma}_{\mathrm{B}} + \cdots (+\text{off-diagonal interaction blocks}) \tag{3.2}$$

where the fluorine $\boldsymbol{\Gamma}_{\mathrm{F}}$ block is 9×9 and the hydrogen $\boldsymbol{\Gamma}_{\mathrm{H}}$ block is 2×2 in the 3-21G AO basis (panel 1 of Figure 3.1). The orbitals that diagonalize the atomic blocks of each angular symmetry type are the *NAOs* [7] (NAOs, $\theta_k^{(\mathrm{A})}$, $\theta_k^{(\mathrm{B})}$, …), satisfying the local eigenvalue equation

$$\boldsymbol{\Gamma}_{\mathrm{A}} \theta_k^{(\mathrm{A})} = n_k^{(\mathrm{A})} \theta_k^{(\mathrm{A})} \tag{3.3}$$

where $n_k^{(\mathrm{A})}$ is the electronic occupancy (population) of orbital $\theta_k^{(\mathrm{A})}$.

Accordingly, the density matrix in the NAO basis (panel 2) has elements $n_k^{(\mathrm{A})}$ along the diagonal that uniquely identify the *natural electron configuration* (NEC) of each atom in the molecular environment,

$$\mathrm{NEC}^{(\mathrm{A})} = \left[\theta_1^{(\mathrm{A})} \left(n_1^{(\mathrm{A})}\right)\right] \left[\theta_2^{(\mathrm{A})} \left(n_2^{(\mathrm{A})}\right)\right] \cdots \tag{3.4}$$

The labels along the left edge (1s, 2s, 2p, etc.) identify the NAOs in the familiar language of Bohr theory; for example, the effective atomic configurations in the present case are found to be (to nearest 0.01e)

$$\mathrm{NEC}^{(\mathrm{F})} = \left[1\mathrm{s}^{2.00}\right] \left[2\mathrm{s}^{1.92}\right] \left[2\mathrm{p}_x^{2.00}\right] \left[2\mathrm{p}_y^{2.00}\right] \left[2\mathrm{p}_z^{1.58}\right] \tag{3.5a}$$

2) Note that Eq. (2.1) somewhat simplifies the full mathematical notation for the kernel of the integral operator (see Ref. [11]), but is adequate for present purposes in suggesting how the density operator depends on the wavefunction.

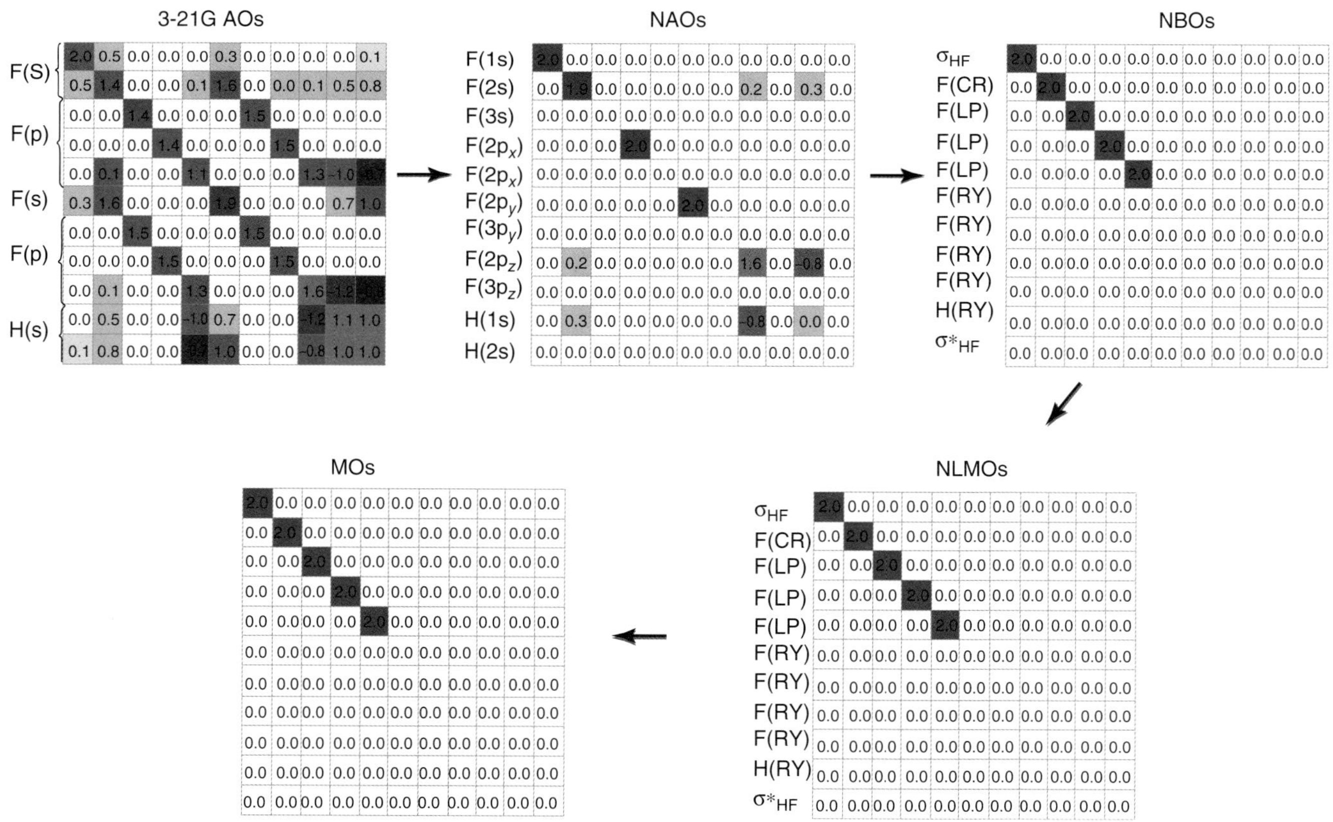

Figure 3.1 The 1e-density matrices for the molecule HF resulting from analysis of the Hartree–Fock self-consistent field computation with the 3-21G basis set. Elements are shaded according to their absolute values, with darkest shading corresponding to density of 2.0 and lightest shading to 0.0. Labels at the top of each matrix identify the basis set and labels along the left indicate the nature of the basis orbitals.

$$\mathrm{NEC}^{(\mathrm{H})} = 1\mathrm{s}^{0.50} \tag{3.5b}$$

Except for the missing "half-electron" of H that appears to have been "transferred" to the valence shell of F, the NEC assignments are remarkably similar to expected Bohr-like configurations of constituent atoms. In contrast, the population distributions of common AO basis functions bear *no* such relationship to basic valency and bonding concepts. Recovery of these sharp Bohr-like properties in the NAO basis allows the subsequent steps of NBO analysis to make easy contact with traditional models of chemical bonding.

The NAOs and their occupancies also underlie the comprehensive *natural population analysis* [7] (NPA) of atomic properties that is provided in NBO output, including the natural charge and effective 1e orbital energies of each atom. From a mathematical viewpoint, the critical feature of the AO → NAO transformation is attainment of strict *orthogonality* between NAOs, which ensures that these orbitals (unlike nonorthogonal basis AOs) can satisfy Hermitian eigenvalue equations such as Eq. (3) that give "orbital population" a well-defined physical meaning. Similar to the NBOs and NLMOs, which are described subsequently, the NAOs form a complete orthonormal set that can be used to *exactly* describe every detail of the computed electron density, while avoiding the many overlap-type artifacts that plague Mulliken population analysis and alternative charge or energy decomposition methods [13]. As can be seen in Figure 3.1, the transformation from basis AOs to NAOs constitutes the most important single step in bringing the density matrix to a simplified localized form. NAOs also provide a low-dimensional *natural minimal basis* (NMB) subset of orbitals that contain the vast majority (commonly, >99.9%) of electron density, consistent with the assumptions of empirical bonding models. From the long-range separation limit to final equilibrium geometry, the numerically determined NAOs (with their occupancies, energies, and other properties) serve to define the effective "natural atoms" [14] that persist as identifiable building blocks of the electron density distribution throughout molecule formation.

By searching similarly for the highest occupied eigenorbitals corresponding to each diatomic block of the density matrix, one obtains the optimal NBOs, as shown in the third panel of Figure 3.1. The labels along the left edge identify the chemical nature of each NBO: CR = core orbital, LP = valence lone pair (nonbonding), LV = lone vacancy (unoccupied borane-like valence orbital), RY = unoccupied Rydberg orbital, BD = 2-center bonding orbital (σ, π, etc.), and BD* = 2-center antibonding orbital (σ^*, π^*, etc.). The pattern of bonds and lone pairs constitutes the *natural Lewis structure* (NLS) for the molecule, which commonly describes the vast majority (>99%) of total electron density as assumed in elementary bonding models. The NBOs can be considered to be the intrinsic "chemist's basis set" by which the wavefunction expresses itself in an optimally condensed Lewis-like form.

Note that each Lewis-type (L-type) NBO of the NLS achieves *near* (but not exact) double occupancy that reflects its exquisite adaptation to the molecular environment. The slight departures from exactly degenerate "electron pair" occupancy (shown as the almost imperceptibly shaded "non-Lewis" blocks in the NBO density matrix) ensure the numerical *uniqueness* of NBOs in solving the diatomic eigenvalue equations analogous to Eq. (3.3). The weak residual occupancies of NL-type NBOs also constitute a rigorous measure of "error" of the NLS description, quantifying the accuracy of a localized Lewis-structural model.

As depicted in the fourth panel of Figure 3.1, incorporation of the weak mixings between Lewis and NL orbitals results in *NLMOs* [15], which achieve full double occupancy while preserving maximum similarity to the original NBOs. Among the infinite number of massively degenerate "MOs" of exact double occupancy (including CMOs), all mathematically *equivalent* for forming a single-determinant wavefunction, the NLMOs are singled out as "least delocalized" or "most NBO-like" by their mathematical construction. The NLMOs also differ fundamentally from alternative "localized molecular orbitals" (LMOs) of Edmiston–Ruedenberg, Boys, or related type [16], in that they are computed much more efficiently by small-matrix or weak-perturbation procedures that never require availability (or existence) of CMOs. Indeed, NLMOs are just as easily calculated from wavefunctions of *any* form, and they bear no intrinsic relationship to CMOs or other LMOs, except in the single-determinant limit where all such "MOs" become unitarily equivalent. NLMOs, unlike CMOs, are not symmetry adapted, nor are they delocalized over the entire molecule. In all but exceptional cases, the NLMOs and parent NBOs look similar, save for the addition of small "delocalization tails" that represent the resonance-type interactions (as described later) of each localized Lewis-type NBO with NL NBOs of the nearby molecular environment.

The final panel of Figure 3.1 depicts the NLMO → CMO transformation that is optionally computed (serving the user-requested "CMO" keyword option) for single-determinant MO or DFT wavefunctions. The matrix elements of this transformation provide a direct "bridge" to conventional CMO representations of molecular properties, allowing one to see the essential underlying VB-like simplicity that is commonly obscured by the "unnecessary delocalizations" (terminology of R. Hoffmann, private communication) of the CMO basis.

Figure 3.1 is somewhat limited in that it refers only to closed-shell systems of "restricted" Hartree–Fock (RHF) character, with spatial orbitals of maximum double occupancy. The *unrestricted* Hartree–Fock (UHF) formulation of MO theory (or correlated post-HF levels) leads more generally to two *distinct* 1e density matrices, one each for α- and β-spins. In such "different orbitals for different spins" (DODS) description [17], one employs spin orbitals of maximum *single* occupancy. However, NBO theory easily generalizes to DODS levels, leading naturally to the concept of "different Lewis structures for different spins" (DLDS) and associated pictures of distinct Lewis-like spin-orbital distributions and resonance corrections (as discussed subsequently). This generalization also provides a direct bridge to higher level correlated wavefunction treatments (VBSCF, GVB, MPx, CASSCF, MRCI, etc.) [12] that can be similarly analyzed by NBO analysis of the one-electron density matrix.

We now illustrate application of NBO methods to atoms and to simple chemical bonds.

3.2.1
NBO Analysis of Free Atoms and Atoms in Molecular Environments

Central to the Bohr/Lewis paradigm is the assumption that atoms retain their identifiability, even in the molecular environment. Let us begin with the "simple" atom, Be, in order to illustrate NBO analysis of restricted, unrestricted, and correlated wavefunctions.

The nominal ground state of Be is a $1s^2 2s^2$ configuration. MO or DFT calculations of the Be ground state using either restricted or unrestricted calculations conform to this description. In contrast, NBO analysis of the correlated UMP2 (unrestricted second-order Møller–Plesset) wavefunction reveals the angular correlation of the two valence electrons in both the electron configuration ($1s^2 2s^{1.94} 2p^{0.06}$) and the orbital shapes, as illustrated in Figure 3.2. Figure 3.2a shows the featureless 2s-type valence NBOs (identical for α and β spin) at simple DFT level, whereas Figure 3.2b shows the more interesting spin polarization arising from the sp-hybridized spin NBOs of correlated UMP2 description, tending to keep α and β electrons on opposite side of the nucleus. Despite such subtle electron correlation effects, the NBOs of free atoms remain closely tied to simple Bohr-type single-configuration concepts.

Although the NBO method leads to uniquely identifiable atoms within molecules, one anticipates that the nature of the atom varies with the nature of the molecular

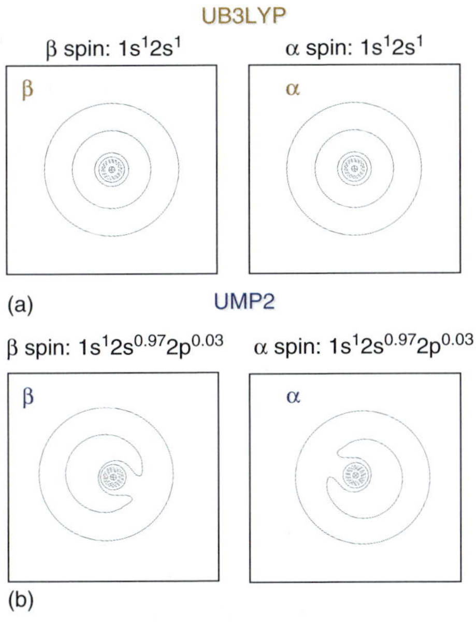

Figure 3.2 (a, b) Contour plots of the NBOs for the Be valence electrons in the atomic ground state obtained from unrestricted DFT (UB3LYP) and MP2 (UMP2) calculations illustrating the effects of angular correlation on the natural electron configurations (NECs) and orbital shapes.

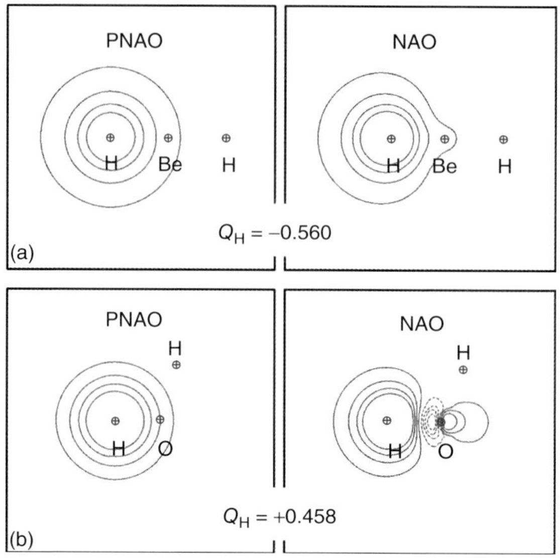

Figure 3.3 Contour plots of the PNAOs and NAOs and atomic charges (Q_H) for H atoms in the BeH_2 (a) and H_2O (b) molecular environments.

environment. Consider BeH_2 and H_2O. One expects significantly different atomic sizes and charges of the H atoms in the two environments, owing to the large differences in the electronegativity of Be and O. As shown in Figure 3.3, these expectations are borne out by the "preorthogonalized" natural atomic orbitals ("pre-NAOs" or PNAOs), NAOs and natural atomic charges (Q_H) of the H atoms obtained by NBO analysis of the DFT results.[3)]

As shown in Figure 3.3, the $1s_H$ PNAO on hydrogen is recognizably "1s-like" (spherical, nodeless) in both species, but with distinctly smaller radius in H_2O (where $Q_H = +0.458$) than in BeH_2 (where $Q_H = -0.560$). The corresponding NAOs show the nonspherical contractions and nodal features that reflect the increased "steric pressure" near the filled orbitals of the oxygen core and opposite hydride bond (as could also be judged from overlap diagrams for the corresponding PNAOs). NAOs therefore reflect exquisite adaptation to details of the H_2O versus the BeH_2

3) (PNAOs and NAOs are identical except for the neglect of a final interatomic orthogonalization step in the former. PNAOs associated with a given atom are therefore orthogonal to one another but not necessarily orthogonal to PNAOs on other atoms, enabling them to serve as "visualization orbitals" for overlap-based estimates of interaction strength. In contrast, the fully orthogonal NAOs include oscillations in the interatomic regions that represent the increased kinetic energy of Pauli-type *steric repulsions* [18] as atomic electron distributions begin to interpenetrate. While the PNAOs resemble common AO depictions in textbooks and are more effective in displaying "orbital overlap" estimates of interaction strength, the absolute orthogonality of the NAOs is required for satisfaction of the Pauli principle in the molecular environment and general consistency with the Hermitian representation of physical properties, as discussed earlier.)

molecular environment, but the essential continuity with free-atom NAOs is easily established in both species.

Note that conventional AO-based analyses would require highly extended numbers of multi-zeta and polarized basis functions to describe the adaptive variations of a *single* NAO. The effective dimensionality of NAO/NBO-based description therefore remains close to *minimal* (NMB level, as assumed in elementary bonding models), no matter how extended the numerical AO basis.

3.2.2 NBO Analysis of Simple Chemical Bonds: LiOH and H_2O

Despite many advances in quantum theoretical methods, the concepts of chemical bonds, Lewis dot structures, and valence orbital hybridization persist as the most fundamental model by which chemists understand the nature of molecules. In this section we demonstrate how the purely mathematical transformations of the reduced density matrix that constitute NBO analysis reconstruct these qualitative concepts as introduced by Lewis, Pauling, and others nearly a century ago [19].

When confronted with formulas such as H_2O and LiOH, even the beginning chemistry student would identify the molecules as intrinsically different, despite the identical number of valence electrons; water comprises two (polar) shared electron-pair bonds, whereas LiOH is ionic with a Li^+ cation and diatomic OH^- anion. In this intuition, the student is informed by numerous rich empirical clues, such as the positions of the elements in the periodic table, conductivity properties, acid–base properties, and so forth. NBO analysis of the gas-phase molecules successfully recovers these basic descriptions, despite the complete independence of *ab initio* quantum mechanical models from any empirical data. Some key NBO metrics of H_2O and LiOH are summarized in Table 3.1.

Do Lewis dot representations of H_2O and LiOH accurately depict the quantum mechanical electron density distributions? The "% Lewis character" metric provides a rigorous, quantitative answer: the NLS dot structures account for >99.5% of the *valence* electron density (disregarding the core orbitals) in both molecules. Such percentages are obtained by summing the diagonal elements (populations) of the VB and lone pair contributions to the reduced density matrix in the NBO basis and dividing by the total number of electrons. Commonly, one finds similarly high "% Lewis character" values for *any* molecule in which atomic valencies can be satisfied by just a single bonding pattern (e.g., NH_3, F_2, HF, CH_4, etc.). Low values of this critical metric indicate that a single Lewis configuration is insufficient to describe the density distribution. Thus, we associate low values of the "% Lewis Character" metric with descriptors such as strong resonance, delocalization, hypervalency, and so on.

Are the OH bonds of water and lithium hydroxide similar (i.e., "transferable" in some sense)? In NBO analysis, each electron-pair bond between atoms A and B (σ_{AB}, π_{AB}, etc.) is a linear combination of atom-centered hybrid orbitals (h_A, h_B). NBO uses the concise and flexible MO-type (Lennard-Jones, Mulliken) formulation of the electron-pair bond, rather than the inflexible VB-type (Heitler–London) or

Table 3.1 Geometrical (d, θ) and NBO-based metrics for gas-phase optimized geometries of H_2O and LiOH, computed at B3LYP/6-311++G** level.

	H_2O	LiOH
NBO-based dot structure	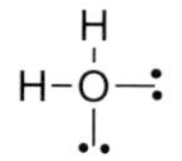	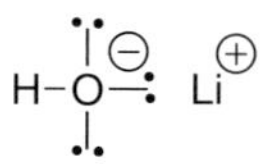
Molecular geometry	$d(\text{O–H}) = 0.962\,\text{Å}$ $\theta(\text{H–O–H}) = 105.1°$	$d(\text{Li–O}) = 1.590\,\text{Å}$ $d(\text{O–H}) = 0.951\,\text{Å}$ $\theta(\text{Li–O–H}) = 180.0°$
Atomic charges	$Q_H = 0.458$ $Q_O = -0.916$	$Q_{Li} = +0.967$ $Q_O = -1.421$ $Q_H = +0.454$
NEC	H: $1s^{0.54}$ O: $1s^2 2s^{1.75} 2p^{5.15}$	H: $1s^{0.54}$ O: $1s^2 2s^{2.00} 2p^{5.58}$ Li: $1s^2 2s^{0.01} 2p^{0.02}$
% Lewis character	99.92%	99.51%
Bonds	2 O–H	1 O–H
Occupancies	1.999e	1.995e
Polarization ($c_O{}^2$, $c_H{}^2$)	73.1% O, 26.9% H	72.9% O, 27.1% H
Hybridization at O	$sp^{3.30}$	$sp^{2.56}$
Lone pairs	2 on O	3 on O
Occupancies	1.997, 1.997	1.991, 1.987, 1.987
Hybridization at O	$sp^{0.87}$, sp^{∞}	$sp^{0.39}$, sp^{∞}, sp^{∞}

most general (Coulson–Fischer) formulation in a given basis of atomic hybrids [20]. Thus, NBO descriptions of density are more similar to GVB functions rather than alternatives such as the apolar Heitler–London bond plus purely ionic structures of classical VB theory. Also, similar to GVB bond functions, NBOs are strictly orthogonal. The general formulation of a sigma bond σ_{AB} is

$$\sigma_{AB} = c_A h_A + c_B h_B \tag{3.6}$$

where c_A and c_B are normalized polarization coefficients whose values reflect the contribution of each hybrid to the bond. The squares of these normalized coefficients define the bond polarizations ($100\% \times c_A{}^2$, $100\% \times c_B{}^2$) and *natural ionicity* (i_{AB}) of the bond

$$i_{AB} = \frac{\left(c_A^2 - c_B^2\right)}{\left(c_A^2 + c_B^2\right)} \tag{3.7}$$

with values ranging from −1 (pure ionic hybrid on B), to 0 (pure covalent), to +1 (pure ionic hybrid on A). According to a fixed NBO numerical threshold, an extremely polarized "bond" function with less than 5% contribution from the opposite center ($|i_{AB}| > 0.95$) is instead identified as a "lone pair" (LP), corresponding to the ionic limit of heterolytic dissociation. Therefore, the basic NBO description of LiOH comprises a cationic Li fragment and a hydroxide anion (two distinct "molecular units"), whereas that for H_2O is a covalently bonded single unit. Nevertheless, the O–H bond ionicities in H_2O and LiOH are seen to be 0.461 and 0.424, respectively, indicating the expected strong transferability of hydroxide bond character between different molecular environments.

The strong similarity of hydroxide bonds in the two species can be further compared in terms of the contributing *natural hybrid orbitals* [8] (NHOs) h_O, h_H at each atomic center. Each NHO is formed from a linear combination of s, p, d, ... NAOs for which the overall mixing ratios can be expressed in conventional $sp^{\lambda}d^{\mu}$ notation; λ is the ratio of squared coefficients for the valence p-orbitals relative to the s-orbital, μ is the ratio of squared coefficients for the valence d-orbitals relative to the s-orbital, and so on. Thus, the angular composition of an sp^4 hybrid orbital is 80% p-character and 20% s-character. The somewhat lower percentage p-character of the O–H bond in LiOH ($sp^{2.56}$) relative to that of H_2O ($sp^{3.30}$) reflects differences in oxygen charge accumulation and molecular geometry (linear vs bent). One should particularly note that the oxygen lone pairs arising from NBO analysis do not have "rabbit ear" orientations, but instead retain the expected pure p-character of MOs in the directions perpendicular to the molecular bonding plane of water, or the linear bonding axis of LiOH.

Figure 3.4 displays the preorthogonalized NHO (PNHO) visualizations of the oxygen bonding hybrids in water ($h_O = sp^{3.30}$, Figure 3.4a) and LiOH ($h_O = sp^{2.56}$, Figure 3.4b), showing the high degree of transferability from one species to the other. Such comparisons exemplify the way in which the qualitative concepts of empirical structure theory are generally recovered with high fidelity in NBO analysis of sufficiently accurate wavefunctions.

3.2.3 Lewis-Like Structures of the P- and D-Block Elements

Early in their experience of chemistry, students learn to apply the octet rule and Lewis dot structures to molecules of the S- and P-block elements. Subsequent introduction to the shell structure of electrons in atoms provides a deeper rationale

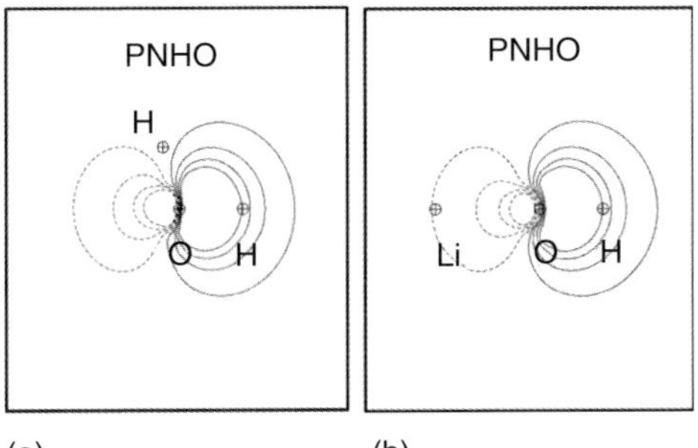

(a) (b)

Figure 3.4 Contour plots of oxygen PNHOs involved in O–H bond formation for water (a) and LiOH (b).

Table 3.2 Graphical depiction of Lewis structure, % Lewis character, hybridization of the central atom bonding hybrid (h_A), and bond angle (θ_{X-A-X}) of representative molecules from the P- and D-block of the periodic table.

P-block examples				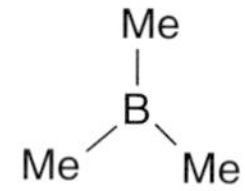	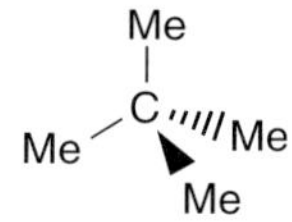	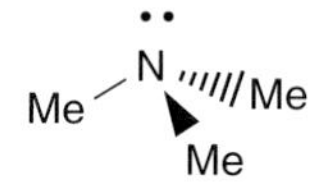
	X = H	X = Me	X = Cl			
% Lewis	99.92	99.31	99.51	98.86	99.35	99.06
h_A	$sp^{3.30}d^{0.01}$	$sp^{2.92}$	$sp^{9.87}d^{0.13}$	$sp^{2.00}$	$sp^{3.00}$	$sp^{2.89}$
θ_{X-A-X}	105.1	113.6	113.6	120.0	109.5	112.9
D-block examples		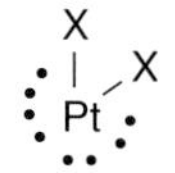		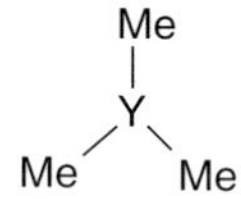	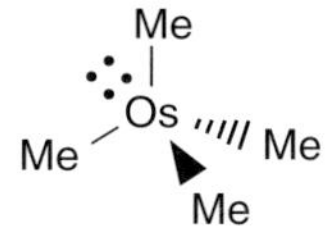	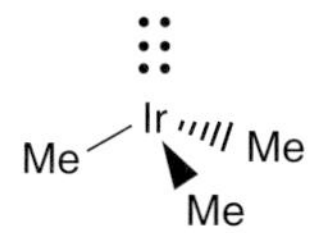
	X = H	X = Me	X = Cl			
% Lewis	99.41	99.31	99.05	99.51	98.98	99.13
h_A	$sd^{1.23}p^{0.01}$	$sd^{1.28}$	$sd^{1.14}p^{0.01}$	$sd^{2.29}p^{0.01}$	$sd^{3.00}$	$sd^{2.19}$
θ_{X-A-X}	85.1	103.2	110.7	110.1	109.5	106.7

for the octet rule and the orbital forces that underlie observed molecular shapes. Do such concepts survive the scrutiny of modern *ab initio* computations? Can Lewis structures form the basis for understanding molecular electronic structures across the periodic table? We now address these questions with a small sampling of simple molecules from the P- and D-block, as shown in Table 3.2.

As judged by the "% Lewis" character of the 12 structures summarized in Table 3.2, simple Lewis (NLS) dot structures of bonds and lone pairs are good descriptors for *both* P- and D-block central atoms. In all cases, the *average* gross hybridization corresponds to either $sp^{(ve/2-1)}$ or $sd^{(ve/2-1)}$ where ve is the valence electron count at the central atom. Thus, the active valence orbitals are the *ns* and *np* orbitals for the P-block and the *ns* and $(n-1)$d orbitals for the D-block. For example, YMe_3 and CMe_4, which use all valence electrons in bond forming with no lone pairs, have approximate hybridizations of sd^2 and sp^3, respectively. As expected, the octet (8e) rule distinguishes between normal-valent (8e; OX_2, CMe_4, NMe_3), hypovalent (<8e; BMe_3), and hypervalent (>8e) molecules in the P-block. Similarly for the D-block, Lewis-like structures conform to a duodectet (12e) rule [21], which enables the fundamental distinction between normal valency (12e; PtX_2, $OsMe_4$, $IrMe_3$), hypovalency (<12e; YMe_3), and hypervalency (>12e).

Consistent with the general theory of directional hybrid bonding, the compositions of valence hybrids lead to certain expectations about bond angles in molecules. For the hybrid orbital representations sp^λ and sd^μ, the values of λ and μ are determined by the %p and %d orbital compositions as follows.

$$\lambda = \frac{(\%p)}{(\%s)} \tag{3.8a}$$

$$\mu = \frac{(\%d)}{(\%s)} \tag{3.8b}$$

The corresponding hybrid orbitals can be written as

$$sp^{\lambda} = (1+\lambda)^{-\frac{1}{2}}\left[s + \lambda^{\frac{1}{2}} p\right] \tag{3.9a}$$

$$sd^{\mu} = (1+\mu)^{-\frac{1}{2}}\left[s + \mu^{\frac{1}{2}} d\right] \tag{3.9b}$$

Application of orthonormalization constraints to a pair of hybrid orbitals h_i and h_j leads to the Coulson directionality theorem [22] (and its analog for sd^{μ} hybrids [23]), which dictates the intrinsic angle, ω_{ij}, between hybrids in terms of their hybridization parameters, viz,

$$\cos\omega_{ij} = -\left(\lambda_i\lambda_j\right)^{-\frac{1}{2}} \quad \text{(P block)} \tag{3.10a}$$

$$\cos\omega_{ij} = \pm\left\{\frac{\left[1-2\left(\mu_i\mu_j\right)^{-\frac{1}{2}}\right]}{3}\right\}^{\frac{1}{2}} \quad \text{for } \mu > 2, \tag{3.10b}$$

$$= 0 \quad \text{for } \mu \leq 2 \ \text{(D block)}$$

The data in Table 3.2 illustrate general trends in hybridization and molecular shapes that are seen over much broader samples of molecules.

Because the hybridization preferences for lone pair orbitals and bond orbitals are not identical, the actual hybridizations differ from the gross average values. For example, the *average* hybridization of water is sp^3, but the individual O–H bond hybridizations ($sp^{3.3}$) and the lone pair hybridizations ($sp^{0.87}$ and sp^{∞}) differ substantially. In the D-block compounds, there is a distinct preference for lone pairs to have pure d character, whereas in P-block compounds, especially those below the second row of the periodic table, the preference of lone pairs is for orbitals of high s character. Such preferences are anticipated by the generally lower energies of atomic $(n-1)$d orbitals versus ns orbitals in the transition series and by the lower energies of ns versus np orbitals of the main group. Hybrid orbital compositions vary with the electronegativity of the atoms bound to the central atom. The large difference in composition of the oxygen bond-forming hybrids of OH_2 ($sp^{3.3}$) versus OCl_2 ($sp^{9.87}$) well illustrate *Bent's rule* [24], which, in its crudest form for the P-block, stipulates that more electronegative ligands favor higher p-character hybrids at the central atom. For PtH_2 ($sd^{1.23}$) and $PtCl_2$ ($sd^{1.14}$), the more electronegative ligand favors *less* d-character (or alternatively, more s-character) in the bonds. Lone pairs can be considered as bonded to the least electronegative entity possible, the vacuum [25]. A consistent extension of Bent's rule to both P-block and D-block species can be formulated in terms of ionic resonance structures [26],

Table 3.3 Intrinsic angles for idealized sp^{λ} and sd^{μ} hybridizations (cf. Eqs. (2.10a–c)).

Hybridization (sp^{λ})	Intrinsic angle (ω)	Hybridization (sd^{μ})	Intrinsic angle (ω)
sp^1	180°	sd^1	90°
sp^2	120°	sd^2	90°
sp^3	109.5°	sd^3	71°, 109°
sp^4	104.5°	sd^4	66°, 114°
sp^{10}	95.7°	sd^5	63°, 117°

providing more robust and accurate structural predictions than the VSEPR (valence shell electron pair repulsions) theory.

To the extent that internuclear axes align with hybrid orbitals, one can directly associate natural bond angles with hybridization parameters. Familiar pairings include sp^2 with 120° and sp^3 with 109.5°, but the association of bond angle with hybridization extends far beyond these textbook examples to include nonintegral hybridizations (e.g., $sp^{3.3}$ for the O–H bonds of water) and hybrids that comprise s and d AO combinations for molecules of the D-block (Table 3.3). Also, note that "bond bending" or the misalignment of hybrid orbital directions and internuclear axes may indicate severe geometrical constraints, such as the 60° C–C–C angle of cyclopropane, or the inability of a single Lewis configuration to describe the full electronic structure (as discussed subsequently).

The striking parallels (Eqs. (3.8)–(3.10)) between P-block and D-block hybridization and Lewis-like bonding were not foreseen by Lewis, Pauling, and other bonding pioneers. However, their subsequent emergence from computational studies of modern wavefunctions [27] bears eloquent testimony to the power and generality of Lewis-like structural concepts and their broad applicability across the periodic table.

3.2.4
Unrestricted Calculations and Different Lewis Structures for Different Spins (DLDS)

So far our discussion has concerned molecules in which pairs of electrons with opposite spins occupy identical orbitals, that is, molecules that are described adequately by restricted wavefunctions. Do Lewis-like structures apply to odd-electron species and higher spin states? We consider a simple, classical problem: the nature of CH_2 (methylene) in its singlet and triplet states as seen through NBO analysis of the Kohn–Sham orbitals.

The lowest energy singlet state of methylene is well described by a Lewis structure comprising one lone pair and two equivalent C–H bonds. At the simplest level, the overall hybridization corresponds to sp^2 hybridization. However, application of Bent's rule suggests that the bonds will have greater p-character than the lone pair, leading to approximately sp^5 (83% p-character) hybridization of the C–H bond-forming hybrids and leaving the lone pair significantly enriched in s-character

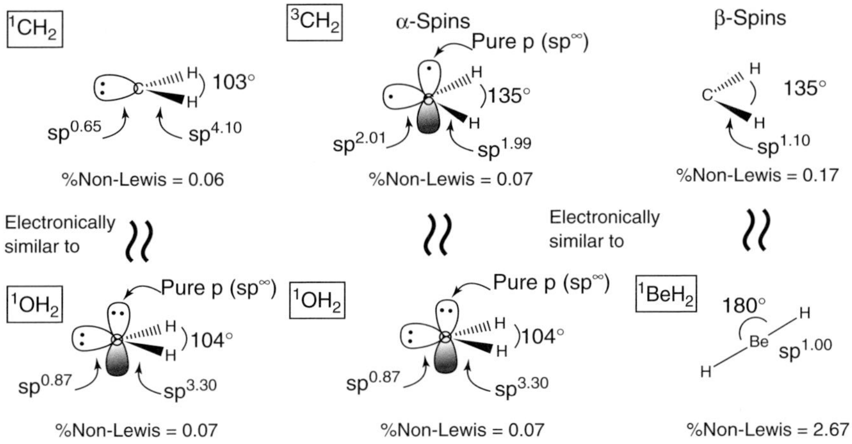

Figure 3.5 Schematic overview of NBO formulations for $^{1,3}CH_2$, illustrating the conceptual usefulness of "different Lewis structures for different spins" in the triplet case (see text).

(67% s or $sp^{0.33}$). The computed bond angle of approximately 102° matches that predicted by simple hybridization schemes.

The triplet state of CH_2 lies about 9 kcal mol^{-1} lower in energy than the singlet state. Within the unrestricted wavefunction formalism, the wavefunctions and density matrices for the α spins and β spins of the triplet state are not identical. Thus, the optimal Lewis structures and NBOs for the two spins differ, leading to a DLDS paradigm. In the case of triplet CH_2, one can represent the Lewis structures as shown in Figure 3.5, keeping in mind that a "bond" line segment represents a single electron shared between two centers.

The NBO Lewis-like structures obtained from unrestricted calculations of triplet methylene suggest that the ideal hybridization of the β spin orbitals should be approximately sp^1, because there are just two active spin orbitals, each making a 1e bond. A reasonable analog from restricted calculations is BeH_2, with two 2e bonds. In contrast, there are four active α spin orbitals for $^3CH_2$ and H_2O is the appropriate analog from restricted calculations. (Although water has a fourth active lone pair orbital, in NBO analysis this is a pure p-orbital perpendicular to the molecular plane, leaving two p-type orbitals to hybridize with the s orbital for in-plane skeletal bonding, as in triplet methylene.) The triplet state C–H bond hybridizations of the α and β spin orbitals ($sp^{1.99}$ and $sp^{1.10}$, respectively) differ significantly, and neither has the directionality expected for the 135° H–C–H bond angle. However, the *average* C–H bond hybridization ($sp^{\lambda(av)}$) can be simply calculated from the average %p-characters of the α and β spin NBOs, as shown in Eqs. (3.11a,b):

$$\%\text{p (av)} = \frac{1}{2}\left[(66.3\%)_{\alpha\text{-spin}} + (55.0\%)_{\beta\text{-spin}}\right] = 60.7\% \tag{3.11a}$$

$$\text{sp}^{\lambda(\text{av})} = \text{sp}^{\left(\frac{60.7}{39.3}\right)} = \text{sp}^{1.54} \tag{3.11b}$$

Application of Coulson's directionality theorem (3.10a) to the *average* C–H bond hybridization for $^3CH_2$ then predicts a bond angle of 130.5°, which lies close to the value of 135° obtained by energy minimization. Therefore, *molecular geometries obtained from unrestricted wavefunctions can be considered to be the result of a balance of forces arising between the geometrical preferences of the α and β spin-orbital hybridizations.*

Before leaving the topic of open-shell descriptions, consider dioxygen. No small molecule has done more to fuel fires – both literally and conceptually – but the analysis of O_2 becomes straightforward when one considers the DLDS paradigm. As shown diagrammatically here, O_2 has seven α spins and five β spins. Remembering that each NBO has just one electron in unrestricted calculations, the α-valence space is filled with just one O–O σ bond and three lone "half-pairs" on each O atom, whereas the β-valence space is completed by an O≡O triple bond. Thus, the *average* bond order is two (= (3+1)/2).

α-spins β-spins

Because two of the lone "half-pairs" on each oxygen of the α-Lewis structure lie in p-orbitals that are spatially aligned with the p-orbitals that make the two π-bonds of the β-Lewis structures, the DLDS model is completely consistent with the "one sigma plus two 3e π-bonds" bonding description of O_2. As in the methylene case, the equilibrium O_2 bond length can be approximated as the average of idealized single (α-spin) and triple bond (β-spin) Lewis-structural values, consistent with empirical bond order–bond length relationships.

3.3
Beyond Lewis-Like Bonding: The Donor–Acceptor Paradigm

As we have already seen, many molecules are not described well by a single Lewis structure configuration. A special vocabulary has arisen to describe departure of electronic structures from the ideal Lewis structure limit, including terms such as *resonance, aromaticity, hypervalence, 3c/2e bonds, 3c/4e bonds, metallic bonding*, and *hydrogen bonding*. The NBO analysis procedure naturally separates the molecular wavefunction Ψ, density ρ, and NBO basis set $\{\Omega_i\}$ into well-defined Lewis (L) and NL components, namely,

$$\Psi = \Psi^{(L)} + \Psi^{(NL)} \tag{3.12a}$$

$$\rho = \rho^{(L)} + \rho^{(NL)} \tag{3.12b}$$

$$\{\Omega_i\} = \left\{\Omega_i^{(L)}\right\} + \left\{\Omega_i^{(NL)}\right\} \tag{3.12c}$$

As we have seen in the previous section, Lewis structures adequately describe many molecules, as judged by the high percentage Lewis character (equivalently, high %-$\rho^{(L)}$). Thus, for these examples, the full wavefunction solution Ψ, on average, must be well approximated by the NLS wavefunction $\Psi^{(L)}$, with only minor corrections from NL components $\Psi^{(NL)}$. However, as resonance effects become more important, the NL $\Psi^{(NL)}$ contributions to the wavefunction will increase, even to the extent of dominating the chemical description as in the case of aromatic or metallic species.

We can think of $\Psi^{(L)}$ as representing the exact solution to the Schrödinger equation in a world that is free of resonance effects. Accordingly, a model Hamiltonian operator $\boldsymbol{H}^{(L)}$ for the idealized "resonance-free world" satisfies the Schrödinger equation

$$\boldsymbol{H}^{(L)}\Psi^{(L)} = E^{(L)}\Psi^{(L)} \tag{3.13a}$$

where $E^{(L)} = \langle\Psi^{(L)}|\boldsymbol{H}|\Psi^{(L)}\rangle$ is the energy of the NLS wavefunction (determined by deletion techniques to be described subsequently). The full Hamiltonian operator in the real world, $\boldsymbol{H}$, is composed of the Lewis model operator $\boldsymbol{H}^{(L)}$ and its NL component $\boldsymbol{H}^{(NL)}$,

$$\boldsymbol{H} = \boldsymbol{H}^{(L)} + \boldsymbol{H}^{(NL)} \tag{3.13b}$$

and the total energy in the real world consists largely of the resonance-free energy with small corrections due to delocalization effects,

$$E = E^{(L)} + E^{(NL)} \tag{3.13c}$$

Thus, the NBO approach is fundamentally perturbative in spirit, with $\boldsymbol{H}^{(L)}$ representing a resonance-free unperturbed operator (with eigenfunctions and energies that are known from calculable NLS properties) and $\boldsymbol{H}^{(NL)}$ representing the perturbation due to the delocalization.

Densities obtained from single-determinant MO or DFT wavefunctions yield Lewis $\{\Omega_i^{(L)}\}$ and NL $\{\Omega_j^{(NL)}\}$ NBOs that satisfy the one-electron eigenvalue equations of the resonance-free operator $h^{(0)}$.

$$h^{(0)}\Omega_i^{(L)} = \varepsilon_i^{(L)}\Omega_i^{(L)}, \quad i = 1, 2, \ldots, N \tag{3.14a}$$

$$h^{(0)}\Omega_j^{(NL)} = \varepsilon_j^{(NL)}\Omega_j^{(NL)}, \quad j = N+1, N+2, \ldots \tag{3.14b}$$

In the resonance-free limit, the Lewis NBOs are filled and the NL NBOs are vacant; thus, we can refer to the Lewis NBOs as "donors" and NL NBOs as "acceptors" of electronic occupancy. In the limit of a resonance-free world, these donor and acceptor NBOs have *zero* interaction. However, application of the *real-world* effective 1e-Hamiltonian operator $\boldsymbol{F}$ (Fock or Kohn–Sham operator) leads to nonvanishing donor–acceptor matrix elements $\boldsymbol{F}_{ij}$:

$$\boldsymbol{F}_{ij} = \int \Omega_i^{(L)}\boldsymbol{F}\Omega_j^{(NL)}\mathrm{d}\tau \neq 0 \tag{3.15}$$

Such donor–acceptor mixings, or *delocalizations*, cause NL orbitals to become partially populated, which is equivalent to saying that other resonance configurations are involved. Thus, NBO analysis expresses all molecular environments as a single resonance structure (usually, but not necessarily, optimal in terms of describing the total density) with delocalizations resulting from interactions between Lewis and NL orbitals.

Because the accuracy (%-$\rho^{(L)}$) of Lewis-like structures for simple molecules often exceeds 99%, it is appropriate to analyze donor–acceptor stabilization energies using the following second-order perturbation treatment [28], where q_i is the occupancy of the Lewis NBO

$$\Delta E_{ij}^{(2)} = \frac{q_i \left| \boldsymbol{F}_{ij} \right|^2}{\left(\varepsilon_j^{(\mathrm{NL})} - \varepsilon_i^{(\mathrm{L})} \right)} \tag{3.16}$$

The "$E^{(2)}$ table" of such donor–acceptor stabilizations is often the first section of output scanned by the accomplished NBO user.

The NBO program also provides a unique alternative facility for gauging the impact of a donor–acceptor interaction on the molecular structure: using deletion ($DEL) keylists[4] [29]. One can instruct the program to *delete* the Fock matrix element $\boldsymbol{F}_{ij}$ corresponding to a particular donor–acceptor interaction, then reoptimize the geometry as though the interaction is physically absent. The variational energy E($DEL) is thereby raised *above* the full energy E (corresponding to loss of the stabilizing donor–acceptor interaction) by roughly the magnitude of the second-order $\Delta E_{ij}^{(2)}$ estimate. Although numerically tedious, E($DEL) optimizations allow one to accurately assess both structural and energetic consequences of a particular donor–acceptor interaction, often allowing one to isolate the "smoking gun" interaction responsible for a particular structural anomaly.

The NBO program also provides a still more powerful option to quantitatively assess resonance effects and fractional bond order changes: *natural resonance theory* [30] (NRT). NRT extends the NBO search from looking for the best *single* Lewis-like density to a search for the best weighted manifold of Lewis-like densities. Thus, the total 1e-density matrix ($\boldsymbol{\Gamma}$) is approximated as the resonance-theoretic superposition $\boldsymbol{\Gamma}_{\mathrm{NRT}}$,

$$\boldsymbol{\Gamma}_{\mathrm{NRT}} = \sum_{\alpha} w_{\alpha} \boldsymbol{\Gamma}_{\alpha} \tag{3.17}$$

where $\boldsymbol{\Gamma}_{\alpha}$ is the 1e-density matrix for a given Lewis-like structure and the w_{α}s are corresponding positive weightings summing to unity. The NRT procedure involves minimization of the rms difference between the full wavefunction density and $\boldsymbol{\Gamma}_{\mathrm{NRT}}$. Because NRT analysis is performed on the densities, it fundamentally differs from multiconfigurational VB methods in which the total *wavefunction* is optimized as a linear combination of determinants in order to achieve the lowest *variational energy*. In VB-RT applications, the configurational coefficients are of varying signs and the question of how to handle nonvanishing cross-terms is an issue, whereas such cross-terms are fundamentally absent in the density-based NRT formulation of Eq. (3.17). As mentioned earlier, another important difference

4) See, for example, "Tutorial on Energetic Analysis with NBO Deletions ($DEL Keylist)" *http://nbo6.chem.wisc.edu/tut˙del.htm* on the NBO6 website (Ref. [10]).

is that the NBO Lewis configurations are based on polarizable MO-type two-center bond functions, whereas multiconfigurational VB methods are commonly restricted to Heitler–London-type ionic and covalent bond functions. Despite the fundamentally different methods of computation, it is remarkable that weightings from NRT and multiconfigurational VB [31] are often qualitatively similar.

Although "donor–acceptor interaction" and "resonance" might initially seem to have little connection, there is in fact a *one-to-one mapping* of any particular NBO donor–acceptor interaction onto a corresponding resonance structure. This mapping can also be visually represented in mnemonic *arrow-pushing* terms as a concerted pattern of charge-transfer shifts from the parent Lewis structure. Figure 3.6 presents a "Rosetta stone" that relates various NBO donor–acceptor motifs to the corresponding arrow-pushing and resonance diagrams of the organic chemist. The almost magical power of resonance structure diagrams to visually convey the structural and charge shifts associated with a particular delocalization provides a powerful complement to the numerical $\Delta E_{ij}^{(2)}$ values. The donor–acceptor paradigm is therefore found to fit "hand in glove" with familiar

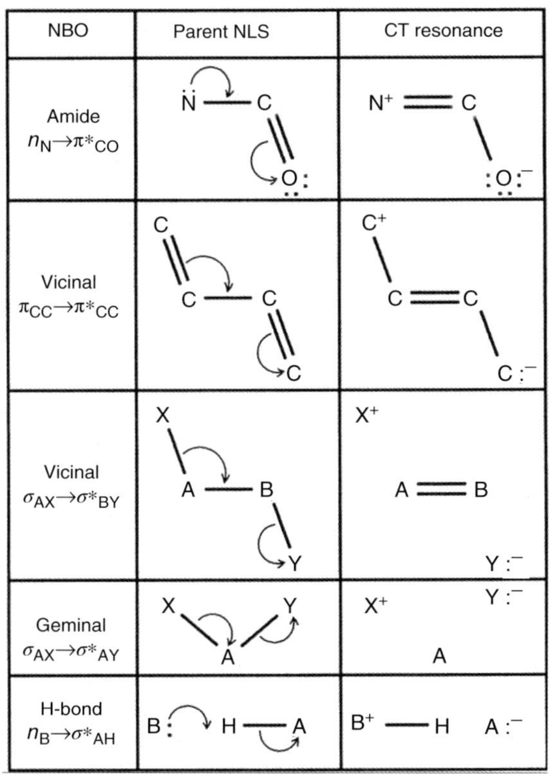

Figure 3.6 Generic arrow-pushing and resonance diagrams, comparing NBO donor–acceptor description (left) with corresponding parent NLS (middle) and secondary charge-transferred structures (right) for representative conjugative and hyperconjugative motifs.

resonance-type concepts of the organic chemist, confirming the close connection of NBO-based descriptors to important empirical insights of the precomputer era.

In summary, important indicators of donor–acceptor delocalizations in a molecular system include the following:

1) Lewis orbital occupancies < 2,
2) NL orbital occupancies > 0,
3) donor–acceptor stabilization energies $\Delta E_{ij}^{(2)} > 0$,
4) significant E($DEL) effects of *deleting* particular $\boldsymbol{F}_{ij}$ delocalizations, and
5) direct evaluation of the weights of other resonance configurations, as determined by NRT.

In the ensuing subsections, we briefly illustrate how such donor–acceptor interactions provide weak hyperconjugative corrections to the idealized NLS properties of simple molecules discussed in Section 3.2, as well as how such interactions become increasingly dominant in cases of strong resonance mixing.

3.3.1
Hyperconjugative Effects in Bond Bending

Let us return to the OCl_2 and $PtCl_2$ species discussed in Section 3.2.3. Although the respective $sp^{9.87}$ and $sd^{1.14}$ bond hybridizations of OCl_2 and $PtCl_2$ are consistent with bent molecular geometries (Table 3.2), the optimized bond angles depart significantly from the near-90° values expected from hybrid orbital orthogonality constraints (Eq. (3.10a,b)). In this case, "bond bending" (the deviation of the bond hybrid directions from the line of centers between the two nuclei) reveals significant divergence from the "perfect" Lewis structure description. What is the origin of bond bending and the larger-than-expected bond angles?

For both OCl_2 and $PtCl_2$, the dominant delocalization interaction is donation from the two in-plane Cl lone pairs into the central atom-Cl *antibonds*, as shown in Figure 3.7. As a result, each of the in-plane Cl lone pairs has occupancy 1.97e (0.03e *less* than 2.00), with most of that density appearing as occupancy of the O–Cl and Pt–Cl antibonds. For both molecules, the donor–acceptor interaction energy can be estimated to be about 6.6 kcal mol^{-1}. Because the energies needed for substantial deformation of bond angles commonly are only a few kilocalories per mole, NBO analysis suggests that forces beyond hybridization are giving rise to the unexpectedly large Cl–O–Cl and Cl–Pt–Cl bond angles. Can this model be tested?

The NBO deletion techniques (as described earlier) are helpful in this context. Upon *deleting* the appropriate $n_{Cl} \rightarrow \sigma^*_{OCl}$ (for OCl_2) or $n_{Cl} \rightarrow \sigma^*_{PtCl}$ (for $PtCl_2$) donor–acceptor interactions, we find that the reoptimized bond angles for OCl_2 and $PtCl_2$ are reduced by more than 10°, to 104.2° and 95.9°, respectively, with minor increases in the bond lengths (0.03 Å). Thus, the donor–acceptor paradigm provides a rational basis for discussing departures from the Lewis structure ideal, and why the internuclear axes might not lie coincident with hybrid orbital direction. As we can show with many similar examples, Lewis structures and donor–acceptor

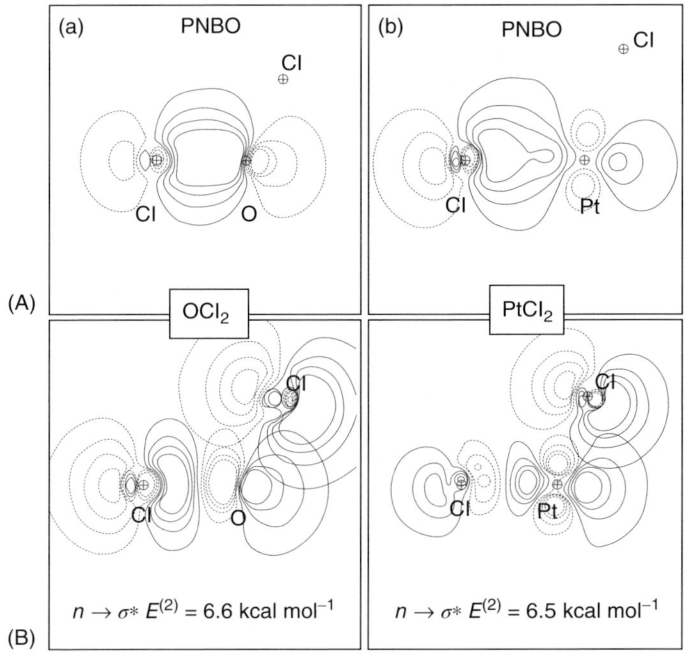

Figure 3.7 (A) Contour plots of pre-natural bond orbitals (PNBOs) of OCl_2 (a) and $PtCl_2$ (b), illustrating the bond bending and the apparently responsible $n_{Cl} \rightarrow \sigma^*_{OCl}$ or $n_{Cl} \rightarrow \sigma^*_{PtCl}$ hyperconjugative donor–acceptor interactions (B).

interactions constitute a powerful method for rationalizing or predicting molecular structures that are applicable across the periodic table, allowing one to identify the effects of even the weak hyperconjugative delocalizations of saturated systems.

3.3.2
C_3H_3 Cation, Anion, and Radical: Aromaticity, Jahn–Teller Distortions, Resonance Structures, and 3c/2e Bonding

The inability of simple resonance theory to treat cyclobutadiene correctly or to distinguish between $C_3H_3^+$ and $C_3H_3^-$ has led to the popular misconception that simple VB theory is unable to describe these molecules. The fundamental problem concerns the nature of the orbital interactions (stabilizing vs destabilizing) that result in electron delocalizations, or departures from simple Lewis concepts. In this section, our focus is on the application of simple NBO metrics to identify strong delocalizations, the application of NRT to describe electronic structures in terms of multiple resonance structures, and the prediction of molecular geometries based on analysis of donor–acceptor interactions. We begin by considering the singlet states of $C_3H_3^+$ and $C_3H_3^-$ at D_{3h} molecular geometries.

Any competent undergraduate chemistry student could devise Lewis configurations for the anions and cations of $C_3H_3^+$, as shown in Figure 3.8. Our focus now

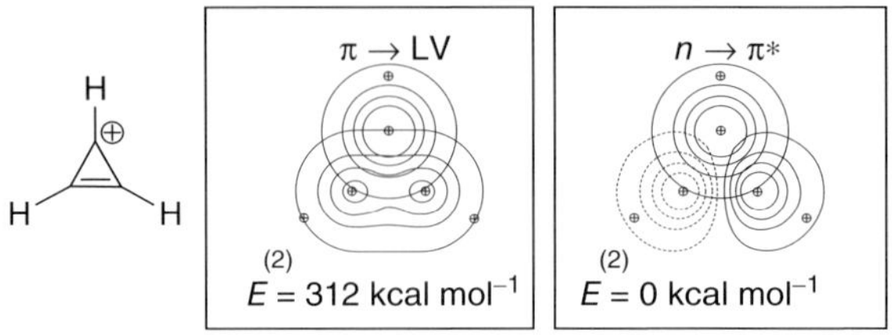

Figure 3.8 NBOs and delocalization energies for the donation of the π electron pair into the empty p-orbital (LV) of $C_3H_3^+$ (a) and donation of the lone pair of electrons (n) into the vacant π* orbital of $C_3H_3^{-++}$ (b). Both molecular geometries correspond to the optimized structure of the cation.

is to examine the donor–acceptor interactions in the NBOs of the singlet cation that lead to strong resonance in the cation and preclude resonance in the anion.

As shown in Figure 3.8, strong delocalization of the $C_3H_3^+$ π-bond electron pair into the empty (LV) carbon p-orbital is indicated by high overlap and substantial perturbative estimates of the π → LV donor–acceptor interaction. Further indications of strong delocalization come from other NBO metrics: low occupancy (1.33e) of the C=C π bond, partial occupancy (0.67e) of the "empty" C p-orbital, and 3.9% NL character in the overall density. Strong, stabilizing delocalizations involving all three C p-π orbitals result in aromatic properties for the cation.

However, at the same fixed D_{3h} geometry, the *anion* $C_3H_3^-$ cannot achieve similar resonance stabilization, because the donor lone pair (n) and the acceptor π* orbitals are orthogonal. This results in "antiaromatic" properties and geometric distortions of the anionic singlet, including pyramidalization at the carbanionic C and lengthening of the two C–C single bonds to distances common for strained σ bonding. (The antiaromaticity of C_4H_4 is similarly accounted for by the vanishing π→π* stabilizations in D_{4h} geometry.) Because the singlet ground state of $C_3H_3^-$ is doubly degenerate, this constitutes a Jahn–Teller distortion [32]. In contrast to the symmetry-based Jahn–Teller rationalization of molecular distortion, the NBO perspective is purely local and makes *no* reference to overall symmetry. Instead, focus is directed to local donor–acceptor interactions and their geometrical consequences.

As we have seen for triplet O_2 and CH_2 diradicals, the neutral C_3H_3 radical is best described by an open-shell wavefunction with different orbitals for α and β spins. This leads to a DLDS representation in the NBO framework. As illustrated in Figure 3.9 for D_{3h} symmetry, the Lewis structure for the two α electrons consists of one "half-bond" and one "half-lone pair" – in other words, it is the half-filled analog of the C_3H_3 *anion*. The single π-β electron makes one C–C π bond, leaving one C p-π orbital empty – in other words, the β Lewis structure is the half-filled analog of the C_3H_3 *cation*. The overall molecular geometry can be viewed as a "balance of forces" originating from the maximization of the bonding interactions in the two spin sets. For the β spins, strong π → LV delocalization is maximized at the planar D_{3h} geometry. But the orbital interactions in the α Lewis structures give strong preference to a localized C=C double bond and pyramidalized carbon

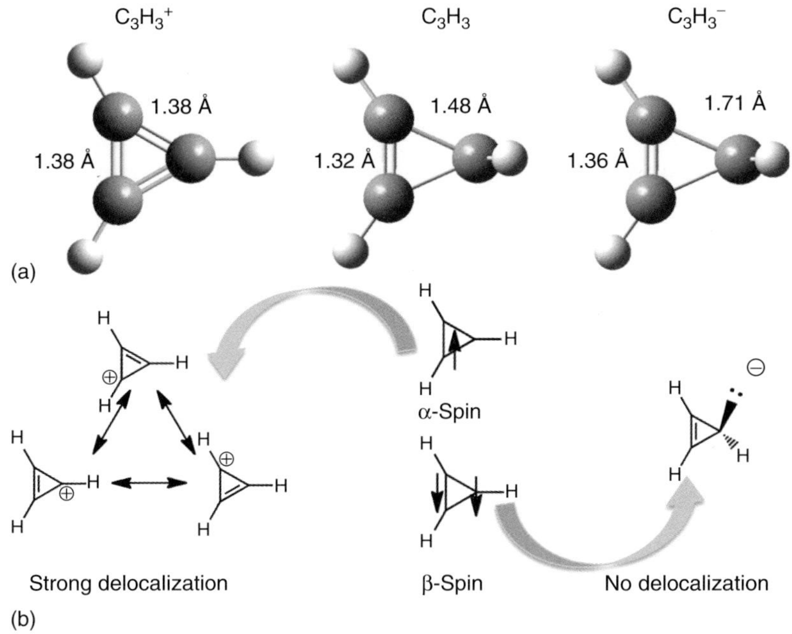

Figure 3.9 Optimized geometries of cyclopropenyl cation, radical, and anion (a) and important resonance contributors (b). The electronic and geometrical structures of the cylclopropenyl radical represent an "averaging" of the cation-like (α spin) and anion-like (β spin) structures.

radical. Thus, the optimized structure is a compromise that lies roughly half-way between the two limiting structures. The significance of this set of examples lies in the simplicity with which the basic Lewis structure concept, combined with quantitative estimates of donor–acceptor interaction energies, can provide insight into both the *direction and size* of Jahn–Teller and related distortions.

Significant occupation of NL orbitals can be equivalently described as the participation of other Lewis-like configurations. For the case of $C_3H_3{}^+$, the obvious resonance configurations can be generated by "arrow pushing" the π electron pair to the cationic center, thus shifting the charge and C=C double bond position. As expected, NRT analysis of $C_3H_3{}^+$ results in the equal population (31.5%) of the three resonance structures produced by permutation of the C=C and carbocationic positions. As expected for a molecule with strong donor–acceptor delocalizations, there is a dramatic improvement in the "fit" between the wavefunction density and the model density when multiple configurations are included. Minor population of resonance structures involving hyperconjugative interactions between C–C and C–H bonds with neighboring C–H and C–C antibonds account for the remaining density. The cumulative effect of populating these different configurations is represented in the NRT resonance-averaged bond orders of 1.33 for the C–C bonds and 0.99 for the C–H bonds.

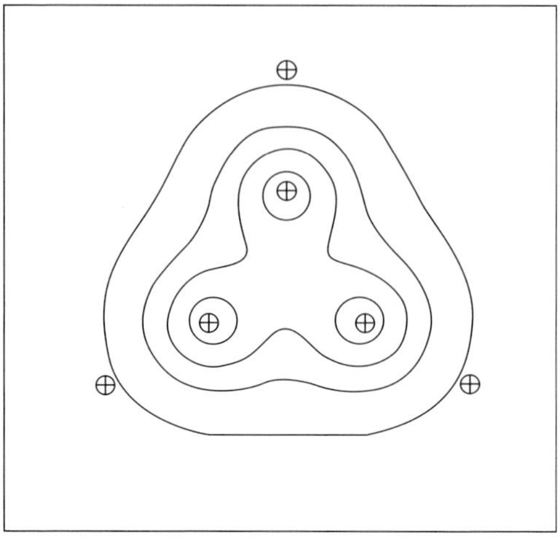

Figure 3.10 Contour diagram of the 3c/2e bonding NBO of $C_3H_3^+$, which is identical to the corresponding CMO and NLMO in this special case.

The very strong donor–acceptor characteristics present in the π-system of $C_3H_3^+$ yield a symmetric geometry with short C–C bonds, even though there are just 2e to share among the three C atoms. Thus, the π-bonding framework in $C_3H_3^+$ has strong parallels with the *hypovalent* bonding in the σ framework of molecules such as diborane (B_2H_6). From the NBO perspective, both of these apparent violations of the octet rule originate from strong donor–acceptor interactions; in diborane, the donors are the B–H bond pairs and the acceptors are the two unoccupied boron 2p orbitals. In recognition of these very strong donor–acceptor interactions, Lipscomb and others [33] adopted the three-center/two-electron (3c/2e) bond as a fundamental unit of molecular structure in order to simplify the description of boranes and related molecules. The current *NBO 6.0* program [34] can search for such strong "electron deficient" 3c/2e interactions automatically, without the "3CBOND" keyword that was required in earlier versions. Indeed, the default NBO6-level description of $C_3H_3^+$ consists of six σ bonds and one 3c/2e bond in the π-system. Inclusion of the 3c/2e bond as an extension to the Lewis description provides a more compact and accurate NLS (the density "error" of $C_3H_3^+$ is 0.79e for the conventional, single reference structure vs just 0.12e for the 3c/2e model). As shown in Figure 3.10, the 3c/2e bond of $C_3H_3^+$ is identical to the bonding orbital of MO or Kohn–Sham type.

3.3.3 3c/4e Hypervalency

Molecules for which the number of electrons around an atom exceeds the nominal atomic valency are common. Such molecules are identified as "hypervalent" [35] and require consideration of bonding interactions beyond the simple Lewis structure to understand their molecular geometries and properties.

We begin our treatment of hypervalency with three molecular anions, each containing two terminal fluorines but with central atoms from different regions of the periodic table: FHF^- (bifluoride), $FBrF^-$, and $FCuF^-$. From the NBO perspective, all three AF_2^- molecules share the characteristics expected of three-center/four-electron (3c/4e) bonding. The Lewis configurations (I–III) shown below

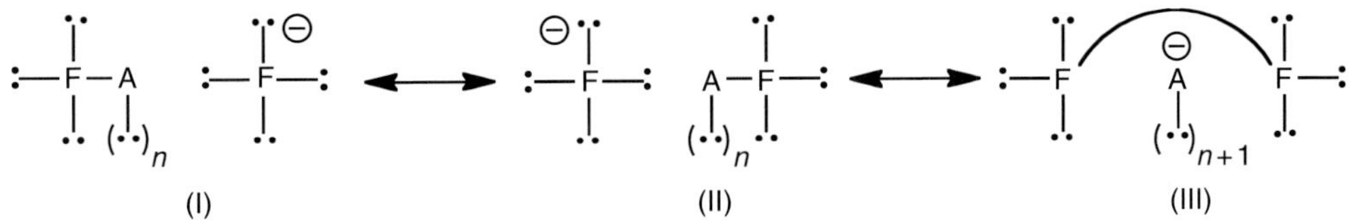

each satisfy normal valency constraints for our three AF_2^- molecular anions. The primary difference among the three anions concerns the number of lone pairs (n) on the central atom: H ($n=0$), Br ($n=3$), Cu ($n=5$). Lewis configurations (I) and (II) capture the H-bonding "arrow-pushing" motif of Figure 3.6, suggesting strong delocalization of the $n_F \rightarrow \sigma_{AF}{}^*$ type. Lewis configuration (III) represents a "long-bond" structure [9]. One should not consider the F–F interaction of the long bond to be a normal bond but should focus on the charge-leveling effect of decreasing charge on the termini by shifting some density to the central atom. For molecules in which the terminal atoms are more electronegative than the central atom, one expects a minor role for resonance structures such as (III).

Plots of the NBOs (Figure 3.11) for the FHF^-, $FBrF^-$, and $FCuF^-$ ions reveal clear similarities in the nature and strength of the $n_F \rightarrow \sigma_{AF}{}^*$ delocalizations. Strong delocalizations of this type characteristically yield *linear* geometry (to maximize $n_F \rightarrow \sigma_{AF}{}^*$ overlap) and *symmetric* structures (the two A–F bond lengths become identical, with equal weighting of the leading two resonance structures).

NRT analysis of the ions using the resonance structures (I–III) gives the following fractional weightings (w_I):

1) FHF^- $w_I = 0.492$, $w_{II} = 0.492$, $w_{III} = 0.016$
2) $FBrF^-$ $w_I = 0.484$, $w_{II} = 0.484$, $w_{III} = 0.032$
3) $FCuF^-$ $w_I = 0.496$, $w_{II} = 0.496$, $w_{III} = 0.008$

As expected, the long-bond structure (III) plays a minor role and both (I) and (II) are equally weighted. Thus, these ions represent strong examples of 3c/4e bonding

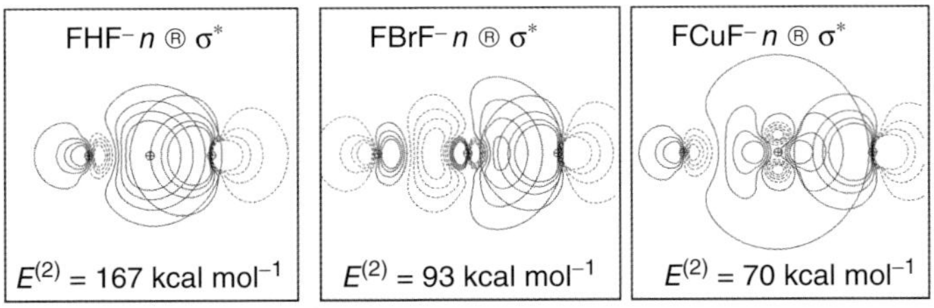

Figure 3.11 Plots of the $n \rightarrow \sigma^*$ donor–acceptor delocalizations and estimates of donor–acceptor interaction energy for FHF^-, $FBrF^-$, and $FCuF^-$ ions. In each case, the donor is an F lone pair and the acceptor is an A–F σ^* orbital.

interactions. Such interactions are so common in D- and P-block bonding that it is convenient to assign a name – the *hyperbond* – and a one-character label – ω – to distinguish the 3c/4e bond from various 2c/2e bonds (σ, π, δ) and the 3c/2e bond (τ). The ω-bond represents strong intermolecular donation ($n_X \rightarrow \sigma_{AY}{}^*$) of a localized pair of electrons of ligand X into the antibonding orbital ($\sigma_{AY}{}^*$) associated with the central atom A and attached ligand Y, and vice versa ($n_Y \rightarrow \sigma_{AX}{}^*$), and is automatically recognized in the "3C HYPERBOND" section of NBO output. Its presence implies a preference for linear arrangements of the three atoms and, relative to 2c/2e bonding, significant lengthening of the bond and increased electron density on the ligands. Related 3c/4e delocalizations are pervasive in chemistry, ranging from the powerful $\pi \rightarrow \pi^*$ delocalizations that enforce the planarity and symmetric C–C bond lengths in the allyl anion, down to the even weaker forms of H-bonding that result from replacing the F atoms of FHF^- by less electronegative ligands [36].

Before turning to more general application of 3c/4e bonding across the periodic table, let us consider another "Jahn–Teller-like" phenomenon: the plasticity of Cu complexes in the +2 oxidation state. Our models are organometallic $CuMe_2$ cation (+3 oxidation number), neutral (+2 oxidation number), and anion (+1 oxidation number). We can readily conclude that $CuMe_2{}^-$ should adopt a linear structure, similar to $CuF_2{}^-$, in order to maximize ω-bonding. In contrast, the fictitious but easily computed +3 oxidation number complex, $CuMe_2{}^+$, should have a strongly bent geometry, close to 90°, because the Lewis-like structure of $CuMe_2{}^+$ comprises four lone pairs and two 2c/2e Cu–C bonds with approximate sd^1 hybridization and, hence, a 90° angle preference. Consistent with these expectations, the energy surfaces for the angular bending of the Me–Cu–Me anion and cation (Figure 3.12) exhibit *deep* wells with minima at 180° and 90°, respectively. What about the surface for the +2 oxidation state? Without resorting to a computation, the DLDS paradigm suggests that α-spin bonding will be similar to $CuMe_2{}^-$ and β-spin bonding will be similar to $CuMe_2{}^+$. Accordingly, the energy surface for the bending coordinate should be the *average* of those for the anion and cation. Indeed, the computed surface for neutral $CuMe_2$ is essentially that average, resulting in a *very soft* surface with minimum close to 135°, as shown in the dotted line of Figure 3.12.

The addition of ω-bonding to the more familiar Lewis structure elements of 2c/2e bonds and lone pairs enables a Lewis-like analysis of computed electron densities to be consistently applied across the periodic table. For example, the "T-shapes" of ClF_3 and $PdH_3{}^-$ have closely related origins. Each molecule is hypervalent "by one pair," and has one 2c/2e σ-bond, one 3c/4e ω-bond, and hybrid orbital orthogonalities at 90° (due to high p-character in the Cl–F bonds and sd^1 hybridization for Pd–H bonds). Effectively, the MX_3 "T-shape" is easily seen as the *trans* addition of a third ligand to a bent, normal valent MX_2 fragment, so as to maximize $n \rightarrow \sigma^*$ delocalization and ω-bonding. Similar considerations apply to many species with square planar (two ω-bonds; e.g., $PtH_4{}^{2-}$ and $BrF_4{}^-$), square pyramidal (one σ bond, two ω-bonds; e.g., BrF_5 and $IrH_5{}^{2-}$), and octahedral (three ω-bonds; $BrF_6{}^-$ and $OsH_6{}^{4-}$) molecular shapes. Even molecules as complex as ferrocene, for which Pauling's formulation with more than 500 resonance

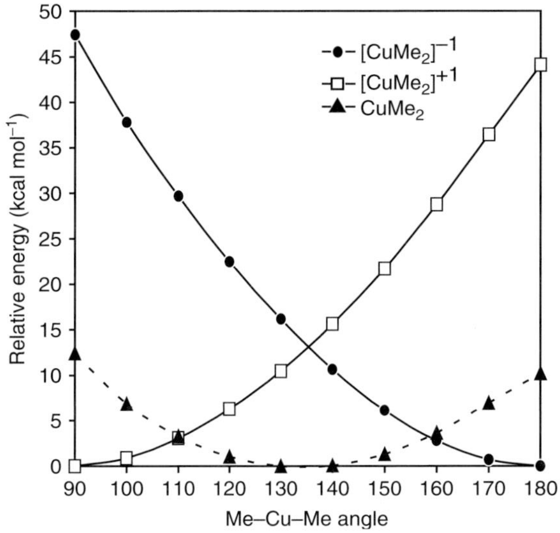

Figure 3.12 Computed energy surfaces for the Me–Cu–Me bending coordinate of neutral, anionic, and cationic $CuMe_2$. These plots illustrate how the neutral, radical surface is the result of "averaging" of the cation-like (α spin) and anion-like (β spin) surfaces, in accordance with the DLDS paradigm.

structures has been described as "repulsive" [37], simplify to compact Lewis-like descriptions: three ω-bonds, sd^2 hybridization, and three lone pairs at iron.

The simple Lewis-like and resonance concepts based on fundamental units of 2c/2e, 3c/4e, and 3c/2e bonding are found to be fully consistent with densities computed by the most accurate current wavefunction methods. We can conclude that the NBO-based Lewis-like conceptual framework serves to (i) highlight the deep chemical kinships between P- and D-blocks of the periodic table, (ii) exhibit the generality of the NBO donor–acceptor paradigm in describing "unusual" bonding motifs, and (iii) provide a powerful pedagogical tool for unifying the description of chemical interaction phenomena, guided by the refinements of modern wavefunction technology.

3.4 Conclusion

As Lewis and others brilliantly observed a century ago, atoms connect to one another according to logical patterns. Even the largest molecules can be broken down to a collection of atoms held together by local interactions involving a relatively small number of bond types. The chemical bond is the ultimate divide-and-conquer representation of matter.

For users of *ab initio* computations who share the bewildered skepticism expressed in the introductory quotations, our advice is, "Get a better basis set!" By this we mean that one should analyze computational results with basis sets that enable fundamental bonding interactions to emerge (we do not suggest simply redoing the calculation with a larger basis set). Localized NAO/NBO basis sets – in effect, selected by the wavefunction as optimal for its own self-description – offer a fundamentally simplified ("VB-like") view of molecular electronic structure that averts the heavy, "unnecessary" delocalization and AO nonorthogonality in conventional linear combination of atomic orbital (LCAO)–MO depictions. The widespread availability of NBO analysis tools now makes such improved wavefunction description readily accessible in many popular electronic structure programs. As illustrated for a variety of simple molecules in this chapter, the results show that traditional localized bonding and resonance concepts of Lewis, Pauling, and other bonding pioneers are indeed alive and well in modern wavefunctions, continually sharpened and strengthened as modern computational technology advances.

References

1. Rocke, A.J. (2010) *Image and Reality: Kekulé, Kopp, and the Scientific Imagination*, University of Chicago Press, Chicago, IL.
2. Pauling, L. (1960) *The Nature of the Chemical Bond and the Structure of Molecules and Crystals*, 3rd edn, Cornell University Press, Ithaca, NY, p. 220.
3. (a) F. Weinhold, in *Encyclopedia of Computational Chemistry*, P.v.R. Schleyer, N. L. Allinger, T. Clark, J. Gasteiger, P. A. Kollman, H. F. Schaefer III, P. R. Schreiner Eds. John Wiley & Sons, Ltd, Chichester 1998), Vol. 3, p. 1792; (b) Glendening, E.D., Landis, C.R., and Weinhold, F. (2012) *WIREs Comput. Mol. Sci.*, **2**, 1.
4. Weinhold, F. and Landis, C.R. (2012) *Discovering Chemistry with Natural Bond Orbitals*, John Wiley & Sons, Inc., Hoboken, NJ.
5. Weinhold, F. and Landis, C.R. (2005) *Valency and Bonding: A Natural Bond Orbital Donor-Acceptor Perspective*, Cambridge University Press.
6. Weinhold, F. (2012) *J. Comput. Chem.*, **33**, 2363.
7. Reed, A.E., Weinstock, R.B., and Weinhold, F. (1985) *J. Chem. Phys.*, **83**, 735.
8. Foster, J.P. and Weinhold, F. (1980) *J. Am. Chem. Soc.*, **102**, 7211.
9. Landis, C.R. and Weinhold, F. (2013) *Inorg. Chem.*, **52**, 5154.
10. NBO6 *http://nbo6.chem.wisc.edu* (accessed 22 November 2013).
11. (a) Löwdin, P.-O. (1955) *Phys. Rev.*, **97**, 1474. (b) Davidson, E.R. (1976) *Reduced Density Matrices in Quantum Chemistry*, Academic Press, London.
12. For explanations of standard acronyms for common electronic structure methods and basis sets used herein, see: Foresman, J.B. and Frisch, A.E. (1996) *Exploring Chemistry with Electronic Structure Methods*, 2nd edn, Gaussian Inc., Pittsburgh.
13. (a) Weinhold, F. and Carpenter, J.E. (1988) *J. Mol. Struct. THEOCHEM*, **165** (189). (b) Weinhold, F. (2003) *Angew. Chem. Int. Ed.*, **35**, 4188 *Angew. Chem.*, (2003), **35**, 4320. (c) Weinhold, F. and Landis, C.R. (2005) *Valency and Bonding: A Natural Bond Orbital Donor-Acceptor Perspective*, Cambridge University Press, pp. 229–234.
14. Weinhold, F. (2012) *J. Comput. Chem.*, **33**, 2440.
15. Reed, A.E. and Weinhold, F. (1985) *J. Chem. Phys.*, **83**, 1736.
16. See, e.g., Kleier, D.A., Halgren, T.A., Hall, J.H., and Lipscomb, W.N. (1974) *J. Chem. Phys.*, **61**, 3905. Note that NLMOs generally resemble alternative LMOs (as much as these LMOs resemble one

another) for fully occupied members, but corresponding localized *virtual* NLMOs are also provided that usually lack practical counterparts in alternative LMO procedures.

17. (a) Carpenter, J.E. and Weinhold, F. (1988) *J. Mol. Struct. THEOCHEM*, **169**, 41. (b) Weinhold, F. and Landis, C.R. (2012) *Discovering Chemistry with Natural Bond Orbitals*, John Wiley & Sons, Inc., Hoboken, NJ, p. 82ff.

18. (a) Badenhoop, J.K. and Weinhold, F. (1997) *J. Chem. Phys.*, **107**, 5406–5422. (b) Weisskopf, V.F. (1975) *Science*, **187**, 605. (c) Weinhold, F. and Landis, C.R. (2012) *Discovering Chemistry with Natural Bond Orbitals*, John Wiley & Sons, Inc., Hoboken, NJ, p. 136ff.

19. (a) Shaik, S. and Hiberty, P.C. (2004) *Rev. Comput. Chem.*, **20**, 1. (b) Gavroglu, K. and Simones, A. (2012) *Neither Physics nor Chemistry: A History of Quantum Chemistry*, MIT Press, Cambridge MA.

20. Coulson, C.A. and Fischer, I. (1949) *Philos. Mag.*, **40**, 386.

21. Weinhold, F. and Landis, C.R. (2005) *Valency and Bonding: A Natural Bond Orbital Donor-Acceptor Perspective*, Cambridge University Press, pp. 365–366.

22. Coulson, C.A. (1952) *Valence*, 2nd edn, Chapter 8, Oxford University Press, London.

23. Weinhold, F. and Landis, C.R. (2005) *Valency and Bonding: A Natural Bond Orbital Donor-Acceptor Perspective*, Cambridge University Press, p. 372ff.

24. Bent, H.A. (1961) *Chem. Rev.*, **61**, 275.

25. Weinhold, F. and Landis, C.R. (2005) *Valency and Bonding: A Natural Bond Orbital Donor-Acceptor Perspective*, Cambridge University Press, p. 132.

26. Weinhold, F. and Landis, C.R. (2005) *Valency and Bonding: A Natural Bond Orbital Donor-Acceptor Perspective*, Cambridge University Press, p. 421ff.

27. Landis, C.R., Cleveland, T., and Firman, T.K. (1995) *J. Am. Chem. Soc.*, **117**, 1859; *Science*, **272**, 179 (1996).

28. (a) Reed, A.E., Curtiss, L.A., and Weinhold, F. (1988) *Chem. Rev.*, **88**, 899–926. (b) Weinhold, F. and Landis, C.R. (2005) *Valency and Bonding: A Natural Bond Orbital Donor-Acceptor Perspective*, Cambridge University Press, pp. 16–19. (c) Weinhold, F. and Landis, C.R. (2005) *Valency and Bonding: A Natural Bond Orbital Donor-Acceptor Perspective*, Cambridge University Press.

29. Weinhold, F. and Landis, C.R. (2012) *Discovering Chemistry with Natural Bond Orbitals*, John Wiley & Sons, Inc., Hoboken, NJ, p. 105ff.

30. Glendening, E.D., Badenhoop, J.K., and Weinhold, F. (1998) *J. Comput. Chem.*, **19**, 628.

31. Braïda, B. and Hiberty, P.C. (2008) *J. Phys. Chem. A*, **112**, 13045.

32. (a) Jahn, H. and Teller, E. (1937) *Proc. R. Soc. (London)*, **A161**, 220–235. (b) Bersuker, I.B. (2001) *Chem. Rev.*, **101**, 1067.

33. (a) Eberhardt, W.H., Crawford, B.L. Jr., and Lipscomb, W.N. (1954) *J. Chem. Phys.*, **22**, 989. (b) Lipscomb, W.N. (1973) *Acc. Chem. Res.*, **8**, 257.

34. (a) Glendening, E.D., Badenhoop, J.K., Reed, A.E., Carpenter, J.E., Bohmann, J.A., Morales, C.M., Landis, C.R., and Weinhold, F. (2013) NBO 6.0, Theoretical Chemistry Institute, University of Wisconsin, Madison; (b) Glendening, E.D., Landis, C.R., and Weinhold, F. (2013) NBO 6.0: natural bond orbital analysis program. *J. Comput. Chem.*, **34**, 1429.

35. (a) Musher, J. (1969) *Angew. Chem. Int. Ed.*, **8**, 54. (b) Weinhold, F. and Landis, C.R. (2005) *Valency and Bonding: A Natural Bond Orbital Donor-Acceptor Perspective*, Cambridge University Press, p. 275ff.

36. Weinhold, F. and Klein, R.A. (2012) *Mol. Phys.*, **110**, 565.

37. Hoffmann, R., Shaik, S., and Hiberty, P.C. (2003) *Acc. Chem. Res.*, **36**, 750.

4
The EDA Perspective of Chemical Bonding

Gernot Frenking and F. Matthias Bickelhaupt

4.1
Introduction

The fundamental approach for understanding the nature of a chemical bond A–B considers the kinetic energy of and electrostatic forces between the nuclei and electrons within and across fragments A and B as elementary quantities of chemical bonding, which are described with quantum mechanical equations. Two types of chemical bonds may then be distinguished. Ionic bonds can be reasonably discussed in terms of classical electrostatic interactions between point charges. Purely ionic bonds exist only in the solid state where the overlap of the valence orbitals between the neighboring atoms is negligible. When the orbital overlap at the equilibrium distances A–B becomes significant, covalent bonding takes place [1]. For example, the equilibrium distance of diatomic CsF in the gas phase is 2.345 Å whereas the Cs−F distance in solid cesium fluoride is 3.005 Å. The cesium−fluorine interactions in the solid state may be reasonably discussed in terms of classical electrostatic forces between Cs^+ and F^- whereas the chemical bond of diatomic CsF in the gas phase, although largely ionic, still has some covalent character that is not negligible. It is only the latter type of bonding that is considered in this work.

The understanding of covalent bonds in terms of basic laws of physics was long an enigma. In 1927, Heitler and London [2] showed that the strongly attractive forces between two hydrogen atoms may be explained when the electrons are considered as waves rather than particles that lead to a consistent description of the electronic structure of H_2. They called it a "characteristic quantum mechanical resonance phenomenon", which is crucial for the understanding of the interactions between neutral atoms. We want to emphasize that covalent bonding and ionic bonding are *both* the result of electrostatic forces but the mathematical description of the former type of bonding is far more complicated than the latter. The reason is that the motion and the spatial distribution of the electrons is dictated by quantum mechanical laws that determine interatomic interactions in molecular systems. A key quantity for understanding the occurrence of covalent bonds is the kinetic energy of quantum mechanical particles, which decreases as the available

The Chemical Bond: Fundamental Aspects of Chemical Bonding, First Edition.
Edited by Gernot Frenking, Sason Shaik.

volume increases, thus leading to a stabilization of the total system (cf. particle in a one-dimensional box).

The different interactions between two electrons when they are described either as point charges or as waves and where they possess the same or opposite spin are shown in Figure 4.1. The dashed curve gives the repulsive interactions between two point charges that is given by the classical equation $1/R$. The dotted curve shows the repulsion between two electrons that is described by 1s hydrogen functions rather than by point charges q. The two curves become very similar at larger distances where the overlap becomes negligible. When the two 1s functions start to overlap, the repulsion becomes smaller and converges toward a finite value at $R = 0$. The solid curve finally gives the interactions between the two electrons in 1s orbitals that have the same spin. It becomes obvious that the latter curve is very sensitive to the size of the overlap S and that it rises sharply as soon as $S \neq 0$. Figure 4.1 makes it clear that the description of electron–electron interactions using point charges in the region of covalent bonding leads to grave errors. It becomes also clear that the Pauli repulsion (or exchange repulsion) between two electrons that possess the same spin may become more important than the electrostatic interactions. We will show in this chapter that the Pauli repulsion is a very important factor for the structure and the bond strength of molecules [3].

It must be emphasized that the "resonance phenomenon" of Heitler and London, which is called *orbital* interactions in modern terminology, is *not* associated with the pairing of two electrons that was believed by Lewis [4] to be the driving force for a covalent bond. Resonance of the wave function takes place even when there is only one electron such as in H_2^+ [5]. It is the mixing of the wave functions of the interacting fragments A and B that leads to the stabilizing resonance term yielding covalent bonding. This is a quantum mechanical phenomenon that cannot be explained by the classical laws of physics. The physics which is connected with the quantum theory of the chemical bond is discussed by Ruedenberg, Schmidt, and Ivanic in Chapter 1 of this book [6].

When the pioneering work by Heitler and London, which paved the way for a true understanding of the chemical bond, was published in 1927, chemists had already developed heuristic bonding models that were amazingly successful as ordering schemes for experimental observations and as helpful guidelines in the search for new compounds and novel reactions [7]. The simple approach and the intuitive concepts in concert with empirical rules made them powerful tools for synthetic chemistry. It is the reason that heuristically developed bonding models are still very popular in the chemical community. This holds in particular for the bonding model of Lewis [4] who suggested that the chemical bond is associated with an electron pair. The pre-quantum chemical electron-pair model of Lewis was underpinned by Pauling [8] with quantum theoretical arguments that were not undisputed, however. Pauling's dogmatic view in favor of valence bond theory and his refusal to recognize molecular orbital (MO) theory as a valid theory of chemical bonding made it deceptively easy for experimental chemists to continue using classical ad-hoc bonding models. Nevertheless, his epochal work "The Nature of

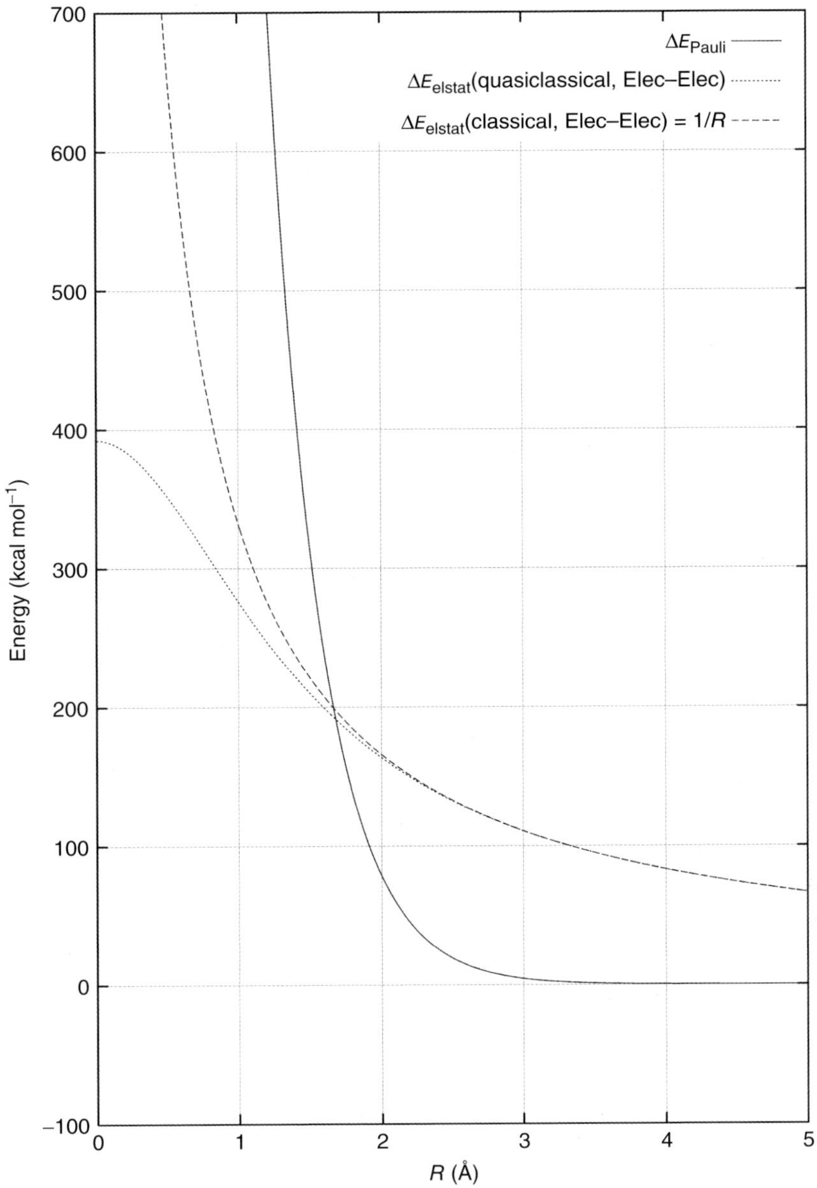

Figure 4.1 Calculated interaction energies between two electrons as a function of their distance r_{12}: calculated classically as point charges $\Delta E_{\text{elstat}}(\text{classical}) = q_1 \times q_2/r_{12}$ (dashed line); quasiclassical repulsion between two electrons in 1s orbitals $\Delta E_{\text{elstat}} = \int \rho_1 \times \rho_2 / r_{12} \, d\tau_1 \, d\tau_2$ (dotted line); Pauli repulsion between two electrons with the same spin in 1s orbitals ΔE_{Pauli} (solid line).

the Chemical Bond'' showed that it is possible to build a bridge between the heuristic bonding models of synthetic chemistry and quantum theory.

In this chapter, we describe a more recently developed method that connects classical concepts of chemical bonding with a rigorous quantum chemical approach. The method is based on theoretical work that was independently carried out by Morokuma [9] and by Ziegler and Rauk [10]. In this work, it was suggested that the energy lowering that is associated with the formation of a chemical bond may be divided into three different terms, which can be interpreted in a physically meaningful way. Although such division is arbitrary like all partitioning schemes are, it is mathematically unambiguously defined and the three terms may be identified with classical chemical concepts but also with quantum chemical terms. The EDA (energy decomposition analysis) focuses on the intrinsic interactions A–B in the molecules AB which must be distinguished from the bond dissociation energy (BDE). The latter term involves the geometrical and electronic relaxation of the fragments A and B from the molecule A–B to the free species A and B in the equilibrium geometry of the electronic ground state. Thus, the EDA gives a faithful representation of the actual bonding situation in the molecule. Since the sum of the various energy terms of the EDA and the relaxation energy give the BDE, they are connected to an observable. This makes the EDA a very attractive and powerful tool for the interpretation and explanation of chemical phenomena such as molecular structure and reactivity. In principle, the EDA can be used in conjunction with any quantum chemical approach. Herein, it is combined with the quantitative MO model contained in the Kohn–Sham approach to density functional theory (DFT), which includes correlation energy. For *ab initio* methods, the EDA has only been developed for Hartree–Fock calculations until today. This is the reason that the EDA method is usually employed in conjunction with DFT calculations.

We wish to emphasize that the breakdown of the interatomic interaction energy into different terms does not have a physical meaning in the sense that the calculated data refer to experimental observables. The EDA is an attempt to provide quantitative expressions for terms that are associated with chemical bonding models. The latter have been called chemical ''unicorns,'' because like the mystical animal, everybody seems to know them although they have never been observed [11]. This is not so much a weakness as a classical human approach to find a perspective on nature that reveals patterns behind observable phenomena. Therefore, ''unicornig'' bonding models play an important role as ordering schemes for molecules and as a useful tool for designing new chemical reactions.

In the following we discuss representative examples of the application of the EDA for the analysis of covalent bonds. We have chosen classical chemical bonds which are typical for various types of molecules. For a more extended collection of EDA results we refer to review articles that were published in recent years [12]. The application of the EDA for understanding particular classes of chemical bonds has also been described in other chapters of this book [13]. We also introduce a recent extension of the method where the EDA is combined with the NOCV (Natural Orbital for Chemical Valence) method [14]. The EDA-NOCV [15] scheme has the advantage that the orbital interactions between two interacting species may be divided into pairwise contributions of the most relevant MOs.

4.2
Basic Principles of the EDA Method

The focus of the bonding analysis in the EDA approach is the instantaneous interaction energy ΔE_{int} of a bond A–B between two fragments A and B in the particular electronic reference state and in the frozen geometry of AB. In the case of diatomic molecules E_2, ΔE_{int} is the energy difference between E_2 and the atoms E in the electronic reference state which may or may not be the electronic ground state. It is very important to recognize the electronic reference states of the fragments in order to provide a meaningful analysis of the interactions A–B. The interaction energy is divided into three main components:

$$\Delta E_{int} = \Delta E_{elstat} + \Delta E_{Pauli} + \Delta E_{orb} \quad (4.1)$$

The term ΔE_{elstat} corresponds to the quasi classical [16][1] electrostatic interaction between the unperturbed charge distributions of the prepared atoms and is usually attractive (vide infra). The Pauli repulsion ΔE_{Pauli} arises as the energy change associated with the transformation from the superposition of the unperturbed electron densities $\rho_{E(\alpha)} + \rho_{E(\beta)}$ of the isolated atoms to the wavefunction $\Psi^0 = N\hat{A}\,[\Psi_{E(\alpha)}\Psi_{E(\beta)}]$, that properly obeys the Pauli principle through explicit antisymmetrization ($\hat{A}$ operator) and renormalization (N constant) of the product wavefunction [12c]. It comprises the destabilizing interactions between electrons on either atom of the same spin. The orbital interaction ΔE_{orb} accounts for charge transfer (i.e., donor–acceptor interactions between occupied orbitals on one moiety with unoccupied orbitals of the other, including the HOMO–LUMO interactions), polarization (empty/occupied orbital mixing on one fragment due to the presence of another fragment), and electron-pair bonding (the stabilization arising from the formation of the electron-pair bonding configuration in which the bonding combination between the SOMOs is formed and doubly occupied) [17]. Likewise, one-electron bonding and three-electron bonding (or repulsion) can be defined and quantified [18]. The ΔE_{orb} term can be decomposed into contributions from each irreducible representation of the point group of the interacting system. This makes it possible to quantitatively estimate the intrinsic strength of orbital interactions from orbitals having σ, π, δ, and so on symmetry.

For molecules where the fragments do not electronically or geometrically relax after bond breaking, the interaction energy ΔE_{int} gives directly (by definition with opposite sign) the BDE D_e. If the two fragments of the chemical bond are in an electronically excited state or if they have more than one atom which means that

1) The term "quasi-classical" indicates that the electron density ρ is calculated quantum theoretically but the interactions between the electrons and nuclei are calculated assuming classical electrostatic interaction between these densities as they are kept frozen, i.e., no quantum interference takes place.

there is a relaxation of the fragments during bond rupture into the equilibrium geometry, the preparation energy ΔE_{prep} must be added to ΔE_{int} in order to obtain the BDE D_e which is by definition the negative value of the total bond energy ΔE:

$$\Delta E(= -D_e) = \Delta E_{int} + \Delta E_{prep} \tag{4.2}$$

Because the atomic fragments that were used in our calculations of the diatomic molecules are in the electronic ground state, it follows that, for the E_2 species, ΔE_{prep} is zero and $\Delta E_{int} = -D_e$. For bonds A–B between polyatomic fragments, the term ΔE_{prep} gives the difference between the intrinsic binding energy ΔE_{int} of the bond A–B in the molecule and the BDE D_e. Thus, the absolute value of ΔE_{int} is a faithful expression for the bond strength whereas the D_e values are affected by the geometrical and possibly the electronic relaxation of the fragments. There are molecules where the fragments A and B are lower in energy than A–B which makes the D_e values useless for estimating the bond strength A–B. This holds in particular for high-energy materials and explosives where bond breaking is a highly exothermic process. It is thus advisable to use the interaction energy ΔE_{int} for comparing the bond strength of molecules.

The EDA involving open-shell fragments does, for technical reasons, neglect the spin-polarization in the fragments yielding slightly too stable bonds (in the order of a few kcal mol^{-1} per unpaired electron). The bond energies have been corrected for the spin-polarization error ΔE_{corr} which is given in the tables. Further details about the EDA can be found in the literature [12c, d].

4.3 The EDA-NOCV Method

The difference between the EDA and the EDA-NOCV methods lies in the expression for the orbital interaction term ΔE_{orb}, which is broken down in the EDA-NOCV approach into pairwise contributions of interacting orbitals of the two fragments. The starting point is the deformation density $\Delta\rho(r)$, which is the difference between the densities of the fragments before and after bond formation. The deformation density $\Delta\rho(r)$ can be expressed in terms of pairs of complementary eigenfunctions $(\psi_k\ \psi_{-k})$ with the eigenvalues υ_k and υ_{-k} that possess the same absolute value but opposite sign [14, 15].

$$\Delta\rho(r) = \sum_k \upsilon_k[-\psi^2_{-k}(r) + \psi^2{}_k(r)] = \sum_k \Delta\rho_k(r) \tag{4.3}$$

The NOCVs ψ_k and the associated eigenvalues υ_k are obtained through diagonalization of the difference density matrix $\Delta P_{\mu\nu}$ of the system [14]. Equation (4.3) makes it possible to express the total charge deformation $\Delta\rho(r)$ that goes along with the bond formation in terms of pairwise charge contributions $\Delta\rho_k(r)$ that come from particular pairs of (NOCV) orbitals. The total orbital interaction ΔE_{orb} may likewise be derived from pairwise orbital interaction energies $\Delta E^k{}_{orb}$ that are associated with $\Delta\rho_k(r)$:

$$\Delta E_{\mathrm{orb}} = \sum_k \Delta E^k{}_{\mathrm{orb}} = \sum_k \upsilon_k \left[-F^{TS}_{-k,-k} + F^{TS}_{k,k}\right] \tag{4.4}$$

The terms $F^{TS}_{-k,-k}$ and $F^{TS}_{k,k}$ are diagonal transition-state (TS) Kohn–Sham matrix elements corresponding to NOCVs with the eigenvalues $-\upsilon_k$ and υ_k, respectively. Here, the term "transition state" (TS) refers to the charge density, which is intermediate between the density of the final molecule AB and the superimposed fragment densities of A and B. The ΔE^{orb}_k term of a particular type of bond is assigned by visual inspection of the shape of the deformation density, $\Delta\rho_k$.

Experience has shown that the ΔE_{orb} term of the EDA-NOCV has usually only a very small number of significant contributions of $\Delta E^k{}_{\mathrm{orb}}$ that makes it possible to identify specific interactions, which lead to a chemical bond. This is particularly interesting for analyzing the TS of a chemical reaction. Two examples are presented and discussed in Section 9. Moreover, the EDA-NOCV makes it possible to estimate the charge donation $\Delta\rho_k$, which is associated with the pairwise orbital interactions, and to visualize its spatial extent. The change in the electron density distribution that comes from bond formation between two fragments or intermolecular interactions during a chemical reaction can be quantitatively expressed through the eigenvalues of the NOCVs.

4.4
Chemical Bonding in H_2 and N_2

The chemical bond in dihydrogen is the classical example for covalent bonding. It is often used in textbooks of chemistry and in introductory curricula at universities for teaching the nature of the covalent bond. Scheme 4.1 schematically shows the difference between the classical and the quantum theoretical approaches for chemical bonding in H_2.

The classical approach for chemical bonding takes the charge distribution ρ of the atoms as central physical quantity for calculating the interatomic interactions. This yields the term E_{elstat} for the energy of the molecule $E(H_2)$, which exhibits only a shallow energy minimum for the bond energy (Figure 4.2). The quantum theoretical approach introduces the wave function Ψ as the representative expression for atoms and molecules, which is related to the charge distribution ρ by the square, for example, $\rho(H_2) = \Psi(H_2)^2$ (Scheme 4.1). The interatomic interactions are described by the plus or minus combination of the atomic wave functions, which gives the molecular wave function $\Psi(H_2)$ as a result that yields $[\Psi(H_2)]^2$ for the charge distribution of the molecule $\rho(H_2)$. Scheme 4.1 shows that the expression for $[\Psi(H_2)]^2$ contains a term $\pm\Psi(H_a)\Psi(H_b)$, which is absent in the classical expression for $\rho(H_2)$. This quantum theoretical term is called *resonance*, because it comes from the mixing ("interference") of the atomic wave functions $\Psi(H)$. The covalent bond comes from the plus combination of $\Psi(H_a)\Psi(H_b)$ where the two electrons have opposite spin while the minus combination of $\Psi(H_a)\Psi(H_b)$ describes the situation when the two electrons have the same spin. The latter wave function

Chemical bonding: Classical approach

$$\rho$$

$$\rho(H_2) = \rho(H_a) + \rho(H_b)$$

$$E(H_2) = E[\rho(H_2)]$$

$$= E_{elstat}$$

Chemical bonding: Quantum chemical approach

$$\Psi$$

$$\rho(H_2) = [\Psi]^2$$

$$\Psi(H_2) = c_1\Psi(H_a) \pm c_2\Psi(H_b)$$

$$[\Psi(H_2)]^2 = [c_1\Psi(H_a) \pm c_2\Psi(H_b)]^2$$

$$= [c_1\Psi(H_a)]^2 + [c_1\Psi(H_a)]^2 \pm 2[c_1\Psi(H_a)c_2\Psi(H_b)]$$

$$q(H_a) \qquad q(H_b) \qquad \textit{Resonance}$$

$$E(H_2)_{\alpha,\beta} = E[\rho(H_2)] \pm E[\Psi(H_a)\,\Psi(H_b)]$$

$$= E_{elstat} \pm E[\Psi(H_a)\,\Psi(H_b)]$$

Scheme 4.1 Schematic representation of the difference between the classical and the quantum theoretical approaches for describing chemical bonding in H_2. (a) The classical description uses the charge ρ as central physical quantity for the bond formation which leads to E_{elstat} as the energy of the molecule $E(H_2)$. (b) The quantum theoretical approach uses the wave function Ψ as central physical quantity where the bond formation comes from the mixing of the wave functions of the two hydrogen atoms yielding $\Psi(H_2)$. This leads to a new term in the expression for the energy of $E(H_2)$ called *resonance* that is absent in the classical approach and which is the very origin of covalent bonding.

yields a repulsive energy curve for H_2. It becomes obvious that the covalent bond formation can only be explained in terms of quantum chemical laws using the wave function Ψ as a fundamental starting point which gives the resonance term as the driving force for the chemical bond. *We wish to point out that the origin of the covalent bond is* not *the formation of an electron pair but the appearance of the resonance term $\Psi(H_a)\Psi(H_b)$ which emerges already when there is only a one-electron bond such as in H_2^+.*

It is reasonable to present the EDA results for H_2 as a starting point of this chapter. Table 4.1 shows the calculated data while Figure 4.2 displays the individual contributions of the EDA to the interaction energy ΔE_{int} as a function of the H–H distance. The ΔE_{int} values give directly the BDE D_e because there is no electronic or geometric relaxation of the fragments. The calculated BDE $D_o = 106.3$ kcal mol^{-1}, which considers zero-point energy (ZPE) contributions, is in good agreement with the experimental value of 103.3 kcal mol^{-1}.

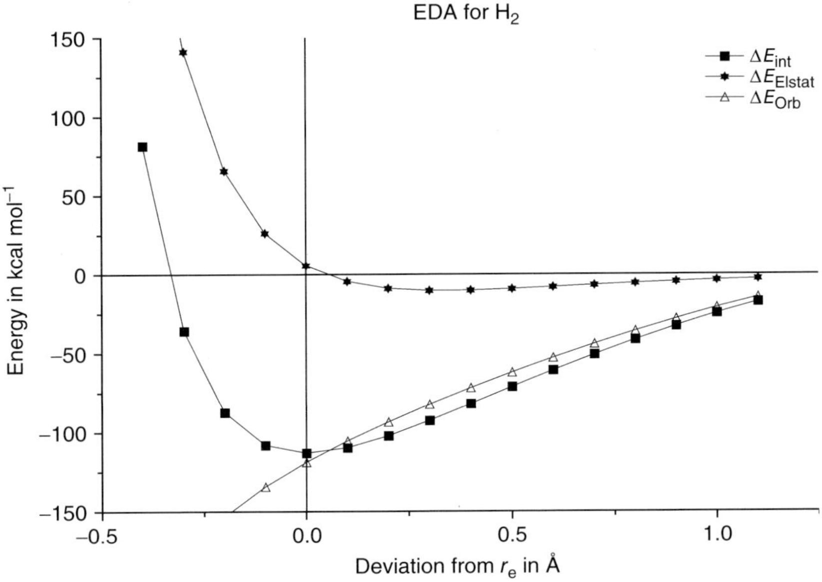

Figure 4.2 Trend of the energy terms terms of the EDA as a function of the interatomic distance for H_2 calculated at RPBE/TZP.

There is one important technical aspect that concerns the use of present DFT functionals for EDA calculations, which are subject to the self-interaction error (SIE). The SIE introduces small errors in the calculated energy terms of the EDA that become obvious when the Pauli repulsion in H_2 is considered that, by definition, is zero. EDA calculations of H_2 using common gradient-corrected functional give small negative values for ΔE_{Pauli}. Since the calculated Pauli repulsion comes from a physically unrealistic correction of the functionals, the ΔE_{Pauli} values have been added to ΔE_{orb}, which lead to the data that are shown in Table 4.1 and Figure 4.2.

The EDA results are in agreement with the present knowledge of the chemical bond in H_2. The covalent bond comes exclusively from the orbital term ΔE_{orb} whereas the electrostatic term ΔE_{elstat} at equilibrium distance is even slightly repulsive. The latter term becomes weakly attractive at longer distances (Figure 4.2) where it has an energy minimum at a rather long distance of ~1.1 Å and a BDE of only ~10 kcal mol^{-1}. The latter result was already obtained in the original work of Heitler and London [2]. There is no Pauli repulsion in H_2 because there are only two electrons that possess opposite spin. But this is not the only anomaly that characterizes bonding in dihydrogen as atypical for a covalent bond. A second anomaly is the unusually weak electrostatic attraction, which is because the hydrogen atom has a nucleus with a rather small positive charge. The atypical feature of the bonding in H_2 comes to the fore when the EDA results are compared with the data for N_2. The latter molecule was chosen, because nitrogen atom in its ^{4}S electronic ground state with the configuration $(1s)^2(2s)^2(1p_x)^1(1p_y)^1(1p_z)^1$ has

Table 4.1 Energy decomposition analysis of the H–H and N–N bonds at RPBE/TZP.[a]

	H_2	N_2
ΔE_{int}	−112.9	−232.2
ΔE_{Pauli}	0.0	791.7
ΔE_{elstat} [b]	+5.8	−308.5 (30.1%)
ΔE_{orb} [b]	−118.7 (100%)	−715.4 (69.9%)
ΔE_{σ} [c]	−118.7 (100%)	−470.0 (65.7%)
ΔE_{π} [c]	0.0	−245.4 (34.3%)
R	0.745 (0.741)	1.105 (1.098)
D_e	112.9	232.2
D_0	106.3 (103.3)	228.8 (225.0)

Energy values in kcal/mol. Bond lengths R in Å. Experimental values are given in parentheses.
[a] Data are taken from Ref. [19].
[b] The values in parentheses give the percentage contribution to the total attractive interactions $\Delta E_{elstat} + \Delta E_{orb.}$
[c] The values in parentheses give the percentage contribution to the orbital interactions $\Delta E_{orb.}$

like hydrogen a spherically symmetrical electron density. The EDA results are thus independent from the orientation of the occupied orbitals of the interacting atoms.

The EDA values for N_2 (Table 4.1) suggest that the largest contribution to the N–N interactions comes from the ΔE_{Pauli} term which amounts to 791.7 kcal mol^{-1}. This is striking evidence for the large role played by the Pauli repulsion in molecules that have more than two electrons. The second qualitative difference between H_2 and N_2 is the very large attractive contribution of the electrostatic term $\Delta E_{elstat} = -308.5$ kcal mol^{-1} in dinitrogen that is even larger than the total interaction energy $\Delta E_{int} = -232.2$ kcal mol^{-1}. The latter result is not unusual for diatomic molecules. Hirshfeld and Rzotkiewicz [20] reported as early as in 1974 that the quasi-classical [16] electrostatic attraction in the heavier first-row diatomic molecules provides a large part of the total binding energy. In 1986, Spackman and Maslen [21] investigated chemical bonding in 148 diatomic molecules and found that electrostatic attraction between two neutral atoms in most homonuclear diatomic molecules and many heteronuclear diatomic molecules is very strong and may even be stronger than the total BDE. *It becomes clear that a point-charge approximation and the use of partial charges for estimating quasi-classical electrostatic interactions in molecules completely fail. The results should be used as a warning against the common tendency to use atomic partial charges as a measure for the ionic character of a bond.*

The EDA results suggest that the orbital (covalent) interactions in N_2 are still stronger than the electrostatic attraction. The contribution of $\Delta E_{orb} = -715.4$ kcal mol^{-1} accounts for 69.9% of the total N–N attraction. The latter term comprises −470.0 kcal mol^{-1} (65.7%) of σ bonding and −245 kcal mol^{-1} (34.3%) of π bonding which agrees with chemical intuition that σ bonding is stronger than π bonding. The strong total attraction of −1023.9 kcal mol^{-1}

and the large Pauli repulsion of 791.7 kcal mol^{-1} give a net attraction of ΔE_{int} ($= -D_e$) = −232.2 kcal mol^{-1} that is in excellent agreement with the experimental BDE.

Pauli repulsion is not only a significant component of the bond strength, it is also a crucial factor for the length of a chemical bond. The equilibrium distance of a covalent bond is often explained in terms of maximum orbital overlap. Figure 4.3 shows schematically the feature of a σ and a π orbital of a diatomic molecule E_2 as is introduced in most chemistry textbooks. The onset of the σ orbital overlap S(σ) occurs at longer E–E distances than the π overlap S(π). The overlap S(σ) has a maximum value at a certain distance R(E–E) whereas at shorter distances S(σ) becomes smaller. The maximum value of S(σ) is then identified with the equilibrium distance of an E–E single bond. By contrast, the maximum value S(π) is reached when R(E–E) = 0. The additional contribution of the π interactions is then suggested as the reason why multiple bonds are shorter than single bonds.

Figure 4.4a shows the orbital overlaps of the σ and π AOs of nitrogen in N_2 as a function of the N–N distance. The figure displays the curves for the total σ and π overlaps S(σ) and S(π) and the σ components ss, sp_σ, $p_\sigma p_\sigma$ as well as the π components $p_\pi p_\pi$. The calculated curves indicate that the maximum value of S(σ) is found at *shorter* distance than the equilibrium bond length r_e. This is clear proof that the bond length in N_2 is *not* determined by the maximum orbital overlap as usually assumed. If this were true, the N–N equilibrium bond length should be ~0.4 Å shorter than experimentally found. There is an interatomic force that prevents a bond length where maximum bond overlap occurs.

Figure 4.4b shows the curves for the various energy terms of the EDA as a function of the N–N distance. It becomes obvious that both attractive forces, for example, the electrostatic term ΔE_{elstat} and the orbital term ΔE_{orb}, further increase when the N–N distance becomes shorter than the equilibrium distance. This holds not only for the π-orbital interactions ΔE_π but also for the σ bonding ΔE_σ, which still becomes stronger when the N–N bond is shortened by 0.4 Å. The electrostatic attraction also increases at shorter bond length by up to ~0.2 Å where it reaches a maximum value. *The crucial force which prevents further shortening below the equilibrium distance and which determines the bond length in* N_2 *is the Pauli repulsion* ΔE_{Pauli}, which sharply increases at a shorter distance (Figure 4.4b). This conclusion is valid for nearly all covalent bonds! An investigation of the trend of the orbital overlap and the EDA terms in diatomic molecules E_2 of all atoms of the

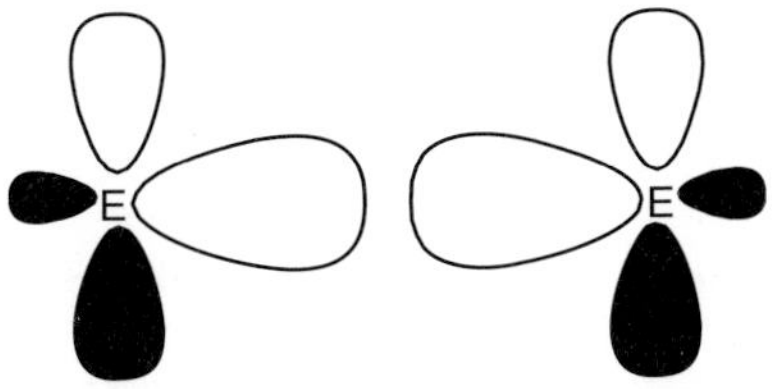

Figure 4.3 Schematic representation of bonding overlap between two sp^x hybrid orbitals and two p(π) AOs.

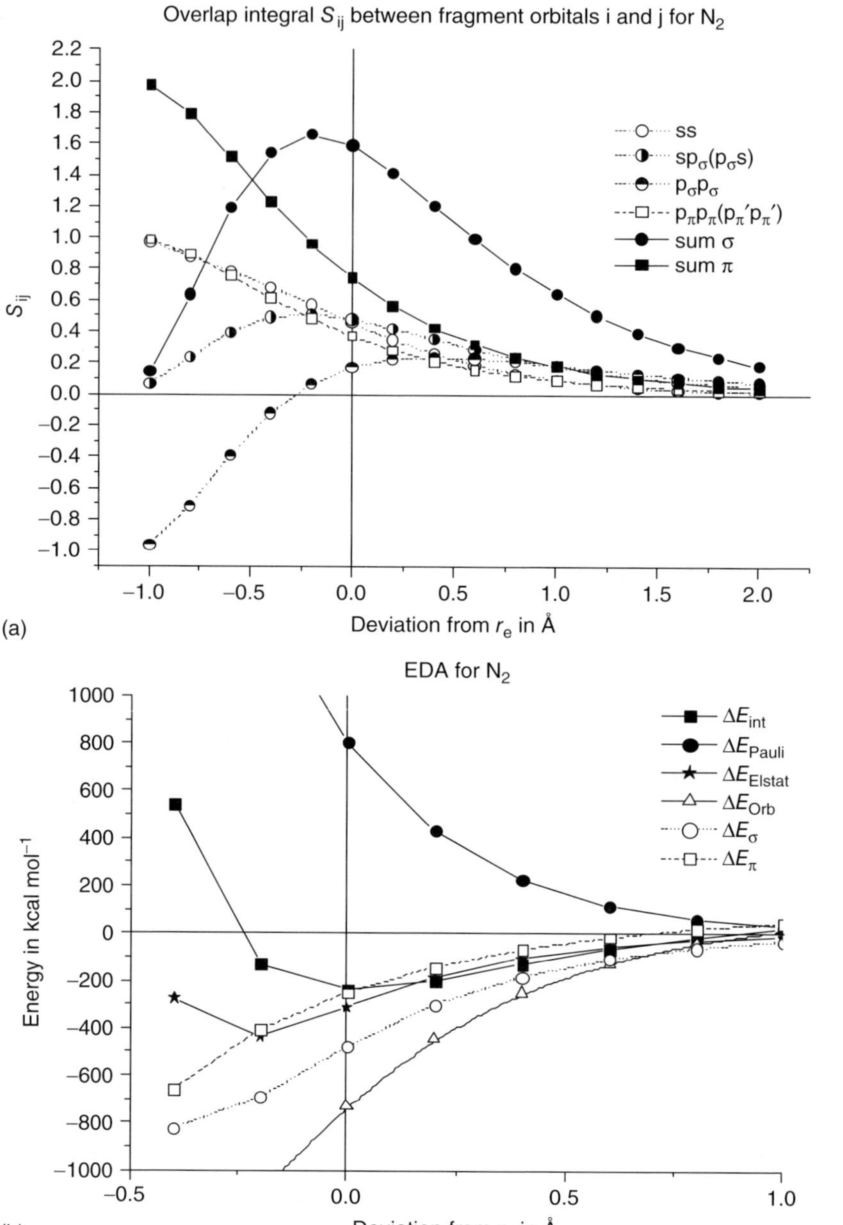

Figure 4.4 (a) Overlap integrals of the atomic 2s and 2p orbitals of N_2 as function of the interatomic interaction. The overlap integral of $p_\sigma p_\sigma$ refers to the bonding combination of the orbitals which, therefore, converges to a value of −1. (b) Calculated EDA values at BP86/TZ2P for N_2 as function of the interatomic distance. The reference value 0.0 is the calculated equilibrium bond length 1.102 Å.

first and second octal rows has shown that in all cases the rising of ΔE_{Pauli} which overcompensates the increase of the attractive forces ΔE_{elstat} and ΔE_{orb} at shorter distances determines the equilibrium bond length of the molecules [22].

4.5
Comparison of Bonding in Isoelectronic N_2, CO and BF

The EDA results may be used to explain trends in the nature and in the bond strength of related molecules. A challenging example is given by the isoelectronic species N_2, CO, and BF [19]. There is an increase in the polarity [23] of the bonds $N_2 < CO < BF$ but the BDE has the order $CO > N_2 > BF$. In the following, we give an explanation for the latter trend using the EDA results of the molecules. At the same time we discuss the problem of choosing the electronic state of the interacting fragments for CO, which is not directly obvious as in cases like H_2 and N_2.

Figure 4.5 shows schematically the electron configurations of the atomic fragments and the alignment of the orbitals as they were chosen for the EDA calculations of the diatomics. The case of N_2 is trivial because of the isotropic electron density of nitrogen atom in the ^4S ground state (Figure 4.5a). However, there are two different ways of combining the carbon and oxygen atoms in their electronic ground states[2)] with the electron configurations $1s^2 2s^2 2p^{\alpha} 2p^{\alpha} 2p^0$ (^3P) carbon and $1s^2 2s^2 2p^2 2p^{\beta} 2p^{\beta}$ (^3P) oxygen yielding ($^1\Sigma^+$) CO. Figure 4.5b displays the approach where the overall symmetry of the atomic configurations is the same as the point group of the CO molecule ($C_{\infty v}$). This is the symmetry adapted fragmentation procedure that leads to a donor–acceptor σ bond and two degenerate electron-sharing π bonds. The doubly occupied $2p^2$ orbital of oxygen and the empty $2p^0$ orbital of carbon have σ symmetry. The second approach shown in Figure 4.5c has only C_{2v} symmetry for the atomic fragments. Here the doubly occupied $2p^2$ orbital of oxygen and the empty $2p^0$ orbital of carbon have π symmetry. The latter approach has one electron-sharing π bond, one donor–acceptor π bond and an electron-sharing σ bond, as suggested by valence bond methods. The choice of the alignment of boron and fluorine atoms in their ^2P ground state for the EDA calculations is straightforward (Figure 4.5d). Note that the Pauli repulsion ΔE_{Pauli} in all systems, which are shown in Figure 4.5, comes from electrons in σ orbitals. There are no π electrons in both fragments that have the same spin.

Table 4.2 gives the EDA results for N_2, CO, and BF. The calculated BDEs of the three molecules are in excellent agreement with experimental values. Since there is no preparation energy ΔE_{prep} of the interacting fragments, the trend of the BDEs is directly related to the change in the strength of the interatomic interactions ΔE_{int}. The EDA data for the different energy terms of N_2 and CO suggest that the stronger bond in the latter molecule does *not* come from stronger interatomic

2) One may also use electronically excited states of C and O for the EDA calculation of CO. There are indeed cases where the electronic reference states of the fragments in a molecule are excited states rather than the ground state. Examples are discussed later in the chapter and in the literature [12]. Excited states of O and C are not relevant for the EDA of CO.

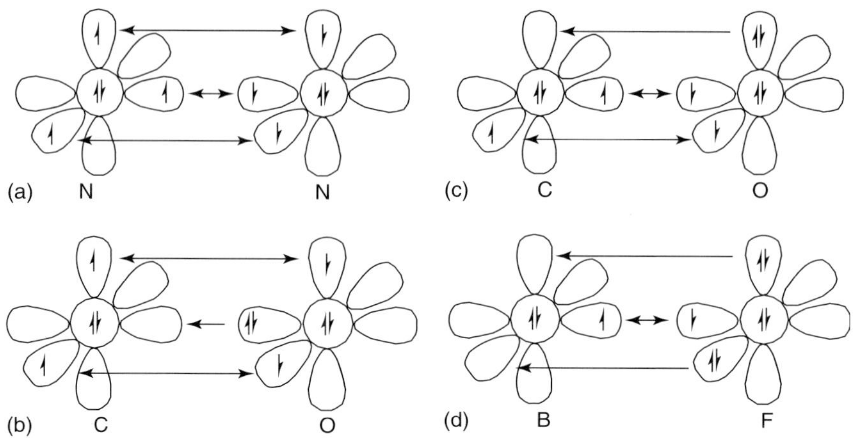

Figure 4.5 Schematic representation showing the different ways of combining the electron configurations of the atoms in some diatomic molecules for the EDA calculations and the associated orbital interactions. (a) Two nitrogen atoms in the ^{4}S ground state yielding N_2. (b) $C_{\infty v}$ symmetry adapted configurations of (^{3}P) C and (^{3}P) O for CO. (c) C_{2v} symmetry adopted configurations of (^{3}P) C and (^{3}P) O for CO. (d) $C_{\infty v}$ symmetry adapted configurations of (^{2}P) B and (^{2}P) F for BF.

attraction. Both attractive components, for example, electrostatic attraction ΔE_{elstat} and orbital interactions ΔE_{orb}, are weaker in CO than in N_2. This holds for both fragmentation schemes of CO (Table 4.2). The calculated data indicate that the stronger bond in CO compared with N_2 comes from weaker Pauli repulsion, which overcompensates the loss of electrostatic attraction and orbital interactions. The data for ΔE_{Pauli} in CO are in both fragmentation schemes smaller than in N_2.

Which of the two fragmentation schemes for CO shown in Figure 4.5b and c gives a more faithful description of the interatomic interactions? The answer is the absolute values of ΔE_{orb}. The orbital interactions ΔE_{orb} signify the change of the electronic structure of the fragments, which is induced by the mixing of the occupied and vacant orbitals within the molecule. A smaller value for ΔE_{orb} indicates that the fragment is better suited for the covalent bonding than other electronic states. Thus, the symmetry adapted configuration shown in Figure 4.5b should be used for EDA calculations of CO. The data suggest that the contribution of π bonding to the total orbital interactions is larger (49.2% of ΔE_{orb}) than in N_2 (34.3%). CO is clearly a molecule with a triple bond that should be written with the Lewis structure $^{(-)}|C\equiv O|^{(+)}$.

The EDA results for BF show (Table 4.2) that all energy terms ΔE_{elstat}, ΔE_{Pauli} and ΔE_{orb} are smaller than in N_2 and in CO, but now the weaker Pauli repulsion does not compensate for the loss of electrostatic attraction and orbital interactions. It seems that the EDA terms do not provide an explanation[3] for the much weaker bond in BF

3) We want to point out that the term "explanation" is used here in the sense of correlation which is typical for addressing chemical matter. A fundamental explanation would address the question in terms of elementary forces of physics which is rarely meant in chemistry.

Table 4.2 Energy partitioning analysis of the N–N, C–O, and B–F Bonds at RPBE/TZP.[a]

	N_2	CO	CO	BF
Symmetry[b]	$D_{\infty h}$	$C_{\infty v}$	C_{2v}	$C_{\infty v}$
ΔE_{int}	−232.2	−258.4	−258.4	−180.8
ΔE_{Pauli}	791.7	575.8	725.9	476.1
ΔE_{elstat}[c]	−308.5 (30.1%)	−240.0 (28.8%)	−291.1 (29.6%)	−210.5 (32.0%)
ΔE_{orb}[c]	−715.4 (69.9%)	−594.2 (71.2%)	−693.2 (70.4%)	−446.4 (68.0%)
ΔE_{σ}[d]	−470.0 (65.7%)	−301.7 (50.8%)	−464.7 (67.0%)	−396.4 (88.8%)
ΔE_{π}[d]	−245.4 (34.3%)	−292.5 (49.2%)	−228.5 (33.0%)	−50.0 (11.2%)
			b_1: −143.8[e]	
			b_2: −84.7[e]	
Overlap σ	1.58	1.51	1.51	1.26
Overlap π	0.74	0.70	0.70	0.55
bond length	1.105 [1.098]	1.144 [1.128]	1.144 [1.128]	1.285 [1.262]
D_e	232.2	258.4	258.4	180.8
D_o	228.8 [225.0]	255.4 [255.7]	255.4 [255.7]	278.9 [179.9]

Energy values are given in kcal/mol. Bond lengths are given in Å. Experimental values are given in brackets.
[a] Data are taken from Ref. [19].
[b] The symmetry is given by the electronic structure of the atoms with respect to the molecule, see Figure 4.5.
[c] Values in parentheses give the percentage of the total attractive interactions $\Delta E_{elstat} + \Delta E_{orb}$.
[d] Values in parentheses give the percentage of the total orbital interactions ΔE_{orb}.
[e] The b_1 value gives the electron-sharing contribution to the π bond and the b_2 value gives the donor–acceptor contribution to the π bond, see Figure 4.5c.

compared with N_2 and CO. However, the breakdown of ΔE_{orb} into contributions of σ and π orbitals reveals that there is a significant difference between BF and the other two diatomics. The strength of ΔE_{π} in BF is only −79.0 kcal mol^{-1} which is 12.9% of ΔE_{orb}. In contrast, the ΔE_{π} values for N_2 (−245.4 kcal mol^{-1}, 34.3% of ΔE_{orb}) and CO (−292.5 kcal mol^{-1}, 49.2% of ΔE_{orb}) indicate much stronger π bonding. Note that the strength of the σ orbital interactions in BF (396.4 kcal mol^{-1}) is even larger than in CO (301.7 kcal mol^{-1}). The EDA data thus suggest that the weaker bonding in BF comes from the much weaker π orbital interactions compared with N_2 and CO. This is in agreement with chemical intuition and with the most important Lewis structure that has a B–F single bond but not a triple bond $^{(2-)}|B{\equiv}F|^{(2+)}$. According to the EDA results, BF is a molecule with a single bond whereas N_2 and CO are species with triple bonds, which explains why there is a weaker bond in BF.

4.6 Bonding in the Diatomic Molecules E_2 of the First Octal Row E = Li–F

A strong feature of the EDA is the ability to connect qualitative bonding models with a quantitative analysis of the actual wave function. This shall be demonstrated

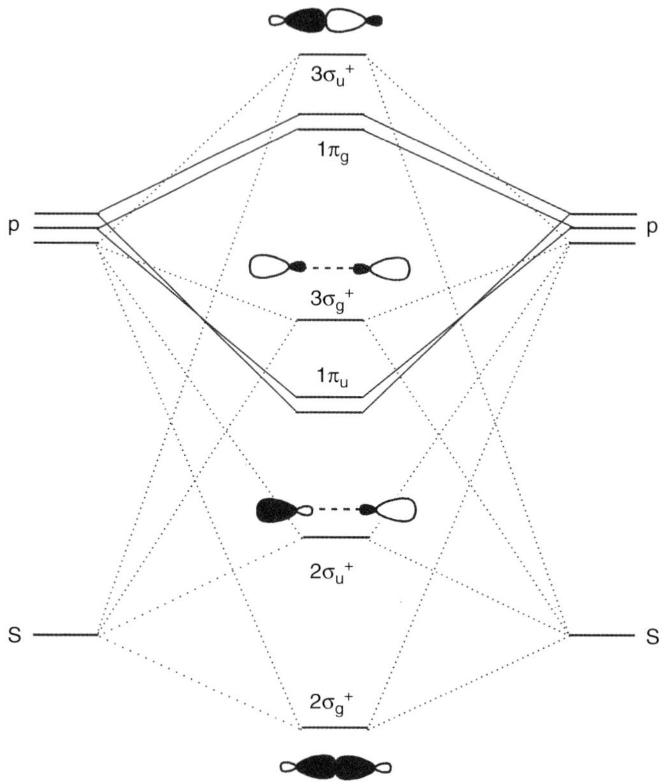

Figure 4.6 Orbital correlation diagram for diatomic molecules of main group elements.

by analysing the chemical bond in a first octal-row sweep of diatomic molecules E_2. Figure 4.6 shows a correlation diagram for the valence s and p orbitals of the main-group atoms E, which give the molecules E_2 [24]. The s/p hybridization lowers the energy of the bonding $2\sigma_g^+$ and antibonding $2\sigma_u^+$ MOs while it raises the energy levels of the bonding $3\sigma_g^+$ and antibonding $3\sigma_u^+$ MOs. Thus, for the lighter species Li_2–N_2 the bonding $3\sigma_g^+$ MO is slightly higher in energy than the $1\pi_u$ orbital. As a result, B_2 has a triplet ($X^3\Sigma_g^-$) ground state whereas C_2 has a singlet ($X^1\Sigma_g^+$) ground state [25].

Figure 4.7 schematically shows the electron configuration and the orientation of the atoms with the chosen orbital populations that are in accord with the electronic ground states of E_2. Note that the atomic fragments, which are used in the EDA calculations of the ($^3\Sigma_g^-$) triplet state of O_2, are two oxygen atoms in the ^3P ground state, which is the symmetry allowed dissociation product [26].[4] Hence, a spin change of one electron in a p(π) orbital must take place during the EDA calculation because a triplet molecule is constructed from two triplet fragments.

4) The generation of O_2 in the $^3\Sigma_g^-$ ground state from two (^3P) oxygen atoms can not be illustrated in a simple picture as shown as in Figure 4.7f which is therefore not quite correct.

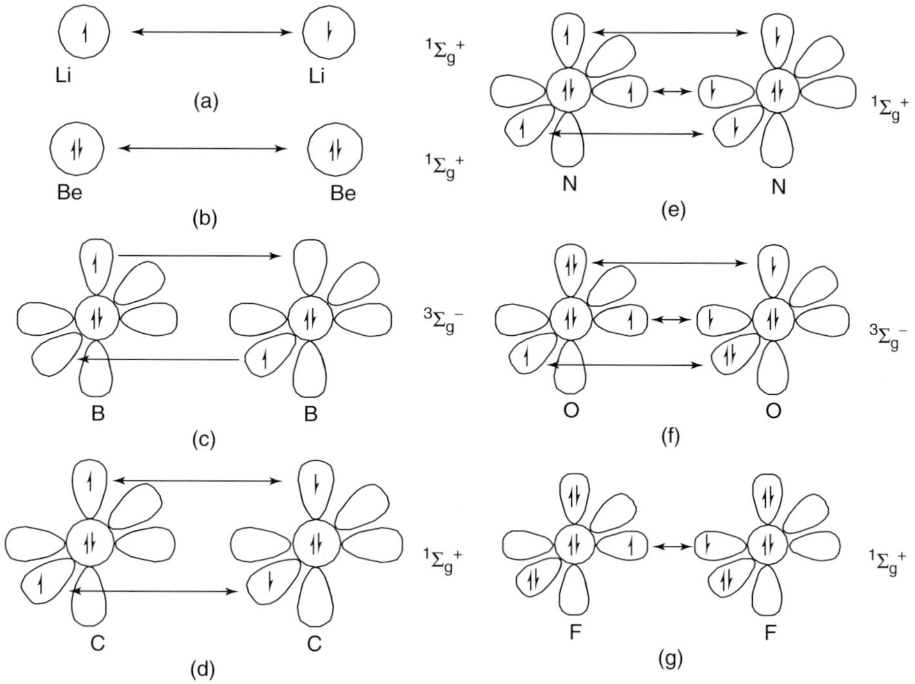

Figure 4.7 Schematic representation of the electron configuration and the orientation of the atoms with the chosen orbital populations for the EDA. Note that for the $^3\Sigma_g^-$ state of O_2 a spin change of one electron in a singly occupied p(π) orbital takes place in the EDA calculation [26].

The two components of the π orbital interactions in the EDA have therefore slightly different values, which is, however, not relevant for the present purpose.

The EDA results for the diatomic molecules Li_2–F_2 are given in Table 4.3. The data of Li_2 suggest that the attractive lithium–lithium interactions have about one third electrostatic character whereas about two thirds come from orbital interactions. The ΔE_{Pauli} value is very small because there is only Pauli repulsion between core–core and core–valence electrons but no valence–valence repulsion. The chemical bond in Li_2 is valence isoelectronic to H_2 but it possesses unusual features. It was previously suggested [27] that induction forces significantly contribute to the chemical bond in Li_2. This can be investigated through an EDA calculation of Li_2 where the p valence AOs and all polarization functions at Li are deleted, which inhibits the polarization of the 2s valence orbital and thus, eliminates induction. Table 4.4 gives the EDA result of Li_2 with the reduced orbital space. The ΔE_{orb} term lowers from $-14.2\,kcal\,mol^{-1}$ to $-9.8\,kcal\,mol^{-1}$. The remaining part comes from genuine orbital interactions, whereas the difference of $-4.3\,kcal\,mol^{-1}$ is polarization and enhanced orbital interaction through hybridization. The EDA calculation of Li_2 where also the higher-order s functions are deleted gives $\Delta E_{orb} = -8.1\,kcal\,mol^{-1}$ (Table 4.4). The small difference of $-1.7\,kcal\,mol^{-1}$ to the ΔE_{orb} value where the

Table 4.3 Energy partitioning analysis of the first row dimers E_2 (E = Li–F) in C_{2v} at BP86/TZ2P (ZORA).[e]

E	Li	Be	B	C	N	O	F
el. State	$^1\Sigma_g^+$	$^1\Sigma_g^+$	$^3\Sigma_g^-$	$^1\Sigma_g^+$	$^1\Sigma_g^+$	$^3\Sigma_g^-$	$^1\Sigma_g^+$
ΔE_{int}	−20.7	−7.9	−74.7	−140.8	−240.2	−141.9	−52.9
ΔE_{Pauli}	1.8	41.6	135.0	252.2	802.4	464.9	146.1
ΔE_{elstat}[a]	−8.3 (36.9%)	−17.9 (36.1%)	−33.1 (15.8%)	−3.2 (0.8%)	−312.9 (30.0%)	−159.7 (26.3%)	−41.2 (20.7%)
ΔE_{orb}[a]	−14.2 (63.1%)	−31.6 (63.9%)	−176.5 (84.2%)	−389.8 (99.2%)	−729.8 (70.0%)	−447.1 (73.7%)	−157.8 (79.3%)
ΔE_{a1}[b] (σ)	−14.2	−31.6	−104.5 (59.2%)	−201.7 (51.8%)	−478.8 (65.6%)	−319.5 (71.5%)	−151.5 (96.0%)
ΔE_{a2}[b] (δ)	0.0	0.0	0.0	0.0	0.0	0.0	0.0
ΔE_{b1}[b] (π)	0.0	0.0	−36.0 (20.4%)	−94.0 (24.1%)	−125.5 (17.2%)	−59.1 (13.2%)	−3.1 (2.0%)
ΔE_{b2}[b] (π)	0.0	0.0	−36.0 (20.4%)	−94.0 (24.1%)	−125.5 (17.2%)	−68.5 (15.3%)	−3.1 (2.0%)
$\Delta E_{corr.}$[d]	0.3	0.0	1.9	3.3	4.2	4.8	2.7
D_e[c]	20.4 (24.6)	7.9 (2.3) 2.7[calc.][d]	72.8 (71.2)	137.5 (145.9)	236.1 (228.4)	137.1 (120.2)	50.2 (38.3)
r(E–E)[c]	2.731 (2.673)	2.442 (2.45)[d]	1.617 (1.590)	1.253 (1.243)	1.102 (1.098)	1.224 (1.208)	1.420 (1.412)

Energies in kcal mol^{-1}, distances r(E–E) in Å.
[a] Values in parenthesis give the percentage contribution to the total attractive interactions $\Delta E_{elstat}+\Delta E_{orb}$
[b] Values in parenthesis give the percentage contribution to the total orbital interactions ΔE_{orb}
[c] Experimental values are given in parentheses..
[d] Correction for the spin polarization
[e] Data are taken from Ref. [22a].

Table 4.4 Energy partitioning analysis of some first row dimers E_2 and related species at BP86/TZ2P using designated fragments.[a]

E_2 El. state	Li_2[e] $^1\Sigma_g^+$		Li_2^+ $^2\Sigma_g^+$			Be_2 $^1\Sigma_g^+$
Virtual space	only s orbitals in virtual space	no virtual orbitals	full	only s orbitals in virtual space	no virtual orbitals	only s orbitals in virtual space
ΔE_{int}	−16.4	−14.6	−27.9	−14.9	−12.3	19.2
ΔE_{Pauli}	1.8	1.8	2.9	2.9	2.9	41.6
ΔE_{elstat}	−8.3	−8.3	+2.1	+2.1	+2.1	−17.9
ΔE_{orb}	−9.8	−8.1	−32.8	−19.9	−17.3	−4.6
ΔE_{a1} (σ)	−9.8	−8.1	−32.8	−19.9	−17.3	−4.6
ΔE_{a2} (δ)	0.0	0.0	0.0	0.0	0.0	0.0
ΔE_{b1} (π)	0.0	0.0	0.0	0.0	0.0	0.0
ΔE_{b2} (π)	0.0	0.0	0.0	0.0	0.0	0.0
ΔE_{corr}[b]			0.1			
D_e[a]			27.7(33.2)			
r(E–E)[a]			3.111			

Energies in kcal/mol, distances r(E-E) in Å
[a]Data are taken from Ref. [22a].
[b]Correction for the spin polarization.

valence p and polarization functions are deleted indicates the energetic effect of mixing in the 3s (and higher) functions yielding a different radius for the Li valence s function in Li_2.

A peculiar feature of the two-electron bond in Li_2 is the finding that removing one electron from the doubly occupied $2\sigma_g^+$ bonding orbital (Figure 4.6) *strengthens* the binding interactions. The BDE of Li_2^+ ($D_e = 33.2\,kcal\,mol^{-1}$) is clearly higher than the BDE of Li_2 ($D_e = 24.6\,kcal\,mol^{-1}$) [25, 28]. This is contrary to dihydrogen where the bond in H_2^+ is about half as strong as in H_2 [25]. Table 4.4 gives the EDA results of Li_2^+ which indicate that the stronger bond comes solely from the ΔE_{orb} term. The Pauli repulsions in Li_2 and Li_2^+ have nearly identical values and the electrostatic interaction in Li_2^+ is even repulsive. Thus, Li_2^+ is, besides H_2, one of the rare examples where the classical electrostatic interaction at equilibrium distance is not attractive. What about induction forces in Li_2^+? Table 4.4 gives also the EDA results of Li_2^+ where the p valence AOs and the polarization functions are deleted. After deleting the latter orbitals the ΔE_{orb} term increases significantly from −32.8 to $-19.9\,kcal\,mol^{-1}$, which shows that the polarization of the valence AO in Li_2^+ is very large. It is much larger than in Li_2 where deleting the p valence AOs reduces ΔE_{orb} from −14.2 to $-9.8\,kcal\,mol^{-1}$. However, the strength of the orbital interactions where only s AOs remain in the valence space in Li_2^+ is still bigger ($-19.9\,kcal\,mol^{-1}$) than in Li_2 ($-9.8\,kcal\,mol^{-1}$). It follows that the stronger bond in Li_2^+ comes from genuine orbital interactions due to the much lower lying

2s AOs of the cation that is further enhanced by strong induction forces[5]. Table 4.4 shows also the ΔE_{orb} value after deleting the higher order (n)s orbitals ($n > 2$) of Li in Li_2^+. The calculated data of −17.3 kcal mol^{-1} suggests that the change of the 2s orbital radius through bond formation in Li_2^+ yields a stabilization of 2.6 kcal mol^{-1}. This is slightly more than in Li_2 where the higher order (n)s orbitals contribute 1.7 kcal mol^{-1} to the bond.

The qualitative MO diagram of Be_2 suggests that the molecules have only a very weak bond because the repulsive interactions of the antibonding $2\sigma_u^+$ MO should partly cancel the bonding contribution of the $2\sigma_g^+$ MO (Figure 4.6). Indeed, the calculation of Be_2 ($^1\Sigma_g^+$) gives a BDE of only 7.9 kcal mol^{-1} which is slightly higher than in the experiment (Table 4.3). The EDA calculation gives a significantly larger ΔE_{Pauli} value of 41.6 kcal mol^{-1} compared with Li_2 (1.8 kcal mol^{-1}). There is Pauli repulsion between the valence electrons in Be_2, which does not occur in Li_2. Somewhat surprisingly, the ΔE_{orb} term contributes −31.6 kcal mol^{-1} to the attractive interactions which is twice as high as in Li_2 (−14.2 kcal mol^{-1}). How is it that the ΔE_{orb} term gives such a large stabilizing value for Be_2 whereas the MO diagram suggests that the orbital interactions should be weakly repulsive? The answer is, that the energy rise of the $2\sigma_g^+$ MO comes from the symmetry constraints that are imposed through the Pauli principle. The energy levels of the valence MOs, which are shown in the correlation diagram in Figure 4.6, consider the Pauli repulsion whereas the ΔE_{orb} term only gives the relaxation of the previously antisymmetrized wave function toward the final SCF solution. The sum of the destabilizing E_{Pauli} term (41.6 kcal mol^{-1}) and ΔE_{orb} (−31.6 kcal mol^{-1}) gives a net destabilization of 10.0 kcal mol^{-1}, which agrees with the rise of the $2\sigma_g^+$ MO.

The stabilizing contribution of ΔE_{orb} in Be_2 comes mainly from the mixing of the 2p(σ) AO of Be into the $2\sigma_g^+$ and $2\sigma_g^-$ MOs which enhances the bonding character of the former and weakens the antibonding character of the latter orbital. This becomes evident through an EDA calculation of Be_2 where the 2p valence orbitals and the polarization functions are deleted. Table 4.4 shows that the value for the ΔE_{orb} term becomes dramatically smaller after deleting the empty orbitals. The strength of the remaining orbital interactions is only −4.60 kcal mol^{-1} that comes from the mixing of the 3s and higher order s functions with the valence orbitals.

Diatomic B_2 has a triplet ($X^3\Sigma_g^-$) ground state where two electrons occupy the degenerate $1\pi_u$ orbital (Figure 4.6). According to the MO correlation diagram, the molecule is bonded solely through its π bond. The EDA results provide a quantitative estimate for the latter interpretation of the boron–boron bond. The strength of the π orbital interactions (−72.0 kcal mol^{-1}) matches nearly exactly the total attraction of $\Delta E_{int} = -74.4$ kcal mol^{-1}. The stabilizing contributions of $\Delta E(\sigma)$ (−104.5 kcal mol^{-1}) and ΔE_{elstat} (−33.1 kcal mol^{-1}) are roughly cancelled by the Pauli repulsion $\Delta E_{Pauli} = 135.0$ kcal mol^{-1}. Diboron B_2 may therefore be considered as a π-bonded molecule. Note that the EDA results for B_2 depend on the alignment of the occupied AOs of the boron atoms in the calculation (Figure 4.7c),

5) The difference of the chemical bonding between Li_2 and Li_2^+ has been discussed in reference [5].

which are not invariant to coordinate transformation. The arrangement shown in Figure 4.7c follows from the Aufbau principle (Figure 4.6).

The qualitative MO diagram for $(X^1\Sigma_g^+)$ C_2 suggests that, like in $(X^3\Sigma_g^-)$ B_2, chemical bonding, which is enhanced because the degenerate $1\pi_u$ orbital is now fully occupied, comes solely from the π orbitals (Figure 4.6). The EDA results are in nice agreement with the qualitative model. The attractive π orbital interactions (−188 kcal mol^{-1}) are even stronger than the total attraction of $\Delta E_{int} = -140.8$ kcal mol^{-1}, because the Pauli repulsion at the rather short C–C distance of 1.253 Å overcompensates ($\Delta E_{Pauli} = 252.2$ kcal mol^{-1}) the contribution of the σ interactions ($\Delta E(\sigma) = -201.7$ kcal mol^{-1}). A peculiar feature of $(X^1\Sigma_g^+)$ C_2 is the unusually weak electrostatic attraction of ΔE_{elstat} −3.22 kcal mol^{-1} (Table 4.3), which can be explained with the occupation of the valence shell of carbon and the orientation of the occupied orbitals for bonding interactions (Figure 4.7d). There are two electrons in the 2s AO, 2 electrons in the p(π) AO but none in the p(σ) AO. A detailed analysis of the electrostatic interactions between atoms, which have occupied s, p(σ) and p(π) valence orbitals, showed that p(π) valence electrons yield repulsion while p(σ) valence electrons yield strong attraction [22]. Valence electrons in s orbitals like in Li_2, Be_2 and B_2 lead to weak electrostatic attraction but the strength of the ΔE_{elstat} term depends on the interatomic distance and particularly on the radius of the valence orbital. Carbon has a more compact 2s orbital than boron there are two electrons in p(π) AOs at carbon which leads to weaker electrostatic attraction in C_2 although it has a shorter bond (1.253 Å) than B_2 (1.617 Å). Like B_2, C_2 is a π bonded molecule.

The EDA results for $(X^1\Sigma_g^+)$ N_2 were already presented in the previous section where the trend of the isoelectronic molecules N_2, CO, and BF were discussed. A comparison of the results for N_2, given in Tables 4.2 and 4.3, shows numerical differences that are due to the different computational levels (BP86/TZ2P and PBE/TZP) that were used in respective theoretical studies. The differences between the absolute values are rather small and the percentage contributions of the different terms to the total interactions are nearly the same. This is an important finding because the EDA results would not be very useful if they depended significantly on the theoretical level of the calculations.

It is intriguing to discuss the EDA data for N_2 in comparison with the other diatomic molecules E_2. Because the nitrogen atom in the ^4S electronic ground state has the electron configuration $1s^22s^22p_x^{\,1}2p_y^{\,1}2p_z^{\,1}$, the p(σ) AO of nitrogen is occupied (Figure 4.7e). This leads to a very large electrostatic attraction of $\Delta E_{elstat} = -312.9$ kcal mol^{-1} for N_2, which is a striking difference to the value of $\Delta E_{elstat} = -3.2$ kcal mol^{-1} for C_2. One could argue that dinitrogen is solely bonded through electrostatic attraction, because the strong Pauli repulsion of $\Delta E_{Pauli} = 802.4$ kcal mol^{-1} overcompensates the total orbital interactions of $\Delta E_{orb} = -729.8$ kcal mol^{-1}. Although the argumentation is valid, it is like all partitioning schemes based on an arbitrary summing up of different terms. A different but equally valid summation would come to the conclusion that N_2 is bonded through its π bonds. The strength of the π orbital interactions $\Delta E(\pi) = -251.0$ kcal mol^{-1} comes close to the total interaction energy

$\Delta E_{int} = -240.2$ kcal mol^{-1}, which means that the other three terms $\Delta E(\sigma)$, ΔE_{elstat}, and ΔE_{Pauli} roughly cancel. The summing up of $\Delta E(\sigma)$ and ΔE_{Pauli} is reasonable, because there is Pauli repulsion only between electrons in the σ orbitals, which means that the σ bond in N_2 is very weak if not repulsive! It has already been suggested in previous analyses of the chemical bond in dinitrogen that the σ bond is rather weak [5, 27].

The large value of $\Delta E_{elstat} = -312.9$ kcal mol^{-1} for N_2 shall be used to examine the EDA approach for estimating the strength of electrostatic interactions that uses the frozen charge densities of the fragments for the calculation of ΔE_{elstat}. This may be criticized, because the mere superposition of charge densities does not consider the Pauli principle and thus, violates a fundamental law of physics. This could lead to values for ΔE_{elstat} being too large, because the antisymmetry constraint of the overlapping orbitals imposed by the Pauli principle leads to removal of electronic charge from the interatomic region. This question has been addressed in a theoretical study where the change in the charge density of N_2 in the ΔE_{Pauli} step (antisymmetrization and renormalization) and in the following ΔE_{orb} step (orbital relaxation) have been investigated [22a].

Figure 4.8a shows that the antisymmetrization and renormalization in the ΔE_{Pauli} step as expected lead to removal of electronic charge from the nitrogen–nitrogen bonding region. However, the following relaxation in the ΔE_{orb} step (Figure 4.8b)

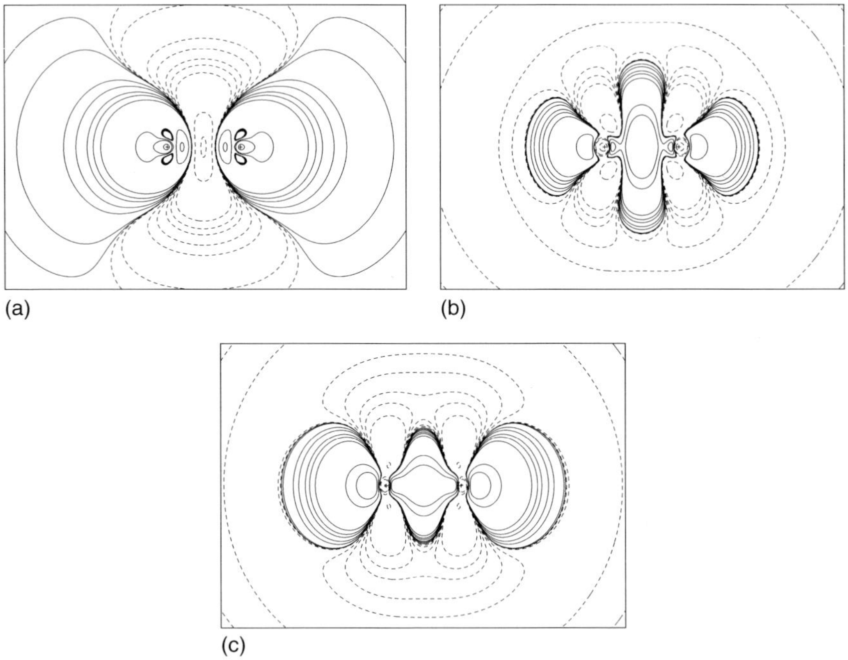

Figure 4.8 Changes in the electron density of N_2 during the EDA. Solid lines indicate charge accumulation, dashed lines indicate charge depletion. (a) Antisymmetrization and renormalization. (b) Orbital relaxation. (c) Total changes during step a and b.

restores electron density in the area of σ and π bonding. It becomes obvious that, in the ΔE_{Pauli} step, electronic charge from the bonding region is withdrawn and becomes accumulated in the backside σ regions of the atoms. In the ΔE_{orb} step electronic charge is removed from the atomic p(π) region into the bonding area of σ and π bonding but the charge concentration in the backside σ regions is also further enhanced. Figure 4.8c shows the total changes that occur in the previous two steps. The net changes in the electron density during the ΔE_{Pauli} step and the ΔE_{orb} step lead to an enhanced accumulation along the internuclear axis in the σ and π bonding interatomic region and in the backside area while electronic charge is removed from the p(π) area of the atoms. This should further enhance the attractive contribution of the electrons in s and p(σ) orbitals to the ΔE_{elstat} term. From this it can be concluded that the quasiclassical electrostatic bonding using the unpolarized charges, which are employed in the ΔE_{elstat} of the EDA, underestimates rather than overestimates the strength of the electrostatic contribution to the chemical bond.

The EDA results for ($X^3\Sigma_g^-$) O_2 indicate (Table 4.3) that all energy terms have smaller absolute values compared with N_2 and that the chemical bond in dioxygen is clearly weaker than in dinitrogen, which concurs with experiment. The MO correlation diagram (Figure 4.6) suggests that the weaker bond in O_2 comes from the occupation of the antibonding $1\pi_g$ orbital. This is supported by the EDA values for $\Delta E(\pi)$ that suggest that π-bonding in O_2 (−127.6 kcal mol^{-1}) is roughly half as strong as in N_2 (−251.0 kcal mol^{-1}). Like for N_2, it could be argued that O_2 is a π-bonded molecule, because the strength $\Delta E(\pi)$ is nearly as large as the total interaction energy $\Delta E_{\text{int}} = -141.9$ kcal mol^{-1}. Note, however, that O_2 is the first molecule in the series where there is Pauli repulsion between π electrons, because the oxygen atoms have a doubly occupied p(π) AO interacting with a singly occupied p(π) orbital of the other atom (Figure 4.7f). The value $\Delta E_{\text{Pauli}} = 464.9$ kcal mol^{-1} may thus not be considered anymore as part of the σ interactions alone although the contribution of the π electrons to the Pauli repulsion will be much smaller than σ Pauli repulsion because the overlap of the π electrons S(π) is much smaller than S(σ). Table 4.3 shows that the quasiclassical electrostatic attraction $\Delta E_{\text{elstat}} = -159.7$ kcal mol^{-1} is slightly stronger as the total interaction energy $\Delta E_{\text{int}} = -141.9$ kcal mol^{-1}. The terms ΔE_{Pauli} (464.9 kcal mol^{-1}) and ΔE_{orb} (−447.1 kcal mol^{-1}) nearly cancel each other. Without conceptual biases, it can be concluded that O_2 would not be a stable molecule without the stabilizing contribution of the quasiclassical attraction ΔE_{elstat}.

The Lewis bonding model for F_2 assigns a single bond to the molecule. This agrees with the MO correlation diagram which suggests that there is no π-bonding in difluorine, because the bonding and antibonding π orbitals $1\pi_u$ and$1\pi_g$ are fully occupied (Figure 4.6). This leaves only the σ orbitals as the source for the bonding orbital interactions, because there are two bonding σ orbitals ($2\sigma_g^+$ and $3\sigma_g^+$) but only one antibonding σ orbital ($2\sigma_u^+$). The EDA results in Table 4.3 appear at first glance to support the view that the F–F bonding comes from the σ orbital interactions. The stabilizing contributions of $\Delta E(\sigma) = -151.5$ kcal mol^{-1} and $\Delta E(\pi) = -6.2$ kcal mol^{-1} imply that the binding interactions come from the σ orbital that are much stronger, however, than the calculated total attraction of $\Delta E_{\text{int}} = -52.9$ kcal mol^{-1}.

The picture changes when the Pauli repulsion is considered. The contribution of $\Delta E_{Pauli} = 146.1\,kcal\,mol^{-1}$ nearly cancels the value of $\Delta E_{orb} = -157.8\,kcal\,mol^{-1}$ that makes the electrostatic attraction $\Delta E_{elstat} = -41.2\,kcal\,mol^{-1}$ to become the dominant term for the chemical attraction.

The comparison of the MO correlation diagram with the EDA results make it clear that the trend of the bond strengths for the diatomic species E_2 from E = Li–F can be discussed in terms of σ and π orbital interactions only when the Pauli repulsion and the electrostatic attraction are sometimes neglected but sometimes considered in an arbitrary way. In Li_2, ΔE_{Pauli} is very small and the Li–Li bond can be identified as σ bond, which is enhanced by electrostatic attraction. In Be_2, Pauli repulsion is important, because it must compensate the σ orbital interactions and most of the electrostatic attraction. The sole π bonds in B_2 and C_2 that are suggested by the MO correlation diagram (Figure 4.6) are recovered from the EDA results when the remaining terms $\Delta E(\sigma)$, ΔE_{Pauli}, and ΔE_{elstat} are either ignored or mutually compensating, which amounts the same. The classical triple-bond model with σ and degenerate π bonding for N_2 is only valid when the very strong Pauli repulsion in the EDA is completely ignored, which yields, however, a bond strength that is far too high. The same holds true for O_2 where the classical double bond, which consists of one σ bond and two semi π bonds, is only recovered when $\Delta E(\sigma)$ and $\Delta E(\pi)$ are considered as exclusive terms for bonding which leads to a bond strength that is too high, however. Finally, the attractive $\Delta E(\sigma)$ and $\Delta E(\pi)$ values for F_2 which suggest a single bond are nearly compensated by the Pauli repulsion. The F–F net attraction is only recovered by electrostatic bonding. All these findings highlight the importance of considering ΔE_{elstat}, ΔE_{Pauli}, and ΔE_{orb} each individually as essential quantities that are needed all together for a true understanding of the bonding mechanism.

4.7 Bonding in the Dihalogens $F_2 - I_2$

Covalent bonds A–B between valence isoelectronic atoms tend to become weaker for heavier atoms A and B. This holds in particular [29] when chemical bonds between atoms of the first octal row are compared with heavier homologues. A well known exception is the dihalogen group of atoms, where the BDE of F_2 is significantly smaller than for Cl_2. This is usually explained with the repulsion between the lone-pair electrons of the halogens that is assumed to become exceptionally large in difluorine because the F–F distance is rather short. This reasoning can be questioned, because the radius of the very electronegative element fluorine is very small, which means that the overlap of the p(π) orbitals in F_2 might not be very large. Also, the EDA calculations of N_2 and O_2 have shown that it is the σ orbitals and not the π orbitals that yield strong Pauli repulsion. A comparison of the EDA data for the dihalogens $F_2 - I_2$, which are given in Table 4.5, provides a plausible basis to check whether unusually strong Pauli repulsion between the lone-pair electrons is responsible for the weak bond in F_2.

Table 4.5 Energy decomposition analysis of the E–E bond for E = F – I at RPBE/TZP.[a]

	F_2	Cl_2	Br_2	I_2
ΔE_{int}	−47.4	−59.1	−47.8	−34.1
ΔE_{Pauli}	140.8	123.2	76.7	45.4
$\Delta E_{elstat}{}^{b}$	−38.2 (20.3%)	−44.1 (24.2%)	−31.9 (25.6%)	−19.8 (24.9%)
$\Delta E_{orb}{}^{b}$	−150.0 (79.7%)	−138.1 (75.8%)	−92.6 (74.4%)	−59.7 (75.1%)
$\Delta E_{\sigma}{}^{c}$	−144.6 (96.4%)	−123.8 (89.6%)	−85.9 (92.8%)	−59.4 (99.5%)
$\Delta E_{\pi}{}^{c}$	−5.4 (3.6%)	−14.3 (10.4%)	−6.7 (7.2%)	−0.3 (0.5%)
R(E–E)	1.424 (1.412)	2.037 (1.987)	2.381 (2.281)	2.860 (2.666)
D_e	−47.4	−59.1	−47.8	−34.1
D_o	−46.0 (−36.9)	−58.3 (−57.2)	−47.4 (−45.4)	−33.8 (−35.6)

Energy values in kcal mol^{-1}. Bond lengths R in Å. Experimental values are given in parentheses.
[a] Data are taken from Ref. [19].
[b] The values in parentheses give the percentage contribution to the total attractive interactions $\Delta E_{elstat} + \Delta E_{orb}$.
[c] The values in parentheses give the percentage contribution to the orbital interactions ΔE_{orb}.

The EDA results suggest that the orbital interactions ΔE_{orb} increase monotonously if one goes from the heavier to the lighter systems that have the regular trend $F_2 > Cl_2 > Br_2 > I_2$. The value for the Pauli repulsion ΔE_{Pauli} exhibits the same trend but there is not an unusually large increase from Cl_2 to F_2. This becomes obvious by inspection of the graphical display for the trend of the energy terms that is shown in Figure 4.9. The increase in ΔE_{orb} and ΔE_{Pauli} exhibits a

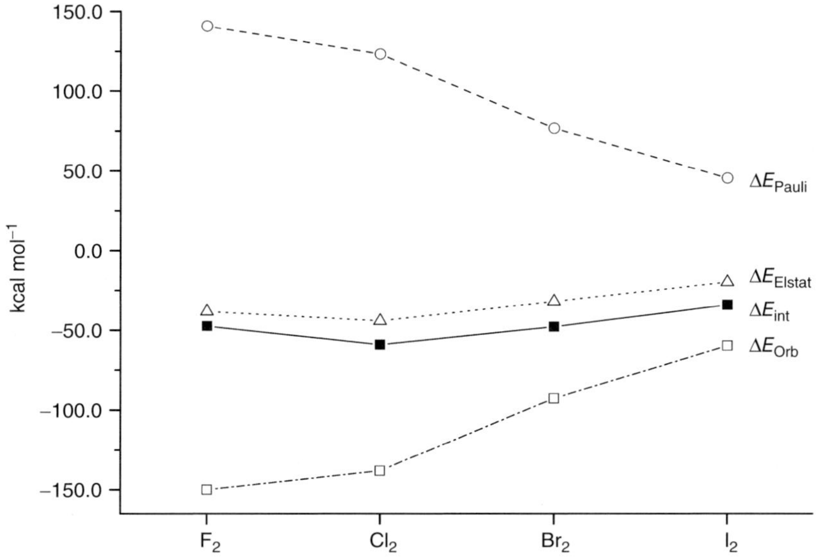

Figure 4.9 Trends of the various terms of the EDA for the dihalogens $F_2 - I_2$ calculated at RPBE/TZP.

straight line from I_2 to Br_2 and Cl_2 but the further rise in the absolute values from Cl_2 to F_2 is less steep than in the previous steps. There is no abnormally large increase in Pauli repulsion for F_2! The EDA results rather suggest that the trend of the electrostatic attraction ΔE_{elstat} shows a peculiar behavior when one goes from Cl_2 to F_2. The electrostatic attraction becomes larger in the order $I_2 < Br_2 < Cl_2$ but then it decreases for F_2. Thus, the energy term which exhibits an abnormal behavior when the chemical bonding in the dihalogens are compared is not the Pauli repulsion but the electrostatic attraction. This can be explained with the very small radius of the atomic valence orbitals of fluorine that yield less attraction in F_2 compared with Cl_2. The question whether Pauli repulsion or electrostatic attraction is the key factor for the weak bonding in F_2 is thus a chicken-and-egg problem. At shorter F–F distance, electrostatic attraction would become stronger but bond shortening is prevented by Pauli repulsion. The weak F–F bond could therefore equally be related to the long F–F bond caused by the increase of Pauli repulsion and to the weak electrostatic attraction at the equilibrium distance.

4.8 Carbon–Element Bonding in CH_3-X

The principles uncovered and discussed above for diatomics apply likewise for archetypal bonds A–B in polyatomic molecules. The only fundamental difference now is that the bond energy ΔE $(=-D_e)$ is smaller than the instantaneous interaction ΔE_{int} because it also considers the preparation energy ΔE_{prep} that is associated with the deformation of the polyatomic fragments A and B through bond formation in the molecule A–B (see Eq. (4.2)). The preparation energy ΔE_{prep} is often much smaller than ΔE_{int} but sometimes it can be rather large. In the latter case, the trend of the bond strength given by the interaction energy ΔE_{int} and the bond energy ΔE $(=-D_e)$ may become very different. Here, we illustrate both situations using the archetypal carbon–element bonds C-X in model systems CH_3X, in which X varies down group 18 (F, Cl, Br, I) [30] or along the second period (F, OH, NH_2, CH_3) of the periodic table.

The former series of halomethanes is an example of the more common situation in which ΔE_{prep} is small but it does not change the trend of the bond strength given by ΔE_{int}. Again, a full understanding requires taking into account ΔE_{elstat}, ΔE_{Pauli}, and ΔE_{orb} but the homolytic bond strength ΔE between a methyl radical and a halogen atom follows the trend in polarity and becomes weaker from C–F to C–Cl, to C–Br to C–I. This reflects the decrease in polarity as the halogen becomes less electronegative. The latter shows up in a valence np$_\sigma$ SOMO that rises in energy and, from below, approaches the sp^3-type SOMO of the methyl radical (see Figure 4.10). The raise in halogen SOMO energy directly translates into less stabilization for the electron stemming from the higher-energy methyl SOMO which leads to the computed (and observed) weakening of the C–X bond.

On the other hand, if X in CH_3X varies along the first octal row, the deformation associated with ΔE_{prep} at a certain point becomes decisive for the trend in C–E bond

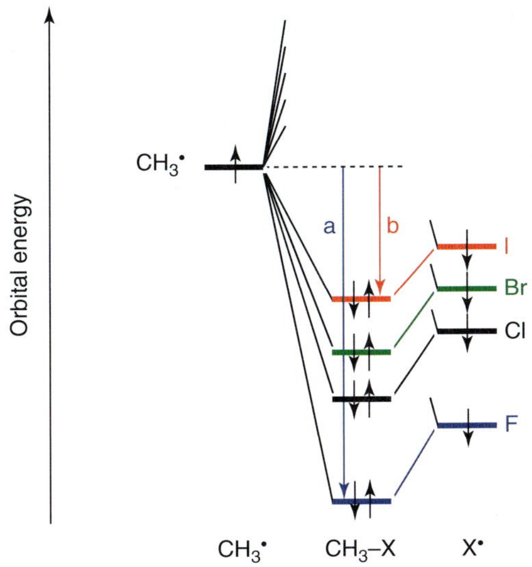

Figure 4.10 Schematic orbital interaction diagram for homolytic carbon–halogen bond formation in halomethanes. The bond strength weakens from CH_3F to CH_3I because the electron that drops from the methyl SOMO into the electron-pair bonding combination with the halogen np_σ SOMO gains less stabilization as the halogen becomes less electronegative from F (stabilization a) to I (stabilization b).

strength [31]. The C–E bond in CH_3X weakens if one goes from C–F to C–O to C–N (from $\Delta E = -114.4$ to -93.8 to -85.4 kcal mol^{-1}), following a similar pattern as for the series in carbon–halogen bonds that become weaker as the bond polarity decreases (cf. Figure 4.10). But then, from C–N to C–C, the C–E bond becomes stronger again (from -85.4 to -89.9 kcal mol^{-1}, Table 4.6). This variation in bond strength ΔE along the different C–E bonds emerges from a similar behavior in the instantaneous interaction ΔE_{int}. Note, however, that the strengthening from C–N to C–C bond is even more pronounced for ΔE_{int} than for ΔE but that it is counteracted, yet not dominated, by a striking increase in the deformation energy ΔE_{prep} from 9.4 to 18.4 kcal mol^{-1} (see Table 4.6).

This phenomenon is a consequence of the fact that the methyl group in ethane adopts a pyramidal geometry. This pyramidalization is associated with a relatively high deformation energy ΔE_{prep} of 18.4 kcal mol^{-1} as compared to 6.3–9.4 kcal mol^{-1} for the other C–E bonds in the series. The overall effect of this deformation is, however, favorable. In this way, Pauli repulsion between vicinal C–H bonds is avoided and, therefore, the C–C bond can contract and the instantaneous interaction ΔE_{int} increases. If one artificially suppresses the pyramidalization of one of the two methyl groups, the term ΔE_{prep} drops to 8.5 kcal mol^{-1} (see CH_3–[CH_3] in Table 4.6). However, the small gain in avoided deformation energy induces an enormous increase in Pauli repulsion at any C–C bond distance [31]. If the geometry of ethane is optimized with the constraint that one methyl group remains planar (CH_3–[CH_3]), the C–C bond becomes substantially longer (1.677 Å)

Table 4.6 Energy decomposition of the C–E bond in CH_3F, CH_3OH, CH_3NH_2, and CH_3CH_3 at ZORA-BLYP/TZ2P.[a]

	R(C–E)	ΔE	ΔE_{prep}	ΔE_{int}
CH_3–F	1.413	−114.4	6.3	−120.7
CH_3–OH	1.443	−93.8	8.1	−101.9
CH_3–NH_2	1.480	−85.4	9.4	−94.8
CH_3–CH_3	1.540	−89.9	18.4	−108.4
CH_3–[CH_3][b]	1.677	−63.0	8.5	−71.4

Energy values in kcal/mol. Bond lengths R in Å.
[a] Data are taken from Ref. [30].
[b] Optimized while keeping one methyl group fixed in a planar geometry.

and weaker ($\Delta E = -63.0\,\text{kcal mol}^{-1}$) than in the equilibrium geometry of ethane (1.540 Å; $\Delta E = -108.4\,\text{kcal mol}^{-1}$).

In the other C–E bonds, there is also Pauli repulsion with E–H bond orbitals as well as E lone pairs. But here, bending away the E–H bonds (which would be the equivalent to methyl pyramidalization) does not yield a reduction in Pauli repulsion. The reason is that the lone-pair orbitals at the central atom of NH_2, OH and also F do not involve any substituent atoms (such as the H atoms in the E–H bonds) that could be bent away. Note that the artificial CH_3–[CH_3] (with one planar methyl group) resembles the other CH_3X (with lone pairs that can notbend) more closely than true ethane. Consequently, along CH_3F, CH_3OH, CH_3NH_2 and $CH_3[CH_3]$, we recover again the trend of a monotonic weakening in ΔE and ΔE_{int} along the entire series that stems (among others) from the reduction in the electronegativity difference across the C–E bond as explained above for the carbon–halogen bonds (see Figure 4.10).

This finding on the trends in carbon–element bonding highlights how bond distance and bond strength are the result of an interplay of, on the one hand, the instantaneous interaction ΔE_{int} between molecular fragments that are connected by that bond and, on the other hand, the geometrical deformation and strain energy ΔE_{prep} that results from this interaction. This interplay and the role of geometrical strain becomes even more important and often dominates trends if one goes from local energy minima (molecular structure and bond strength) to saddle points (transition states and reaction barriers); [32] this is, however, beyond the scope of this chapter.

4.9 EDA-NOCV Analysis of Chemical Bonding in the Transition State

At present, the most common approach for investigating the interactions between two molecules that involve bond formation/bond breaking rests on the inspection of the energetically highest lying occupied and lowest lying vacant orbitals of the species. The frontier molecular orbital (FMO) theory was introduced into chemistry by Fukui [33] in the 1960s and it has become a cornerstone for understanding and predicting chemical reactions that involve covalent bond formation. The FMO model considers the energy levels and the spatial distribution of HOMO and LUMO of the reacting species as key factors for the reaction course [34]. The success of the FMO model is somewhat surprising considering the fact that the frontier orbitals of the separated fragments at their equilibrium geometries are considered whereas the critical step of the reaction is the TS which often invokes a substantial distortion of the interacting species. This issue is solved in the activation strain model of chemical reactivity in which the fragment approach is transported from stable molecules (local minima) to TSs (saddle points) and entire reaction profiles (non-stationary points) [32, 35]. Thus, the change in energy along the reaction coordinate is analyzed in terms of the two separate reactants as they approach from infinity and begin to interact and deform each other more and more on their way to the products. In the simplest form of this model, the activation energy $\Delta E^{\ddagger}$ associated with the TS is decomposed into the strain energy $\Delta E^{\ddagger}{}_{\text{strain}}$ and the interaction energy $\Delta E^{\ddagger}{}_{\text{int}}$:

$$\mathrm{D}E^{\ddagger} = \mathrm{D}E^{\ddagger}{}_{\text{strain}} + \mathrm{D}E^{\ddagger}{}_{\text{int}} \tag{4.5}$$

The activation strain $\Delta E^{\ddagger}{}_{\text{strain}}$ in Eq. (4.5) is the energy associated with deforming the reactants from their equilibrium geometry into the geometry they acquire in the activated complex. It can be divided into a contribution stemming from each of the reactants. The TS interaction $\Delta E^{\ddagger}{}_{\text{int}}$ is the actual interaction energy between the deformed reactants in the transition state, and can be further decomposed using either the regular EDA or EDA-NOCV. The EDA-NOCV method makes it possible to quantify the charge flow and the pairwise orbital interactions between two species even in the absence of any symmetry. This is why a particular value of the EDA-NOCV lies in the information that can be obtained about the bonding between molecular fragments that are connected by multiple bonds or in species that loose (some of) their symmetry elements along a geometrical deformation or reaction.

We present the results of EDA-NOCV calculations for two elementary steps of addition reactions that demonstrate the power of the method for gaining insight to the bond formation/bond breaking reaction, which goes far beyond the information given by the FMO model. We wish to emphasize that the EDA-NOCV method is not suggested as replacement for the FMO approach, but rather as a significant progress in gaining more detailed insight into chemical reactions.

The two reactions concern the addition of H_2 and CO_2 to the amido-substituted digermyne R_2N-GeGe-NR_2:

$$\mathrm{R_2N - GeGe - NR_2 + H_2 \rightarrow R_2N - GeGe(H)_2 - NR_2} \tag{4.6}$$

$$\mathrm{R_2N - GeGe - NR_2 + CO_2 \rightarrow R_2N - Ge(CO)_2Ge - NR_2} \tag{4.7}$$

We do not discuss the complete mechanisms of reactions 6 and 7 which involve several steps. Here we are only concerned with the transition states of the first step of the reactions. Details of the two reactions are given in the literature [36].

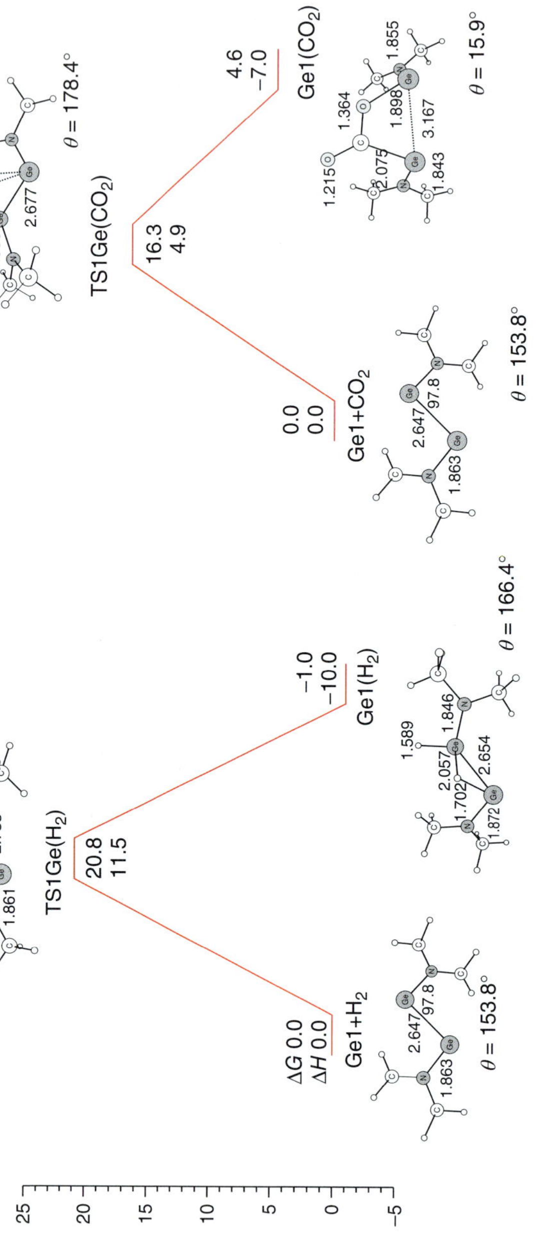

Figure 4.11 Calculated reaction profiles at BP86/TZVPP for the first steps of the addition of (a) H_2 to Me_2N-GeGe-NMe$_2$ and (b) CO_2 to Me_2N-GeGe-NMe$_2$.

Figure 4.11 shows the calculated reaction profiles at BP86/TZVPP of the first steps of the addition reactions 6 and 7 of **1Ge** where R = CH_3. The transition state for the hydrogenation reaction suggests that H_2 approaches only one Ge but one hydrogen atom is tilted toward the second Ge (Figure 4.11a). The reaction product of the first step of the reaction **1Ge(H$_2$)** has one hydrogen atom that is bridging the germanium atoms while the second H is bonded to one Ge. The first step of the CO_2 addition proceeds as insertion reaction yielding **1Ge(H$_2$)** where the Ge–Ge bond is broken (Figure 4.11b).

The question addressed concerns the dominant orbital interactions in the transition states **TS1Ge(H$_2$)** and **TS1Ge(CO$_2$)**. The FMO model suggests that the mixing of HOMO and LUMO of the reacting species should be the dominant contribution to ΔE_{orb}. Table 4.7 gives the numerical results of the EDA-NOCV calculations of the transition states where **1Ge** + H_2 and **1Ge** + CO_2 are taken as interacting fragments. It becomes obvious that, in the hydrogenation reaction, there are two major components $\Delta E_{orb}(1)$ and $\Delta E_{orb}(2)$ which contribute with $-37.0\,\text{kcal mol}^{-1}$ and $-35.1\,\text{kcal mol}^{-1}$ to the total orbital term ΔE_{orb}. In contrast, there is only one major component $\Delta E_{orb}(1)$ which provides the major contribution of $-32.8\,\text{kcal mol}^{-1}$ to the orbital interactions in the CO_2 addition.

Figure 4.12 graphically displays the most important orbital interactions in the transition state of the hydrogenation reaction 6. The left hand side (a) qualitatively sketches the FMO picture of the reaction. The middle part (b) shows the HOMO of H_2 and the LUMO of **1Ge** which refer to the H_2→**1Ge** donation and the HOMO of **1Ge** and the LUMO of H_2 which are related to the H_2←**1Ge** backdonation. Finally, the right hand side (c) shows the deformation densities $\Delta\rho(1)$ and $\Delta\rho(2)$ which are associated with the orbital components $\Delta E_{orb}(1)$ and $\Delta E_{orb}(2)$ which contribute

Table 4.7 EDA-NOCV calculations at BP86/TZVP+ of the transition states **TS1Ge(H$_2$)** and **TS1Ge(CO$_2$)** for the addition of H_2 and CO_2 to the amido-substituted digermynes Me_2N-GeGe-NMe_2 (see Figures 4.12–4.14).

Fragments	TS1Ge(H$_2$) Me$_2$N-GeGe-NMe$_2$ + H$_2$	TS1Ge(CO$_2$) Me$_2$N-GeGe-NMe$_2$ + CO$_2$
ΔE_{int}	−13.9	−11.2
ΔE_{Pauli}	131.1	69.1
ΔE_{elstat}[a]	−60.0 (41.4%)	−37.4 (46.6%)
ΔE_{orb}[a]	−84.9 (58.6%)	−42.9 (53.4%)
$\Delta E(1)_{A\rightarrow LGeGeL}$[b]	−37.0 (43.5%)[c]	−6.4 (14.9%)[d]
$\Delta E(2)_{A\leftarrow LGeGeL}$[b]	−35.1 (41.3%)[c]	−32.8 (76.5%)[d]
ΔE_{Rest}	−12.8 (15.2%)	−3.7 (8.6%)

Energy values in kcal/mol.
[a]The values in parentheses give the percentage contribution to the total attractive interactions $\Delta E_{elstat} + \Delta E_{orb.}$
[b]The values in parentheses give the percentage contribution to the orbital interactions $\Delta E_{orb.}$
[c]Fragment A = H_2
[d]Fragment A = CO_2

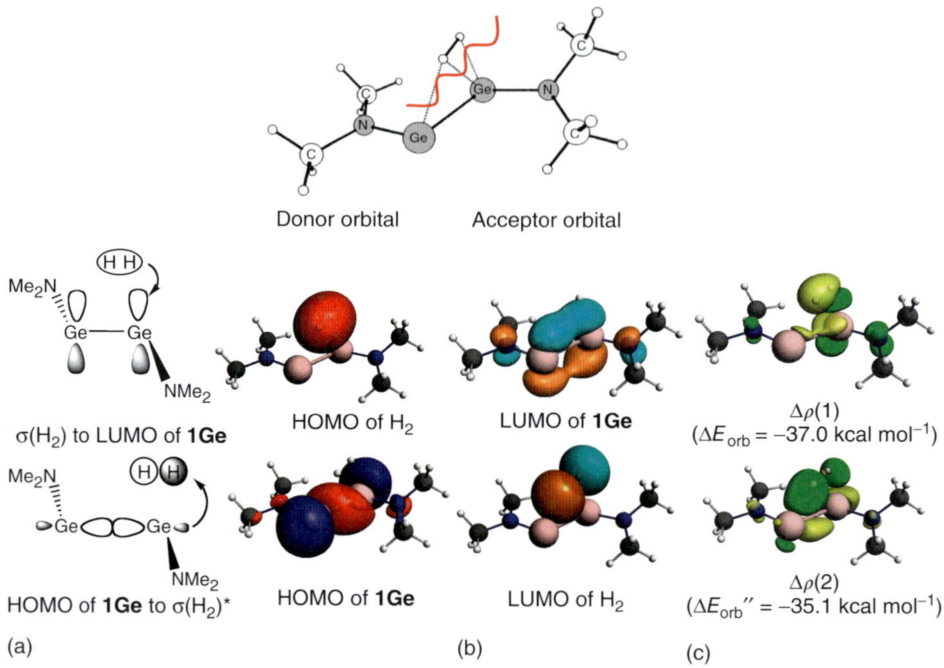

Figure 4.12 (a) Qualitative sketch of the orbital interactions for the addition reaction of H_2 to Me_2N-GeGe-NMe_2. (b) Frontier orbitals of the reacting species H_2 and Me_2N-GeGe-NMe_2. (c) EDA-NOCV results of the two most important pairs of interacting orbitals in the transition state. The colour coding for the orbitals indicates the sign of the MOs while the colour coding of the deformation densities $\Delta\rho(1)$ and $\Delta\rho(2)$ gives the charge flow from the yellow to the green areas.

with $-37.0\,\text{kcal mol}^{-1}$ and $-35.1\,\text{kcal mol}^{-1}$ to ΔE_{orb}. Note that the colour coding for the orbitals indicates the sign of the MOs whereas the colour coding of the deformation densities gives the charge flow from the yellow space to the green area. The shape of the charge flow suggested that the slightly larger component of $-37.0\,\text{kcal mol}^{-1}$ can be assigned to the $H_2\rightarrow$**1Ge** donation (top) whereas the second major contribution of $-35.1\,\text{kcal mol}^{-1}$ comes from the $H_2\leftarrow$**1Ge** backdonation. But it also becomes obvious that the donation and the backdonation involve not only the HOMO and LUMO of the reactants. Inspection of the NOCV pairs Ψ_k/Ψ_{-k} ($k=1, 2$), which are associated with the deformation densities $\Delta\rho(1)$ and $\Delta\rho(2)$ (Eq. (4.3), shows that there are indeed further orbitals of the reactants that become important for the orbital interaction in the transition state.

Figure 4.13a displays the NOCV pair Ψ_1/Ψ_{-1} which is connected to the deformation density $\Delta\rho(1)$. The occupied NOCV Ψ_1 has large contributions from the HOMO of H_2 but also from the HOMO-2 and HOMO-4 of **1Ge**. Also, the vacant NOCV Ψ_{-1} is a linear combination of the LUMO of H_2 with large contributions of the LUMO and LUMO+1 of **1Ge** that leaves the p(π) AO of one germanium atom and the 1s AO of only one hydrogen atom as dominant parts of the NOCV Ψ_{-1}.

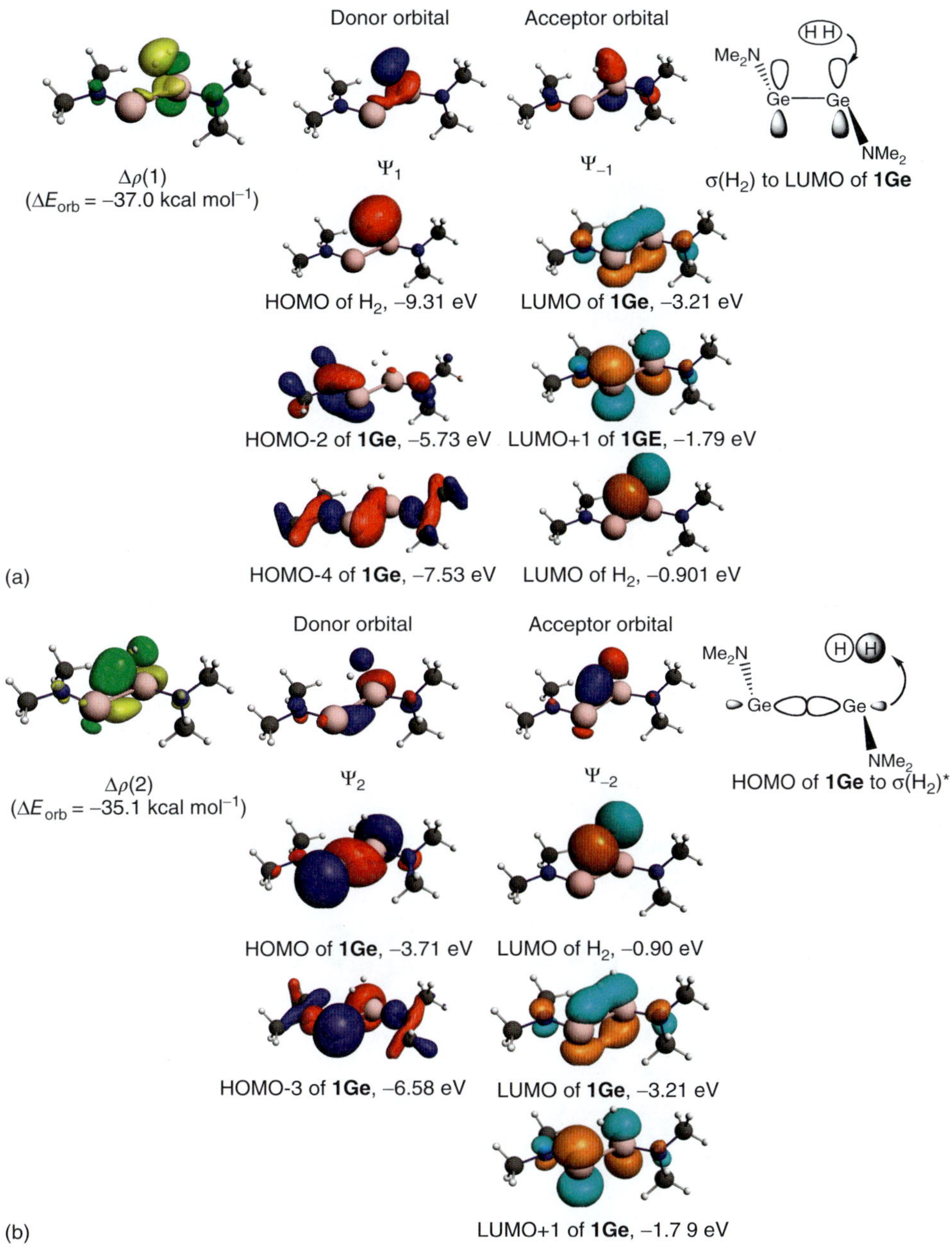

Figure 4.13 (a) Breakdown of the NOCV pair Ψ_1/Ψ_{-1} which is associated with the orbital interaction $\Delta E_{orb} = -37.0\,kcal\,mol^{-1}$ in the transition state **TS1Ge(H$_2$)** into the most important occupied and vacant orbitals of the reactants. (b) Breakdown of the NOCV pair Ψ_2/Ψ_{-2} which is associated with the orbital interaction $\Delta E_{orb} = -35.1\,kcal\,mol^{-1}$ in the transition state **TS1Ge(H$_2$)** into the most important occupied and vacant orbitals of the reactants.

The significant participation of occupied orbitals of **1Ge** to the NOCV Ψ_1 and of vacant orbitals of **1Ge** to the NOCV Ψ_{-1} comes from the large deformation of the electronic charge due to the geometrical changes from the equilibrium structure to the transition state. A similar situation is found for the NOCV pair Ψ_2/Ψ_{-2} (Figure 4.13b) which is connected to the deformation density $\Delta\rho(2)$. The occupied NOCV Ψ_2 has two large contributions from the HOMO and HOMO-3 of **1Ge** whereas the vacant NOCV Ψ_{-2} has three major components from the LUMO of H_2 and from the LUMO and LUMO+1 of **1Ge**.

The breakdown of the orbital interactions in the transition state of the CO_2 addition to **1Ge** into contributions of the donor and acceptor orbitals is shown in Figure 4.14. The EDA-NOCV results give a simpler interpretation of the MO terms than for the hydrogenation reaction. The single dominant deformation density $\Delta\rho(1)$ is easily associated with the $CO_2\leftarrow$**1Ge** backdonation which provides $-32.8\,\text{kcal mol}^{-1}$ to the orbital interactions. The comparison of the spatial distribution of $\Delta\rho(1)$ with the HOMO of **1Ge** and the LUMO of CO_2 makes it easy to identify the latter frontier orbitals as the most important MOs for the interactions in the transition state **TS1Ge(CO$_2$)**. We do not analyze the orbital interactions

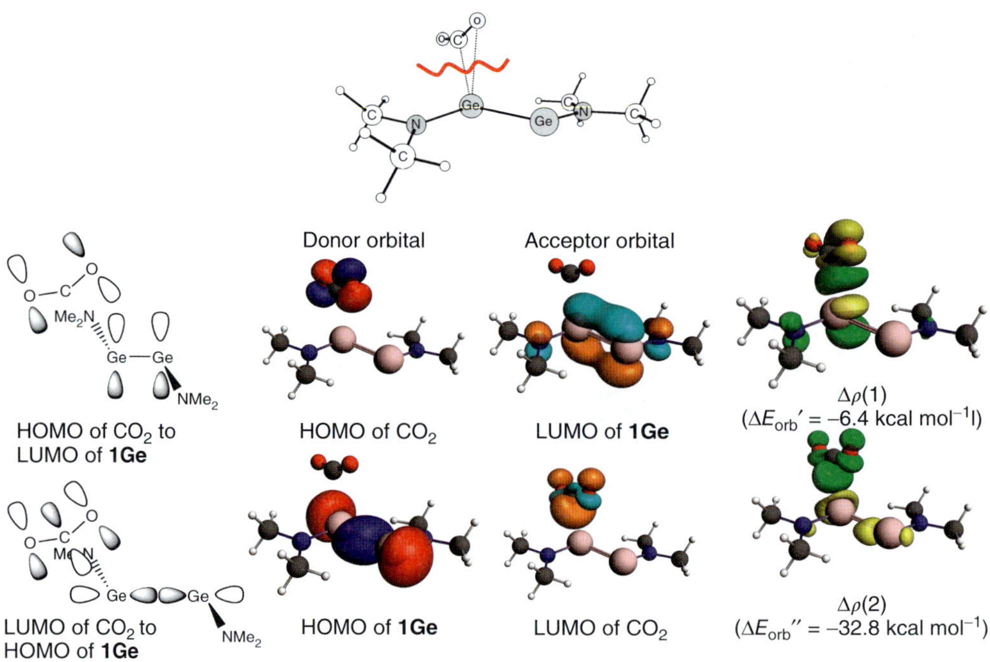

Figure 4.14 Frontier orbitals of the reacting species for the reaction of CO_2 to Me_2N-GeGe-NMe_2 and EDA-NOCV results of the two most important pairs of interacting orbitals in the transition state. The color coding for the orbitals indicates the sign of the MOs while the colour coding of the deformation densities $\Delta\rho(1)$ and $\Delta\rho(2)$ gives the charge flow from the yellow to the green areas.

that are associated with the next largest deformation density $\Delta\rho(2)$ that is shown in Figure 4.14. It comes from the $CO_2 \rightarrow$ **1Ge** donation, which is much weaker ($-6.4\ \text{kcal mol}^{-1}$) than the $CO_2 \leftarrow$ **1Ge** backdonation. The numerical data agree with the chemical intuition that CO_2 is a better electron acceptor than donor.

In summary, the EDA-NOCV method has the advantage over the standard EDA approach that the orbital term ΔE_{orb} can be broken down into pairwise contributions for the interactions between the reacting species even if the model system has no symmetry (C_1 point group symmetry). The associated energies give the strength of the orbital interactions. In agreement with the FMO theory, there is usually only a very small number of orbital pairs that make a large contribution to ΔE_{orb}. Unlike the FMO model, the EDA-NOCV method uses the actual orbitals of the reacting species for the description of the transition state. Of course, the EDA-NOCV can also be used for analyzing the chemical bond. Examples are discussed in the chapter "Donor–Acceptor Complexes of Main-Group Elements" in Vol. 2 of this book.

4.10 Summary and Conclusion

The results of the EDA calculations of archetypical types of chemical bonds which are discussed in this chapter demonstrate that it is possible to address the question about the nature of the chemical bond in terms of familiar concepts that can be identified and quantitatively expressed using accurate quantum chemical methods. The three energy terms ΔE_{elstat}, ΔE_{Pauli}, and ΔE_{orb} which are uniquely defined in the EDA method can be interpreted in a plausible way that connects heuristic bonding models with the physical mechanism of the chemical bonds. The EDA results show that orbital interactions are not always the crucial terms for determining the trend of the bond strength. The electrostatic attraction, which can be very large even in nonpolar bonds, and the repulsive forces, which come from the Pauli exclusion principle, can be very important for understanding the strength of the interatomic interactions. The analysis of the bonding in the diatomic molecules $Li_2 - F_2$ shows that the equilibrium distance of covalent bonds is determined by the increase of the Pauli repulsion but not by the maximum overlap of the valence orbitals. Another advantage of the EDA partitioning method is that the instantaneous interaction energy of a chemical bond is considered. The ΔE_{int} values can be very different from the BDEs because the preparation energy of the interacting fragments may be large. The advancement to the EDA-NOCV method makes it possible to estimate the charge flow and the associated energy contribution of the pairwise orbital interactions even in the absence of symmetry. This makes the EDA a very powerful method for analyzing interatomic interactions at equilibrium structures and in transition states. The EDA can be considered as a bridge between the classical heuristic bonding models of chemistry and the physical mechanism of chemical bond formation.

Acknowledgements

We thank the following organizations for financial support: the Deutsche Forschungsgemeinschaft (DFG), the Alexander von Humboldt Foundation (AVH), the National Research School Combination – Catalysis (NRSC-C), and the Netherlands Organization for Scientific Research (NWO-CW and NWO-EW).

References

1. Bickelhaupt, F.M., Solà, M., and Fonseca Guerra, C. (2007) *J. Comput. Chem.*, **28**, 238.
2. Heitler, W. and London, F. (1927) *Z. Phys.*, **44**, 455.
3. (a) A bonding model which considers *only* Pauli repulsion as the crucial factor for determining molecular geometries is the VSEPR (Valence Shell Electron Pair Repulsion) scheme: Gillespie, R.J. and Popelier, P.L.A. (2001) *Chemical Bonding and Molecular Geometry. From Lewis to Electron Densities*, Oxford University Press, New York. (b) For a criticism of the VSEPR model see: Frenking, G. (2003) *Angew. Chem. Int. Ed.*, **42**, 143. (c) Gillespie, R.J. (2003) *Angew. Chem. Int. Ed.*, **42**, 3331. (d) Frenking, G. (2003) *Angew. Chem. Int. Ed.*, **42**, 3335.
4. (a) Lewis, G.N. (1916) *J. Am. Chem. Soc.*, **38**, 762. (b) Lewis, G.N. (1923) *Valence and the Structure of Atoms and Molecules*, American Chemical Society Monograph Series, The Chemical Catalog Company, New York.
5. Kutzelnigg, W. (1990) The physical origin of the chemical bond, in *The Concept of the Chemical Bond* (ed. Z.B. Maksic), Springer, Berlin, Heidelberg.
6. Schmid M. W., Ivanic J., and Ruedenberg K., The physical origin of covalent bonding, *The Chemical Bond*, Vol. 1, Chapter 1, Wiley-VCH Verlag GmbH, Weinheim.
7. Mierzecki, R. (1985) *The Historical Development of Chemical Concepts*, Kluwer Academic Publishers, Dordrecht, p. 202.
8. Pauling, L. (1960) *The Nature of the Chemical Bond*, 3rd edn, Cornell University Press, Ithaca, NY.
9. Morokuma, K. (1971) *J. Chem. Phys.*, **55**, 1236.
10. Ziegler, T. and Rauk, A. (1977) *Theor. Chim. Acta*, **46**, 1.
11. Frenking, G. and Krapp, A. (2007) *J. Comput. Chem.*, **28**, 15.
12. (a) Lein, M. and Frenking, G. (2005) in *Theory and Applications of Computational Chemistry: The First 40 Years* (eds C.E. Dykstra, G. Frenking, K.S. Kim, and G.E. Scuseria), Elsevier, Amsterdam, p. 367. (b) Frenking, G., Wichmann, K., Fröhlich, N., Loschen, C., Lein, M., Frunzke, J., and Rayón, V.M. (2003) *Coord. Chem. Rev.*, **238-239**, 55. (c) te Velde, G., Bickelhaupt, F.M., Baerends, E.J., van Gisbergen, S.J.A., Fonseca Guerra, C., Snijders, J.G., and Ziegler, T. (2001) *J. Comput. Chem.*, **22**, 931. (d) Bickelhaupt, F.M. and Baerends, E.J. (2000) in *Reviews in Computational Chemistry*, vol. 15 (eds K.B. Lipkowitz and D.B. Boyd), John Wiley & Sons, Inc., New York, p. 1.
13. See the chapters (a) (Frenking, G., Chemical bonding in transition metal complexes. *The Chemical Bond*, Vol. 2, Chapter 7 Wiley-VCH Verlag GmbH, Weinheim; (b) Frenking, G. and Tonner, R., Donor-acceptor complexes of main-group elements. *The Chemical Bond*, Vol. 2, Chapter 4, Wiley-VCH Verlag GmbH, Weinheim; (c) Fernandez, I., Conjugation, hyperconjugation and aromaticity. *The Chemical Bond*, Vol. 2, Chapter 12, Wiley-VCH Verlag GmbH, Weinheim; (d) Frenking, G., Multiple bonding of heavy main-group elements. *The Chemical Bond*, Vol. 2, Chapter 2, Wiley-VCH Verlag GmbH, Weinheim.
14. Mitoraj, M. and Michalak, A. (2007) *J. Mol. Model.*, **13**, 347.

15. Mitoraj, M.P., Michalak, A., and Ziegler, T. (2009) *J. Chem. Theory Comput.*, **5**, 962.

16. It has first been suggested by Ruedenberg: Ruedenberg, K. (1962) *Rev. Mod. Phys.*, **34**, 326.

17. Bickelhaupt, F.M., Nibbering, N.M.M., van Wezenbeek, E.M., and Baerends, E.J. (1992) *J. Phys. Chem.*, **96**, 4864.

18. Bickelhaupt, F.M., Diefenbach, A., de Visser, S.P., de Koning, L.J., and Nibbering, N.M.M. (1998) *J. Phys. Chem. A*, **102**, 9549.

19. Esterhuysen, C. and Frenking, G. (2004) *Theor. Chem. Acc.*, **111**, 381; Erratum (2005) *Theoret. Chem. Acc.*, **113**, 294.

20. Hirshfeld, F.L. and Rzotkiewicz, F.L.S. (1974) *Mol. Phys.*, **27**, 1319.

21. Spackman, M.A. and Maslen, E.N. (1986) *J. Phys. Chem.*, **90**, 2020.

22. (a) Krapp, A., Bickelhaupt, F.M., and Frenking, G. (2006) *Chem. Eur. J.*, **12**, 9196. (b) See also: Bickelhaupt, F.M., DeKock, R.L., and Baerends, E.J. (2002) *J. Am. Chem. Soc.*, **124**, 1500.

23. (a) The polarity of a bond does not always correlate with the overall polarity of the electronic structure and the dipole moment of the molecule. The C-O σ and π bonds of CO are clearly polarized towards the oxygen end but the dipole moment is surprisingly small (0.11 D) and it has its negative end at the carbon atom: Muenter, J.S. (1975) *J. Mol. Spectrosc.*, **55**, 490. (b) The bonding situation in CO and the correlation with the dipole moments has been discussed in: Frenking, G., Loschen, C., Krapp, A., Fau, S., and Strauss, S.H. (2007) *J. Comput. Chem.*, **28**, 117. (c) Bickelhaupt, F.M., Nagle, J.K., and Klemm, W.L. (2008) *J. Phys. Chem. A*, **112**, 2437.

24. For a thorough discussion of the molecular orbital of diatomic molecules $Li_2 - F_2$ see: Albright, T.A., Burdett, J.K., and Whangbo, M.H. (2013) *Orbital Interactions in Chemistry*, 2nd edition John Wiley & Sons, Inc., New York.

25. Huber, K.P. and Herzberg, G. (1979) *Molecular Spectra and Molecular Structure IV. Constants of Diatomic Molecules*, Van Nostrand-Reinhold, New York.

26. For a further explanation see: Herzberg, G. (1989) *Molecular Spectra and Molecular Structure. I. Spectra of Diatomic Molecules*, 2nd edn, Reprint of the, Krieger Publishing, Malabar, FL, p. 25 and p. 318.

27. Kutzelnigg, W. (1990) in *The Concept of the Chemical Bond*, vol. 1 (ed. Z.B. Maksic), Springer, Berlin/Heidelberg, p. 1.

28. Dupuis, M. and Liu, B. (1978) *J. Chem. Phys.*, **68**, 2902.

29. For very heavy elements holds that a chemical bond between heavier atoms may be stronger than between lighter homologues because of relativistic effects. For relativistic effects on chemical bonding see Schwerdtfeger P., Relativity and chemical bonding. *The Chemical Bond*, Vol. 1, Chapter 11, Wiley-VCH Verlag GmbH, Weinheim.

30. (a) Bickelhaupt, F.M., Hermann, H.L., and Boche, G. (2006) *Angew. Chem. Int. Ed.*, **45**, 823. (b) Deng, L., Branchadell, V., and Ziegler, T. (1994) *J. Am. Chem. Soc.*, **116**, 10645. See also (c) Bickelhaupt, F.M., Ziegler, T., and Schleyer, P.v.R. (1996) *Organometallics*, **15**, 1477.

31. van Zeist, W.-J. and Bickelhaupt, F.M. (2009) *Phys. Chem. Chem. Phys.*, **11**, 10317.

32. van Zeist, W.-J. and Bickelhaupt, F.M. (2010) *Org. Biomol. Chem.*, **8**, 3118.

33. (a) Fukui, K. (1971) *Acc. Chem. Res.*, **4**, 57. (b) Fukui, K. (1975) *Theory of Orientation and Stereoselection*, Springer-Verlag, Berlin.

34. Fleming, I. (2010) *Molecular Orbitals and Organic Chemical Reactions*, John Wiley & Sons, Inc., Hoboken, NJ.

35. Bickelhaupt, F.M. (1999) *J. Comput. Chem.*, **20**, 114.

36. (a) H_2 addition: Li, J., Schenk, C., Goedecke, C., Frenking, G., and Jones, C. (2011) *J. Am. Chem. Soc.*, **133**, 18622. (b) CO_2 addition: Li, J., Hermann, M., Frenking, G., and Jones, C. (2012) *Angew. Chem. Int. Ed.*, **51**, 8611.

5
The Valence Bond Perspective of the Chemical Bond

Sason Shaik, David Danovich, Wei Wu, and Philippe C. Hiberty

5.1
Introduction

One of the fundamental paradigms of chemistry is the chemical bond, *the building block from which an entire chemical universe can be constructed* [1]. This theoretical notion serves as the means of understanding the apparent magic of chemistry: one molecule disappears and is replaced by a new one. The magic is accordingly represented in terms of molecules changing into one another by breaking some bonds and making new ones. As such, the bond is the basis of the grand scheme of "*Lego*" by which practicing chemists devise and control the formation of new molecules of ever-increasing complexity and beauty.

With the birth of quantum mechanics, the chemical bond was described by both valence bond (VB) and molecular orbital (MO) theories. This was done in a manner that seemed entirely different and at times with bitter rivalry, which in the long run has caused more damage than good [2]. Our chapter on the "*Bridging Cultures*" (Chapter 2 in Vol. 1) shows that the VB and MO descriptions are mutually transformable, and hence are *not* different. There are quite a few other chapters in the monograph that describe the MO perspective of bonding. This chapter recounts the VB perspective of the bond paradigm and focuses on the most ubiquitous bonding form, the two-electron bond.

The electron-pair bond has been traditionally described in terms of two bond families, the covalent (inclusive of polar covalent) and ionic bonds. We shall describe these traditional classes, but as we shall show, the articulation of the bond paradigm requires that alongside the covalent and ionic bonds, there should exist a third and a distinct class of bonding, the so-called charge-shift (CS) bonding [3], which has typical characteristics and some unique experimental signatures. We shall describe these three classes in this chapter. Furthermore, we shall form bridges to other theories by demonstrating that CS bonding can be articulated also from theories based on electron density consideration, specifically the electron localization function (ELF) [4] and atoms in molecules (AIM) [5] theories, which are also reviewed in this book (Chapters 10 and 8). There are of course other bond types, such as the odd-electron and hypervalent bonds, which also belong to the

The Chemical Bond: Fundamental Aspects of Chemical Bonding, First Edition.
Edited by Gernot Frenking, Sason Shaik.

CS-bonding family. However, our choice is to focus here on electron-pair bonding, while these other bond types are mentioned in passing.

This chapter introduces initially the Pauling application of VB theory to derive the classical bond families, and discusses the major deficiency of his scheme. The second part presents the modern VB theory [6] and derives the three bond families. Because this is not a chapter on the techniques of VB theory, we have added a short appendix, which outlines the key elements of the modern VB theory. More details can be found in review sources [6, 7]. In the third part of the chapter, we create bridges to ELF and AIM theories and show how CS bonding emerges also from these theories. Then, we discuss the physical origins of CS bonding. And, subsequently, we describe experimental manifestations of this new bond family. Finally, we try to trace the potential territory of CS bonding. We hope that the experimental community will find novel experimental articulation of the theory of CS bonding.

5.2
A Brief Historical Recounting of the Development of the Chemical Bond Notion

The concept of the electron-pair bond was formulated in a stroke of genius by Gilbert Newton Lewis (Figure 5.1) in his famous 1916 JACS article, "*The Atom and the Molecule*" [8]. Lewis was seeking an understanding of the behavior of strong and weak electrolytes in solution [9]. And as with the Biblical King, Saul, who meant to seek for his father's donkeys and instead found a kingdom, so did Lewis find the concept of the electron-pair bond as an intrinsic property that stretches between the covalent and ionic situations. This work has eventually had its greatest impact in chemistry through the work of Langmuir [10], who very ably articulated the Lewis concept, coining new and catchy terms [11]. Another important influence of

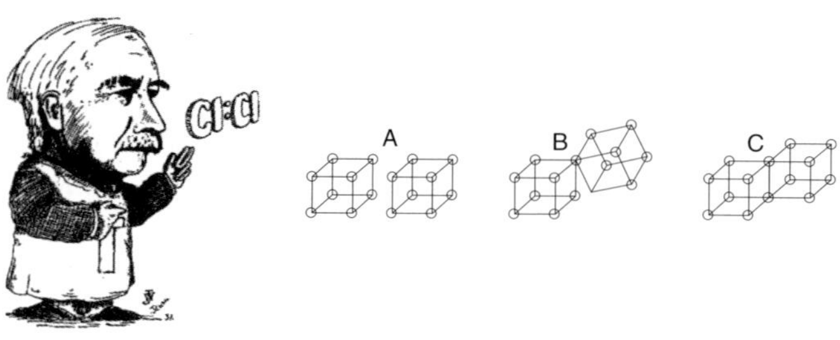

(a) (b)

Figure 5.1 The development of the electron-pair bonding description by Lewis in his seminal paper [8]. (a) A Lewis caricature with his electron-pair bonding cartoon appearing in the second half of the paper. (Reproduced with permission of W.B. Jensen.) (b) The description of the electron-pair bond in terms of atoms depicted as boxes with octet (C). Note that the bond is a dynamic entity, stretching between covalent (C) and ionic forms (A). (Part (b) is reproduced with permission from Ref. [8]. Copyright 2011 American Chemical Society.)

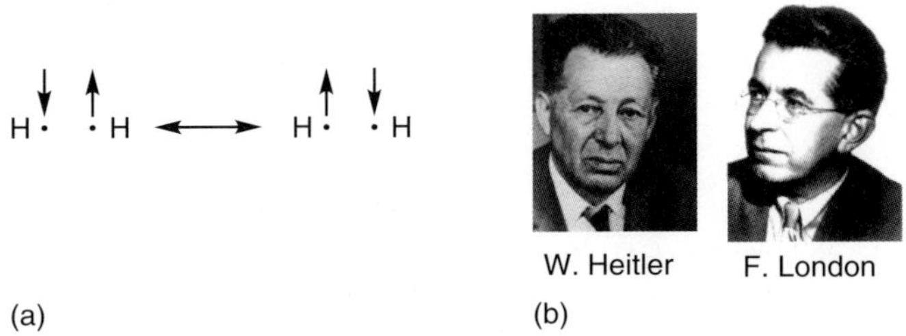

Figure 5.2 (a) A pictorial representation of the Heitler–London wave function. (b) Photos of Heitler and London.

the Lewis ideas was on the understanding of reaction mechanisms as established by the English School of Lapworth, Robinson, Lowry, and Ingold [12]. This has eventually led to the development of the science of physical organic chemistry [13], which rules mechanistic thinking to this day.

The overwhelming chemical evidence that supported the Lewis model presented an exciting agenda for research directed toward understanding the mechanism whereby the electron pair could constitute a bond. This, however, remained mysterious until 1927, when Walter Heitler and Fritz London went to Zürich to work with Schrödinger. In the summer of the same year, they published their seminal paper [14], which calculates the bonding energy in H_2. Using modern terminology, the bond energy in H_2 was shown to arise from the resonance interaction between the two spin arrangement patterns, $H^{\uparrow}H^{\downarrow}$ and $H^{\downarrow}H^{\uparrow}$, required to generate a singlet electron pair (Figure 5.2). In the winter of 1928, London [15] drew the basic principles of the nonionic bond, and his theory was, in essence, a quantum mechanical articulation of Lewis' covalent bond.

The Heitler–London papers mark the birth of the VB theory. Lewis is not cited in these papers; this omission may simply reflect the rift between the two communities. Nevertheless, Pauling was a chemist who was a resident in the two communities, and he wrote to Lewis [16] in 1928 about the exciting verification of his ideas by Heitler and London. This recognition propelled Pauling to develop a quantum mechanical version of the Lewis model. This quantum mechanical articulation of Lewis's shared-pair model has culminated in a generalizing intellectual construct [17], which described the electron-pair bond A–X as a superposition of covalent (Φ_{cov}) and ionic forms, $\Phi_{A^+X^-}$ and $\Phi_{A^-X^+}$ (Figure 5.3), and thereby enabled a unified description of bonding in any molecule, in terms of the VB theory. Around the same time, Slater and van Vleck [18] showed that an MO treatment followed by a complete configuration interaction is equivalent to the VB-based covalent-ionic scheme of Pauling. In Pauling's work, the long sought for *Allgemeine Chemie* of Ostwald [19], the father of physical chemistry, was finally achieved.

In retrospect, reading Lewis' paper shows that he anticipated the ideas that underlie the physical organic chemistry school [11, 12b, 16] of Ingold, and the resonance concept [12b, 20] expounded by Pauling. Indeed, Pauling himself dedicated his monograph [17] to Lewis, thereby expressing the link and influence of Lewis' work on his own. As noted by Hager [21], Pauling's biographer, Pauling

$$\mathrm{A\cdot \!-\!\!-\! \cdot X \longleftrightarrow A^{+}\ :X^{-} \longleftrightarrow A:^{-}\ X^{+}}$$

(a)

L. Pauling

(b)

Figure 5.3 (a) A schematic representation of Pauling's covalent-ionic superposition scheme. (b) Pauling's photo.

discovered Lewis' 1916 paper [8] by reading Langmuir's [10] 1919 paper. Until reading these two papers in 1920, Pauling had been teaching a chemistry course at Oregon Agricultural College in which he used the image of a chemical bond as consisting of hooks and eyes, for example, with the sodium atom having an eye and chorine having a hook. Thus, the birth of the VB theory in chemistry was an ingenious quantum chemical dressing of the Lewis seminal idea by Pauling, and can be referred to as the *Pauling–Lewis VB theory*.

5.3 The Pauling–Lewis VB Perspective of the Electron-Pair Bond

Figure 5.4 describes the key elements of the Pauling–Lewis perspective of electron-pair bonding in terms of VB mixing diagrams [3g]. Three structures, one covalent, Φ_{cov}, and two ionic ones, $\Phi_{A^+X^-}$ and $\Phi_{A^-X^+}$, describe any A–X bond, which may either be homo or heteronuclear.

The covalent structure (Figure 5.4a) is stabilized by spin pairing due to the resonance of the $A^{\uparrow}X^{\downarrow}$ and $A^{\downarrow}X^{\uparrow}$ spin arrangement forms, and contributes the covalent bond energy due to spin pairing, denoted as D_{cov}. For a dominantly covalent bond, where Φ_{cov} is the lowest VB structure (Figure 5.4b), this stabilization energy D_{cov} is the covalent contribution to the total bonding energy. On the other hand, ionic structures are stabilized by electrostatic interactions, relative to the separated atoms by an amount D_{ion}. When an ionic form, for example, $\Phi_{A^+X^-}$, is the lowest among the VB structures (Figure 5.4c), the bond is ionic and the electrostatic stabilization energy is the ionic contribution to the bond energy. The covalent-ionic mixing results in a resonance energy contribution that augments, in principle, the bonding of either covalent or ionic bonds. In the original literature, we referred to this quantity as the *charge-shift resonance energy*, RE_{CS} [3], because the pair density inherent in the VB wave function shows that *covalent-ionic mixing is associated with fluctuation of the electron pair from the average electron population.* As we shall see later, the RE_{CS} quantity figures prominently in the CS-bonding motif.

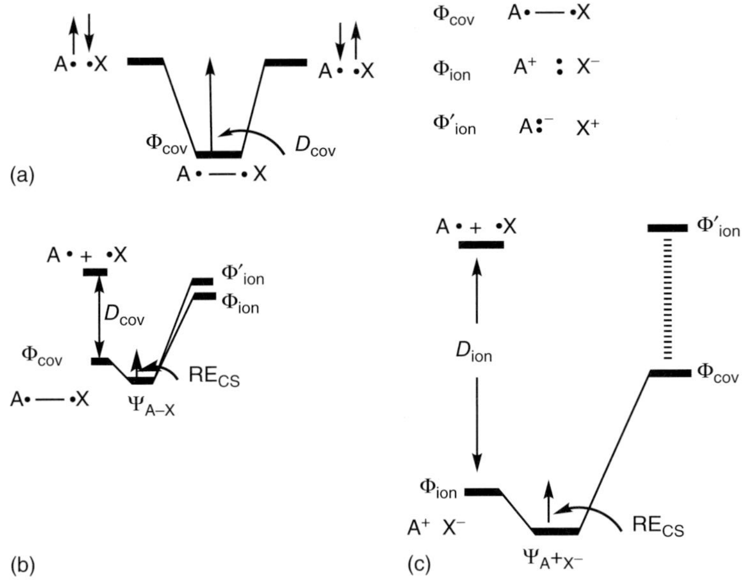

Figure 5.4 VB mixing diagrams for an A–X bond: (a) The stabilization of the covalent structure by D_{cov} due to resonance between the two spin arrangement patterns. (b) The covalent-ionic mixing for a polar-covalent A–X bond. (c) The covalent-ionic mixing for an ionic $A^+ :X^-$ bond. D_{ion} is the net electrostatic stabilization of the ionic structure relative to the separate atoms. In both (b) and (c), RE_{CS} is the covalent-ionic resonance energy, so-called in this chapter, the *charge-shift resonance energy*.

To use the scheme to calculate bond energies, Pauling assumed that for homonuclear bonds $RE_{CS} = 0$.[1] This assumption enabled him to estimate D_{cov} as the average of the bond energies of the two homonuclear bonds, A–A and X–X. For example, the geometric average as in Eq. (5.1):

$$D_{cov} = (D_{AA} \cdot D_{XX})^{1/2} \tag{5.1}$$

Using Eq. (5.1), the remaining contribution to the actual bond energy, D_e, was considered to be the resonance energy due to covalent-ionic mixing, and this value was used to gauge the electronegativity scale, as shown in Eq. (5.2),

$$D_e - D_{cov}(\text{kcal mol}^{-1}) = 23(\chi_x - \chi_A)^2 \tag{5.2}$$

where χ is the electronegativity. Furthermore, once the electronegativity is known, the *bond polarity* (δ) can be quantified as in Eq. (5.3), thereby providing the extent

1) The use of RE_{CS} (A–A) ≈ 0 appears as a working assumption, for example, on pages 73–100 (see also footnote 13 on page 73), in Ref. [17] where it is estimated that the ionic structures in, for example, Cl_2 will contribute <2% to the total bond energy. A stricter assumption is used in Sanderson's treatment [22] later, which neglects the resonance energy altogether.

of ionic $A^+ X^-$ character in the bond:

$$\delta = 1 - \exp[-0.25(\chi_X - \chi_A)^2] \quad (5.3)$$

Despite the successful performance of the Pauling construct in Figure 5.4 and Eqs. (5.1)–(5.3), this construct has a major problem that originates in the assumption that $RE_{CS} = 0$ for the homonuclear bond. This assumption undermines the role of RE_{CS} and overestimates the magnitude of D_{cov}. As shown later, this assumption is incorrect, and its implementation leads to a loss of an entire bond family, the so-called *CS bonds*.

As such, in practice, the Pauling covalent-ionic superposition scheme has traditionally become associated with two bond families, based on a criterion of static charge distribution; these are the covalent bond and ionic bond families in Scheme 5.1. In the first family, the major contribution to bonding comes from spin pairing. And let us recall that in homopolar bonds the RE_{CS} contribution was assumed – in Pauling's original scheme[1] and in subsequent treatments based on it – to be very small and *was set to zero*. In heteropolar bonds, the primary contribution to bonding is normally considered to be the D_{cov} quantity,[1] while the RE_{CS} is of secondary importance, except for very polar bonds involving the very electronegative atoms. Furthermore, as shown by Eq. (5.2), the magnitude of RE_{CS} is considered to vary in proportion to the electronegativity difference of the fragments, A and X, much like the charge distribution that is, the "bond polarity" in Eq. (5.3).

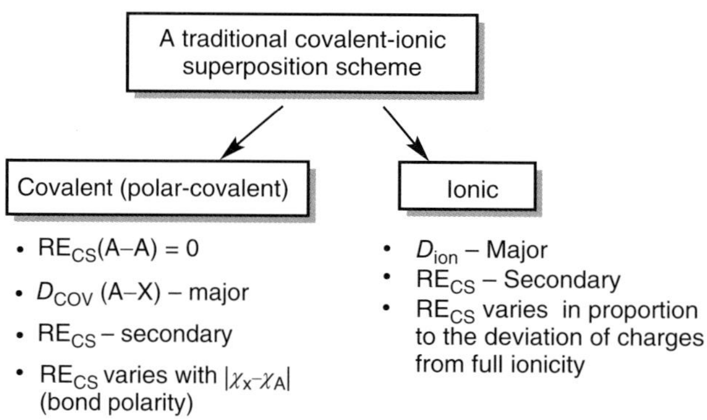

Scheme 5.1 The traditional covalent and ionic bond families based on Pauling's covalent-ionic superposition scheme. (Reproduced from Ref. [3g] with permission of Wiley-VCH.)

In the second family, the major bonding contribution comes from the electrostatic energy in the dominant ionic structure, while RE_{CS} is a minor factor; its magnitude is supposed to vary in proportion to the deviation of the charge distribution from full ionicity.

As such, in the traditional classification of both bonding types, it is assumed that one can deduce the magnitude of the covalent-ionic resonance energy *by simply inspecting the static charge distribution of the molecule.*

Using MO theory, it is possible to transform the delocalized canonical MOs to a set of localized molecular orbitals (LMOs) that describe two-center bonds [23] (see also the chapter 3 on NBO). The LMOs retrieve the covalent-ionic superposition scheme as follows: the electron-pair bond is the LMO itself, while the covalent-ionic superposition can be quantified from the charge polarization of the LMO, namely, the relative size of the coefficients of the hybrids, of the contributing fragments, to the LMO determines the bond polarity. Accordingly, MO theory leads, in principle, to the same electron-pair bonding picture as the classical covalent-ionic paradigm of Pauling. In fact, both VB and MO descriptions support the Lewis formulation of electron-pair bonding.

Thus, our bonding paradigm is now 98 years old [8], and yet even a cursory search in the literature suggests that this is perhaps not the whole story. Just consider the bonds of silicon to electronegative atoms. By the criterion of the static charge distribution, these bonds are virtually as ionic as, for example, LiF or NaCl (e.g., $H_3Si^{+0.85}F^{-0.85}$ vs $Li^{+0.94}F^{-0.94}$, $Na^{+0.91}Cl^{-0.91}$, etc.) [24]. However, while Li^+F^- and Na^+Cl^- behave as genuine ionic bonds, the Si^+X^- bonds behave chemically as covalent bonds [25, 26]. The bonds look so similar, yet they are so very different in their chemical behavior. Indeed, all Si–X bonds are more ionic than the corresponding C–X bonds [25a], according to static charge distribution, and, nevertheless, these are the C–X bonds that exhibit ionic chemistry in condensed phases, whereas the ionic Si–X chemistry is extremely rare, with a handful of exceptions [25, 26, 26a, 27]. For example, trityl perchlorate is an ionic solid, $Ph_3C^+ClO_4^-$, similar to NaCl [25c], while the silicon analog, is a covalent solid, with a short Si–O bond [25d]. It is apparent, therefore, that the static charge distribution is not a reliable indicator of the nature of bonding. *There must be an additional property of the bond that is missing in the traditional covalent-ionic superposition scheme.*

5.4
A Preamble to the Modern VB Perspective of the Electron-Pair Bond

These and many similar puzzles prompted us in 1990 to reexamine the classical covalent-ionic paradigm using the tools of modern VB theory [3]. The reader interested in modern VB theory can consult Appendix 5.A and a recent review [7].

Our first intriguing findings [3a,b] concerned the F–F bond, which has been intensely debated, and which requires quite a high level of configuration interaction to yield a quantitative bond energy [28].[2] By any known measure, the F–F bond would be defined as a "covalent bond." First, it is a homonuclear bond, where

2) As seen from Table 10.2 of Ref. [2], CASSCF/6-31G* or GVB/6-31G* leads to BDE values of ~16.0 kcal mol^{-1}, compared with the experimental value of 38.3 kcal mol^{-1}. Using here MRCI/cc-pVTZ the value is 35.9 kcal mol^{-1}.

ionicity should not matter. Second, the weight of its covalent structure is as large as that for the H–H bond [3a,b,g]. Is the F–F bond really covalent as the H–H bond?

Figure 5.5 displays the dissociation energy curves of a few bonds showing the dominant VB structure of the bond alongside the "exact" ground state, which is a resonating combination of the covalent and ionic components, calculated by means of the modern VB theory. Inspection of Figure 5.5a,b makes it apparent that the bonding natures of the H–H and F–F bonds are very different. While in H_2 the covalent VB structure displays by itself a potential well, which is already a good approximation of the exact curve (Figure 5.5a), the covalent component of F_2 is, on the contrary, purely *repulsive* (Figure 5.5b), and *the bonding is in fact sustained by the very large* RE_{CS} due to the mixing of the ionic structures (not shown in the figure) into the repulsive covalent structure. Thus, although F–F may be formally a covalent bond according to its zero static charge distribution, this definition cannot tag its true nature; the F–F bond is in fact a CS bond, because *the bonding exists as a result of the ionic-covalent fluctuation of the electron-pair density*. This F–F case shows that the assumption underlying the classical Pauling scheme (Scheme 5.1) is incorrect; the covalent bonding by itself is not necessarily stabilizing even for homopolar bonds, and even in cases where the covalent structure clearly dominates the wave function, as in F–F. Importantly, *homopolar bonds can have very large* RE_{CS}. Moreover, the result of a repulsive covalent structure shows that these structures can be repulsive when the covalent spin-pairing energy is frustrated by Pauli repulsion. We shall elaborate this point later when we discuss the physical origins of CS bonding, and we shall see that once the assumption in the Pauling scheme is removed this enables recharting of the mental map of the chemical bond.

However, CS bonding is not restricted to F–F, as can be seen by comparing Figure 5.5d to c. It is clear that the B–H bond is classically covalent, while the F–H bond is not; its covalent structure is weakly bonded, while the majority of bonding energy in the exact VB wave function arises from the RE_{CS} due to the mixing of the ionic structures. Thus, the repulsion here weakens the covalent bonding very much, but not to the extent found in Figure 5.5b for F–F.

Finally, Figure 5.5e,f shows the NaF and NaCl bonds. It is clear that for both bonds, the dominant VB structure is ionic, and it is very close to the exact covalent-ionic superposition curve, with a negligible RE_{CS} contribution. These two bonds are classically ionic.

As such, Figure 5.5 reveals very clearly the two classical bond families, covalent and ionic, but alongside them it reveals also the presence of another bond type *wherein the bonding does not arise from any one of the structures but rather from the resonance interaction,* RE_{CS}. This is the CS-bonding family. In recent years, a variety of σ- and π-bonds, both homo and heteronuclear, were shown to share this property, thereby forming a growing family of CS bonds [3], which we are going to focus on in this chapter.

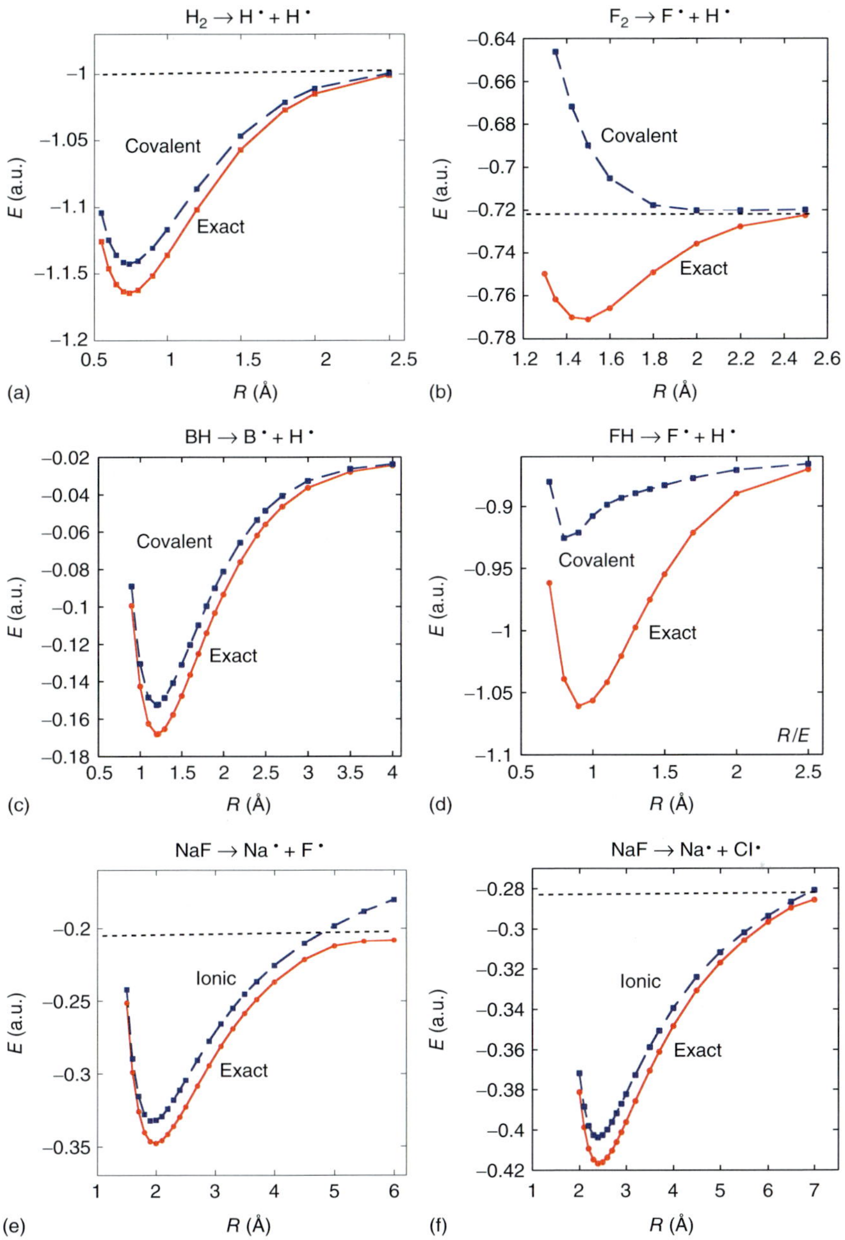

Figure 5.5 Dissociation energy curves for (a) H_2, (b) F_2, (c) BH, (d) FH, (e) NaF, and (f) NaCl. The blue lines, with squares, show the purely covalent VB structures in a–d, and the purely ionic VB structures in e,f. Red lines, with squares, show the optimized covalent + ionic "exact" ground state. (Reproduced from Ref. [3j] with permission of Nature Chemistry.)

5.5
Theoretical Characterization of Bond Types by VB and Other Methods

The emergence of the CS-bond family and its eventual acceptance would revise the electron-pair bond concept in chemistry. One prerequisite for such a revision is having alternative and independent theoretical derivations of the CS-bonding phenomenon, and, most importantly, drawing links to experimental data. We shall start by finding alternative theoretical probes for bonding and especially for the signatures of CS bonding.

5.5.1
VB Characterization of Bond Types

As discussed, the emergence of three bonding families, covalent, ionic, and CS bonds, was originally derived from modern VB calculations [3a,b]. Thus, the VB wave function of an A–X bond is computed as a combination of the covalent form Φ_{cov} (A•–•X), and the two ionic forms, Φ_{ion} (A^+X^-) and Φ'_{ion} (A^-X^+), shown in Eq. (5.4):

$$\Psi(\text{VB}) = c_1\Phi_{cov} + c_2\Phi_{ion} + c_3\Phi'_{ion} \tag{5.4}$$

Equation (5.4) is identical to the Pauling wave function (Figure 5.3), but all the terms are now computed by means of the modern VB theory (Appendix 5.A). The principal VB structure is the one having the lowest energy, and hence also the largest coefficient among the three structures in Eq. (5.4). Its contribution to the total bond dissociation energy (BDE, D_e) is referred to as D_{cov} or D_{ion}, wherein the subscript specifies the dominant VB structure. In all cases, the RE_{CS} is determined by reference to the bonding energy of the principal VB structure, as expressed by Eq. (5.5a) and (b):

$$D_e(\text{polar-covalent}) = D_{cov} + RE_{CS} \tag{5.5a}$$

$$D_e(\text{ionic}) = D_{ion} + RE_{CS} \tag{5.5b}$$

Note that Eq. (5.5a) covers both covalent and CS bonds, depending on the magnitude of RE_{CS} vis-à-vis the total bond energy. These quantities characterize the bonding type as could already be gleaned from Figure 5.5. Thus, in Figure 5.5a,c, the principal VB structure for both H–H and B–H is Φ_{cov}, while the RE_{CS} quantity is small and much less significant than the large D_{cov}. In accord, these bonds are classical and polar-covalent types, respectively. By contrast, F–H in Figure 5.5d displays a weakly bound principal structure Φ_{cov}, and the major contribution to the bond comes from RE_{CS}. An extreme case is the F–F bond, in Figure 5.5b, in which the principal structure Φ_{cov} is not even bonded, that is, D_{cov} is *negative*, while RE_{CS} is even larger than the total bonding energy. In accord, F–H and F–F are both CS bonds. Finally, in Na–F and Na–Cl (Figure 5.5e,f) the principal VB structure is now Φ_{ion}, and the RE_{CS} quantity is a minor contributor, making both classical ionic bonds, where most of the bonding energy arises from the ionic structure.

5.5.2
ELF and AIM Characterization of Bond Types

An alternative way to characterize bonding uses electron density theories, such as ELF [4] and AIM [5]. Will ELF and AIM show these three families? What will be the ELF and AIM signatures for CS bonds?

5.5.2.1 ELF Characterization of Bond Types

The ELF approach (described in detail in Chapter 10 in Vol. 1) uses a function related to the Pauli repulsion to carry out a partition of the molecular space into basins that correspond to the volumes occupied by core inner shells, bonds, and lone pairs. As in the Lewis model, a valence basin may either belong to a single atomic shell or be shared by several ones. In the first case, the basin is called *monosynaptic* and corresponds to a lone pair region, and in the second case it is polysynaptic, and specifically bisynaptic for a two-center bond that is of interest in this chapter.

A given basin can be characterized by a statistical analysis of the density. Thus, it is possible to calculate the basin population, $\overline{N}$, and its variance, σ^2, by integrating the one-electron and the pair density over the volumes of the corresponding basins. In the statistical theory of the basin populations, the variances measure the electron fluctuation in a given basin, and the covariances [3g, 29] are thought to gauge directly the covalent-ionic fluctuations in terms of weights of ionic structures. However, as the covariance values exhibit trends similar to those of the variance [3g], we shall focus only on the latter quantities.

For a classical covalent bond, the basin is disynaptic, its population is close to 2.0, and the variance (and covariance) is significantly smaller than the population, while a classical ionic bond such as NaCl has only core and monosynaptic basins [3g, 4, 30]. These characteristics are summarized in Scheme 5.2. Any bond with very different values, of the population and the corresponding fluctuation index,

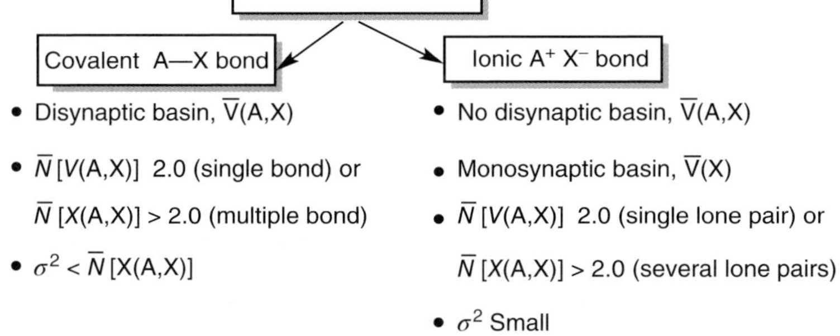

Scheme 5.2 Expected ELF properties for the ideal covalent and ionic bonds. (Reproduced from Ref. [3g] with permission of Wiley-VCH.)

will not qualify as either covalent or ionic. As will be shown, the CS bonds have typical and new ELF characteristics.

5.5.2.2 AIM Characterization of Bond Types

In AIM theory (see Chapter 8 in Vol. 1 for details), a bond is generally characterized by a bond path, which defines a maximum density path connecting the bonded atoms. The point of the path at which the density is at minimum is called the *bond critical point* (*BCP*), and the values of the density, $\rho(\mathbf{r}_c)$, and its Laplacian, $\nabla^2\rho(\mathbf{r}_c)$ at this point are characteristic of the interaction type in the bonding region. According to AIM, a classical covalent bond is typified by a significant $\rho(\mathbf{r}_c)$ value, and a large negative $\nabla^2\rho(\mathbf{r}_c)$. By contrast, closed-shell interactions, suffering from Pauli repulsions (also known as *overlap repulsion* or *exchange repulsion*), as in ionic bonds, or the He–He interaction, have characteristically a small critical density *and a positive Laplacian.*

The Laplacian is especially a telling quantity [5, 31], as it is connected to the kinetic and potential energy densities at BCP, $G(\mathbf{r}_c)$ and $V(\mathbf{r}_c)$, respectively, by the following local-virial theorem expression:

$$\frac{\hbar^2}{4m}\nabla^2\rho(\mathbf{r}_c) = 2G(\mathbf{r}_c) + V(\mathbf{r}_c) \tag{5.6}$$

Thus, a negative Laplacian means that the bonding region is dominated by lowering of the potential energy, while a positive Laplacian means that the interaction in the bonding region is typified by excess kinetic energy, and is hence repulsive. All the AIM parameters for bonds in a molecule can be either calculated or derived from experimental density determination, and are used by experimental chemists to characterize interactions within molecules (See Chapter 9 in Vol. 1) [32]. As such, we might expect AIM to reveal the presence of CS bonds.

5.6 Trends of Bond Types Revealed by VB, AIM and ELF

5.6.1 VB and AIM Converge

In order to provide a global picture of the various categories of bonds, we collected in Table 5.1 27 bonds [3] that are organized into three groups, labeled as (**I–III**). The first group involves homonuclear bonds starting from H–H all the way to the "inverted" C–C bond in [1.1.1]propellane (Scheme 5.3) [3i]. Groups (**II**) and (**III**) involve heteronuclear bonds, starting from C–H all the way to Si–F.

Each bond in the table is characterized by five VB properties; the weight of the principal VB structure (ω_{cov} or ω_{ion}), the bonding energy of that structure (D_{cov} or D_{ion}), the full BDE (D_e), the RE_{CS}, and the relative resonance energy (%RE_{CS}), which is the percentage ratio of RE_{CS} to D_e. For some of the bonds, we show AIM-derived quantities ρ and $\nabla^2\rho$ as well as the Laplacian components in the BCP

Table 5.1 A collection of bonds with their VB and AIM properties: group (**I**) corresponds to homonuclear covalent and CS bonds, (**II**) to heteronuclear covalent and CS bonds, and (**III**) to ionic bonds.

I.	**A–A**	ω_{cov}	D_{cov}	D_e	RE_{CS}	$\%RE_{CS}$	ρ	$\nabla^2\rho$	$\nabla^2\rho_{cov}$	$\nabla^2\rho_{res}$
1	H–H	0.76	95.8	105.0	9.2	8.8	0.27	−1.39	−0.70	−0.31
2	Li–Li	0.96	18.2	21.0	2.8	13.1	0.01	−0.01	−0.01	0.00
3	Na–Na	0.96	13.0	13.0	0.0	0.2	0.01	0.00	0.00	0.00
4	$H_3C–CH_3$	0.55	63.9	91.6	27.7	30.2	0.25	−0.62	−0.26	−0.36
5	$H_2N–NH_2$	0.62	22.8	66.6	43.8	65.7	0.29	−0.54	−0.02	−0.68
6	HO–OH	0.64	−7.1	49.8	56.9	114.3	0.26	−0.02	+0.46	−0.75
7	F–F	0.69	−28.4	33.8	62.2	183.9	0.25	+0.58	+1.00	−0.83
8	Cl–Cl	0.64	−9.4	39.3	48.7	124.1	0.14	+0.01	+0.14	−0.26
9	Br–Br[a]	0.71	−15.3	44.1	59.4	143.8	—	—	—	—
10	$C–C_i$(prop)[b]	0.62	−2.2	~70	72.2	>100	0.19[c]	+0.43[c]	—	—
11	C–C(prop)[b]	~0.55	—	—	—	—	0.25[c]	−0.51[c]	—	—
II.	**A–X**	ω_{cov}	D_{cov}	D_e	RE_{CS}	$\%RE_{CS}$	ρ	$\nabla^2\rho$	$\nabla^2\rho_{cov}$	$\nabla^2\rho_{res}$
12	$H_3C–H$[a]	0.69	90.2	105.7	15.1	14.3	—	—	—	—
13	$H_3Si–H$[a]	0.65	82.5	93.6	11.1	11.9	—	—	—	—
14	B–H	0.71	78.2	89.2	11.0	12.3	0.19	−0.61	−0.59	−0.04
15	Cl–H	0.70	57.1	92.0	34.9	37.9	0.26	−0.81	−0.33	−0.42
16	F–H	0.52	33.2	124.0	90.8	73.2	0.38	−2.52	−1.82	−0.52
17	$H_3C–F$[a]	0.45	28.3	99.2	70.9	71.5	—	—	—	—
18	$H_3C–Cl$[a]	0.62	34.0	79.9	45.9	57.4	—	—	—	—
19	$H_3Si–Cl$[a]	0.57	37.0	102.1	65.1	63.8	—	—	—	—
20	$H_3Ge–Cl$[a]	0.59	33.9	88.6	54.7	61.7	—	—	—	—
21	F–Cl[a]	0.59	−39.7	47.9	87.6	182.9	—	—	—	—
22	Cl–Br[a]	0.69	−9.2	40.0	49.2	123.0	—	—	—	—
III.	$A^+ X^-$	ω_{ion}	D_{ion}	D_e	RE_{CS}	$\%RE_{CS}$	ρ	$\nabla^2\rho$	$\nabla^2\rho_{ion}$	$\nabla^2\rho_{res}$
23	Li–F	0.76	93.3	104.5	11.2	10.7	0.07	+0.62	+0.51	−0.01
24	Na–F	0.72	77.0	86.0	9.0	10.4	0.05	+0.37	+0.27	+0.02
25	Li–Cl	0.56	76.8	88.5	11.7	13.3	0.04	+0.24	+0.16	0.00
26	Na–Cl	0.63	71.4	79.5	10.1	8.1	0.03	+0.18	+0.13	0.00
27	$H_3Si–F$[a]	0.36	103.8	140.4	36.6	26.1	—	—	—	—

[a] From Refs [3g, j]. All other data are from Ref. [3h] unless noted otherwise. Energies in kcal mol^{-1}, ρ in au (e. a_o^{-3}), $\nabla^2\rho$ in au (e. a_o^{-5}). The partial Laplacians are based on weights of the corresponding VB structures (see Eqs. 9–16 in Ref. [3h]).

[b] $C–C_i$ (prop) is the inverted bond in [1.1.1]propellane, the C–C(prop) is one of the wing bonds of the same molecule. The VB data are from Ref. [3h,i].

[c] Experimental data [33] for a substituted [1.1.1]propellane derivative. The values for the wing bonds are averaged. Energies in kcal mol^{-1}, ρ in au (e. a_o^{-3}), $\nabla^2\rho$ in au (e. a_o^{-5}).

for bonding due to the principal structure of the bond ($\nabla^2\rho_{cov}$ or $\nabla^2\rho_{ion}$) and the covalent-ionic resonance ($\nabla^2\rho_{res}$) [3h].

Let us first inspect the homonuclear bonds in (**I**), which by all definitions could not possess static bond ionicities. The bond energies in entries 1–4 are dominated by the covalent component, with RE_{CS} being the minor bonding contribution

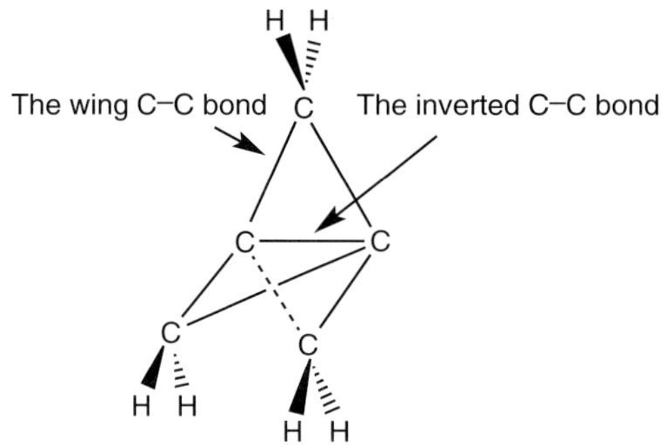

Scheme 5.3 The molecule [1.1.1]propellane and its two C–C bond types.

(%RE_{CS} < 50%). By contrast, the bonds in entries 6–10 all have a bonding energy dominated by RE_{CS} (%RE_{CS} > 100%), while the covalent structure is repulsive ($D_{cov} < 0$). The N–N bond, entry 5, is a borderline case, with %RE_{CS} accounting for 66.6% of the total bonding energy. Leaving aside the weak Na–Na and Li–Li bonds for which all AIM parameters are close to zero, there is an excellent correlation between the RE_{CS} quantities and the AIM parameters, especially within the same row of the periodic table. Thus, from C–C all the way to F–F (entries 4–7), the resonance component of the Laplacian ($\nabla^2\rho_{res}$) is more and more negative, in line with the increase in RE_{CS}, while the covalent component ($\nabla^2\rho_{cov}$) goes from negative to positive values, in line with the repulsive nature of the covalent structure in CS bonds. As a result, the total Laplacian $\nabla^2\rho$ is large and negative for classically covalent bonds and small or positive for CS bonds. Note that, according to the computed RE_{CS} and the experimentally-derived $\nabla^2\rho$ values [33], the [1.1.1]propellane molecule embodies the two categories of bonds, classically covalent for the wing bonds (entry 11) and CS bond for the "inverted" central bond (entry 10).

These relationships are illustrated more vividly in Figure 5.6, which plots the covalent part of the Laplacian against the covalent bond energies, D_{cov}, for homonuclear bonds [3h]. In the right lower quadrant, where $D_{cov} > 0$ and $\nabla^2\rho_{cov} < 0$, there are the bonds with stabilized covalent bonding. The second group, in the upper left quadrant, involves *electronegative and lone-pair-rich atoms and "inverted carbons"* that undergo CS bonding. It can be seen that this bonding type is associated with weakened covalent spin pairing ($D_{cov} < 0$), owing to lone-pair repulsion, which raises the kinetic energy as seen from the positive sign of $\nabla^2\rho_{cov}$. Clearly, the AIM and VB theories converge; both revealing the presence of a CS-bonding family.

Turning to heteropolar bonds in **(II)** in Table 5.1, we note the following trends. While the covalent VB structure is the principal one for all these bonds, the bonds still fall into two distinct groups. Specifically, the bonds in entries 12–15 belong to the classical polar-covalent bond family based on their %RE_{CS} that is well below 50%. By contrast, the bonds in entries 16–22 all have weakly bonded covalent structures, and large %RE_{CS} exceeding 50% and in some cases >100%.

In part **(III)** of the table, the principal VB structure of all bonds is ionic. The bonding energies in entries 23–26 are all dominated by the electrostatic

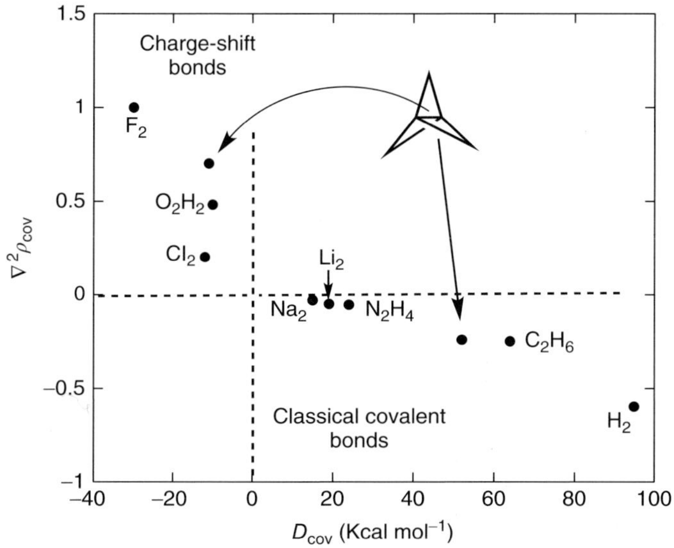

Figure 5.6 Correlation of the covalent Laplacian with the value of D_{cov} for homonuclear bonds. The charge-shift bonds are concentrated in the upper left quadrant, while the classical covalent bonds are in the lower right quadrant. Note the two bond types of [1.1.1]propellane; each belonging to a different quadrant. (Reproduced from Ref. [3h] with permission of Wiley-VCH.)

contribution to bonding (D_{ion}), with small RE_{CS} contributions. These are classical ionic bonds. Finally, the Si–F bond in entry 27 is special; its principal VB structure is ionic; its static ionicity is large, but its RE_{CS} is significant, much larger than that in the classical ionic bonds in **(III)**. VB theory predicts that this bond will be very different from ionic bonds. As already alluded to earlier, the Si–X bonds behave as though they were covalent despite their large ionic characters in terms of charge distribution. Here, in **(II)** and **(III)**, these bonds and their heavier analogs are clearly marked either as CS bonds (Si–Cl, Ge–Cl) [3d] or as bonds with a large CS character (Si–F) [3g].

The AIM analysis of the heteropolar bonds in **(II)** does not distinguish between the covalent and CS bonds, but the Laplacian components in the BCP show that the CS bonds have more pronounced $\nabla^2\rho_{res}$ values [3h], compared with the classical covalent bonds, in line with the dominant RE_{CS} quantity. Finally, the AIM analysis of the classical ionic bonds in **(III)** [3h] shows the expected characteristics from closed-shell interactions; all have positive Laplacians that are dominated by the ionic component, $\nabla^2\rho_{ion}$.

5.6.2
VB and ELF Converge

Table 5.2 collects the ELF results for a group of single bonds, reported in the original literature [3g]. For covalent and CS bonding, we show the population of the

Table 5.2 Populations $\overline{N}(\Omega)$ and population variance σ^2 of ELF basins [3g].

Entry	Molecule	Basin	$\overline{N}(\Omega)$	σ^2	Bond type
1	H–H	V(H,H)	2.0	0.0	cov.
2	Li–Li	V(Li,Li)	2.0	0.17	cov.
3	CH_4	V(C,H)	1.97	0.63	cov.
4	SiH_4	V(Si,H)	2.0	0.46	cov.
5	C_2H_6	V(C,H)	2.0	0.63	cov.
		V(C,C)	1.81	0.96	cov.
6	NaCl	C(Na)	10.02	0.11	ion.
7	NaF	C(Na)	10.01	0.12	ion.
8	F_2[a]	V(F,F)	0.44	0.42	CS
9	Cl_2[a]	V(Cl,Cl)	0.73	0.59	CS
10	Br_2	V(Br,Br)	0.81	0.68	CS
11	FCl	V(F,Cl)	0.39	0.35	CS
12	FBr	V(F,Br)	0.28	0.26	CS
13	ClBr	V(Cl,Br)	0.67	0.54	CS
14	H_2O_2	V(O,O)	0.49	0.41	CS
15	N_2H_4	V(N,N)	1.16	0.77	CS
16	HF	V(H,F)	1.22	0.68	CS
17	CH_3F	V(C,F)	0.86	0.64	CS
19	SiH_3F	V(Si,F)	0.27	0.24	Ion-CS

[a]V(F,F) and V(Cl,Cl) are the unions of two monosynaptic basins.

disynaptic basin that corresponds to the A–X bond, $\overline{N}[V(\text{A},\text{X})]$ and its variance σ^2, which is a measure of the charge density fluctuation of the bonding electrons. For ionic bonds (entries 6, 7), the core population of the most electropositive atom A is reported instead of $\overline{N}[V(\text{A},\text{X})]$.

Table 5.2 exhibits three groups of bonds. In entries 1–5, we find bonds with almost 2.0 electrons in the disynaptic basin, with weak-to-moderate fluctuation compared to the total population. These are the classical covalent bonds.

In entries 6 and 7 of Table 5.2, we show two bonds, which do not exhibit any disynaptic basin, but possess a basin for the bare core of sodium. The other basin, which is not shown, resides on the electronegative atoms Cl and F. In both entries, the population of the core basin of sodium is close to $10e^-$, and the variance is rather small (0.11–0.12). These are classical ionic bonds.

The largest group in Table 5.2 corresponds to entries 8–17. In all of these bonds, the population of the A–X basin is of the order of $1e^-$ or less, and the variances of these populations are large, almost of the same order as the population. The small populations of the disynaptic basins indicate that *these are not classical covalent bonds*, in which two spin-paired electrons are expected to provide the bonding. In fact, at higher levels of calculations, in the cases of F–F and Cl–Cl, the disynaptic basins are split into two monosynaptic ones that are 0.2 Å apart, and the electrons in the bonding region behave as though the bonds were "dissociated," with significant Pauli repulsion between the electrons. This, together with the large variance, signifies that the bonding in these molecules is dominated by fluctuation of the charge density. This last group of bonds corresponds therefore to the same CS-bonding type that emerges from the VB calculations.

The last entry in Table 5.2 corresponds to the Si–F bond. With the 6-31G* basis set, this bond does not possess a disynaptic *V*(Si, F) basin. Adding diffuse functions to the basis set (6-31+G*), leads to an emergence of a disynaptic Si–F basin, with a weak population, and large variance, mostly due to the delocalization involving the fluorine lone pairs. Thus, the improvement of the basis set increases the covalent contribution at the expense of the static ionic one, and confirms that although this bond has high static ionicity, it is a borderline case lying in between the groups of ionic bonds and CS bonds.

Clearly, therefore, much like the VB picture, the ELF analysis predicts the same distinction between the covalent and CS bond groups [3g]. Bonds such as H–H, C–C, Li–Li possess disynaptic basins with a population close to 2.0 electrons and small variances, whereas bonds like F–F, Cl–Cl, O–O, Br–Br, N–N, and the inverted C–C bond of [1.1.1]propellane possess small basin populations ($\leq$1.0) [34], with variances as large as the population.

The match between the predictions of the two methods is made vivid by inspecting the homonuclear bonds in Figure 5.7, which shows a plot of the basin population bonds vis-à-vis the RE_{CS}. The correlation is apparent; the smaller the basin population, the larger the RE_{CS}. Furthermore, the heavy red circles in Figure 5.7 show how the population variance approaches the basin's population as the RE_{CS} quantity increases and vice versa. It is apparent that the variance is less than half of the basin population for the covalent bond (C–C), but it gets gradually closer to the population for the CS bonds (from Br–Br to F–F), thereby showing again the connection between repulsive covalency, the fluctuating electron density and the large RE_{CS}. Recalling (Table 5.1) that all the bonds in Figure 5.7 have very

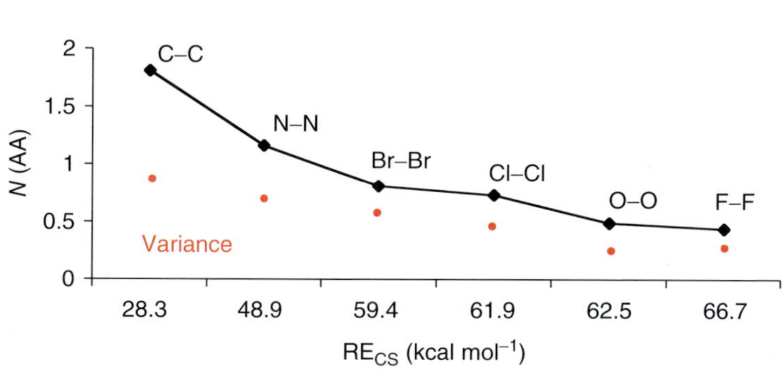

Figure 5.7 A correlation between the population of the disynaptic basin, calculated by ELF, and the charge-shift resonance energy, calculated by VB, for a series of homonuclear bonds. The heavy red circles show the change of the population variances (σ^2) of the bonds. (Reproduced from Ref. [3g] with permission of Wiley-VCH.)

similar weights of covalent and ionic contributions to bonding, it is clear that the major feature of bonding that distinguishes this group is the RE_{CS} quantity that arises from the covalent-ionic fluctuation of the pair density.

The correlation in Figure 5.7 indicates that both theories converge to the same conclusion, thereby substantiating the classification of CS bonding as a distinct bonding type that is supported by a dominant RE_{CS} quantity, owing to covalent-ionic fluctuation. Furthermore, both theories show that *this group of bonds transcends considerations based on static charge distribution, and is more concerned with the "dynamic bond ionicity"* [35].

5.6.3
Convergence of VB, ELF and AIM

Figure 5.8 projects the distinction of the covalent and CS-bond families by the three theoretical approaches. The figure depicts the ELF molecular basins for $H_3C{-}CH_3$, F–F and the two C–C bond types in [1.1.1]propellane, alongside their VB and AIM properties. It is seen that the C–C bond of ethane in Figure 5.8a and the wing C–C bond in [1.1.1]propellane in Figure 5.8c have good cylindrical ELF basins with populations close to 2.0, highly negative Laplacians, and a small or moderate RE_{CS}. These are classical covalent bonds. By contrast, Figure 5.8b,c shows that the disynaptic basins of F–F and the inverted C–C bond of [1.1.1]propellane are in fact two monosynaptic basins, much like dissociated bonds. The corresponding basin populations are tiny, with variances being as large as the populations; the Laplacian is highly positive, indicating repulsive covalent structures; and the RE_{CS} quantities are very large. Thus, the three methods diagnose the same classification of homonuclear bonds into two families. ELF and AIM diagnose the attractive/repulsive nature of the covalent "shared densities," while VB brings additional energetic insight that highlights the dominant role of the RE_{CS} energy in the CS-bond group.

5.6.4
The Three Bonding Families

In summary, CS bonding emerges as a distinct class alongside the covalent and ionic bonds. In VB theory [3], CS bonding is typified by large covalent-ionic resonance energy, RE_{CS}. In ELF, it is typified by a depleted basin population with large variance and covariance [3g]. In addition, homonuclear CS bonding is characterized in AIM by a positive or small Laplacian of the electron density [3i, 36]. It should be noted that the characterizations of CS bonding by AIM and ELF electron density analyses are independent of the level of calculation used to compute the wave function or electron density, for example, MO bonding theory or density functionals [3g,i], showing that the non-VB-based methods effectively account for CS bonding, even if not in the explicit way achieved by VB theory.

There is of course a relationship between the VB method and MO- or DFT-based methods of energy decomposition analysis (EDA) [37], which is described in

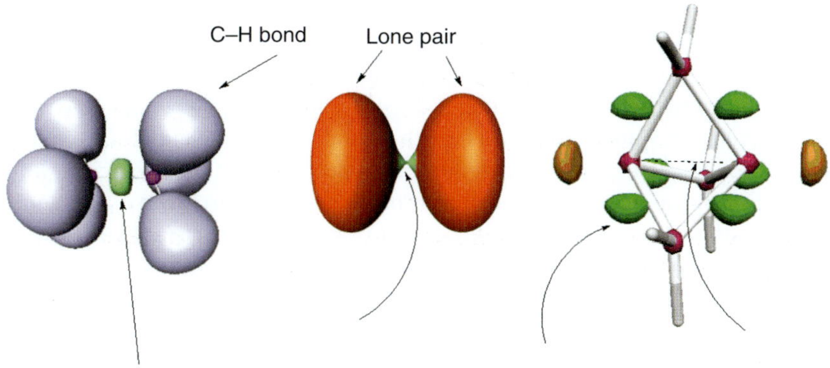

	C–C bond:	F–F bond:	Wing bond:	Central bond:
	$RE_{CS} = 27.7$	$RE_{CS} = 62.2$	—	$RE_{CS} = 72.2$
	$\overline{N}_{CC} = 1.81$	$\overline{N}_{FF} = 0.44$	$\overline{N}_{CC} = 1.76$	$\overline{N}_{CC} = 0.13$
	$\sigma^2 = 0.96$	$\sigma^2 = 0.42$	—	$\sigma^2 = 0.12$
	$\rho = 0.25$	$\rho = 0.25$	$\rho = 0.25$	$\rho = 0.19$
	$\nabla^2\rho = -0.62$	$\nabla^2\rho = +0.58$	$\nabla^2\rho = -0.51$	$\nabla^2\rho = +0.43$
(a)		(b)	(c)	

Figure 5.8 Some ELF representations of electron density in a few typical cases: (a) the ELF disynaptic basin[3g] for $H_3C{-}CH_3$; (b) the monosynaptic basins for the F–F bond; (c) disynaptic basins for the wing bonds of [1.1.1]propellane, and two monosynaptic basins for the central inverted bond [34]. The bond type is further characterized by its covalent-ionic resonance energy RE_{CS}, the ELF basin population $\overline{N}$ and its variance σ^2, the density ρ at the bond critical point and the corresponding Laplacian $\nabla^2\rho$ (energies are in kcal mol^{-1}, densities in e a_o^{-3}, Laplacians in e a_o^{-5}). For $H_3C{-}CH_3$ and F–F, the ELF and AIM parameters are taken from Refs [3g,i], respectively. For [1.1.1]propellane, the AIM parameters are experimental values [33] from the study of a substituted [1.1.1]propellane derivative. (The ELF drawings in Figures 8a and 8b are reproduced with permission of Wiley-VCH from Figures 3 and 4 in Ref. [3g]: Copyright Wiley-VCH Verlag & Co. KGaA. The ELF drawing in Figure 8c is reproduced with permission from Figure 1a in Ref. [34]. Copyright Wiley-Interscience.)

Chapter 4 in Vol. 1. While the EDA methods do not include as yet provisions to characterize CS bonding, they share with the VB model a few essential features: the major one is the Pauli repulsion. Recalling that large Pauli repulsions originate the large covalent-ionic resonance energies, one may hope that energy partition methods would eventually characterize the difference between bonds such as H_2 and F_2.

5.7 Physical Origins of CS Bonding

While the phenomenon of CS bonding is derived from three independent theoretical treatments, one would still like to base this bonding type on some fundamental principles. All the methods show that the emergence of CS bonding coincides with poor bonding by the shared-electron density of the electron pair: in VB calculations, this manifests as a repulsive or weakly attractive covalent structure (Figure 5.5b); in ELF, this is shown by the escape of the shared density from the disynaptic basin (Figures 5.7 and 5.8); and in AIM one finds that the shared density has a positive Laplacian (Figure 5.8) much as in cases that exhibit closed-shell repulsion, for example, He–He. Coupling the AIM technique into VB shows that CS bonds have large resonance Laplacians (entry 7 in Table 5.1). Furthermore, AIM shows also that the positive Laplacian is associated with excess kinetic energy in the bonding region and the same is implied by the ELF definition. As such, we would like to articulate a mechanism that reveals the root causes of weakly bonded or repulsive covalent structure, and that reveals the link of this repulsion to the increase in kinetic energy of the electrons, and the role of RE_{CS}. The following discussion relates the phenomenon to fundamental properties of atoms and bonding mechanisms.

5.7.1 The Role of Atomic Size

The large RE_{CS} quantity of CS bonds is an outcome of the mechanism necessary to establish equilibrium and optimum bonding during bond formation. This mechanism, based on the virial theorem, has been analyzed in the original literature in detail [3b,g, 38], while here we present a simpler analysis.

By comparing the atomic and covalent radii in the periodic table, one finds that as a rule $r_{atom} < r_{cov}$. This means that as atoms (fragments) bind they shrink, and for reasons that are detailed in Appendix 5.B. The shrinkage causes a steep increase in the fragments' kinetic energy, which exceeds the lowering of the potential energy due to the diminished size [39]. Thus, the shrinkage tips the virial ratio of the kinetic (T) versus potential (V) energies off-equilibrium. At equilibrium, the ratio has to be $V/T = -2$ (Appendix 5.B). All resonance energy terms are dominated by kinetic energy lowering (Appendix 5.C), and therefore *the covalent-ionic resonance is the mean whereby the kinetic energy can be reduced to restore the virial ratio* [3b,g, 38, 39], and this is true in all bonds. The kinetic energy rise due to shrinkage is proportional to the compactness of bonding partners, and therefore, as the fragments in bonding become more compact, the kinetic energy rise as a result of the shrinkage will get steeper, and a larger RE_{CS} quantity will be required to restore the equilibrium.

A simple demonstration of the atom compactness effect is Figure 5.9 [3b], which models the effect of orbital compactness by calculating the CS-resonance energy of an H_2' molecule, where H′ is an atom with a variable orbital exponent, ζ. It is seen that as the orbital exponent increases, the CS-resonance energy increases, leading

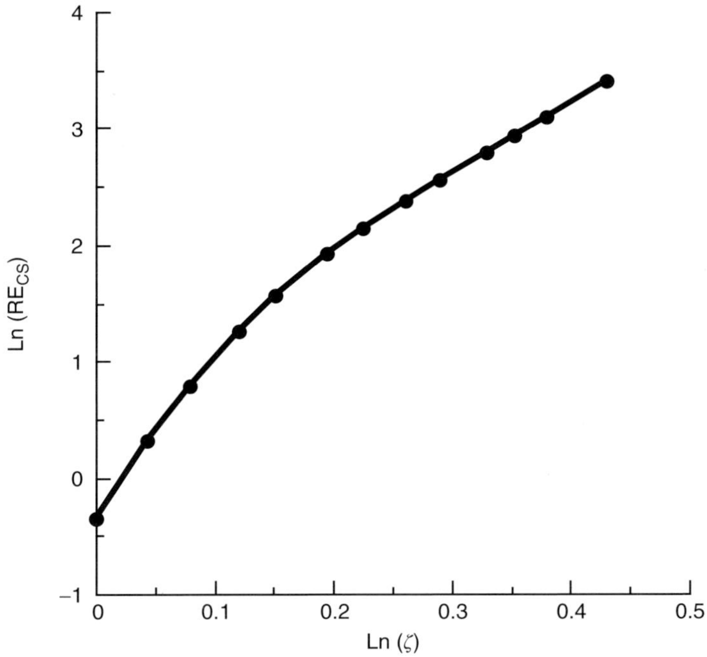

Figure 5.9 A natural logarithm plot of the VB computed [3b] CS-resonance energy against the orbital exponent, ζ, for a pseudo H_2' molecule where the 1s orbitals of H′ have modified orbital exponent, ζ. (Adapted with permission from Ref. [3b]. Copyright 1992 American Chemical Society.)

to the following approximate relationship:

$$\mathrm{Ln}(\mathrm{RE_{CS}}) = 8.9\mathrm{Ln}(\zeta) - 0.4 \tag{5.7}$$

The result in Figure 5.9 and its expression Eq. (5.7) is easy to understand based on this discussion. Thus, *as ζ increases, the orbital of H′ becomes more compact, leading to an increase in the kinetic energy of the atoms, and hence, the CS-resonance energy increases in order to restore equilibrium for the bond.*

5.7.2
The Role of Pauli Repulsion Pressure

Real atoms (fragments) are not merely pseudo hydrogens. They bear also lone pairs or other electron pairs, which can interact with the "test bond" by Pauli repulsion. As we saw, in some bonds such as the F–F, Cl–Cl, and O–O, the spin pairing leads to a covalent structure that is either destabilized or only weakly stabilized relative to the dissociated atoms (Table 5.1). This failure of the covalent structure to provide significant bonding was quantified in VB terms [3a,b,c,d,g,j], and found to originate primarily in *the Pauli repulsion between the bonding electrons and the lone pairs that have the same symmetry as the bond.* The repulsion between the lone pairs themselves contributes too, but to a lesser extent [3f]. Scheme 5.4 illustrates this repulsive interaction that counteracts the spin-pairing stabilization in F–F [3g, 28], using a cartoon.

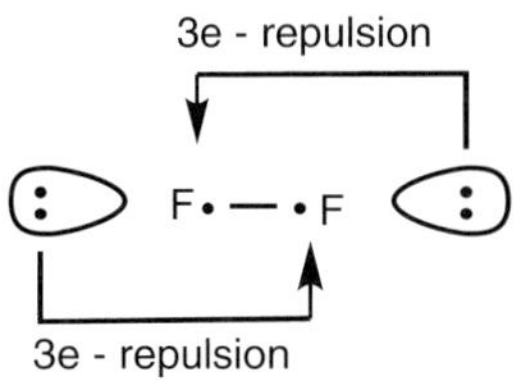

Scheme 5.4 The lone-pair bond repulsion in the covalent structure of F–F. The spin pair is depicted as two electrons (dots) connected by a line.

Thus, as described in Textbox 5.1, both the spin coupling energy as well as each of the three-electron repulsion terms have the same expression but with opposite signs [6, 40]. Hence, the two Pauli repulsive terms add up to a larger quantity than the spin pairing energy, and the covalent structure of F–F is repulsive (Figure 5.5b) with a negative D_{cov} quantity [3k]. The same considerations apply to other bonds, which bear lone pairs. In any case, the Pauli repulsion pressure will weaken the corresponding covalent bonding energy.

Textbox 5.1

Using the Hückel resonance integral β and the overlap S between the two hybrid atomic orbitals (HAOs), which participate in spin pairing, the covalent bonding in the F–F bond leads to the following stabilization energy: [6, 40]

$$\Delta E_{cov} = 2\beta S; \beta < 0 \tag{T.1}$$

Each of the Pauli repulsion terms in Scheme 5.4 has the same expression but with a different sign:

$$\Delta E_{3e\text{-Pauli}} = -2\beta' S' \tag{T.2}$$

Therefore, the net D_{cov} becomes:

$$D_{cov} = -[\Delta E_{cov} + 2\Delta E_{3e\text{-Pauli}}] = -2\beta S + 4\beta' S' < 0 \tag{T.3}$$

D_{cov} is net destabilizing if βS and $\beta' S'$ have comparable magnitudes.

This Pauli repulsion was pointed out originally by Sanderson [22], who termed this as the *lone-pair bond-weakening effect* (*LPBWE*). Because the Pauli repulsion has the same expression as the bonding energy, but with a different sign (Eq. T.2 in Textbox 5.1), its presence in a bond will raise the kinetic energy of the bond, and the effect will become more severe as the number of lone pairs on the atom increases. As the atoms (fragments) are brought together, the LPBWE augments the kinetic energy rise as all βS terms are dominated by the increase in the kinetic energy (Appendix 5.C). This tips the virial ratio off-balance, and hence the only way for the molecule to restore the virial ratio, and achieve equilibrium bonding, is to augment the ionic component and increase thereby the CS-resonance energy [3b, 38].[3)]

Because electronegative fragments are compact and also lone-pair rich, we might expect that in bonds of such fragments, the resonance energy that is required to restore the virial ratio will become necessarily very large, generating thereby bonds with weakened bonding in the covalent structures and large RE_{CS} quantities. Thus, CS bonding is a fundamental mechanism that is necessary to adjust the kinetic and potential energy to the virial ratio at equilibrium, in response to the Pauli repulsive strain exerted on the bond and the shrinkage of the atoms (fragments) that occurs during bonding.

5.8 Global Behavior of Electron-Pair Bonds

Understanding the roots of CS bonding allows us to outline some global correlations for the CS-resonance energy and bonding in general. Recalling that electronegative atoms have small valence orbitals and are lone-pair rich, we might expect that the electronegativity of the atom or fragment (χ_A) will be an organizing quantity for RE_{CS} and, in the absence of LPBWE, also of D_e.

These global relationships are illustrated in Figure 5.10 for RE_{CS}. Figure 5.10a shows the RE_{CS} quantities for homonuclear A–A bonds, plotted against the electronegativity (χ_A) of A. It is seen that in each period, RE_{CS} increases as the electronegativity increases. Figure 5.10b shows a plot of RE_{CS} versus the sum of electronegativities of the fragments using both homonuclear and heteronuclear bonds [3g], while Figure 5.10c,d shows, respectively, the same trend for π-bonds of doubly bonded and triply bonded molecules [3e, 41]. It is apparent that the RE_{CS} quantity of the bond generally increases as the molecular electronegativity of the bond partners increases. In this respect, we note that what determines the RE_{CS} quantity is not the simple orbital overlap of the atoms making the bond. In fact, the RE_{CS} is large for cases where the overlap is small [3b,g]. For example, the orbital overlap in H_2 is much larger than in F_2, while the covalent-ionic resonance energy behaves oppositely. *This underscores the relationship of RE_{CS} to the kinetic energy decrease in the bonding region rather than to the simple "sharing of density" as in covalency.*

3) As seen from footnote 2 here, the F-F bond energy arising from CASSCF or from GVB wave functions is rather poor, 16.0 kcal mol^{-1}. These methods treat the covalent and ionic structures of the bond in a mean-field approximation. Only further extensive CI of the GVB and CASSCF wave functions the bond energy gets closer to experiment. The modern BOVB methods lead to the correct bond energy, about 36 kcal mol^{-1}, by explicit treatment of the ionic structures, which are allowed to take on their particular set of orbitals. Thus bonding in F–F originates from *the response of the electronic structure to the fluctuation of the electron pair density from the average density.*

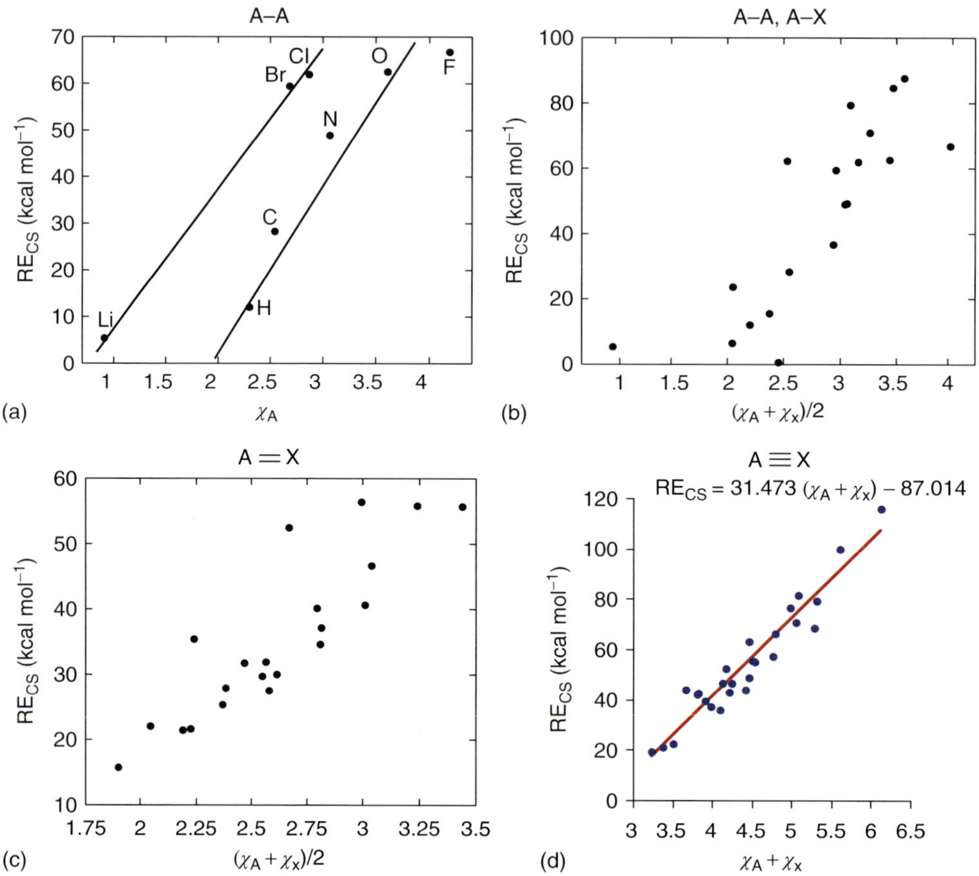

Figure 5.10 Correlation of the charge-shift resonance energy (RE_{CS}) of a bond with the electronegativities (χ) of the bonded atoms or fragments. (a) A plot of RE_{CS}(A–A) versus the electronegativity of A (χ_A). Reproduced from Ref. [3g] with permission of Wiley-VCH. (b) A plot of RE_{CS} for A–A and A–X bonds versus the average electronegativity of the bond. (Reproduced from Ref. [3g] with permission of Wiley-VCH. (c) A plot of RE_{CS} for π-bonds (of doubly bonded molecules, A=X) versus the average electronegativity of the bond, reproduced from Ref. [3g] with permission of Wiley-VCH. (d) A plot of RE_{CS} for π-bonds (of triply bonded A≡X molecules) versus the sum of electronegativities of the bond constituents. Permissions for (a)–(c) are by Copyright Wiley-VCH Verlag & Co. KGaA. Part (d) is reproduced with permission from Ref. [41]. Copyright 2011 American Chemical Society.)

We note that the scatter in the plots in Figure 5.10b–d reflects in part the effect of the electronegativity difference, namely, the classical Pauling effect on the covalent-ionic resonance energy (Eq. 5.2). Thus, for a given molecular electronegativity ($\chi_X + \chi_A$), the RE_{CS} quantity increases, to some extent, with increase in the electronegativity difference ($\chi_X - \chi_A$), thereby reflecting an incremental increase of RE_{CS} due to the stabilization of the ionic structure, A^+X^-, and its stronger mixing

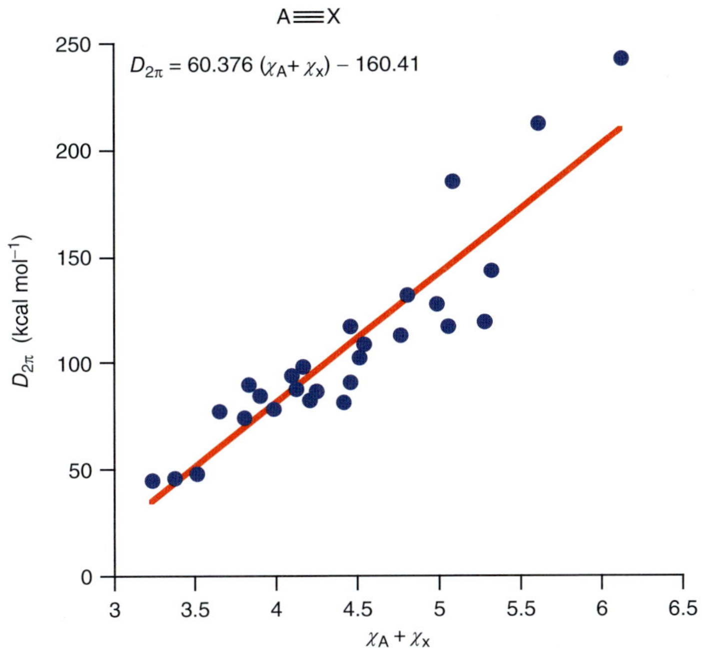

Figure 5.11 A plot of VB calculated sum of π-bond energies, $D_{2\pi}$, for A≡X molecules versus the molecular electronegativity, $(\chi_A + \chi_X)$. (Reproduced with permission from Ref. [41]. Copyright 2011 American Chemical Society.)

into the covalent structure. However, the electronegativity difference constitutes only a secondary influence. Indeed, in contrast to the behavior in Figure 5.10b–d, where a global correlation with $(\chi_X + \chi_A)$ is apparent, no correlation whatsoever is observed when the RE_{CS} data is plotted against $(\chi_X - \chi_A)$ alone. *The fundamental correlation is with the sum of electronegativities*, which gauges both the effects of the fragment compactness and the Pauli repulsion pressure on the shared density.

Because π-bonds do not suffer from LPBWE, even their D_{cov} quantity correlates quite well with the sum of electronegativities of the fragments. This, along with the dominance of bonding by the RE_{CS}, means that one may expect that the total π-bond energy will also correlate with the sum of the fragment electronegativities rather than with their difference. Figure 5.11 shows this global correlation for the total π-bonding energy in triply bonded molecules, A≡X. Thus, the bonding gets stronger as the molecular electron affinity increases.

5.9
Additional Factors of CS Bonding

The Pauli repulsion pressure associated with the lone pairs of electronegative fragments is not the only factor that can promote CS bonding. A recently identified additional factor [3b,c,d,g] was expressed in bonds between metalloids of Group 14 and electronegative groups, similar to all the Si–F, Si–Cl and Ge–Cl bonds in Table 5.1. The VB calculations for these bonds show that the corresponding ionic curve for the $Me_3Si{-}Cl$ bond, for example, is much deeper than that for the corresponding $Me_3C{-}Cl$ bond [42]. Moreover, the ionic curve $Me_3Si^+Cl^-$ has a tighter minimum than Me_3C^+Cl in harmony with the fact that the charge is completely localized on Si in Me_3Si^+, while highly delocalized in Me_3C^+. This causes the ionic and covalent structures to be close in energy in Me_3SiCl, thus leading to a high RE_{CS} quantity, which is apparent from Table 5.1 for the Si–Cl bond.

5.10
Can a Covalent Bond Become CS Bonds by Substitution?

We have discussed the molecule [1.1.1]propellane and its curious inverted C–C bond, which was identified as a CS bond, by contrast to the classical covalent wing C–C bonds [3i, 33]. This is an interesting finding because it shows that bonds having the same atomic constituents can be either covalent or CS bonds, depending on their molecular environment. Understanding the origins of CS bonding in the inverted C–C bond of [1.1.1]propellane is therefore important.

Figure 5.12 shows that the inverted bond in [1.1.1]propellane is embedded inside a cage of six wing C–C bonds, labeled as σ_W. If we make symmetry-adapted combinations of these six localized bonds, we shall find that two of the combinations possess the same symmetry as the inverted C–C bond and their electron densities are projected on the inverted bond. One of these is a cage orbital (σ_{cage}), which consists of three lobes pointing from the CH_2 moieties of the propellane to the center of the inverted bond, and the second one is the all-positive combination of the wing orbitals that projects on the axis (σ_{axis}). These two electron pairs repel the covalent structure of the inverted bond, making it repulsive much like in F–F, and eliciting thereby a large RE_{CS}, which makes the inverted bond a CS-bond.

Understanding this principle we can now set out to design a series of C–C bonds, which exhibit an excursion from classical covalent bonds to CS bonds [3k]. The molecules are depicted in Scheme 5.5, and the target bond computed with VB theory is shown in a bold line.

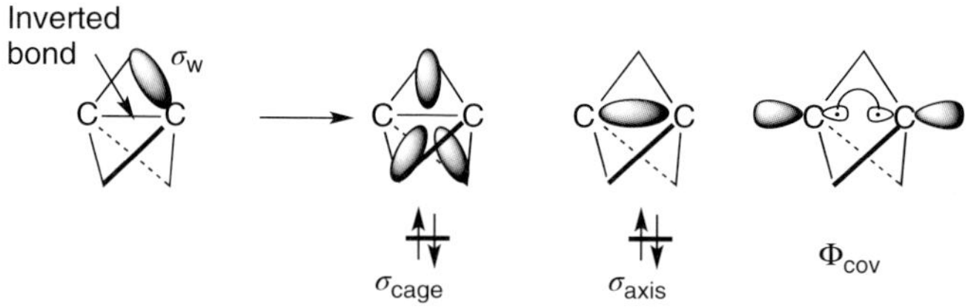

Figure 5.12 Covalent bond-weakening repulsion exerted by the wing C–C bonds on the inverted central bond of [1.1.1]propellane [3k].

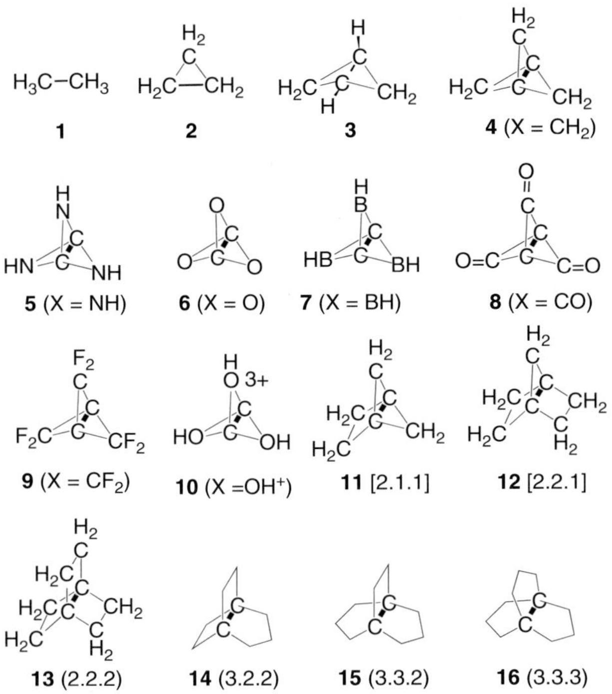

Scheme 5.5 Target C–C bonds highlighted in bold in a series of molecules. The molecules **4–10** are labeled by the wing fragment X. (Adapted from Ref. [3k] with permission of Wiley-VCH.)

For all these bonds we computed by means of VB theory the bond energy and the RE_{CS} quantity. For propellanes one cannot really calculate a BDE, and therefore we calculated for all the molecules the *in situ* bond energy, $D_{in\ situ}$, which gauges the bond energy of the molecule relative to a reference non-bonded structure, called the *quasi-classical* (QC) state [43]. The QC state has a single VB determinant where the spins are not allowed to exchange, and hence its energy does not include any bonding term due to spin pairing (Textbox 5.2). To treat all the bonds in Scheme 5.5 on equal footing, all the bond energies, even for ethane, were calculated as $D_{in\ situ}$. Scheme 5.6 shows the QC state and the various bond quantities that can be calculated by modern VB theory.

Textbox 5.2

The QC state is a single determinant, which includes localized up- and down-spins that are not permitted to exchange. Hence, the entire bonding energy that arises from spin pairing of the covalent structure is lost in the QC state (see the energy diagram in Scheme 5.6). By allowing the spin pairing and ionic structures in, we get the bonded state and the difference relative to the QC state

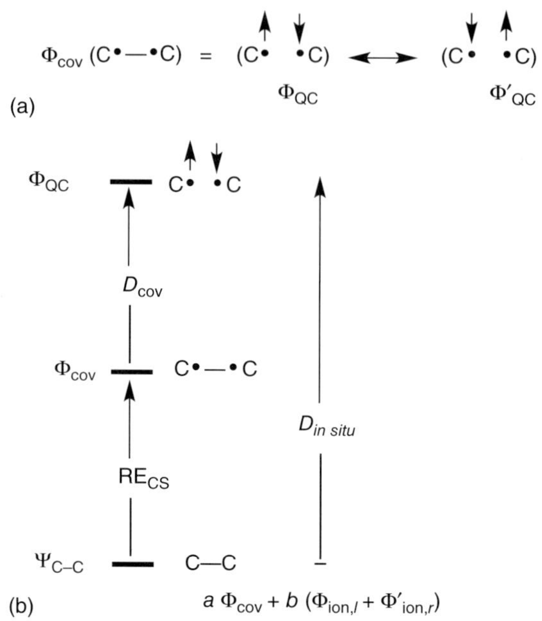

Scheme 5.6 VB calculations of $D_{in\ situ}$ relative to the QC reference. (Adapted from Ref. [3k] with permission of Wiley-VCH.)

is the *in situ* bond energy. In H_2, the QC state is virtually flat when plotted against the internuclear distance [43]. In other cases, where the atoms bear other electron pairs, for example, in $H_3C{-}CH_3$, or F–F, the QC state is repulsive (due to Pauli repulsions). In all cases, the QC state is devoid of any bonding and can serve as a reference for calculating $D_{in\ situ}$ values. Generally, $D_{in\ situ} \geq$ BDE.

Figure 5.13 shows a plot of the ratio of $RE_{CS}/D_{in\ situ}$ vis-à-vis the Laplacian (L) at the BCP of the target bonds in Scheme 5.5. It is seen that the molecules fall into two families: at the lower left quadrant of the plot we find the C–C bonds of ethane, propane, and the large [2.2.2]propellane. All these bonds have a low $RE_{CS}/D_{in\ situ}$ ratio and negative L. Thus, in this family most of the bond energy arises from the covalent spin pairing, and in accord with that, the Laplacian is negative as expected for classical covalent bonds. By contrast, in the upper right quadrant of Figure 5.13, we find the inverted C–C bonds of the smaller propellanes and the wing-substituted ones. All these bonds are typified by a high $RE_{CS}/D_{in\ situ}$ ratio, which means that most of the bonding arises because of CS resonance and, in accordance, the Laplacian is positive, indicating that these bonds suffer from Pauli repulsion pressure. It is conceivable that such two families of the same homopolar bond may exist for other atoms (fragments).

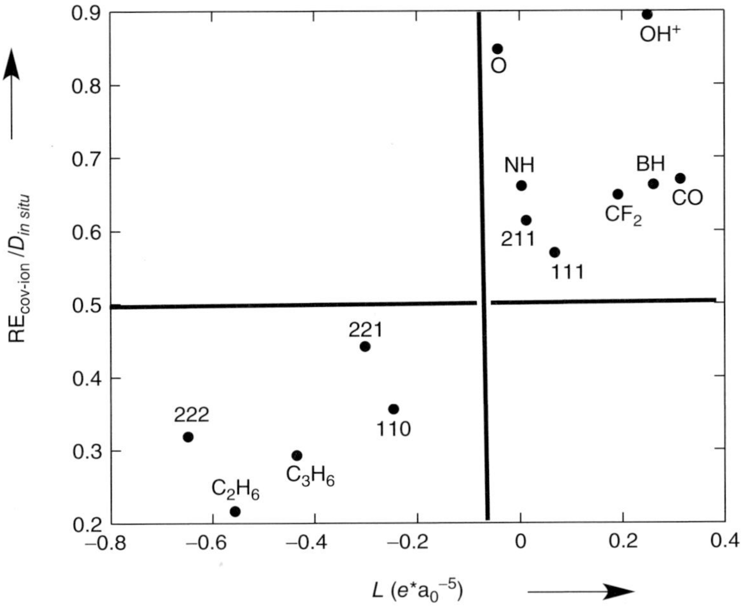

Figure 5.13 A two-dimensional plot of $RE_{cov\text{-}ion}/D_{in\ situ}$ (calculated by VB theory) versus the Laplacian L (calculated by AIM, in atomic units) at the BCP for the target C–C bonds in **1–13** (Scheme 5.5), with vertical lines drawn to emphasize the separation into two families of C–C bonds. The substituted [1.1.1]-X_3-propellanes (**4–10** in Scheme 5.5) are specified by their wing fragment X (X = CH_2, NH, etc.). (Reprinted from Figure 5 in Ref. [3k] with permission of Wiley-VCH.)

5.11 Experimental Manifestations of CS Bonding

Having shown the emergence of CS bonding and its promoting factors, here we follow with some evidence for the signature of this bond type in the chemical behavior.

5.11.1 Marks of CS Bonding from Electron Density Measurements

The existence of the CS bond family will eventually be consolidated by experimental determination of the Laplacian of various bonds, as already done for propellane [33], N_2O_4 [44], $(Mg–Mg)^{2+}$ cores [45], and so on [32]. Thus, as we already alluded to, for [1.1.1]propellane derivatives [33], the experimental Laplacian clearly shows the existence of a classical covalent C–C bond in the wing positions, vis-à-vis a CS C–C bond in the inverted bond region. Such experimental characterization for other bonds will further show the importance of this new bond family.

In the meantime, the existence of two distinct families has already emerged from electron density difference maps (available experimentally), which plot the difference between the actual molecular density and the density of a reference state made from spherical atoms ($\Delta\rho = \rho_{Mol} - \rho_{Ref}$), placed at the same geometry as the molecule. These data [46] clearly show a group of bonds (e.g., Li–Li, C–C, Si–Si, C–H) with $\Delta\rho > 0$, which coincides with the classical covalent bond, and a second group (e.g., F–F, Cl–Cl, O–O, S–S, N–N, N–O, C–F, C–O, etc.) of "*no-density bonds*" with $\Delta\rho \leq 0$, which coincides with the CS-bonding family outlined in this chapter. While the deformation density depends on the definition of the reference atomic state [47], the example of [1.1.1]propellanes [46] is virtually free of this limitation as the $\Delta\rho$ quantity is determined by comparing two different bonds in the same molecule, and using the same ρ_{Ref}. The findings clearly show that the C–C bonds in the wings possess $\Delta\rho > 0$, while the "inverted" (C–C) has $\Delta\rho < 0$. Furthermore, these bond types were identified also by the experimental Laplacians, which revealed a fundamental difference between the wing and inverted C–C bonds [33].

5.11.2
Marks of CS Bonding in Atom Transfer Reactivity

We would expect to see manifestations of CS bonding on reactivity for cases that involve cleavage of CS bonds. Our studies [48] showed that one of these manifestations is the computational [49] and experimental [50] result that halogen transfer reactions (and especially of fluorine), Eq. (5.8a), have much larger barriers (by >20 kcal mol^{-1} for X=F) than the corresponding hydrogen transfer processes, Eq. (5.8b).

$$H\cdot + X - H' \rightarrow H - X + \cdot H' \tag{5.8a}$$

$$X\cdot + H - X' \rightarrow X - H + \cdot X' \tag{5.8b}$$

Thus, as we demonstrated recently by means of VB calculations [48], the two processes have almost identical barriers for the covalent structures, but they differ in the status of the ionic VB structures in the vicinity of the transition state. In reaction (8b), the key combination of ionic structures, H• $X:^-$ H^+ and H^+ $X:^-$ •H, is destabilized by two repulsive 3-electron interactions [6, 40, 51] between H• and the X^- fragment (Scheme 5.7a). By contrast, the ionic combination $X:^-$ H^+ •X and X• H^+ $X:^-$, for Eq. (5.8a), is devoid of repulsive interactions (Scheme 5.7b vs 5.7a). The destabilization of the ionic structures during X-transfer results in a loss of RE_{CS} in the respective transition state. Becuse the H–X bonds, and especially so H–F, have large RE_{CS} to begin with, the loss is significant and the barrier is higher for the X transfer reaction. The largest destabilization occurs for X=F, as H–F has the largest RE_{CS} quantity among the hydrogen halides.

Interestingly, the barrier difference between the two series was found [48] to follow a very simple relationship, as 1/4 of the RE_{CS} quantity of the H–X bond that undergoes cleavage during the two processes. Therefore, RE_{CS} is given as

$$RE_{CS(H-X)} = 4[\Delta E^{\ddagger}{}_{H/XH} - \Delta E^{\ddagger}{}_{X/HX}] \tag{5.9}$$

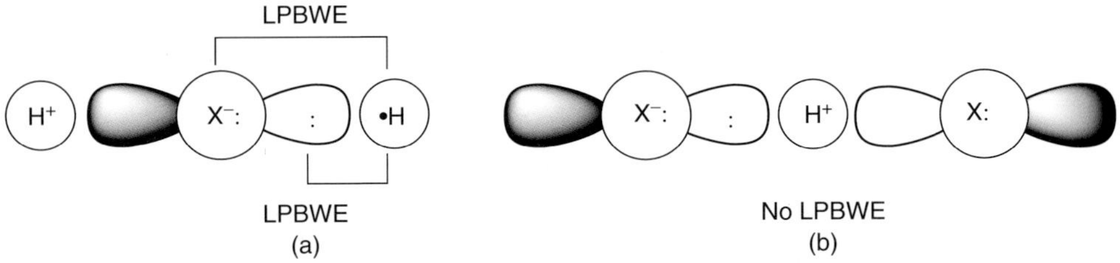

Scheme 5.7 Lone-pair bond-weakening effect (LPBWE) in the ionic structure for halogen (X) transfer reactions and the lack of LPBWE in the ionic structure of H transfer reactions. (Adapted from Ref. [3g] with permission of Wiley-VCH.)

As such, measurement of the barrier difference for the two series enables to quantify the CS resonance energy from experimental barriers. This is exciting!

5.11.3
Marks of CS Bonding in the Ionic Chemistry of Silicon in Condensed Phases

A large CS-resonance energy typifies also bonds with a high static ionicity, similar to H–F, C–F, Si–F, Si–Cl, Ge–Cl, and so on (Table 5.1). This arises because of a combination of effects, one being the atomic shrinkage and LPBWE of the lone-pair-bearing heteroatom, and the second is the strong covalent-ionic interaction due to the decreased energy gap between the two structures [3d,j]. In the case of Si–X bonds, the ionic VB structure undergoes a special stabilization that can be appreciated from the calculated charge distribution of the ionic structures, in Figure 5.14, for Si–Cl versus C–Cl [3c,d]. It is seen that the positive charge of H_3Si^+ is concentrated on silicon, while in the case of $CH_3{}^+$ the charge is delocalized over all the atoms, placing only a small charge on the carbon. This in turn causes much stronger electrostatic interactions in the ionic structure $H_3Si^+Cl^-$ compared with $H_3C^+Cl^-$. *The result is that the minimum of the ionic curve becomes very deep for* $H_3Si^+Cl^-$ *and it coincides with the minimum of the covalent structure,* leading thereby to a strong covalent-ionic mixing and large RE_{CS} compared with the carbon analog (Table 5.1). The same situation carries over to any R_3Si–Cl versus R_3C–Cl, R = alkyl, and so on; in each case the $R_3Si^+Cl^-$ structure will have a minimum coincident with the covalent structure and a large resultant RE_{CS}, some 21 kcal mol^{-1} larger than that of the carbon analog [3d]. In a condensed phase, the ionic structure is stabilized by the environment, but because the Si^+Cl^- minimum is tight, the stabilization will be only moderate; hence, the ionic curve should remain close to the covalent curve, thereby retaining the large RE_{CS} interaction of the bond. Thus, in a condensed phase, the covalent-ionic mixing remains large, giving rise to Si–X bonds that stay intact owing to the large CS-resonance energy.

Indeed, as discussed, our recent VB study showed [42] that the $Me_3Si^+Cl^-$ structure in aqueous solution retains the tight ion-pair minimum, and thus mixes strongly with the covalent structure and acquires large RE_{CS}. This large RE_{CS} is the major reason why the bond will not undergo heterolysis in solution (but will prefer

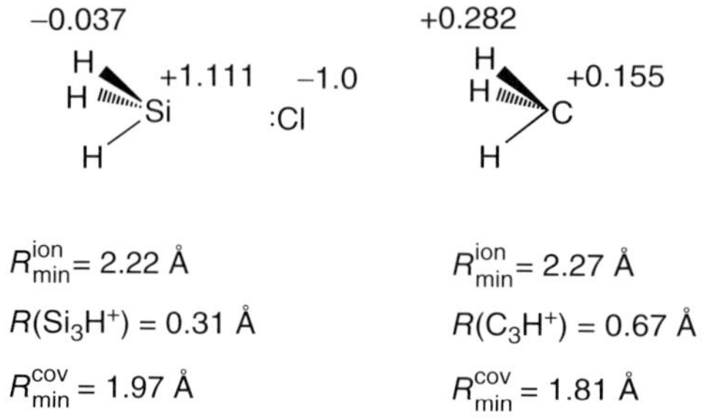

Figure 5.14 Charge distribution and geometric parameters of the ionic structures $H_3C^+Cl^-$ and $H_3Si^+Cl^-$, calculated by VB theory. (Reproduced from Ref. [3g] with permission of Wiley-VCH.)

associative processes), and why in the solid state even $Ph_3Si{-}OClO_3$ is a covalent solid [25c] in contrast to the carbon analog, which has an Na^+Cl^--type lattice with Ph_3C^+ and ClO_4^- ions [25d], and so on [25–27].

5.12 Scope and Territory of CS Bonding

The territory of CS bonding for electron pair bonding is in fact larger than we describe in this chapter. In the area of electron-pair bonds, we should mention the recent VB study [52] by Galbraith which showed that coordinate bonding such as the one between tetravalent boron and amines, $R_3B{-}NR'_3$, is dominated by CS-resonance energy. Similarly, Coote *et al.* [53] found that the dependence of the relative bond strengths of R–X bonds (R = Me, Et, *iso*-Pr, *tert*-Bu; X = H, F, OH, OCH_3) follows the CS-resonance energy. Our recent study of M–H bonds, where M is a first-row transition metal [54], showed that the CS-resonance energy is quite significant despite the apparent "covalency" of most of the bonds. Furthermore, RE_{CS} was found to increase from left to right in the period, and to be affected by the presence of the $2s^22p^6$ core electron pairs, which behave as lone pairs on an atom. The analyzed propensity of atoms (fragments) to generate CS bonding suggests that bonds of first-row transition metals will tend to be CS bonds, especially when the bonding partner is an electronegative and/or lone-pair-rich atom. More such CS bonds should be looked for among the bonds between the heavier elements of the periodic table.

We have not discussed in this chapter the odd-electron bonds [55]. In odd-electron bonds such as in F_2^- or He_2^+, H_2^+ , the entire bonding arises from the RE_{CS} because of the mixing of the odd-electron structures, as depicted in Scheme 5.8, for an archetypical odd-electron bond. Thus, for example, the 3-electron structures of F_2^- are by themselves repulsive partly because of the 3-electron repulsion between the two F moieties [56], and in part because of the Pauli repulsion from the σ-lone pairs. Nevertheless, F_2^- is significantly bonded by about 29.5 kcal mol^{-1}, which arises because of a large RE_{CS} quantity that overcomes the repulsive interaction in the individual VB structures.

$$\mathrm{F}{:}^{-}\ {\bullet}\mathrm{F} \longleftrightarrow \mathrm{F}{\bullet}\ {:}\mathrm{F}^{-} \qquad D_e = 30\ \text{kcal mol}^{-1}$$

Scheme 5.8 The resonance of the two 3-electron VB structures that constitute F_2^- and its bond dissociation energy [a CCSD(T)/aug-cc-pV5Z datum].

Another bond type that was not discussed in the chapter is hypercoordination, which may well be invariably bonded by CS bonding. For example, the symmetrically hydrogen-bonded species $(FHF)^-$ has been computed by means of modern VB theory [57], and was shown to be stabilized by $RE_{CS} = 92$ kcal mol^{-1}, which contributes to its stability, relative to the separated $F^- + HF$. Large RE_{CS} values might also be the root cause for the stabilities of other species such as PCl_5, SF_n ($n =$ 4, 6), and bonded noble-gas-containing molecules, which form hypercoordination species, for example, XeF_n (n = 2, 4, 6), and all hypercoordinated species [58a] could well be CS bonded. A most recent VB study [58b] shows that XeF_2 is indeed CS bonded.

Future directions for articulating this bonding form are many. A fruitful direction is hypercoordination and aggregation. Thus, for example, the small size of R_3Si^+, and heavier analogs, means that they will tend to form hypercoordinated compounds; in solution, in the solid state [59] and even in the gas phase, where some unusual molecules have been reported [60], and bridged systems, $(Si{-}X{-}Si)^+$, which participate in catalytic bond exchange reactions [61, 62]. Metal–metal bonds in some bimetallic complexes could well be CS bonds, as in M_2(formamidinate)$_4$ complexes (M = Nb, Mo, Tc, Ru, Rh, Pd) where large positive values of $\nabla^2\rho(\mathbf{r}_c)$ have been reported [63]. Other directions involve the generation of [1.1.1]propellane in which the CH_2 wings are substituted by heteroatoms that exert exchange repulsion pressure on the inverted C–C bond, for example, HN, and so on [3k]. The in-plane π-type bond in *ortho*-benzyne is another bond that suffers from exchange repulsion pressure. Protonation or methylation (by Me^+) of C–N bonds may convert them into CS bonds [58a], a fact that may concern DNA bases, and may have mechanistic effects, as in the protonated arginine in the mechanism of nitric oxide synthase [64]. Bonds under immense external pressure [65] are likely to be CS bonds, and encapsulated highly positive ions may well be CS bound [66, 67], and so on. A growing territory lies ahead for exploration.

5.12.1 Concluding Remarks

CS bonding originates from the equilibrium condition of the bond, defined by the virial ratio. It is promoted by two main factors:

1) By Pauli repulsion that weakens the covalency of the bond and induces large covalent-ionic resonance energies (RE_{CS}). This excessive exchange repulsion is typical to electronegative and lone-pair-rich atoms, or bonds weakened by

exchange-repulsion pressure, as the bridgehead C–C bond in [1, 1, 1]propellane, and other small ring propellanes.

2) Fragments that form extremely small cations, which resemble a proton, with all the positive charge located at the central atom, much like in silicenium cation, R_3Si^+ and so on, will promote CS bonding and hypercoordination, especially with electronegative and lone-pair-rich atoms.

With these promoters, CS bonding forms a distinct group of bonding that transcends consideration of static charge distribution, and that possesses unique chemical signatures. Thus, CS bonding is not merely an academic abstraction. As new examples or experimental manifestations of CS bonding start to accumulate and become recognized, the concept of CS bonding will gradually find more articulations [68] and ultimately be accepted by the chemical community.

Appendix

5.A
Modern VB Methods

The principles of modern VB theory have been recently reviewed [7]. The modern methods differ on whether they use semilocalized orbitals, such as GVB (generalized valence-bond) [69] and spin-coupled (SC) VB [70], or strictly local HAOs in a multistructure VB scheme. The latter approach has been used successfully for calculating bond energies with accuracy that matches the Gold standard method, CCSD(T) [7].

In the localized HAO approach, a bond, A–X is described by the following wave function:

$$\Psi_{A-X} = c_{COV}\Phi_{COV} + c_{A^+X^-}\Phi_{A^+X^-} + c_{A^-X^+}\Phi_{A^-X^+} \tag{5.A1}$$

where the cs are structural coefficients and the Φs are the covalent and ionic VB structures that distribute the electron pair between the orbitals of the two bonded fragments; these orbitals are called the *active orbitals*. The remaining electrons (inactive shell) in each VB structure are arranged in doubly occupied orbitals. There are a few modern methods for handling the calculations, and we describe briefly the main three methods.

1) The basic method is called the *valence bond self-consistent field* (*VBSCF*) method [71]. It performs optimizations of both orbital and VB-structural coefficients. If all the structures within the active shell are taken into account, the VBSCF method is nearly equivalent to a CASSCF (complete active space self-consistent field) method. Thus, VBSCF takes care of the static correlation but does not include any dynamic correlation.
2) A VB method that incorporates dynamic correlation is the breathing orbital valence bond (BOVB) method [72], which allows the structures in Eq. (5.A1) to assume different orbitals, while performing optimizations of orbitals and

structural coefficients. Most of the data in this chapter are computed with BOVB [3g].

3) The valence bond configuration interaction (VBCI) method [73], initially performs a VBSCF calculation. Subsequently, dynamic correlation is introduced by singles and doubles configuration interaction (CI) from the optimized orbitals of the fundamental structures into virtual orbitals. The virtual orbitals are constrained to be localized on the same bonds as the occupied orbitals. In this manner, the excited VB structures keep the same spin pairing and charge characters as the fundamental structures, and it is therefore possible to contract each fundamental structure and its excited ones into a single structure, covalent or ionic, as in Eq. (5.A1). VBCI is used in some case in Table 5.1 [3g,j].

The weights of the VB structures are determined by use of the Coulson–Chirgwin [7] formula, Eq. (5.A2), which is the equivalent of a Mulliken population analysis in VB theory.

$$W_K^{\mathrm{CC}} = C_K^2 + \sum_{L \neq K} C_K C_L \langle \Phi_K | \Phi_L \rangle \tag{5.A2}$$

5.B The Virial Theorem

The virial theorem expressed in Eq. (5.B1),

$$-\frac{R\mathrm{d}E}{\mathrm{d}R} = 2T + V \tag{5.B1}$$

where E is the total energy, V and T are respectively the potential and kinetic components, and R is the interatomic distance between the two atoms (or fragments).

The term on the left of Eq. (5.B1) (the force acting on the molecule times the respective length, R) is zero at equilibrium, that is, for $R = R_{\mathrm{eq}}$. This means that any properly optimized wave function at its own equilibrium distance R_{eq} must obey the following virial ratio of the kinetic to the potential energy:

$$-\frac{2T}{V} = 1 \tag{5.B2}$$

Achieving this ratio results in energy lowering, namely, in bonding, which is expressed in Eq. (5.B3):

$$\Delta E = -\Delta T = 0.5\Delta V; (\Delta V < 0) \tag{5.B3}$$

However, far from equilibrium, the virial ratio ($-2T/V$) will be off-balance: larger than unity for $R < R_{\mathrm{eq}}$, (i.e., the kinetic energy T being too large to satisfy Eq. 5.B2), and smaller than unity for $R > R_{\mathrm{eq}}$.

Let us now bring together two atoms (fragments): at infinite separation, the atoms (fragments) obey individually the virial ratio (Eq. 5.B2). Simply shortening the distance, without any other change, will cause a decrease in the kinetic energy, and will thereby put the virial ratio of the species off-balance. The wave function will then have to relax in some way in order to restore this ratio and decrease the total energy. At the purely covalent level, this is accomplished by orbital shrinkage, which lowers the potential energy but raises the kinetic energy steeply. At the optimal distance of the covalent structure, R_{eq}^{cov}, the virial ratio will be obeyed. However, the purely covalent wave function yields, as a rule, too small bonding energies and too long bonding distances relative to accurately calculate spectroscopic parameters; this wave function is insufficient. To optimize the bonding, *the orbital shrinkage mechanism has to be augmented by an additional mechanism*. This mechanism allows the wave function to include ionic structures, thereby resulting in an optimized covalent-ionic wave function Ψ_{A-X} with shorter equilibrium distance R_{eq}^{A-X}, where the virial ratio is obeyed. At this distance, which is shorter than R_{eq}^{cov}, the covalent component of the wave function no longer obeys the virial ratio, which now exceeds unity, and it is the mixing of the ionic structures that restores the ratio to unity. Therefore, it is clear that the effect of ionic structures is to diminish the $-2T/V$ ratio, that is, to reduce the excess of kinetic energy relative to potential energy.

Thus, CS resonance due to the mixing of ionic structures into the covalent one is expected to be responsible, along with orbital shrinkage, for the adjustment of the kinetic and potential energy terms to the virial ratio. This effect of the CS resonance energy was ascertained by us using BOVB calculations. The following trends were found for H_2 at equilibrium distance: (i) upon binding, the orbitals shrink, (ii) the shrinkage of the orbitals in the covalent structure lowers the potential energy, but excessively raises the kinetic energy, T, thereby tipping the virial ratio off-balance; and (iii) adding the ionic structures lowers T without having much of an effect on V, thus restoring the correct virial ratio [38]. Generalizing to typical *classical covalent bonds*, such as H–H or C–C bonds, the mechanism by which the virial ratio is obeyed during bond formation is primarily orbital shrinkage, and therefore the RE_{CS} is only a small corrective effect.

Let us now consider a bond that bears adjacent electron pairs, as in F_2, Cl_2, and so on. The presence of lone pairs creates additional Pauli repulsions (LPBWE), which raise the kinetic energy. An extreme case is that of the F_2 molecule, in which the covalent component of the bond is repulsive, having excessive kinetic energy, thus making the virial ratio greater than unity at any distance in the covalent wave function. Therefore, by reference to H_2, which lacks these repulsive interactions, the additional increase of kinetic energy brought by LPBWE must be compensated for by greater participation of covalent-ionic mixing. Moreover, the necessity of large RE_{CS} is further reinforced by the fact that the atoms that bear lone pairs are also compact, and shrinkage of a compact atom (fragment) raises its kinetic energy more steeply than in the case of a diffuse atom (fragment)[4]. It follows therefore, from this discussion, that the smaller the atom (fragment) and the more lone-pair rich it is, the larger will be the excess kinetic energy, in the covalent structure, and greater RE_{CS} would be required to restore the viral ratio. This is the reason why the homonuclear bonds of F, O, N, Cl, and so on, are CS bonds, and why many heteronuclear bonds containing F, Cl, O, S, N, P, will have a propensity for CS

4) For a valence orbital in a central field, the potential energy varies in proportion to ζ (the orbital exponent), while the kinetic energy changes as ζ^2.

bonding. This is also the reason why RE_{CS} peaks for F, which is the most compact and lone-pair-rich atom in its period. This property comes inherently with the atom (fragment). It follows therefore that CS bonding is the outcome of a fundamental mechanism of bonding.

5.C
Resonance Interaction and Kinetic Energy

Kutzelnigg [39a] showed that the resonance integral β is generally made of a dominant negative kinetic component, β_T, and a smaller and positive potential component, βv. It follows, therefore, that the RE_{CS}, which results from the mixing of ionic structures into the covalent structure, acts mainly to lower the kinetic energy in the bonding region. Thus, the large RE_{CS} quantity that typifies CS bonding is associated with the lowering of the kinetic energy in the bonding region.

Kutzelnigg has also given an approximate expression for β_T as a function of ζ, the orbital exponent, and the internuclear distance R:

$$\beta_T = -\frac{1}{3}[\zeta^4 R^2 e^{-\zeta R}] \tag{5.C1}$$

Assuming a perturbation expression for the RE_{CS}, it would be proportional to the square of β_T, and will therefore be expected to vary with the eigth power of ζ. This expectation is well substantiated in Figure 5.9, where the slope of the log–log plot of RE_{CS} versus ζ is 8.89, thus showing indeed quite a great sensitivity of the RE_{CS} to the compactness of the valence orbital.

References

1. Shaik, S. (2007) *J. Comput. Chem.*, **28**, 51–61.
2. Shaik, S. and Hiberty, P.C. (2008) *A Chemist's Guide to Valence Bond Theory*, John Wiley & Sons Inc., Hoboken, NJ, pp. 1–25.
3. (a) Sini, G., Maitre, P., Hiberty, P.C., and Shaik, S.S. (1991) *J. Mol. Struct. THEOCHEM*, **229**, 163–188; (b) Shaik, S., Maitre, P., Sini, G., and Hiberty, P.C. (1992) *J. Am. Chem. Soc.*, **114**, 7861–7866; (c) Lauvergnat, D.L., Hiberty, P.C., Danovich, D., and Shaik, S. (1996) *J. Phys. Chem.*, **100**, 5715–5720; (d) Shurki, A., Hiberty, P.C., and Shaik, S. (1999) *J. Am. Chem. Soc.*, **121**, 822–834; (e) Galbraith, J.M., Blank, E., Shaik, S., and Hiberty, P.C. (2000) *Chem. Eur. J.*, **6**, 2425–2434; (f) Lauvergnat, D. and Hiberty, P.C. (1995) *J. Mol. Struct. THEOCHEM*, **338**, 283–291; (g) Shaik, S., Danovich, D., Silvi, B., Lauvergnat, D., and Hiberty, P.C. (2005) *Chem. Eur. J.*, **11**, 6358–6371; (h) Zhang, L., Ying, F., Wu, W., Hiberty, P.C., and Shaik, S. (2009) *Chem. Eur. J.*, **15**, 2979–2989; (i) Wu, W., Gu, J., Song, J., Shaik, S., and Hiberty, P.C. (2009) *Angew. Chem. Int. Ed.*, **48**, 1407–1410; (j) Shaik, S., Danovich, D., Wu, W., and Hiberty, P.C. (2009) *Nat. Chem.*, **1**, 443–449; (k) Shaik, S., Chen, Z., Wu, W., Stanger, A., Danovich, D., and Hiberty, P.C. (2009) *ChemPhysChem*, **10**, 2658–2669.
4. Silvi, B. and Savin, A. (1994) *Nature*, **371**, 683–686.
5. Bader, R.F.W. (1990) *Atoms in Molecules: A Quantum Theory*, Oxford University Press, Oxford.

6. Shaik, S. and Hiberty, P.C. (2008) *A Chemist's Guide to Valence Bond Theory*, Chapter 3, 9, 10, John Wiley & Sons Inc., Hoboken, NJ.
7. Wu, W., Su, P., Shaik, S., and Hiberty, P.C. (2011) *Chem. Rev.*, **111**, 7557–7593.
8. Lewis, G.N. (1916) *J. Am. Chem. Soc.*, **38**, 762–785.
9. Servos, J.W. (1990) *Physical Chemistry from Ostwald to Pauling*, Princeton University Press, Princeton, NJ, pp. 4–5, 135.
10. Langmuir, I. (1919) *J. Am. Chem. Soc.*, **41**, 868–934.
11. Jensen, W.B. (1984) *J. Chem. Educ.*, **61**, 191–200.
12. (a) Davenport, D.A. (1996) *Bull. Hist. Chem.*, **19**, 13–18; (b) Calvin, M. (1984) *J. Chem. Educ.*, **61**, 14–18.
13. Nye, M.J. (1993) *From Chemical Philosophy to Theoretical Chemistry*, University of California Press, Los Angeles, CA, p. 186.
14. (a) Heitler, W. and London, F. (1927) *Z. Phys.*, **44**, 455–472; (b) English Translation Hettema, H. (2000) *Quantum Chemistry Classic Scientific Paper*, World Scientific, Singapore, pp. 140–155.
15. London, F. (1928) *Z. Phys.*, **46**, 455–477.
16. Saltzman, M.D. (1996) *Bull. Hist. Chem.*, **19**, 25–32.
17. Pauling, L. (1939) *The Nature of the Chemical Bond*, 3rd edn, 1960, Cornell University Press, Ithaca, New York.
18. (a) Slater, J.C. (1965) *J. Chem. Phys.*, **43**, S11–S17; (b) van Vleck, J.H. and Sherman, A. (1935) *Rev. Mod. Phys.*, **7**, 167–228.
19. Servos, J.W. (1984) *J. Chem. Educ.*, **61**, 5–10.
20. Gavroglu, K. and Simoes, A. (1994) *Stud. Biol. Phys. Sci.*, **25**, 47–110.
21. Hager, T. (1995) *Force of Nature: The Life of Linus Pauling*, Simon and Schuster, New York, pp. 62–63.
22. Sanderson, R.T. (1983) *Polar Covalence*, Academic Press, New York.
23. For pioneering localization methods, see: (a) Edmiston, C. and Ruedenberg, K. (1963) *Rev. Mod. Phys.*, **35**, 457–464; (b) Rutledge, R.M. and Saturno, A.F. (1965) *J. Chem. Phys.*, **43**, 597–602; (c) Boys, S.F. (1960) *Rev. Mod. Phys.*, **32**, 296–299; (d) Foster, J.M. and Boys, S.F. (1960) *Rev. Mod. Phys.*, **32**, 300–302; (e) Weinhold, F. and Landis, C.R. (2005) *Valency and Bonding: A Natural Bond Orbital Donor-Acceptor Perspective*, Chembrige Univeristy Press, Chembridge; (f) Weinhold, F. and Landis, C.R. (2012) *Discovering Chemistry with Natural Bond Orbitals*, John Wiley & Sons Inc., Hoboken, NJ.
24. (a) Silvi, B. (2005) *Chem. Eur. J.*, **11**, 6358–6371; AIM analysis, of SiH_3-F and Li-F using B3LYP/6-31+G* and B3LYP/6-31G* levels cited in Ref. 3g above; (b) Bader, R.F.W. and Nguyen-Dang, T.T. (1981) *Adv. Quantum Chem.*, **14**, 63–124; (c) For other highly ionic bonds to silicon, see: Henn, J., Ilge, D., Leusser, D., Stalke, D., and Engles, D. (2004) *J. Phys. Chem. A*, **108**, 9442–9452.
25. (a) Apeloig, Y. (1989) in *The Chemistry of Organic Silicon Compounds*, Vol. 1, Chapter 2 (eds Y. Apeloig and Z. Rappoport), John Wiley & Sons, Ltd, Chichester; (b) For a variety of crystalized carbocationic salts, see: Laube, T. (1995) *Acc. Chem. Res.*, **28**, 399–405; (c) Gomes de Mesquita, A.H., MacGillavry, C.H., and Eriks, K. (1965) *Acta Crystallogr.*, **18**, 437–443; (d) Prakash, G.K.S., Keyaniyan, S., Aniszfeld, S.K.R., Heiliger, L., Olah, G.A., Stevens, R.C., Choi, H.-K., and Bau, R. (1987) *J. Am. Chem. Soc.*, **109**, 5123–5126; (e) For a recent "bottling" of *tert*-Butyl cation, see: Kato, T. and Reed, A. (2004) *Angew. Chem. Int. Ed.*, **43**, 2908–2911.
26. (a) For a few reviews on the elusive R_3Si^+ cation in condense phases, see: Apeloig, Y. and Stanger, A. (1987) *J. Am. Chem. Soc.*, **109**, 272–273; (b) Lambert, J.B., Kania, L., and Zhang, S. (1995) *Chem. Rev.*, **95**, 1191–1201.
27. (a) For a recent exception, see: Lambert, J.B. and Zhao, Y. (1997) *Angew. Chem., Int. Ed. Engl.*, **36**, 400–401; (b) Kim, K.-C., Reed, C.A., Elliott, D.W., Mueller, L.J., Tham, F., Lin, L., and Lambert, J.B. (2002) *Science*, **297**, 825–827.
28. (a) See also, Cremer, D. and Kraka, E. (1984) *Angew. Chem., Int. Ed. Engl.*, **23**, 627–628; (b) Low, A.A. and Hall, M.B. (1990) in *Theoretical Models of Chemical Bonding, Part 2*

(ed. Z.B. Maksic), Springer-Verlag, New York, pp. 544–591; (c) Schwarz, W.H.E., Valtazanos, P., and Ruedenberg, K. (1985) *Theor. Chim. Acta*, **68**, 471–506.

29. Lusar, R., Beltrán, A., Andrés, J., Noury, S., and Silvi, B. (1999) *J. Comput. Chem.*, **20**, 1517–1526.

30. Silvi, B. (2003) *J. Phys. Chem. A*, **107**, 3081–3085.

31. Kraka, E. and Cremer, D. (1990) in *Theoretical Models of Chemical Bonding, Part 2* (ed. Z.B. Maksic), Springer-Verlag, New York, pp. 457–543.

32. (a) Coppens, P. (2005) *Angew. Chem. Int. Ed.*, **44**, 6810–6811; (b) Coppens, P. (1997) *X-ray Densities and Chemical Bonding*, Oxford University Press, New York.

33. Messershmidt, M., Scheins, S., Grubert, L., Patzel, M., Szeimies, G., Paulmann, C., and Luger, P. (2005) *Angew. Chem. Int. Ed.*, **44**, 3925–3928.

34. Polo, V., Andres, J., and Silvi, B. (2007) *J. Comput. Chem.*, **28**, 857–864.

35. For "dynamic ionicity", see: Maynau, D. and Malrieu, J.-P. (1988) *J. Chem. Phys.*, **88**, 3163–3173.

36. Rincon, L. and Almeida, R. (1998) *J. Phys. Chem. A*, **102**, 9244–9254.

37. (a) Kitaura, K. and Morokuma, K. (1976) *Int. J. Quantum Chem.*, **10**, 325–340; (b) Ziegler, T. and Rauk, A. (1977) *Theor. Chim. Acta*, **46**, 1–10; (c) Velde, G.T., Bickelhaupt, F.M., Baerends, E.J., van Gisbergen, S.J.A., Fonseca Guerra, C., Snijders, J.G., and Ziegler, T. (2001) *J. Comput. Chem.*, **22**, 931–967; (d) Krapp, A., Bickelhaupt, F.M., and Frenking, G. (2006) *Chem. Eur. J.*, **12**, 9196–9216.

38. Hiberty, P.C., Ramozzi, R., Song, L., Wu, W., and Shaik, S. (2006) *Faraday Discuss.*, **135**, 261–272.

39. (a) Kutzelnigg, W. (1990) in *Theoretical Models of Chemical Bonding, Part 2* (ed. Z.B. Maksic), Springer-Verlag, New York, pp. 1–44; (b) Ruedenberg, K. (1962) *Rev. Mod. Phys.*, **34**, 326–376; (c) Feinberg, M.J. and Ruedenberg, K. (1971) *J. Chem. Phys.*, **54**, 1495–1591; (d) Wilson, C.Q. and Goddard, W.A. III, (1972) *Theor. Chim. Acta*, **26**, 195–210; (e) Rozendaal, A. and Baerends, E.J. (1985) *Chem. Phys.*, **95**, 57–91; (f) Ruedenberg, K. and Schmidt, M. (2007) *J. Comput. Chem.*, **28**, 391–410; (g) Bickelhaupt, F.M. and Baerends, E.J. (2000) *Rev. Comput. Chem.*, **15**, 1–86.

40. Shaik, S.S. (1989) in *New Theoretical Concepts for Understanding Organic Reactions*, NATO ASI Series C267 (eds J. Bertran and G.I. Csizmadia), Kluwer Publisher, Dordrecht, pp. 165–217.

41. Ploshnik, E., Danovich, D., Hiberty, P.C., and Shaik, S. (2011) *J. Chem. Theor. Comput.*, **7**, 955–968.

42. Su, P., Song, L., Wu, W., Shaik, S., and Hiberty, P.C. (2008) *J. Phys. Chem. A*, **112**, 2988–2997.

43. Hiberty, P.C., Danovich, D., Shurki, A., and Shaik, S. (1995) *J. Am. Chem. Soc.*, **117**, 7760–7768.

44. Messerschmidt, M., Wagner, A., Wong, M.W., and Luger, P. (2002) *J. Am. Chem. Soc.*, **124**, 732–733.

45. Platts, J.A., Overgaard, J., Jones, C., Iversen, B.B., and Stasch, A. (2011) *J. Phys. Chem. A*, **115**, 194–200.

46. (a) Dunitz, J.D. and Seiler, P. (1983) *J. Am. Chem. Soc.*, **105**, 7056–7058; (b) Dunitz, J.D., Schweizer, W.B., and Seiler, P. (1983) *Helv. Chem. Acta*, **66**, 123–133; (c) Coppens, P., Yang, Y.W., Blessing, R.H., Cooper, W.F., and Larsen, F.K. (1977) *J. Am. Chem. Soc.*, **99**, 760–766; (d) Savariault, J.-M. and Lehmann, M.S. (1980) *J. Am. Chem. Soc.*, **102**, 1298–1303.

47. (a) Ruedenberg, K. and Schwarz, W.H.E. (1990) *J. Chem. Phys.*, **92**, 4956–4969; (b) Generally, a better reference state can be made from deformed rather than spherical atoms. See, Schwarz, W.H.E., Ruedenberg, K., and Mensching, L. (1989) *J. Am. Chem. Soc.*, **111**, 6926–6941.

48. Hiberty, P.C., Megret, C., Song, L., Wu, W., and Shaik, S. (2006) *J. Am. Chem. Soc.*, **128**, 2836–2843.

49. (a) O'Neal, S.V., Schaefer, H.F. III,, and Bender, C.F. (1974) *Proc. Natl. Acad. Sci. U.S.A.*, **71**, 104–106; (b) Dunning, T.H. Jr., (1984) *J. Phys. Chem.*, **88**, 2469–2477; (c) Bender, C.F., Garrison, B.J., and Schaefer, H.F. III, (1975) *J. Chem. Phys.*, **62**, 1188–1190; (d) Voter, A.F. and Goddard, W.A. III, (1981) *J. Chem. Phys.*, **75**, 3638–3639; (e) Dobbs, K.D.

and Dixon, C.A. (1993) *J. Phys. Chem.*, **97**, 2085–2091.

50. Bartoszek, F.E., Manos, D.M., and Polanyi, J.C. (1978) *J. Chem. Phys.*, **69**, 933–935.
51. Shaik, S. and Hiberty, P.C. (2004) *Rev. Comput. Chem.*, **20**, 1–100.
52. Fiorillo, A.A. and Galbraith, J.M. (2004) *J. Phys. Chem. A*, **108**, 5126–5130.
53. Coote, M.L., Pross, A., and Radom, L. (2003) *Org. Lett.*, **5**, 4689–4692.
54. Galbraith, J.M., Shurki, A., and Shaik, S. (2000) *J. Phys. Chem.*, **104**, 1262–1270.
55. Hiberty, P.C., Humbel, S., Danovich, D., and Shaik, S. (1995) *J. Am. Chem. Soc.*, **117**, 9003–9011.
56. (a) Hiberty, P.C., Humbel, S., and Archirel, P. (1994) *J. Phys. Chem.*, **98**, 11697–11704; (b) Wu, W. and Shaik, S. (1999) *Chem. Phys. Lett.*, **301**, 37–42.
57. Shaik, S. and Shurki, A. (1999) *Angew. Chem. In. Ed.*, **38**, 586–625.
58. (a) For a discussion of CS-bonding in XeF_2 due to mixing of, $F:^-$ $^+Xe–F$ and $F–Xe^+$ $F:^-$ into the no-bond structure F• Xe: •F, and the $N–CH_3$ bond energy change due to methyl cation attachment to the nitrogen atom, see, Shaik, S.S. (1991) in *An Encomium to Linus Pauling. Molecules in Natural Science and Medicine* (eds Z.B. Maksic and M.E. Maksic), Ellis Horwood, London, 12; (b) Braida, B. and Hiberty, P.C. (2013) *Nat. Chem.*, **5**, 417–422.
59. Kost, D. and Kalikhman, I. (2009) *Acc. Chem. Res.*, **42**, 303–314.
60. Dávalos, J.Z., Herrero, R., Abboud, J.-L.M., Mó, O., and Yáñez, M. (2007) *Angew. Chem. Int. Ed.*, **46**, 381–385.
61. Panisch, R., Bolte, M., and Müller, T. (2006) *J. Am. Chem. Soc.*, **128**, 9676–9682.
62. Lühmann, N., Panisch, R., Hirao, H., Shaik, S., and Müller, T. (2011) *Organometallics*, **30**, 4087–4096.
63. Llusar, R., Beltran, A., Andrés, J., Fuster, F., and Silvi, B. (2001) *J. Phys. Chem. A*, **105**, 9460–9466.
64. Cho, K.B., Carvajal, M.A., and Shaik, S. (2009) *J. Phys. Chem. B*, **113**, 336–346.
65. Grochala, W., Hoffmann, R., Feng, J., and Ashcroft, N.W. (2007) *Angew. Chem. Int. Ed.*, **46**, 3620–3642.
66. Dognon, J.-P., Clavaguéra, C., and Pyykkö, P. (2009) *J. Am. Chem. Soc.*, **131**, 238–243.
67. Rupar, P., Staroverov, V.N., and Baines, K.M. (2008) *Science*, **322**, 1360–1363.
68. Gershoni-Porrane, R. and Stanger, A. (2012) *Chem. Phys. Chem.*, **13**, 2377–2381.
69. (a) Goddard, W.A. III,, Dunning, T.H., Hunt, W.J., and Hay, P.J. (1973) *Acc. Chem. Res.*, **6**, 368–376; (b) Goodgame, M.M. and Goddard, W.A. III, (1985) *Phys. Rev. Lett.*, **54**, 661–664.
70. (a) Cooper, D.L., Gerratt, J., and Raimondi, M. (1991) *Chem. Rev.*, **91**, 929–964; (b) Sironi, M., Raimondi, M., Martinazzo, R., and Gianturco, F.A. (2002) in *Valence Bond Theory* (ed. D.L. Cooper), Elsevier, Amsterdam, pp. 261–277.
71. For VBSCF, see: (a) Verbeek, J. and van Lenthe, J.H. (1991) *J. Mol. Struct. (THEOCHEM)*, **229**, 115–137; (b) van Lenthe, J.H., Dijkstra, F., and Havenith, W.A. (2002) in *Valence Bond Theory* (ed. D.L. Cooper), Elsevier, Amsterdam, pp. 79–116.
72. For the leading BOVB references, see: (a) Hiberty, P.C., Flament, J.P., and Noizet, E. (1992) *Chem. Phys. Lett.*, **189**, 259–265; (b) Hiberty, P.C. (1997) in *Modern Electronic Structure Theory and Applications in Organic Chemistry* (ed. E.R. Davidson), World Scientific, River Edge, NY, pp. 289–367; (c) Hiberty, P.C. and Shaik, S. (2002) *Theor. Chem. Acc.*, **108**, 255–275.
73. For leading VBCI references, see: (a) Wu, W., Song, L., Cao, Z., Zhang, Q., and Shaik, S. (2002) *J. Phys. Chem. A*, **106**, 2721–2726; (b) Song, L., Wu, W., Zhang, Q., and Shaik, S. (2004) *J. Comput. Chem.*, **25**, 472–478.

6
The Block-Localized Wavefunction (BLW) Perspective of Chemical Bonding

Yirong Mo

6.1
Introduction

Modern quantum chemistry, in which molecular orbital (MO) based methods prevail, has allowed us to compute bonding energies within the chemical accuracy, namely, a few kilojoules per mole. Yet, the desire to understand the bonding nature in intuitive ways persists. While atoms are bound together by chemical bonds [1], the definition and understanding of these bonds are critically related to their theoretical formulation. Langmuir [2] first introduced the term covalent bond into chemistry, and Lewis [3] provided more details to this term by envisioning that "The chemical bond is at all times and in all molecules merely a pair of electrons held jointly by two atoms." Retrospectively, Pauling [4] interpreted the covalent bond as involving "a pair of electrons shared between two atoms, and occupying two stable orbitals, one of each atom." When the two bonding atoms differ considerably in their electronegativity, however, one atom would lose its electron to the other, and subsequently the bond between these two atoms is severely polarized and termed as *ionic bond*. Thus, conceptually covalent bonds arise from shared-electron interactions while ionic bonds come from closed-shell interactions between separated charged species. Contemporary chemical models and theories are broadly dependent on this Lewis [5] concept of electron pair bonding initially proposed nearly 100 years ago. A Lewis structure can be drawn for a bonded molecule, and when a single Lewis structure is not enough to describe the physical and chemical properties, for example, for a conjugated molecule, more Lewis structures (i.e., resonance structures) can be introduced as profoundly illustrated in the resonance theory [1, 6].

The theoretical description of chemical bond was rooted in the work by Heitler and London [7] who conducted the first quantum mechanical calculation of the hydrogen molecule H_2 and presented a mathematical formula for a bonding orbital. It was found that a covalent bond between two atoms originates from the overlap of half-filled valence atomic orbitals of each atom containing one unpaired electron. Since a Lewis structure or resonance structure consists of many covalent bonds each of which is strictly localized between two bonding atoms, Slater and Pauling

The Chemical Bond: Fundamental Aspects of Chemical Bonding, First Edition.
Edited by Gernot Frenking, Sason Shaik.

generalized the Heitler-London equation for the H_2 bond into the Heilter-London-Slater-Pauling (HLSP) wavefunction which corresponds to a particular resonance structure [8]. For example, for a closed system of $N=2n$ electrons, a HLSP can be expressed as

$$\begin{aligned}\Psi_K &= \hat{A}\left[\varphi_1(1)\varphi_2(2)\cdots\varphi_N(N)\prod_{(ij)} 2^{-\frac{1}{2}}[\alpha(i)\beta(j)-\beta(i)\alpha(j)]\right] \\ &= \hat{A}(\phi_{1,2}\phi_{3,4}\cdots\phi_{2n-1,2n}) \end{aligned} \tag{6.1}$$

which corresponds to a resonance structure with a chemical bond between orbitals ϕ_{2i-1} and ϕ_{2i} (or a lone pair if $\phi_{2i-1}=\phi_{2i}$). In the above equation, $\hat{A}$ is the antisymmetrizer and $\{\varphi\}$ are bond functions. The proposal of the HLSP function laid the foundation for *ab initio* valence bond (VB) theory, as the mathematical translation of the resonance theory is that a wavefunction for a molecular state is a linear combination of several HLSP functions.

Soon after the VB theory, an alternative theory called MO theory was developed with the introduction of the linear combination of atomic orbitals (LCAOs) approximation for the MOs [9]. The adoption of a single Slater determinant at the Hartree-Fock (HF) theoretical level and the constraint of orbital orthogonalization dramatically reduce the computational cost, and MO-based methods have flourished since the sixties and now overwhelmingly predominate in the computational chemistry field. In contrast, the non-orthogonality of one-electron orbitals $\{\phi\}$ in Eq. (6.1) which is essential for the fundamental notion that a chemical bond results from the overlap of two bonding orbital, greatly complicates the computational efforts. Even more, Eq. (6.1) is composed of 2^n Slater determinants. As a consequence, the VB theory has been overshadowed by the MO theory for a long time. Still, concepts such as Lewis structure, resonance, electronegativity, hybrid orbital, covalent and ionic bonds are ubiquitous in textbooks and literature and essentially the pillars for chemical models and theories, as they provide intuitive yet qualitative pictures for the understanding of molecular structures and properties. A continuing challenge in modern theoretical chemistry thus has been how to accommodate the conventional theories and concepts into the ab initio MO computations. The scrutiny and even quantification of these concepts at the modern quantum mechanical level have far-reaching implications for the further development of chemistry theory. The discovery and/or synthesis of novel molecules and materials with peculiar properties also stimulate researches in elucidating the chemical bonding, structures and stability of these species. However, MO-based methods, which nowadays are readily available for molecules of a few hundred atoms, are not ideal for this task due to the delocalization nature of MOs. Nevertheless, enormous efforts have been made to bridge MO computation results with VB concepts [10], with the development of numerous post-(SCF) Self-Consistent Field approaches such as localization schemes of canonical MOs [11], atoms in molecule (AIM) [12], natural bond orbital (NBO) [13], adaptive natural density partitioning [14], electron localization function (ELF) [15] and various intermolecular energy decomposition schemes [16]. Notably, Dederichs *et al.* [17] proposed a general density functional theory (DFT) method with arbitrary constraints, for example, a number of electrons constrained in a certain volume. This constrained DFT (CDFT) can effectively define electron-localized states and probe the electron transfer (ET) processes [18]. Similarly, Warshel and coworkers [19] developed the frozen DFT (FDFT) algorithm to model a solute or a chemical reaction system in condensed states.

As many of the chemical concepts have their roots in VB theory, the development of ab initio VB methods can provide chemists innovative tools to examine and justify these concepts and obtain unique and complementary information for high-level MO calculations. During the past two decades, remarkable advances in ab initio VB theory have been made with a few practical software available [5b, 20], including the XMVB program [21]. Although these ab initio VB codes are still limited in their applicability compared with popular MO methods, applications have generated interesting and unique information and provided novel insights into the bonding features in molecules [22]. Further development of ab initio VB approaches is challenging but they can noticeably complement the MO methods and establish the essential chemical concepts on a solid ground.

A promising strategy is to combine the advantages of both MO and VB theories. We have proposed the block-localized wavefunction (BLW) method which is highly efficient and can be run at the HF and DFT levels [16j, 23]. The fundamental assumption in the BLW method is that the total electrons and primitive basis functions can be partitioned into subgroups called *blocks* or *fragments*. Orbitals in the same subspace (block) are subject to the orthogonality constraint, which retains the computational efficiency of MO theory, but orbitals belonging to different subspaces are nonorthogonal, which is a characteristic feature in VB theory. Thus, the BLW method takes the advantages of both MO and VB theories, and can uniquely obtain the wavefunctions for diabatic states self-consistently and express the adiabatic states with a few (usually two or three) diabatic state wave functions. Currently the BLW code has been ported to in-house version of GAMESS software [24].

The BLW method was initially proposed for the study of intramolecular electron delocalization such as the conjugation effect in unsaturated planar molecules [23b]. Later it was extended to probe the intermolecular ET and a unique energy decomposition scheme subsequently was developed [16j]. Considering the one-to-one correspondence between BLWs and diabatic (electron-localized, or resonance) states, we also formulated the whole reaction profiles with a few BLWs within the framework of the combined quantum mechanical and molecular mechanical (QM/MM) theory to simulate chemical processes in condensed states [25]. Similarly, a BLW-based two state model was used for the computations of electronic coupling energy and molecular reorganization energy, which are key quantities in the ET theory [26]. In the following, we will introduce the essence of the BLW method, followed by a few applications in the elucidation of molecular bonding natures based on BLW computations. Finally, an outlook will be provided.

6.2
Methodology Evolutions

6.2.1
Simplifying *Ab Initio* VB Theory to the BLW Method

In VB theory, the bond function in Eq. (6.1) is defined as

$$\phi_{2i-1,2i} = \hat{A}\{\varphi_{2i-1}\varphi_{2i}[(\alpha(2i-1)\beta(2i) - \alpha(2i)\beta(2i-1))]\} \tag{6.2}$$

which can be expanded to two Slater determinants. In MO theory, however, a bond function is formally reduced to one Slater determinant (or HF wavefunction)

$$\phi_{2i-1,2i} = \hat{A}\{\varphi_i'\varphi_i'[\alpha(2i-1)\beta(2i)]\} \tag{6.3}$$

where $\{\varphi_i'\}$ are delocalized and orthogonal, in contrast to $\{\phi_i\}$ in Eq. (6.2) which are localized and nonorthogonal. The hybrid use of Eqs. (6.2) and (6.3) in Ψ_K (Eq. (6.1)) leads to Goddard's generalized VB (GVB) method [27], which retains the VB form for one or a few focused bonds (perfect-pairs) but accommodates the remaining electrons with orthogonal and doubly occupied MO's. The further introduction of the strong orthogonality between bond orbitals and overlap-enhanced orbitals (OEOs), which are essentially MO's, greatly reduces the computational demand for GVB calculations.

An alternative combination of VB and MO methods is to represent bond orbitals with nonorthogonal doubly occupied localized orbitals [16l, 28]. Since 1996 (with the first publication in 1998), we have been working on the proposal of a BLW for each diabatic state [16j, 23]. The purpose of the BLW method is to get an efficient and effective solution to electron-localized diabatic states which correspond to classical resonance structures and can be used as references for the measure of the intramolecular electron delocalization effect. In the BLW approach it is assumed that the total electrons and primitive basis functions are partitioned into k subgroups, in line with the conventional VB ideas. Each subspace consists of a number of pre-determined basis functions and accommodates a preset number of electrons. Obviously, for a Lewis resonance structure every two electrons form a subspace. However, we can generalize the definition of resonance structures and allow a subspace to have any number of electrons, as very often our focus is on the electron movement in a particular region among particular molecular fragments, rather than the overall electron delocalization within the whole molecular system. By the use of block-localized MOs (BL-MOs) with doubly occupations (or singly occupations in open-shell cases), we finally can reduce the original HLSP function to a BLW of one Slater determinant form, where orbitals in the same subspace are subject to the orthogonality constraint, but orbitals belonging to different subspaces are nonorthogonal. Thus, the BLW method combines the advantages or features of both MO and VB theories. Figure 6.1 shows how we construct BLWs for the Kekulé structure of benzene and the three resonance structures used to describe the whole chloride exchange S_N2 reaction profile, where $\Phi(X)$ denotes all block-localized orbitals in a block X.

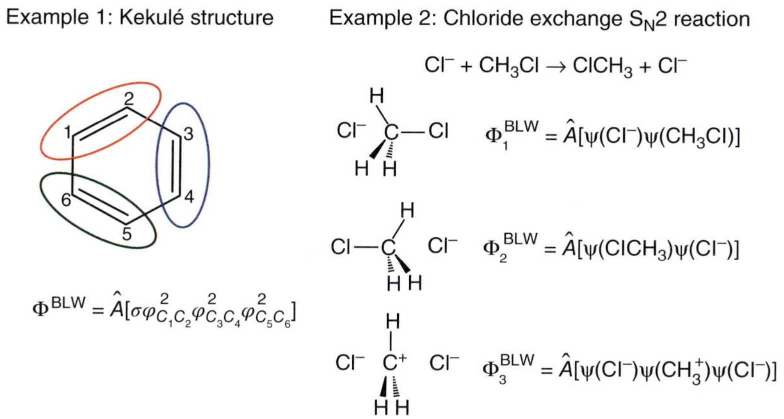

Figure 6.1 Definition of blocks in the examples of Kekulé structure and S_N2 reaction.

The optimization of orbitals in BLW can be accomplished using successive Jacobi rotation [23b] or Gianinettia *et al.*'s algorithm [28j–l, 29]. The latter is highly efficient as it generates coupled Roothann-like equations and each equation corresponds to a block. Furthermore, as the first derivative of the energy with respect to nuclear coordinates directly takes the form in conventional HF theory [28j], the second-derivatives can be computed numerically. This would allow us to derive the optimal geometry and vibrational frequencies and other properties for a hypothetical electron-localized state, and compare the results with experimental data for reference molecules without such electron delocalization effect. This kind of comparison is critical to justify the BLW results and quantify not only the energetic but also the geometrical impacts from the electron delocalization effect.

We have written an independent BLW code at the HF level with high efficiency, and numerous applications have endorsed its usefulness, as electron correlation is mostly self-cancelled when we take the energy difference between BLW and HF as the stabilization energy due to electron delocalization (transfer). For instances, we have studied the charge transfer in BH_3NH_3 [30], probed the nature of the ethane rotation barrier [31] and the cation-π interactions in δ-opioid receptor binding [32], proposed an energetic measure of aromaticity and antiaromaticity based on the Pauling-Wheland resonance energies [22f], and analyzed the charge transfer between solute and solvent with up to 1202 basis functions [33].

It should be noted that the BLW method, which is a quantum chemical model after all, may not work well if basis functions lose atomic characteristics, for example, when a complete basis on a single center for a molecular system is used. This unphysical basis set artifact complicates not only the ab initio VB methods, but also MO-based analyses on atomic properties in molecules. We have extensively tested mid-size basis sets from 6-31G(d) to 6-311+G(d,p) and cc-pVTZ and found the basis set dependence is generally trivial with this range of basis sets [22f,g, 23b, 34]. In addition, the reliability of the BLW method is documented by its computed structural parameters, vibrational frequencies, and NMR data [35].

We also note that the Head-Gordon group has decomposed the total delocalization energy into quasi-perturbative occupied-virtual orbital interaction components with nonorthogonal orbitals called "absolutely localized molecular orbitals" (ALMOs) [16l, 28m], which are actually identical to our BL-MOs.

6.2.2 BLW Method at the DFT Level

Due to the low computational cost and partial incorporation of electron correlation [16f, 36], DFT methods provide a sound basis for the development of computational strategies for studying potential energy surfaces, dynamics, various response functions and spectroscopy, excited states, and more [36c, 37]. In DFT, the self-consistent Kohn-Sham (KS) procedure is strictly analogous to the Hartree-Fock-Roothaan SCF procedure, except that the HF exchange potential is replaced by a DFT exchange-correlation (XC) potential. And the orbital equations of DFT have the same forms as those in HF theory except with a different Fock matrix

$$\boldsymbol{F}^{\alpha} = \boldsymbol{h} + \boldsymbol{J} + \boldsymbol{F}^{XC\alpha} \tag{6.4}$$

where $\boldsymbol{h}$ is the one-electron Hamiltonian matrix and $\boldsymbol{J}$ is the Coulomb matrix. The elements of α exchange-correlation matrix $\boldsymbol{F}^{XC\alpha}$ can be evaluated by a one-electron integral involving the local electron spin densities (LSD methods), or by an integral involving electron densities and their gradients (generalized gradient approximation, or GGA methods). Thus, it is fairly straightforward to implement the BLW idea into DFT as long as we keep all the algorithm unchanged except that the Fock matrix therein is replaced by a DFT one ($\boldsymbol{F}^{XC\alpha}$).

As the density matrix satisfies the symmetry, rank and idempotency conditions, the electron density is given as follows [23c, 38]

$$\rho(\mathrm{r}) = \rho_1(\mathbf{r}) + \cdots + \rho_k(\mathbf{r}) \tag{6.5}$$

where the integration of the fragmental density for block i satisfies the charge constraint by construction of the non-orthogonal block-localized KS orbitals:

$$\int \rho_i(\mathbf{r})\mathrm{d}\mathbf{r} = n_i \tag{6.6}$$

Compared with the BLW-DFT method, the CDFT method involved a set of constraints in KS-DFT calculations by the means of Lagrange multiplier to derive charge-localized states [17, 18]. The CDFT method has been used in numerous applications. For instances, Behler *et al.* [39] studied the important nonadiabatic effects in the adsorption of triplet oxygen on aluminum surface and reactions that are forbidden by the spin selection rule by confining electrons to subspaces of the Hilbert space. Wu and van Voorhis [18, 40] developed an approach to evaluate the coupling matrix element in long-range ET reactions and formulated a configuration interaction (CI) method by using the Complete Active Space SCF (CASSCF) inspired charge and spin constrained configurations. Our BLW-DFT method, however, adopts a different strategy from CDFT. We block-localize the KS orbitals and hence the associated electron density is constrained by construction (e.g., the Mulliken population constraint is inherently imposed). The electron density for each diabatic state is derived from an antisymmetric wave function, consisting of both orthogonal and nonorthogonal block-localized KS orbitals, a feature distinct from the CDFT approach of Dederichs *et al.* [17], but of characteristic to VB theory [10]. Importantly, the BLW method is applicable to both short-range (or strong-coupling) and long-range (or weak-coupling) ET processes [26b].

6.2.3
Decomposing Intermolecular Interaction Energies with the BLW Method

Intermolecular interactions play a central role in chemistry as it is usually the initial step for subsequent chemical reactions. A better understanding of the nature of intermolecular interactions can help chemists tune and eventually control chemical reactions. So far a large variety of theoretical approaches have been proposed to probe the nature of intermolecular interactions by decomposing the interaction energy into various physically meaningful components such as electrostatic, exchange, dispersion, relaxation or polarization, charge transfer, and so on [16]. Although certain arbitrariness is unavoidable in these energy decomposition schemes as these energy terms are not experimentally measurable, important consensus can be drawn from the analysis of a group of similar complexes and significant insights into the origin of intermolecular interactions can be garnered. This kind of studies is also illuminating for the development of next generation force fields in the simulations of large systems. Among the various energy decomposition schemes, the BLW energy decomposition (BLW-ED) approach has the unique advantage of defining the hypothetical electron-localized state self-consistently [16j, 23b, 32, 34a, 35b,c]. Moreover, the BLW has the geometry optimization capability and recently been extended to the DFT level [23a,c]. Thus, both the structural and energetic impacts by the charge transfer among interacting species can be quantitatively evaluated.

The BLW-ED approach involves a number of successive steps as illustrated in Figure 6.2. For the sake of simplicity, our discussion is limited to two-body interactions though it can be easily extended to many-body cases [35c]. We start from two monomers **A** and **B** at their isolated and optimal states. The optimal monomer geometries are more or less different from their geometries in the supermolecule **AB** due to the mutual interactions. Thus, we first need deform these monomers from their optimal geometries to the geometries in the optimal supermolecule. This part of energy cost (penalty) is usually denoted as the deformation energy ΔE_{def}. Afterwards, we bring monomers together to form a supermolecule with the monomer electron densities undisturbed. Obviously, the only contribution to the interaction energy here is the electrostatic interaction ΔE_{ele}. The amount of electrostatic energy can be derived by building a Hartree product of the monomer wavefunctions, and comparing the expectant energy of this Hartree product with the sum of monomer energies. Since the Hartree product is not well-behaved wavefunction as it does not satisfy the antisymmetry requirement, we can impose

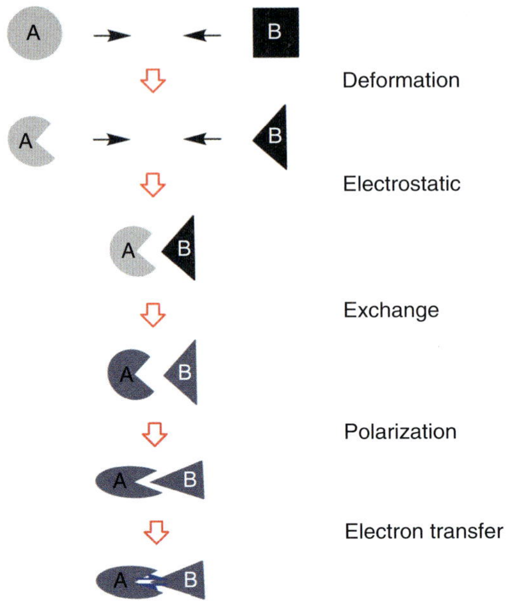

Figure 6.2 Flowchart for the definition of energy terms in the BLW energy decomposition (BLW-ED) approach.

an antisymmetrizer on the Hartree product, but still retain the monomer orbitals, or the monomer electron densities, unchanged as in their isolated states. This actually results in the initial non-optimal BLW and the energy change from the Hartree product to the antisymmetrized initial BLW corresponds to the Pauli exchange repulsion energy ΔE_{ex}.

It should be noted here that both definitions of the electrostatic and exchange energies in the BLW-ED approach are identical to those in many other approaches [16b,f,i,p]. While the electrostatic forces can be well described by classical physics, the Pauli exchange correlation interaction is a quantum mechanical effect. Often we combine these two energies into a single term, called the *Heitler-London energy* ΔE_{HL}.

The approaching of two monomers will inevitably disturb each other's electron density due to the external field imposed by the other monomer. This phenomenon is called *polarization*, which is an energy-lowering (stabilizing) process for the complex. In this step, there is no mutual penetration of electron densities between two monomers, and the BLW is self-consistently optimized. The corresponding energy change from the initial to the optimal BLW is defined as the *polarization energy* ΔE_{pol}. By restricting the relaxation of BL-MOs in only one monomer, we can further define individual polarization energies for monomers as well. However, the sum of individual polarization energies will not be exactly equal to the total polarization energy since polarization interactions are many-body effects. But our extensive computations show that the coupling effect is insignificant and secondary, compared with individual polarization energies.

At the last step, we extend the electron distribution from block-localized orbitals into the basis space of the whole supermolecule **AB**. This process can be characterized as electronic delocalization or simply charge transfer, which further lowers the total energy of the complex relative to the electron-localized state. As the initial BL-MOs are now replaced by delocalized MOs which expand in the whole basis space of the supermolecule, the basis set superposition error (BSSE) is introduced. Consequently, we assign the BSSE correction to the charge transfer energy term:

$$\Delta E_{CT} = E(\Psi) - E(\Psi^{BLW}) + BSSE \tag{6.7}$$

So far the BLW-ED scheme is described at the HF or DFT levels, where electron correlation (ΔE_{corr}) is either completely or partially ignored. This can be remedied by performing high-level quantum mechanical calculations that include correlation effects. In such a way, the overall intermolecular binding energy can be decomposed into a set of physically meaningful contributions as

$$\begin{aligned}\Delta E_b &= \Delta E_{def} + \Delta E_{ele} + \Delta E_{ex} + \Delta E_{pol} + \Delta E_{CT} + \Delta E_{corr} \\ &= \Delta E_{def} + \Delta E_{HL} + \Delta E_{pol} + \Delta E_{CT} + \Delta E_{corr}\end{aligned} \tag{6.8}$$

6.2.4
Probing Electron Transfer with BLW-Based Two-State Models

Electron transfer (ET) effect is ubiquitous in chemical and biological processes and plays a fundamental role in the understanding of enormous chemical and biochemical reactions. According to the general Marcus-Hush model [41], the ET process is often described with two electron-localized diabatic states (**1** and **2** as shown in Figure 6.3). One corresponds to the pre-ET state and the other refers to the after-ET state. With the harmonic approximation, the potential energy for each

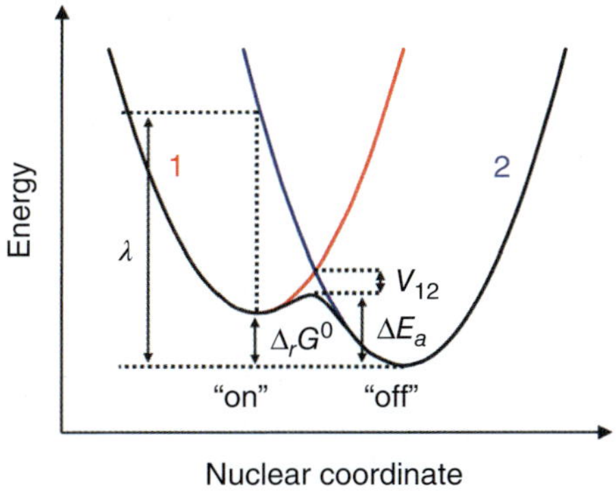

Figure 6.3 Illustration of the general Marcus–Hush two-state model.

diabatic state is represented by a parabolic curve and the adiabatic (ground) state, which is a superposition of **1** and **2**, is characterized by a double-well potential, whose two minima correspond to "on" and "off" states in molecular electronics. The most intriguing issue in this two-state model is the seam of the crossing between the two energy profiles of diabatic states, where charge transfer occurs in line with the Frank-Condon principle. Due to the electronic coupling which is largely dictated by orbital overlap, the ground state energy will be lowered by V_{12}, or the electronic coupling energy, compared with the diabatic state energies. The significance of V_{12} lies in its direct relevance to the ET rate [42], and the evaluation of the electronic coupling term V_{12} and reorganization energy λ thus is the focus of the ET theory.

So far there are a number of approaches toward the computations of the electronic coupling energy [43], including the energy-splitting method based on Koopmans' theorem [43c, 44] and the generalized Mulliken-Hush (GMH) formalism [43e, 45]. But a general feature in these approaches is that the electronic coupling between the diabatic states stems from chosen adiabatic states. A direct evaluation of V_{12} requires the proper definition of diabatic states with the transported electron localized. While this remains a challenge within the MO theory due to the delocalization nature of MOs, for certain systems whose wavefunction cannot be properly described with one Slater determinant, it is possible to use one Slater determinant to represent a charge-localized state at the unrestricted HF (UHF) level [44c, 46]. Of course, the CDFT discussed in the pretext represents the most recent advance in the derivation of the wavefunctions for diabatic states within the framework of the MO theory [17, 18, 40].

Since the BLW method can well define the wavefunctions for electron-localized diabatic states self-consistently and derive their energy profiles at the ab initio level, we can easily derive the electronic coupling energy V_{12} [26a, 47]. Assuming the adiabatic state wavefunction Ψ_0 which is a superposition of two BLW's as

$$\Psi_0 = c_1\Psi_1^{\text{BLW}} + c_2\Psi_2^{\text{BLW}} \tag{6.9}$$

we can obtain a 2×2 secular equation by minimizing the energy of Ψ_0. The solution of the secular equation leads to two energy eigenvalues E_g (lower one) and E_x. At the crossing point of energy curves of diabatic states when their energies are identical, the electronic coupling strength V_{12} is defined as the difference between the diabatic state energy and E_g.

Alternatively, we recognize that the energy of the adiabatic state Ψ_0 can be easily computed with any MO method, and the energies for the diabatic states can be obtained with the BLW method. Based on this information, we can reversely derive the configurational coefficients c_1 and c_2 [Eq. (6.9) and subsequently the electronic coupling energy. Similar to empirical VB approaches, we assume that the overlap integral between Ψ_1^{BLW} and Ψ_2^{BLW} is zero, and subsequently the electronic coupling energy V_{12} is equal to the off-diagonal Hamiltonian matrix element between the two diabatic states whose value is determined via the so-called Reverse CI (configuration interaction) procedure as

$$V_{12} = \sqrt{(\varepsilon_0 - \varepsilon_1)(\varepsilon_0 - \varepsilon_2)} \tag{6.10}$$

6.3 Exemplary Applications

6.3.1 Benzene: Evaluating the Geometrical and Energetic Impacts from π Conjugation

Benzene is a stellar molecule in chemistry and a top example in the introduction of the conjugation concept with the resonance theory [1, 6, 48]. The extraordinary stability exhibited by benzene, later coined as aromaticity, can be well explained with the rapid equilibrium or resonance between two equivalent Kekulé resonance structures

I II

Experimentally, it was found that the hydrogenation heat of 1,3-cyclohexadiene is about two times more than that of cyclohexene [49]. As benzene has three π bonds, its hydrogenation heat was extrapolated to be about three times of that of cyclohexene (see Figure 6.4). However, experimental data showed that the hydrogenation heat of benzene is 36 kcal mol^{-1} lower than what expected, and this energy gap has been used as the experimental estimation of the resonance energy (RE) in benzene. In the matter of fact, Figure 6.4 can be condensed to the following reaction

$$\text{benzene} + 2\,\text{cyclohexane} \rightarrow 3\,\text{cyclohexene} + 36\,\text{kcal}\,\text{mol}^{-1}(\text{expt.}) \qquad (6.11)$$

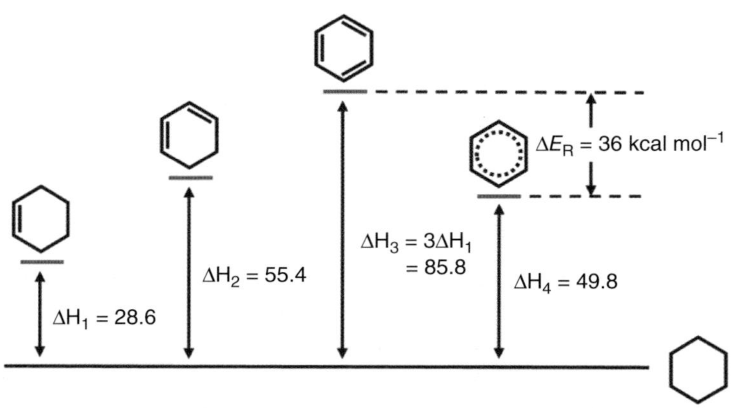

Figure 6.4 Definition of the empirical resonance energy (ΔE_R) of benzene based on hydrogenation heats (kcal mol^{-1}).

However, assigning the reaction heat of Eq. (6.11) to the conjugation stabilization in benzene is problematic, as considerable intramolecular hyperconjugative interaction from the adjacent methylene groups to the π bond in cyclohexene reduces the endothermicity of reaction. This effect can be clearly revealed by the comparison of the hydrogenation heats of ethylene and cyclohexene, which are 32.58 and 28.59 kcal mol^{-1}, respectively [49b, 50]. Thus, the historical resonance stabilization energy of benzene (36 kcal mol^{-1}) is remarkably underestimated [22f, 51]. Pople's "bond separation energy (BSE)" evaluation overcomes this problem by using the following isodesmic reaction [52].

$$3\,\text{ethane} + 3\,\text{ethene} \rightarrow \text{benzene} + 6\,\text{methane} - 65.0\,\text{kcal mol}^{-1}(\text{expt.}) \quad (6.12)$$

Based on Eq. (6.12), the true empirical RE of benzene should be about 65 kcal mol^{-1}.

We must note that when the VB theory and the resonance theory were proposed in the first half of last century, *ab initio* computations were out of reach. Wheland once summarized the empirical rules to identify the important resonance structures for a conjugated molecule [6b], and stated that the RE can be derived "by subtracting the actual energy of the molecule in question from that of the most stable contributing structure." [6a] The total number of independent resonance structures, however, is determined by the Rumer spin coupling patterns [53]. For benzene, there would be 175 resonance structures in total, of which 170 are ionic. Unexpectedly, using minimal basis sets and all possible 175 resonance structures, Norbeck *et al.* [54] and Tantardini *et al.* [55] demonstrated that the five covalent Kekulé and Dewar structures make even less contribution to the ground state of benzene than the rest 170 ionic structures.

The key here lies in the understanding of the Kekulé structure, or the form of one-electron orbitals $\{\varphi\}$ in Eq. (6.1). In the computations by Norbeck *et al.* [54] and Tantardini *et al.* [55], one-electron orbitals are pure atomic orbitals. This echoes the classical notion that a real bond, polar or non-polar bond like H–H, should be described with three VB structures, one covalent and two ionic ones. The Kekulé structure thus was taken as a classical covalent structure in early computations. But we would appreciate the reference Kekulé molecule, or hypothetical 1,3,5-cyclohexatriene, with three "real" double bonds comparable to ethylene and three single bonds comparable to ethane. Consequently, it should be described with 3^3=27 VB structures. Norbeck and Gallup [54] first recognized that the covalent functions cannot represent the benzene molecule in a satisfactory way, and calculated the RE by taking into account all relevant VB structures, 27 for the reference Kekulé structure and 175 for the delocalized ground state. They eventually got a RE of 61.4 kcal mol^{-1}. Later, Hiberty [56] further showed that not only the covalent-only description of benzene is very poor, but also delocalized systems are, as a rule, more ionic than localized ones. This means that if only covalent structures are retained, the description of benzene is even poorer than that of the reference Kekulé structure.

Compared with the classical ab initio VB method where atomic orbitals are used as one-electron orbitals, modern *ab initio* methods allow the expansion of one-electron orbitals to the whole system like MOs with the expansion coefficients optimized self-consistently [20b, 27b, 57]. This kind of semi-delocalized orbitals are called *overlap-enhanced orbitals (OEOs)* which are a direct extension of the Coulson-Fischer orbitals for diatomic molecules [58]. The significant advantage of this type of MO-like orbitals is that most of the correlation energy can be recovered with only a small number of VB structures [20b]. Thus, OEOs play a significant role in the construction of compact wavefunction independent of the growing size of basis sets [20a]. But we need remind that the advantages of OEOs are applicable to the delocalized state only. For a reference electron-localized resonance state such as the hypothetical 1,3,5-cyclohexatriene in the case of benzene, the use of OEOs will introduce additional VB structures other than the 27 VB structures. In other words, the supposedly localized state wavefunction is actually something intermediate between the reference VB structure and the ground state, and will result in unrealistically low resonance energies. That is why we usually call OEOs as "semi-delocalized orbitals." To remedy this drawback, we proposed a kind of localized one-electron orbitals, namely, the bond-distorted orbitals (BDOs), which are expanded with the basis functions centered on two bonding atoms only [59]. Thus, whereas delocalized OEOs allow us to derive a compact expression for the ground state with majority of electron correlation considered, localized BDOs are essential to derive the electron-localized reference state which refers to a Lewis structure whose bonds are comparable to those in actual substances. Similar self-consistent localized orbitals have been adopted by Hiberty and coworkers in the study of resonance effect using *ab initio* VB method [22b,c].

The BLW method, which limits the expansion of block-localized orbitals within a particular physical region of the system, obviously carries the same spirit of the BDOs. The present BLW code even can provide us optimal geometries of electron-localized structures such as the hypothetical 1,3,5-cyclohexatriene (Kekulé) structure. Since in the IUPAC Gold Book, resonance energy is defined as "the difference in potential energy between the actual molecular entity and the contributing structure of lowest potential energy. The RE cannot be measured, but only estimated, since contributing structures are not observable molecular entities," (Muller [60], 1994 #918) the adoption of different geometries for the reference resonance structure could lead to difference values of the RE. Wheland [6b], Mulliken and Parr [28a] and Coulson and Altmann [61] pointed out that in order to compare computational data with the empirical value, geometries of benzene and the hypothetical 1,3,5-cyclohexatriene (Kekulé) should not be the same, and a "compression" energy of the bonds (ΔE_C) would be involved if the calculations are performed at the benzene geometry. Consequently, two types of RE have been defined, one is the vertical resonance energy (VRE [22a, 28a], or quantum mechanical resonance energy, QMRE [22i, 61, 62]) where the Kekulé structure retains the geometry of benzene structure, and the other is the adiabatic resonance energy (ARE [22f, 28e], or equivalently thermochemical resonance energy TRE [22a,h,i,k, 63]) where the localized Kekulé structure takes an optimal bond-alternating geometry. Both kinds of resonance energies are correlated as shown in Figure 6.5, and

$$\mathrm{VRE} = \mathrm{ARE} + \Delta E_C \qquad (6.13)$$

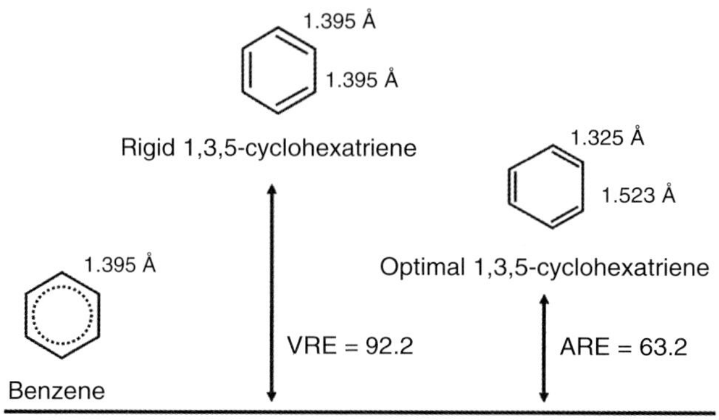

Figure 6.5 Vertical resonance energy (VRE) and adiabatic resonance energy (ARE) in the case of benzene (computed at the B3LYP/6-311+G(d,p) level in kcal mol^{-1}).

It should be noted that the compression energy ΔE_C in Eq. (6.13) can be easily computed with molecular mechanical and quantum mechanical methods. By assuming that the single and double carbon-carbon bond lengths in the optimal Kekulé structure are around 1.54 Å and 1.34 Å, respectively, corresponding to the bond lengths in ethane and ethylene, Wheland [6b] and Mulliken and Parr [28a] estimated the compression energy in benzene about 30 kcal mol^{-1}. By performing quantum mechanical computations of the stretching or compressing of the carbon-carbon bonds in ethane or ethylene, we derived the value 29.4 kcal mol^{-1} at B3LYP/cc-pVTZ level or 27.5 kcal mol^{-1} at the MP2/cc-pVTZ level [34b].

We performed BLW computations of benzene at the B3LYP level with the basis sets of 6-31G(d), 6-311+G(d,p), cc-pVTZ, and aug-cc-pVTZ, and Table 6.1 summarized the key structural parameters and resonance energies.

First of all, we examine the basis set dependency. VB theory and its concepts are essentially based on chemical intuitions, particularly the view that any molecule is composed of individual fragments. One persistent challenge is what kind of

Table 6.1 Optimal structural parameters (bond lengths in Å and bond angles in degree) of benzene and 1,3,5-cyclohexatriene and resonance energies (kcal mol^{-1}) at the DFT(B3LYP) level.

Basis Set	Structure	R_1	R_2	VRE	ARE	ΔE_C
6-31G(d)	benzene	1.397	1.397	88.8	61.8	27.0
	1,3,5-cyclohexatriene	1.329	1.528			
6-311+G(d,p)	benzene	1.325	1.395	92.2	63.2	29.0
	1,3,5-cyclohexatriene	1.325	1.523			
cc-pVTZ	benzene	1.391	1.391	89.1	62.4	26.6
	1,3,5-cyclohexatriene	1.322	1.523			
aug-cc-pVTZ	benzene	1.392	1.392	80.9	57.8	23.1
	1,3,5-cyclohexatriene	1.324	1.513			

mathematical forms should be used for atomic orbitals. The Slater-type orbitals (STOs) derived from the exact solution of Schrödinger equation to hydrogen atom seems the best candidate. But the computational obstacle in the evaluations of four-index integrals is yet to be overcome. Later, Boys introduced the Gaussian functions where the radial decays of the original STOs change from e^{-r} to e^{-r^2} and approximated STOs as linear combinations of Gaussian-type orbitals (GTOs) [64]. The use of GTOs as basis sets for atoms, which is artificial, revolutionized the computational chemistry due to the high computational efficiency by 4–5 orders of magnitude compared to STOs. According to the variational principle, in general the larger the basis sets are, the lower the molecular energy. Thus, with the computational power grows, larger and larger basis sets have been developed. However, when the basis sets get large enough, the atomic individuality would gradually melt down, and ultimately the departure from the atomic concepts with extremely large basis sets diminishes the advantages of VB theory versus MO theory. The completeness of basis sets would diffuse the conventional picture by chemists that molecules are composed of atoms, and bring difficulties in the interpretation of computational results in a physically intuitive way. Mulliken realized this dilemma in the very beginning of MO theory and insightfully pointed out that "the more accurate the calculations become, the more the concepts tend to vanish into thin air." [65]

Table 6.1 shows that from 6-31G(d) to cc-pVTZ, both the structural and energetic results are quite stable, but as expected, the use of aug-cc-pVTZ generates lower resonance energies which theoretically would be zero in the use of infinite basis sets. Optimizations of the fully delocalized benzene result in negligibly shorter carbon-carbon bonds than given by experiment (1.399Å). For the optimal 1,3,5-cyclohexatriene structure, the carbon-carbon double bond lengths are essentially the same as in ethylene at the same theoretical level with the same basis sets, while the single Csp^2–Csp^2 bond lengths, which fluctuate from 1.513 to 1.528 Å, are only slightly shorter than the Csp^3–Csp^3 lengths in ethane. These structural parameters are in good agreement with our chemical intuitions that the double bonds in the hypothetical 1,3,5-cyclohexatriene should be like the ethylene bond and the single Csp^2–Csp^2 bonds should be comparable to or slightly shorter than the ethane bond. In addition, similar BLW optimizations of linear alkenes consistently result in double bond lengths about 1.32 Å and single bond lengths about 1.52 Å, which are in accord with the results and findings on benzene [22f, 34b].

Apart from significant structural changes, π electron delocalization also remarkably stabilizes benzene more than many people initially expected. At the optimal geometry of benzene, the VRE fluctuates slightly around 90 kcal mol^{-1}. After costing about 27 kcal mol^{-1} to distort the σ-frame to a bond-alternating geometry which is close to the classical estimation of 30 kcal mol^{-1} [6b, 28a, 61], benzene exhibits an ARE of around 63 kcal mol^{-1}, which is very much close to Pople's BSE evaluation (65 kcal mol^{-1}, Eq. (31)).

We also note that the VREs from the HF and DFT are very close with the same basis set, suggesting the light impact from the electron correlation, which contributes to both the Kekulé structure and benzene, thus plays a minor role in RE estimates [34c].

While the structural and energetic data in the above are consistent with experimental evidences and thus support the BLW method, it is desirable to obtain NMR data for the Kekulé structure as proton NMR chemical shifts have long been employed as primary experimental indicators of aromaticity in unsaturated organic rings [66]. We have combined the BLW method with the Kutzelnigg's IGLO methodology [67] and introduce the BLW-IGLO method [35a, 68], which enables the direct quantitative evaluation of π-electron delocalization effects on the NMR chemical shifts of polycyclic conjugated and hyperconjugated systems generally.

6.3.2
Butadiene: The Rotation Barrier Versus the Conjugation Magnitude

Butadiene is another typical example of the conjugation effect, which can be described by two resonance structures as **III** and **IV**.

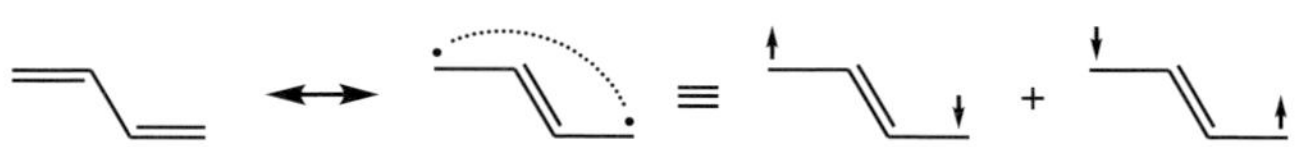

III **IV** **IV**$_a$ **IV**$_b$

The first primary one corresponds to a hypothetical state where the two double bonds are not "conjugated," while the second is essentially a biradical state. The delocalized (Ψ) and strictly localized forms (Ψ_{III} and Ψ_{IV}) of butadiene can be expressed as

$$\Psi = \hat{A}\left[\sigma a_u^2 b_g^2\right] \tag{6.14a}$$

$$\Psi_{\mathrm{III}} = \hat{A}\left[\sigma \pi_{12}^2 \pi_{34}^2\right] \tag{6.14b}$$

$$\Psi_{\mathrm{IV}} = \hat{A}\left[\sigma \pi_{23}^2 \pi_{14}^2\right] \tag{6.14c}$$

where σ represents all orbitals and electrons of σ symmetry. MOs a_u and b_g in Eq. (6.14a) are fully delocalized and expanded in the whole space, while π_{ij} is strictly localized between atoms i and j. As resonance structure **III** is much more stable than **IV**, we choose it as the reference state and performed geometry optimization and RE computations.

Table 6.2 listed the computational results at the DFT(B3LYP) level with several mid-sized basis sets including 6-31G(d), 6-311+G(d,p), cc-pVDZ and cc-pVTZ. Like the case of benzene, structural data clearly showed that the carbon–carbon double bond lengths in the optimal localized 1,3-butadiene structure **III** are essentially of the same values as in ethylene at the same theoretical level, while the single Csp^2–Csp^2 bond lengths (1.518–1.526 Å) are only slightly shorter than the Csp^3–Csp^3 lengths in ethane. The correspondence between the single Csp^2–Csp^2

Table 6.2 Optimal structural parameters (bond lengths in Å) of planar 1,3-butadiene and its primary resonance structure **III** with structural weight (Wgt, %) and resonance energies (kcal mol^{-1}) at the DFT(B3LYP) level.

Basis Set	Structure	R_{12}	R_{23}	VRE	ARE	ΔE_C	Wgt
6-31G(d)	1,3-butadiene	1.340	1.458	13.9	12.2	1.7	89.8
	III	1.329	1.522				
6-311+G(d,p)	1,3-butadiene	1.338	1.456	14.5	12.6	1.9	89.1
	III	1.326	1.526				
cc-pVDZ	1,3-butadiene	1.342	1.458	14.3	12.5	1.8	89.2
	III	1.331	1.524				
cc-pVTZ	1,3-butadiene	1.334	1.453	13.9	12.3	1.6	88.8
	III	1.322	1.518				

and Csp^3–Csp^3 bond lengths implies that the hybridization mode of carbon atoms only modestly influences bond lengths. The π conjugation stabilizes 1,3-butadiene by 12.2–12.6 kcal mol^{-1}, and lengthens the double bonds by only 0.011–0.012 Å, but shortens the central single bond by 0.064–0.070 Å. The difference between VRE and ARE (ΔE_C=1.6–1.9 kcal mol^{-1}), or σ-compression energy, reflects the energy required to distort the optimal geometry of butadiene with a localized wavefunction. Similar to the benzene case [22f, 34b,c] the computed RE can be justified by experimental data and other computational results.

Using the BLW-based two-state model [26, 47], we can also estimate the structural weights of resonance structures **III** and **IV** to the ground state of 1,3-butadiene, which are compiled in Table 6.2. It is clear that the primary structure dominates the ground state wavefunction, while the biradical resonance structure **IV** contributes about 10% to the ground state. We can further define the electron density difference (EDD) between the ground (delocalized) state (Ψ) and the primary resonance state (Ψ_{III}) which can visually describe the electron movement due to the bond delocalization. Figure 6.6 illustrates the electron flow due to conjugation at the B3LYP/cc-pVDZ level, where the lighter color highlights the increasing of the electron density, while the darker color denotes the decreasing of the electron density. Figure 6.6a manifests that the bond delocalization in 1,3-butadiene shifts the π electron density from the two double bonds to the central bond, though this shift is lightly offset by the reverse movement of the σ electrons due to the polarization effect. This picture fits our qualitative chemical intuition very well.

One particularly interesting issue for butadiene is the insignificant variation of the central carbon–carbon bond with a low rotation barrier (the CBS-Q calculated value is 6.0 kcal mol^{-1} [69]) [28f]. At the DFT level with various basis sets, we optimized the perpendicular conformation of butadiene where the π conjugation is completely quenched. The optimal central carbon–carbon bond length is about 1.48 Å (see Table 6.3), which is very close to the value 1.45 Å (Table 6.2) in

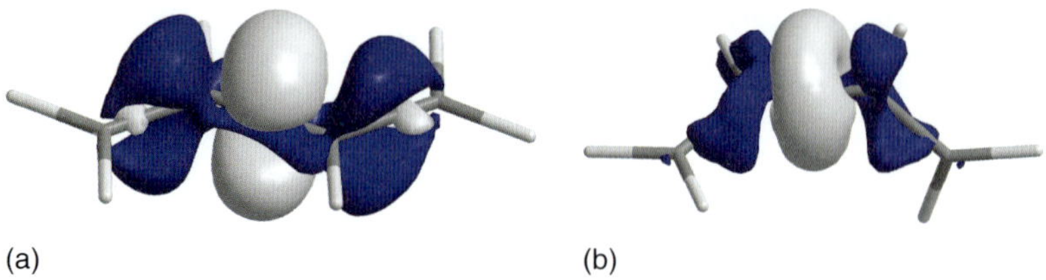

Figure 6.6 Electron density difference (EDD) map showing the electron movement due to the bond delocalization in 1,3-butadiene, (a) Planar and (b) Perpendicular, where grey shows the density enhancement and dark the density reduction (isodensity 1.0×10^{-3} a.u.).

Table 6.3 Optimal structural parameters (bond lengths in Å) of perpendicular 1,3-butadiene and vertical resonance energy (kcal/mol) at the DFT(B3LYP) level.

Basis Set	R_{12}	R_{23}	Barrier	VHCE
6-31G(d)	1.334	1.487	7.5	10.3
6-311+G(d,p)	1.331	1.485	7.0	11.5
cc-pVDZ	1.336	1.486	7.5	11.7
cc-pVTZ	1.327	1.482	7.2	10.4

the planar butadiene. In the optimization of 1,3,5-cyclohexatriene (Table 6.1), we already know that the optimal Csp^2–Csp^2 should be about 1.52 Å. Thus, the shortening of the central bond in the perpendicular butadiene indicates that there is some effect other than the π conjugation. That effect is now recognized as the hyperconjugative interactions between the CC π orbital and the antibonding orbitals of the two central carbon–carbon bond and can be roughly labeled as $\pi_{CC}\rightarrow\sigma_{CC}^*$. Daudey *et al.* first probed the influence of π conjugation and hyperconjugation in the shortening of the central carbon–carbon bond in butadiene by substituting the delocalized π MOs in butadiene with the ethylenic π MO derived from the calculation of ethylene with the same basis set [28f]. They found that the hyperconjugation effect in the perpendicular structure is comparable to the conjugation effect in the planar structure, thus well explained the origin of the central bond shortening.

We derived the optimal BLW (Eq. (6.14b)) for the perpendicular structure to estimate the vertical hyperconjugation energy (VHCE) at the DFT optimal geometries. Table 6.3 listed the data. Our results confirmed the findings by Daudey *et al.* [28f], showing that the hyperconjugation stabilization energy is about 11 kcal mol^{-1}, only 3 kcal mol^{-1} lower than the conjugation energy in the planar structure. Since the rotaion barrier is about 7 kcal mol^{-1}, the remaining 4 kcal mol^{-1} may come from the steric effect. Figure 6.6b further plots the electron density change due to the hyperconjugation effect. Clearly, the carbon–carbon double bonds lose π electron density to the central bond, and the interaction thus can be defined as $\pi_{CC}\rightarrow\sigma_{CC}^*$.

6.3.3
Ethane: What Force(s) Governs the Conformational Preference?

So far, we have demonstrated the construction and optimization of electron-localized state wavefunction by limiting the expansion of BL-MOs in terms of atomic orbitals (or basis functions). In certain cases of molecular symmetries, it is possible (and interesting) to partition the group basis functions with the BLW method. Ethane is such a prominent example [70], as it maintains the D_3 symmetry in the rotation about the carbon-carbon bond and there have been significant controversies over the origin of the corresponding rotational barrier [31, 71].

It has been well known that the staggered structure of ethane is about 3 kcal mol^{-1} more stable than the eclipsed structure [72]. Classically, it was envisioned that the ethane rotation barrier originates from the Pauli exchange repulsion between the two methyl groups [1]. This intuitive, steric repulsion explanation fits to our simple stick-ball model for molecules very well, and was quantitatively confirmed by Sover *et al.*'s analyses by means of bond-orbital model in the sixties [28b], and still remains a popular and illuminating example in the description of steric effect in textbooks. More recently, there have been a few quantum mechanical energy decomposition analyses which further confirm that the repulsive interaction governs the staggered conformation of ethane [72e, 73].

Apart from the classical repulsion model, however, there has been an alternative explanation existing for the origin of ethane rotation barrier, namely, the hyperconjugation model [72f, 74]. The concept of hyperconjugation, introduced by Mulliken in 1939 [75], is used in the description of the interaction between an occupied bond orbital σ_{ij} and a vicinal unoccupied antibond orbital $\sigma_{kl}{}^*$, which results in an occupied delocalized orbital $\sigma_{ij}{}'$ ($\lambda \ll 1$)

$$\sigma'_{\rm ij} = \sigma_{\rm ij} + \lambda\sigma^*_{\rm kl} \tag{6.15}$$

and the stabilization of the system. Based on the approximation of MOs as linear combinations of bond orbitals at the semi-empirical level, Brunck and Weinhold first demonstrated that the vicinal hyperconjugative interaction between the $\sigma_{\rm CH}$ occupied orbitals in one methyl group and the $\sigma_{\rm CH}{}^*$ antibonding orbitals in the other methyl group stabilizes the staggered conformation and thus is the dominant force responsible for the rotational barrier in ethane [74c]. Later Weinhold *et al.* developed the NBO approach at the ab initio level that verified their initial semi-empirical results [13b, 76].

Thus, there are two major competing models to interpret the origin of the ethane rotation barrier, one is the steric repulsion model, and the other is hyperconjugation model. This unsettled issue regained enormous attention in 2001. Based on the NBO method, Pophristic and Goodman carried out a "flexing" analysis in terms of energies associated with structural, steric, exchange and hyperconjugative interactions during methyl rotation. By removing the $\sigma_{\rm CH}$–$\sigma_{\rm CH}{}^*$ hyperconjugative interaction, they found that steric repulsion favors the eclipsed conformation, thus repulsive forces have no effect on the preference for a staggered conformation [71a]. Instantly, this unusual finding inspired more studies [31, 71c, 77].

Bickelhaupt and Baerends first evaluated the Pauli and electrostatic interactions explicitly using a wavefunction composed of fragmental MOs of methyl groups and showed that although hyperconjugation does favor the staggered ethane conformer, Pauli exchange repulsions are the dominant force responsible for the rotational barrier in ethane [71c]. Differently, using *ab initio* VB method, we explicitly computed the hyperconjugation energy in ethane and demonstrated that, in accord with Bickelhaupt and Baerends's finding, although the hyperconjugation effect indeed favors the staggered conformation, its contribution to the barrier is only secondary [31].

Quite interestingly, the best computational strategy for the solution to the nature of the ethane rotation barrier within the MO theory was proposed by Mulliken himself when he coined the hyperconjugation concept in 1939, though *ab initio* calculations were not available at that time. Still, Mulliken insightfully cautioned that "hyperconjugation in ethane should have little or no direct effect in restricting free rotation" since it is "only of second order" [75].

Mulliken first recognized that "Hyperconjugation in ethane involves interaction aming eight electrons which in the absence of any conjugation may be assigned to two sets of "[πe]" C–H bonding MO's, one set localized in each CH_3 group" [75]. We consider the point group D_3 which is unchanged in the ethane rotation. There are two kinds of irreducible bases for D_3, one is the fully symmetry "a" orbitals which are irrelevant to the rotation and the other is twofold degenerate "e" orbitals which are the "[πe]" orbitals in Mulliken's paper. Thus, hyperconjugation in ethane involves eight electrons which occupy the degenerate *e*-symmetric orbitals, as the rest electrons occupy the fully symmetric orbitals and thus are unaffected by the rotation. In the absence of any hyperconjugation, each set of *e* orbitals with four electrons are localized on in each CH_3 group. Figure 6.7 shows the *e* orbitals for methyl groups, where 1π (differentiated with $1\pi'$ and $1\pi''$ for two methyl groups) are occupied with four electrons for each methyl group and 2π are unoccupied. Based on the definition of hyperconjugation in MO theory [78], hyperconjugative interaction results from the charge transfer from occupied group-localized orbitals to vicinal unoccupied group-localized orbitals (i.e., $1\pi'+2\pi''$ and $1\pi''+2\pi'$, see Figure 6.7a), which consequently stabilizes the system. The interaction between neighboring occupied orbitals ($1\pi'$ and $1\pi''$), however, involves the quantum mechanical Pauli exchange repulsion and thus unfavorable. We can term this part of interaction as steric effect (Figure 6.7b).

Once we finish the transformation of atomic basis functions to irreducible basis functions for each methyl group, we can construct the electron-localized reference state by limiting the expansion of *e*-symmetric group orbitals to only one methyl group, and express the corresponding BLW as [70]

$$\Phi_L = \hat{A}(1a_1^2 2a_1^2 3a_1^2 4a_1^2 5a_1^2 1\pi'^4 1\pi''^4) \tag{6.16}$$

where $1\pi'$ and $2\pi''$ are group-localized orbitals confined to only each of the two methyl groups, respectively, as Mulliken initially suggested. In contract, the conventional HF wavefunction for ethane

$$\Psi = \hat{A}(1a_1^2 2a_1^2 3a_1^2 4a_1^2 5a_1^2 1e^4 2e^4) \tag{6.17}$$

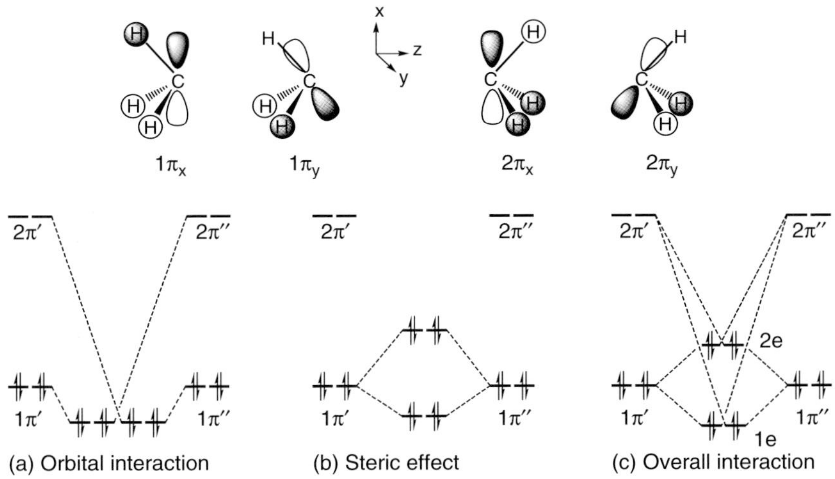

Figure 6.7 The "e"-symmetric group-localized orbitals of methyl groups in ethane. $1\pi_x$ and $1\pi_y$ are degenerate and occupied, whereas the degenerate $2\pi_x$ and $2\pi_y$ are unoccupied. Orbital interactions in ethane: (a) hyperconjugative interaction; (b) steric interaction and (c) overall interactions. (Reprinted with permission from Ref. [70]. Copyright 2007 American Chemical Society)

defines an electron-delocalized state. The energy difference between the delocalized (Ψ) and localized states (Φ_L) reflects the hyperconjugative stabilization (E_{hc}).

Apart from the estimation of the hyperconjugation energy, it is equally essential to quantify the steric effect independently. For this purpose, we proposed a unique solution, namely freezing the group-localized orbitals and monitoring the energy variation in the process of rigid rotations where all structural parameters except the torsional angle are fixed and ascribing this energy variation as the steric energy change (ΔE_s). The re-optimization of group-localized orbitals along the rotation can be easily defined as the electronic relaxation energy (ΔE_s), and the energy change due to the structural change is labeled as the geometric relaxation energy (ΔE_g). In the way as shown in Figure 6.8, we decompose the rotation barrier into a sum of contributions from the hyperconjugation, steric, electronic and geometric relaxation terms

$$\Delta E_b = -E_{hc}(s) + \Delta E_s + \Delta E_e + \Delta E_g + E_{hc}(e) = \Delta E_{hc} + \Delta E_s + \Delta E_e + \Delta E_g \quad (6.18)$$

Obviously the energy decomposition scheme [Eq. (6.18)] relies on the starting geometry. Figure 6.8 describes the rotation from the optimal staggered structure to the optimal eclipsed structure. It would be interesting to examine the energy terms in the rotation from the optimal eclipsed structure to the optimal staggered structure. Tables 6.4 and 6.5 listed the energy contributions to the barriers of ethane, disilane, digermane and methylsilane from different staring geometries with the basis sets of 6-31G(d) and 6-311+G(d,p). Starting from different geometries do have impacts on the energy values, but the changes are very limited and do not alter

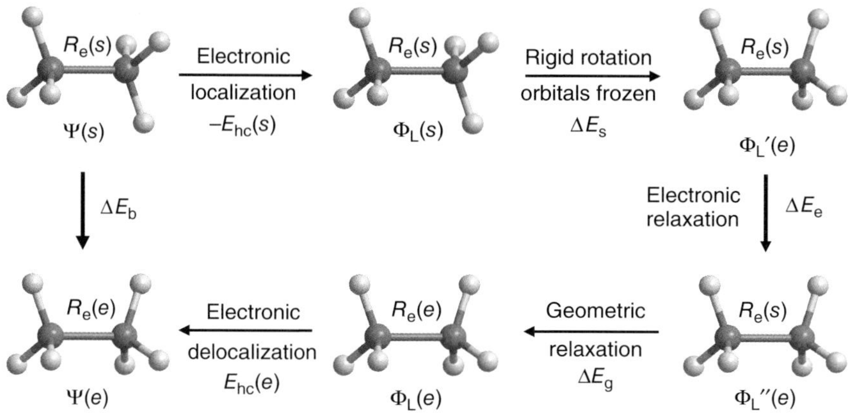

Figure 6.8 A decomposition scheme to explore the geometric impact on the rotational barrier. (Reprinted with permission from Ref. [70]. Copyright 2007 American Chemical Society)

Table 6.4 Energy decomposition of the rotational barriers from staggered structures to eclipsed structures for ethane congeners (kcal/mol).

Molecule	Basis set	ΔE_{hc}	ΔE_s	ΔE_e	ΔE_g	ΔE_b
CH_3CH_3	6-31G(d)	0.76	2.73	−0.01	−0.50	2.98
	6-311+G(d,p)	0.76	2.87	−0.03	−0.54	3.06
SiH_3SiH_3	6-31G(d)	0.30	0.71	0.00	−0.06	0.95
	6-311+G(d,p)	0.26	0.77	−0.01	−0.04	0.98
GeH_3GeH_3	6-31G(d)	0.09	0.78	−0.01	−0.05	0.81
	6-311+G(d,p)	0.14	0.67	0.00	−0.08	0.73
CH_3SiH_3	6-31G(d)	0.38	1.16	−0.01	−0.13	1.40
	6-311+G(d,p)	0.37	1.27	−0.02	−0.16	1.46

the conclusion at all. The geometric relaxation is a combined structural response to both the steric repulsion and hyperconjugative attraction. Compared with the analysis from staggered to eclipsed structures, we observe a slight reduction of both steric energy (ΔE_s) and geometric relaxation energy (ΔE_g), with their sum unchanged, in rotations from eclipsed to staggered structures. In all cases, we found that the electronic relaxation causes trivial energy changes in the rotation, while the geometric relaxation which lengthens the central bonds from staggered to eclipsed structures slightly stabilizes the systems, notably in ethane. This demonstrates that a small change of the central bond in the rotation makes very modest energy variation to the rotation barrier, and as a consequence, both rigid and relaxed rotations have the same mechanism which is dominated by the steric effect, as clearly shown in Tables 6.4 and 6.5.

Table 6.5 Energy decomposition of the rotational barriers from eclipsed structures to staggered structures for ethane congeners (kcal/mol).

Molecule	Basis Set	ΔE_{hc}	ΔE_s	ΔE_e	ΔE_g	ΔE_b
CH_3CH_3	6-31G(d)	−0.76	−2.45	−0.02	0.25	−2.98
	6-311+G(d,p)	−0.76	−2.55	−0.03	0.28	−3.06
SiH_3SiH_3	6-31G(d)	−0.30	−0.66	0.00	0.01	−0.95
	6-311+G(d,p)	−0.26	−0.73	−0.01	0.02	−0.98
GeH_3GeH_3	6-31G(d)	−0.09	−0.75	−0.01	0.04	−0.81
	6-311+G(d,p)	−0.14	−0.62	−0.03	0.06	−0.73
CH_3SiH_3	6-31G(d)	−0.38	−1.08	−0.01	0.07	−1.40
	6-311+G(d,p)	−0.37	−1.16	−0.02	0.09	−1.46

It is of interest to note that the hyperconjugative interaction indeed favors staggered structures, as suggested by the hyperconjugation model, but its magnitude is no more than 30% of the total rotation barriers. In the matter of fact, the geometric relaxation and hyperconjugative interaction partially cancel out, and consequently the steric repulsion accounts for over 80% of rotation barriers in ethane and its analogous.

6.3.4 H_3B-NH_3: Quantifying the Electron Transfer Effect in Donor-Acceptor Complexes

BH_3NH_3 is a textbook example for electron donor-acceptor (EDA) complexes whose bond strengths range from about the average of strong covalent bond to weak van der Waals bond. The boron–nitrogen dative bond, which is the classical example to demonstrate the ET from the lone nitrogen pair to the vacant orbital on boron, has important physiological activities [79]. The simplest system of the boron-nitrogen dative bond is BH_3NH_3, which have been extensively studied experimentally [80] and computationally [16h, 30, 81].

Although BH_3NH_3 is isoelectronic with ethane, its bond strength (31.1 kcal mol^{-1}) is only one third of the latter. Based on the resonance theory, BH_3NH_3 can be well described by the following two resonance structures (or diabatic states)

V VI

The contribution from the ionic state is highlighted by the high dipole moment (5.216 Debye) [80a], suggesting that there is a significant amount of charge transferred from NH_3 to BH_3. The consequence of the intermolecular ET is, the hydrogen atoms in the NH_3 moiety carry a fraction of positive charge whereas those in the BH_3 moiety carry negative charges. The interaction between hydride atoms and protons forms unconventional B-$H^{\delta-}\cdots{}^{\delta+}H$–N bonds [80c, 82], which may play a primary role in crystal packing and supramolecular assembly in molecular aggregations. Using the BLW-ED approach [16j, 35c], we can probe the nature of the boron-nitrogen dative bond in BH_3NH_3.

We first explore the neutral resonance state **V**, where the lone nitrogen pair does not delocalize to the BH_3 moiety, and thus there are only nonbonded interactions including the conventional van der Waals (dispersion and exchange) and electrostatic interactions. Here we used both B3LYP and PBE-D which is a nonempirical density functional PBE plus Grimme's empirical dispersion correction [83], to perform geometry optimizations and BLW-ED analysis. With the B3LYP functional, the BLW optimization shows that the boron–nitrogen distance ranges from 2.151 to 2.528 Å and the binding energy ($\Delta E_b{}'$) fluctuates from −3.5 to −4.6 kcal mol^{-1} after the BSSE correction using the counterpoise method [84] with the basis sets of 6-31G(d), 6-311+G(d,p) and cc-pVTZ. In a previous work, we obtained 2.33 Å and −5.2 kcal mol^{-1} at the VBSCF/6-31G(d) level without BSSE correction [30]. With the PBE-D functional, however, the boron-nitrogen distance reduces to 2.025–2.321 Å and the binding energy increases from −5.3 to −7.8 kcal mol^{-1} with the same three basis sets. The seemingly large variation implies two facts. One is the shallow and flat energy profile for weak van der Waals attraction between NH_3 and BH_3 which can be well fitted by many potential functions such as the Lennard-Jones form. The other is the relative significance of the BSSE effect due to the low binding energy, for example, at the BLW optimal geometries with B3LYP functional, the BSSE amounts to 2.0, 0.7, and 1.5 kcal mol^{-1} with the 6-31G(d), 6-311+G(d,p) and cc-pVTZ basis sets, suggesting that the geometry optimization need be done with BSSE corrections.

The boron-nitrogen distance in the hypothetical van der Waals complex may suggest the atomic van der Waals radius, which is not available for boron due to the lack of experimental data [85]. For comparison, the covalent radii for nitrogen and boron are 0.75 and 0.82 Å, respectively, and the van der Waals radius for nitrogen is 1.55 Å. However, it seems certain that the equilibrium distance in the neutral resonance structure **V** must be shorter than the sum of van der Waals radii for nitrogen and boron due to the electrostatic interactions as nitrogen is an electronegative atom and boron is an electropositive atom compare with hydrogen atoms (Table 6.6).

We then conduct the energy decomposition analysis at the optimal BH_3NH_3 complex whose boron–nitrogen distance is 1.65–1.67 Å, compared with the experimental value 1.657 Å [80a]. The B3LYP binding energy is close to the QCISD/aug-cc-pVTZ data (−29.4 kcal mol^{-1} [81m]), but higher than the experimental measure 23.9 kcal mol^{-1}. With the PBE-D functional, however, the binding energy is quite remarkably overestimated compared with the benchmark value. From the optimal monomer geometries to the optimal complex, considerable structural changes occur for the BH_3 moiety, which is planar in the isolated state but pyramidal in the complex. Calculations show that the total deformation energy cost is about 13 kcal mol^{-1}, which dominantly comes from the BH_3 moiety. The positive

Table 6.6 Optimal nitrogen-boron distance (Å) and energy decomposition analysis (kcal mol^{-1}) for BH_3NH_3

Theoretical Level	R(DFT)	R(BLW)	ΔE_{def}	ΔE_{HL}	ΔE_{pol}	ΔE_{ct}	$\Delta E_b{}'$	ΔE_b
B3LYP/6-31G(d)	1.671	2.151	12.7	9.5	−20.9	−30.9	−4.6	−29.6
B3LYP/6-311+G(d,p)	1.668	2.528	13.0	26.0	−31.0	−35.7	−3.5	−27.7
B3LYP/cc-pVTZ	1.661	2.338	13.2	24.4	−36.0	−30.4	−3.7	−28.8
PBE-D/6-31G(d)	1.663	2.025	12.5	6.2	−22.3	−33.4	−7.8	−37.0
PBE-D/6-311+G(d,p)	1.660	2.321	12.6	24.1	−33.3	−38.4	−5.3	−35.0
PBE-D/cc-pVTZ	1.654	2.116	13.0	22.1	−38.7	−32.4	−6.4	−36.0

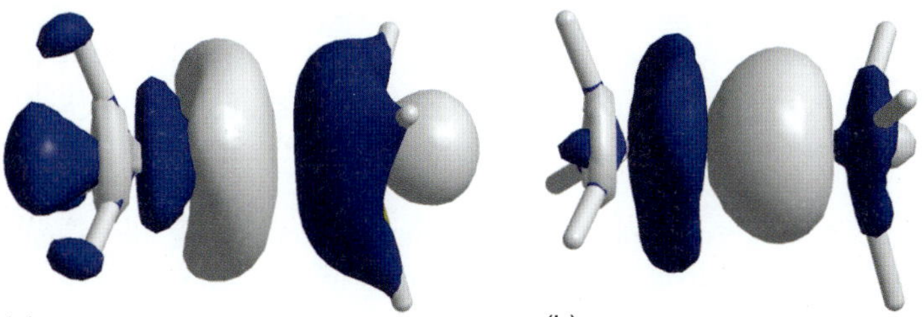

(a) (b)

Figure 6.9 Electron density difference (EDD) map showing (a) the polarization effect and (b) electron transfer from NH_3 (left) to BH_3 (right) (isodensity 5.0×10^{-3} a.u.).

Heitler–London energy indicates that the Pauli repulsion contributes more than the attractive electrostatic interaction. However, there is large discrepancy between the 6-31G(d) and the rest two larger basis sets for this energy term as well as the polarization energy. But the charge transfer stabilization energy is reasonably stable, and almost the same as the final total binding energy. Figure 6.9 exhibits the electron density changes due to both the polarization and ET. It is interesting to view that the electron density in NH_3 moves to the nitrogen side, while the density in BH_3 moves away from the boron atom. These movements are a preparation for the subsequent ET, which moves the electron density from the nitrogen atom to the boron atom, as Figure 6.9b clearly shows.

6.4 Conclusion

Intramolecular electron delocalization, including conjugation and hyperconjugation, plays an important role in molecular structures, properties and reactivity. Conventionally, resonance theory has been developed to elucidate the correlations among the delocalization, structures, and properties at the qualitative level, though delocalization stabilization energy has been well defined as a measure of the magnitude of contributions from resonance structures other than the principal Lewis structure to the ground state of a delocalized molecule. Since the mainstream methods in computational chemistry are MO theory-based and generally unable to define individual electron-localized (resonance) structures, numerous indirect approaches have been designed to estimate the relative delocalization energy with other real reference molecules. But the arbitrariness in the choice of references results in a large variation of the delocalization energy for the same molecule, notably benzene [34b,c]. Thus, the development of ab initio VB methods, which explicitly define resonance structures with HLSP functions, is highly expected.

The BLW method represents the simplest, most efficient ab initio VB method. It partitions all electrons and basis function into several blocks (groups, or fragments), and restricts the expansion of any block-localized MO to only one block. Apart from the orthogonality constraint for BL-MOs within each block, it defines the corresponding electron-state with one Slater determinant. As a consequence, the BLW retains the characteristics of VB theory, but possesses the high efficiency of the MO theory.

It should be noted that experimentally, electron-localized states are usually inaccessible, and the lack of direct experimental proofs makes several significant controversies such as the nature of ethane rotation barrier unsettled and no computational method accepted universally. However, we stress that still there are plenty of experimental evidences existing which can be used to justify computational results. For instance, the geometry optimization of an electron-localized state for a conjugated molecule should generate structural parameters as well as NMR, vibrational frequencies etc comparable with the data for nonconjugated molecules. We have extensively analyzed the resonance in benzene and the BLW computations of the hypothetical Kekulé structure (namely the 1,3-cyclohexadiene) produce results in accord with all experimental proofs. The comparison of the conjugation in the planar structure and the hyperconjugation in the perpendicular structure of butadiene confirms the early findings by Daudey *et al.* [28f] that both have similar magnitudes. Thus, simply using the rotation barrier over a conjugated bond, which disables the conjugation, to evaluate the conjugation effect is often problematic.

The BLW method can also be used to partition irreducible basis function instead of atomic basis functions. As an interesting example, we analyzed the rotation barrier in ethane by partitioning the *e*-symmetric group orbitals in order to estimate the hyperconjugation stabilization energy among the *e* group orbitals. Our computations showed that there are notable hyperconjugative interactions. However, the interactions change only slightly in the process of rotation. In other words, the hyperconjugation effect does not have a decisive role for the barrier, which largely originates from the steric effect, as the conventional wisdom holds. The BLW method can be further applied to the study of intermolecular interactions, and the ET processes.

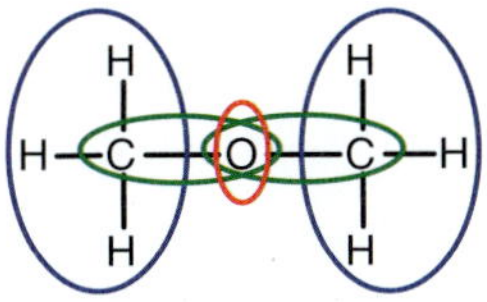

Figure 6.10 Definition of blocks for dimethoxymethane.

6.5 Outlook

The present version of the BLW method allows the assignment of a basis function to only one sub-group of the system and there is no redundancy of members in sub-groups. However, in the construction of electron-localized state for ET within a molecule, such as the hyperconjugation effect, in general there is no way to clearly allocate some basis functions to one block. For instances, in the exploration of the hyperconjugative interactions from the lone pairs on oxygen atoms to adjacent methyl or methylene groups such as in dimethoxymethane shown by Figure 6.10, it is necessary to strictly localize the lone oxygen pairs (shown by the red cycle). But the oxygen atomic orbitals also participate in the σ bonding with carbon atoms (shown by the green cycles). Thus, some basis functions may appear in several blocks. Currently we are developing the generalized BLW method which will allow the definition of blocks more flexible by removing the limit that any basis function belongs to only one block. As such, the applicability of the BLW method will be substantially broadened. It should be noted that such kind of general block-diagonalization approach was first adopted by Stoll *et al.* in a few simple molecules [28c,d]. Lately, Sironi, and coworkers [86] demonstrated that orbitals strictly localized on pre-defined molecular fragments are transferable and thus provide an efficient and effective way to construct the electronic structure of large molecules.

We have tested the generalized BLW idea with a few simple examples such as the anomeric effect in dimethoxymethane and substituted tetrahydropyrans [87] and the conformational preference of formic acid for the *trans* (*Z*) conformer over the *cis* (*E*) conformer [87], and confirmed the applicability of the generalized BLW method.

Acknowledgements

This work was supported by the US National Science Foundation under Grants CHE-1055310 and CNS-1126438.

References

1. Pauling, L.C. (1960) *The Nature of the Chemical Bond*, 3rd edn, Cornell University Press, Ithaca, NY.
2. Langmuir, I. (1919) *J. Am. Chem. Soc.*, **41**, 868–934.
3. Lewis, G.N. (1923) *Valence and the Structure of Atoms and Molecules*, The Chemical Catalog Co., Inc., New York.
4. Pauling, L. (1988) *General Chemistry*, Dover Publications, New York.

5. (a) Lewis, G.N. (1916) *J. Am. Chem. Soc.*, **38**, 762–785. (b) Frenking, G. (2007) *J. Comput. Chem.*, **28** (1), 1–466.
6. (a) Wheland, G.W. (1944) *The Theory of Resonance*, John Wiley & Sons, Inc., New York.(b) Wheland, G.W. (1955) *Resonance in Organic Chemistry*, John Wiley & Sons, Inc., New York.
7. Heitler, W. and London, F. (1927) *Z. Phys.*, **44**, 455–472.
8. (a) Pauling, L. (1931) *J. Am. Chem. Soc.*, **53**, 3225–3237. (b) Pauling, L. (1931) *J. Am. Chem. Soc.*, **53**, 1367–1400. (c) Slater, J.C. (1931) *Phys. Rev.*, **37**, 481–489. (d) Slater, J.C. (1931) *Phys. Rev.*, **37**, 1109–1144.
9. (a) Lennard-Jones, J.E. (1929) *Trans. Faraday Soc.*, **25**, 668–686. (b) Hall, G.G. (1991) *Adv. Quantum Chem.*, **22**, 1–6. (c) Slater, J.C. (1929) *Phys. Rev.*, **34**, 1293–1322.
10. Shaik, S. (2007) *New J. Chem.*, **31**, 1981–2128.
11. (a) Edmiston, C. and Ruedenberg, K. (1963) *Rev. Mol. Phys.*, **35**, 457–465. (b) Foster, J.M. and Boys, S.F. (1960) *Rev. Mod. Phys.*, **32**, 300–302. (c) Pipek, J. and Mezey, P.G. (1989) *J. Chem. Phys.*, **90**, 4916–4926.
12. (a) Bader, R.F.W. (1991) *Chem. Rev.*, **91**, 893–928. (b) Bader, R.F.W. (1994) *Atoms in Molecule: A Quantum Theory*, Oxford University Press.
13. (a) Foster, J.P. and Weinhold, F. (1980) *J. Am. Chem. Soc.*, **102**, 7211–7218. (b) Reed, A.E., Curtiss, L.A., and Weinhold, F. (1988) *Chem. Rev.*, **88**, 899–926.
14. Zubarev, D.Y. and Boldyrev, A.I. (2008) *Phys. Chem. Chem. Phys.*, **10**, 5207–5217.
15. (a) Bader, R.F.W. and Nguyen-Dang, T.T. (1981) *Adv. Quantum Chem.*, **14**, 63–124. (b) Silvi, B. and Savin, A. (1994) *Nature*, **371**, 683–686.
16. (a) Cybulski, S.M. and Scheiner, S. (1990) *Chem. Phys. Lett.*, **166**, 57–64. (b) Kitaura, K. and Morokuma, K. (1976) *Int. J. Quantum Chem.*, **10**, 325–340. (c) Moszynski, R., Heijmen, T.G.A., and Jeziorski, B. (1994) *Mol. Phys.*, **88**, 741–758. (d) Stevens, W.J. and Fink, W.H. (1987) *Chem. Phys. Lett.*, **139**, 15–22. (e) Chalasinski, G. and Gutowski, M. (1988) *Chem. Rev.*, **88**, 943–962. (f) Bickelhaupt, F.M. and Baerends, E.J. (1999) in *Reviews in Computational Chemistry*, vol. 15 (eds K.B. Lipkowitz and D.B. Boyd), John Wiley & Sons, Inc., New York, p. 1. (g) Erhardt, S. and Frenking, G. (2006) *Chem. Eur. J.*, **12**, 4620–4629. (h) Glendening, E.D. and Streitwieser, A. (1994) *J. Chem. Phys.*, **100**, 2900–2909. (i) Lein, M., Szabo, A., Kovacs, A., and Frenking, G. (2003) *Faraday Discuss.*, **124**, 365–378. (j) Mo, Y., Gao, J., and Peyerimhoff, S.D. (2000) *J. Chem. Phys.*, **112**, 5530–5538. (k) van der Vaart, A. and Merz, K.M. Jr., (1999) *J. Phys. Chem. A*, **103**, 3321–3329. (l) Khaliullin, R.Z., Cobar, E.A., Lochan, R.C., Bell, A.T., and Head-Gordon, M. (2007) *J. Phys. Chem. A*, **111**, 8753–8765. (m) Su, P. and Li, H. (2009) *J. Chem. Phys.*, **131**, 014101. (n) Wu, Q., Ayers, P.W., and Zhang, Y.K. (2009) *J. Chem. Phys.*, **131**, 164112. (o) Bagus, P.S., Hermann, K., and Bauschlicher, C.W. Jr., (1984) *J. Chem. Phys.*, **80**, 4378–4386. (p) Chen, W. and Gordon, M.S. (1996) *J. Phys. Chem.*, **100**, 14316–14328. (q) Morokuma, K. (1977) *Acc. Chem. Res.*, **10**, 294–300. (r) Ziegler, T. and Rauk, A. (1977) *Theor. Chem. Acc.*, **46**, 1–10. (s) Michalak, A., Mitoraj, M., and Ziegler, T. (2008) *J. Phys. Chem. A*, **112**, 1933–1939. (t) Mitoraj, M.P., Michalak, A., and Ziegler, T. (2009) *J. Chem. Theor. Comput.*, **5**, 962–975. (u) Heßelmann, A. and Jansen, G. (2002) *Chem. Phys. Lett.*, **357**, 464–470. (v) Jeziorski, B., Moszynski, R., and Szalewicz, K. (1994) *Chem. Rev.*, **94**, 1887–1930. (w) Stone, A.J. (1996) *The Theory of Intermolecular Forces*, Oxford University Press, New York. (x) Szalewicz, K. and Jeziorski, B. (1979) *Mol. Phys.*, **38**, 191–208. (y) Williams, H.L. and Chabalowski, C.F. (2001) *J. Phys. Chem. A*, **105**, 646–659. (z) Angeli, C., Cimiraglia, R., and Malrieu, J.P. (2008) *J. Chem. Edu.*, **85**, 150–158. (aa) Guihéry, N., Malrieu, J.P., Evangelisti, S., and Maynau, D. (2001) *Chem. Phys. Lett.*, **349**, 555–561. (ab) Malrieu, J.P., Guihéry, N., Calzado, C.J., and Angeli, C. (2006) *J. Comput. Chem.*, **28**, 35–50.

17. Dederichs, P.H., Bluegel, S., Zeller, R., and Akai, H. (1984) *Phys. Rev. Lett.*, **53**, 2512–2515.

18. Wu, Q. and Van Voorhis, T. (2006) *J. Chem. Theor. Comput.*, **2**, 765–774.

19. (a) Hong, G.Y., Rosta, E., and Warshel, A. (2006) *J. Phys. Chem. B*, **110**, 19570–19574. (b) Wesolowski, T.A. and Warshel, A. (1993) *J. Phys. Chem.*, **97**, 8050–8053.

20. (a) Cooper, D.L. (2002) *Valence Bond Methods, Theoretical and Computational Chemistry*, Elsevier, Amsterdam. (b) Cooper, D.L., Gerratt, J., and Raimondi, M. (1986) *Nature*, **323**, 699–701. (c) Cooper, D.L., Gerratt, J., and Raimondi, M. (1991) *Chem. Rev.*, **91**, 929. (d) Gallup, G.A., Vance, R.L., Collins, J.R., and Norbeck, J.M. (1982) *Adv. Quantum Chem.*, **16**, 229–272. (e) Zhang, Q. and Li, X. (1989) *J. Mol. Struct.*, **189**, 413–425. (f) Dijkstra, F. and van Lenthe, J.H. (2000) *J. Chem. Phys.*, **113**, 2100–2108. (g) Hiberty, P.C. (1997) *THEOCHEM*, **398-399**, 35–43. (h) Mcweeny, R. (1999) *Int. J. Quantum Chem.*, **74**, 87–96. (i) Shaik, S. and Shurki, A. (1999) *Angew. Chem. Int. Ed.*, **38**, 586–625. (j) Thorsteinsson, T. and Cooper, D.L. (1998) *J. Math. Chem.*, **23**, 105–126. (k) Truhlar, D.G. (2007) *J. Comput. Chem.*, **28**, 73–86. (l) Shaik, S.S.H. and Philippe, C. (2008) *A Chemist's Guide to Valence Bond Theory*, Wiley-Interscience, Hoboken, NJ. (m) Wu, W., Su, P., Shaik, S., and Hiberty, P.C. (2011) *Chem. Rev.*, **111**, 7557–7593. (n) Karadakov, P.B. and Cooper, D.L. (2009) *Int. J. Quantum Chem.*, **109**, 2447–2455. (o) Su, P., Wu, W., Kelly, C.P., Cramer, C.J., and Truhlar, D.G. (2008) *J. Phys. Chem. A*, **112**, 12761–12768.

21. (a) Song, L., Mo, Y., Zhang, Q., and Wu, W. (2005) *J. Comput. Chem.*, **26**, 514–521. (b) Wu, W., Wu, A., Mo, Y., Lin, M., and Zhang, Q. (1998) *Int. J. Quantum Chem.*, **67**, 287–297. (c) Song, L., Chen, Z., Ying, F., Song, J., Chen, X., Su, P., Mo, Y., Zhang, Q., and Wu, W. (2012) An Ab Initio Nonorthogonal Valence Bond Program, Xiamen University, Xiamen.

22. (a) Mo, Y., Wu, W., and Zhang, Q. (1994) *J. Phys. Chem.*, **98**, 10048–10053. (b) Hiberty, P.C. and Byrman, C.P. (1995) *J. Am. Chem. Soc.*, **117**, 9870–9880. (c) Lauvergnat, D. and Hiberty, P.C. (1997) *J. Am. Chem. Soc.*, **119**, 9478–9482. (d) Lauvergnat, D., Maitre, P., Hiberty, P.C., and Volatron, F. (1996) *J. Phys. Chem.*, **100**, 6463–6468. (e) Mo, Y., Lin, Z., Wu, W., and Zhang, Q. (1996) *J. Phys. Chem.*, **100**, 6469–6474. (f) Mo, Y. and Schleyer, P.v.R. (2006) *Chem. Eur. J.*, **12**, 2009–2020. (g) Mo, Y., Schleyer, P.v.R., Wu, W., Lin, M., Zhang, Q., and Gao, J. (2003) *J. Phys. Chem. A*, **107**, 10011–10018. (h) Shaik, S., Shurki, A., Danovich, D., and Hiberty, P.C. (2001) *Chem. Rev.*, **101**, 1501–1539. (i) Shaik, S.S., Hiberty, P.C., Lefour, J.M., and Ohanessian, G. (1987) *J. Am. Chem. Soc.*, **109**, 363–374. (j) Shurki, A. and Crown, H.A. (2005) *J. Phys. Chem. B*, **109**, 23638–23644. (k) van Lenthe, J.H., Havenith, R.W.A., Dijkstra, F., and Jenneskens, L.W. (2002) *Chem. Phys. Lett.*, **361**, 203–208. (l) Shaik, S., Danovich, D., Silvi, B., Lauvergnat, D.L., and Hiberty, P.C. (2005) *Chem. Eur. J.*, **11**, 6358–6371. (m) Hiberty, P.C., Megret, C., Song, L., Wu, W., and Shaik, S. (2006) *J. Am. Chem. Soc.*, **128**, 2836–2843. (n) Linares, M., Humbel, S., and Braida, B. (2008) *J. Phys. Chem. A*, **112**, 13249–13255. (o) DeBlase, A., Licata, M., and Galbraith, J.M. (2008) *J. Phys. Chem. A*, **112**, 12806–12811. (p) Wu, W., Gu, J., Song, J., Shaik, S., and Hiberty, P.C. (2009) *Angew. Chem. Int. Ed.*, **48**, 1407–1410. (q) Braida, B., Prana, V., and Hiberty, P.C. (2009) *Angew. Chem. Int. Ed.*, **48**, 5724–5728. (r) Wu, W., Ma, B., Wu, J., Schleyer, P.v.R., and Mo, Y. (2009) *Chem. Eur. J.*, **15**, 9730–9736.

23. (a) Mo, Y. (2003) *J. Chem. Phys.*, **119**, 1300–1306. (b) Mo, Y. and Peyerimhoff, S.D. (1998) *J. Chem. Phys.*, **109**, 1687–1697. (c) Mo, Y., Song, L., and Lin, Y. (2007) *J. Phys. Chem. A*, **111**, 8291–8301.

24. Schmidt, M.W., Baldridge, K.K., Boatz, J.A., Elbert, S.T., Gordon, M.S., Jensen, J.J., Koseki, S., Matsunaga, N., Nguyen,

K.A., Su, S., Windus, T.L., Dupuis, M., and Montgomery, J.A. (1993) *J. Comput. Chem.*, **14**, 1347–1363.

25. (a) Mo, Y. and Gao, J. (2000) *J. Comput. Chem.*, **21**, 1458–1469. (b) Mo, Y. and Gao, J. (2000) *J. Phys. Chem.*, **104**, 3012–3020.

26. (a) Mo, Y., Wu, W., and Zhang, Q. (2003) *J. Chem. Phys.*, **119**, 6448–6456. (b) Mo, Y., Song, L., Liu, M., Lin, Y., Cao, Z., and Wu, W. (2012) *J. Chem. Theor. Comput.*, **8**, 800–805.

27. (a) Goddard, W.A.I. (1967) *Phys. Rev.*, **157**, 73–80. (b) Bobrowicz, F.W. and Goddard, W.A. III, (1977) in *Methods of Electronic Structure Theory* (ed. H.F. Schaefer III,), Plenum Press, New York, pp. 79–127.

28. (a) Mulliken, R.S. and Parr, R.G. (1951) *J. Chem. Phys.*, **19**, 1271–1278. (b) Sovers, O.J., Kern, C.W., Pitzer, R.M., and Karplus, M. (1968) *J. Chem. Phys.*, **49**, 2592–2599. (c) Stoll, H. and Preuss, H. (1977) *Theor. Chim. Acta*, **46**, 11–21. (d) Stoll, H., Wagenblast, G., and Preuss, H. (1980) *Theor. Chim. Acta*, **57**, 169–178. (e) Kollmar, H. (1979) *J. Am. Chem. Soc.*, **101**, 4832–4840. (f) Daudey, J.P., Trinquier, G., Barthelat, J.C., and Malrieu, J.P. (1980) *Tetrahedron*, **36**, 3399–3401. (g) Mehler, E.L. (1977) *J. Chem. Phys.*, **67**, 2728–2739. (h) Mehler, E.L. (1981) *J. Chem. Phys.*, **74**, 6298–6306. (i) Fülscher, M.P. and Mehler, E.L. (1981) *J. Comput. Chem.*, **12**, 811–828. (j) Famulari, A., Gianinetti, E., Raimondi, M., and Sironi, M. (1998) *Int. J. Quantum Chem.*, **69**, 151–158. (k) Gianinetti, E., Raimondi, M., and Tornaghi, E. (1996) *Int. J. Quantum Chem.*, **60**, 157–166. (l) Gianinetti, E., Vandoni, I., Famulari, A., and Raimondi, M. (1998) *Adv. Quantum Chem.*, **31**, 251–266. (m) Khaliullin, R.Z., Bell, A.T., and Head-Gordon, M. (2006) *J. Chem. Phys.*, **124**, 204105.

29. Famulari, A., Specchio, R., Gianinetti, E., and Raimondi, M. (2002) in *Valence Bond Theory*, vol. 10 (ed. D.L. Cooper), Elsevier, Amsterdam, p. 313.

30. Mo, Y., Song, L., Wu, W., and Zhang, Q. (2004) *J. Am. Chem. Soc.*, **126**, 3974–3982.

31. Mo, Y., Wu, W., Song, L., Lin, M., Zhang, Q., and Gao, J. (2004) *Angew. Chem. Int. Ed.*, **43**, 1986–1990.

32. Mo, Y., Subramanian, G., Ferguson, D.M., and Gao, J. (2002) *J. Am. Chem. Soc.*, **124**, 4832–4837.

33. Mo, Y. and Gao, J. (2006) *J. Phys. Chem. B*, **110**, 2976–2980.

34. (a) Mo, Y. and Gao, J. (2001) *J. Phys. Chem. A*, **105**, 6530–6536. (b) Mo, Y. (2009) *J. Phys. Chem. A*, **113**, 5163–5169. (c) Mo, Y., Hiberty, P.C., and Schleyer, P.v.R. (2010) *Theor. Chem. Acc.*, **127**, 27–38.

35. (a) Steinmann, S.N., Jana, D.F., Wu, J.I., Schleyer, P.v.R., Mo, Y., and Corminboeuf, C. (2009) *Angew. Chem. Int. Ed.*, **48**, 9828–9833. (b) Steinmann, S.N., Corminboeuf, C., Wu, W., and Mo, Y. (2011) *J. Phys. Chem. A*, **115**, 5467–5477. (c) Mo, Y., Bao, P., and Gao, J. (2011) *Phys. Chem. Chem. Phys.*, **13**, 6760–6775.

36. (a) Kohn, W. (1998) *Rev. Mod. Phys.*, **71**, 1253–1266. (b) Kohn, W. and Sham, L.J. (1965) *Phys. Rev.*, **140**, 1133–1138. (c) Parr, R.G. and Yang, W. (1989) *Density-Functional Theory of Atoms and Molecules*, Oxford University Press.

37. Frenking, G. (1999) *J. Comput. Chem.*, **20** (1), 1–183.

38. Cembran, A., Song, L., Mo, Y., and Gao, J. (2009) *J. Chem. Theor. Comput.*, **5**, 2702–2716.

39. (a) Behler, J., Delley, B., Lorenz, S., Reuter, K., and Scheffler, M. (2005) *Phys. Rev. Lett.*, **94**, 036104. (b) Behler, J., Delley, B., Reuter, K., and Scheffler, M. (2007) *Phys. Rev. B*, **75**, 115409. (c) Behler, J., Reuter, K., and Scheffler, M. (2008) *Phys. Rev. B*, **77**, 115421.

40. (a) Wu, Q., Chen, C.-L., and Van Voorhis, T. (2007) *J. Chem. Phys.*, **127**, 164119. (b) Wu, Q. and Van Voorhis, T. (2005) *Phys. Rev. A: At. Mol. Opt. Phys.*, **72**, 024502. (c) Wu, Q. and Van Voorhis, T. (2006) *J. Phys. Chem. A*, **110**, 9212–9218. (d) Wu, Q. and Van Voorhis, T. (2006) *J. Chem. Phys.*, **125**, 164105.

41. (a) Marcus, R.A. (1956) *J. Chem. Phys.*, **24**, 966–978. (b) Marcus, R.A. (1964) *Annu. Rev. Phys. Chem.*, **15**, 155–196. (c) Hush, N.S. (1961) *Trans. Faraday*

Soc., **57**, 557–580. (d) Hush, N.S. (1968) *Electrochim. Acta*, **13**, 1005–1023.

42. (a) Marcus, R.A. and Sutin, N. (1985) *Biochim. Biophys. Acta*, **811**, 265–322. (b) Marcus, R.A. (1993) *Angew. Chem. Int. Ed.*, **32**, 1111–1121.

43. (a) Curtiss, L.A. and Miller, J.R. (1998) *J. Phys. Chem. A*, **102**, 160–167. (b) Voityuk, A.A. and Rösch, N. (2002) *J. Chem. Phys.*, **117**, 5607–5616. (c) Newton, M.D. (1991) *Chem. Rev.*, **91**, 767–792. (d) Prezhdo, O.V., Kindt, J.T., and Tully, J.C. (1999) *J. Chem. Phys.*, **111**, 7818–7827. (e) Cave, R.J. and Newton, M.D. (1997) *J. Chem. Phys.*, **106**, 9213–9226. (f) Pacher, T., Cederbaum, L.S., and Köppel, H. (1988) *J. Chem. Phys.*, **89**, 7367. (g) Larsson, S. (1981) *J. Am. Chem. Soc.*, **103**, 4034–4040. (h) Ratner, M.A. (1990) *J. Phys. Chem.*, **94**, 4877.

44. (a) Paddon-Row, M.N. and Jordan, K.D. (1988) in *Modern Models of Bonding and Delocalization*, vol. 6 (eds J.F. Liebman and A. Greenber), John Wiley & Sons, Inc., New York, p. 115. (b) Jordan, K.D. and Paddon-Row, M.N. (1992) *Chem. Rev.*, **92**, 395–410. (c) Rodriguez-Monge, L. and Larsson, S. (1996) *J. Phys. Chem.*, **100**, 6298–6303. (d) Lu, S.-Z., Li, X.-Y., and Liu, J.-F. (2004) *J. Phys. Chem. A*, **108**, 4125–4131. (e) Prytkova, T.R., Kurnikov, I.V., and Beratan, D.N. (2005) *J. Phys. Chem. B*, **109**, 1618–1625.

45. (a) Cave, R.J. and Newton, M.D. (1996) *Chem. Phys. Lett.*, **249**, 15–19. (b) Newton, M.D. (2000) *Int. J. Quantum Chem.*, **77**, 255–263. (c) Improta, R., Barone, V., and Newton, M.D. (2006) *ChemPhysChem*, **7**, 1211–1214.

46. (a) Newton, M.D. (1980) *Int. J. Quantum Chem.*, **14**, 363–391. (b) Farazdel, A., Dupuis, M., Clementi, E., and Aviram, A. (1990) *J. Am. Chem. Soc.*, **112**, 4206–4214. (c) Zhang, L.Y. and Friesner, R.A. (1998) *Proc. Natl. Acad. Sci. U.S.A.*, **95**, 13603–13605. (d) Li, X. and He, F. (1999) *J. Comput. Chem.*, **20**, 597–603. (e) Pati, R. and Karna, S.P. (2001) *J. Chem. Phys.*, **115**, 1703–1715. (f) Rosso, K.M., Smith, D.M.A., and Dupuis, M. (2003) *J. Chem. Phys.*, **118**, 6455–6466.

47. Mo, Y. (2007) *J. Chem. Phys.*, **126**, 224104.

48. Pauling, L.C. and Wheland, G.W. (1933) *J. Chem. Phys.*, **1**, 362–374.

49. (a) Kistiakowsky, G.B., Ruhoff, J.R., Smith, H.A., and Vaughan, W.E. (1936) *J. Am. Chem. Soc.*, **58**, 146–153. (b) Kistiakowsky, G.B., Ruhoff, J.R., Smith, H.A., and Vaughan, W.E. (1936) *J. Am. Chem. Soc.*, **58**, 137–145.

50. Kistiakowsky, G.B., Romeyn, H. Jr.,, Ruhoff, J.R., Smith, H.A., and Vaughan, W.E. (1936) *J. Am. Chem. Soc.*, **57**, 65–75.

51. Schleyer, P.v.R. and Puhlhofer, F. (2002) *Org. Lett.*, **4**, 2873–2876.

52. Hehre, W.J., Radom, L., Schleyer, P.v.R., and Pople, J.A. (1986) *Ab Initio Molecular Orbital Theory*, John Wiley & Sons, Inc., New York.

53. (a) Pauncz, R. (1979) *Spin Eigenfunctions – Construction and Use*, Plenum Press, New York.(b) McWeeny, R. (1989) *Methods of Molecular Quantum Mechanics*, 2nd edn, Academic Press, London.

54. Norbeck, J.M. and Gallup, G.A. (1974) *J. Am. Chem. Soc.*, **96**, 3386–3393.

55. Tantardini, G.F., Raimondi, M., and Simonetta, M. (1977) *J. Am. Chem. Soc.*, **99**, 2913–2918.

56. (a) Hiberty, P.C. and Ohanessian, G. (1985) *Int. J. Quantum Chem.*, **27**, 245–257. (b) Hiberty, P.C. and Ohanessian, G. (1985) *Int. J. Quantum Chem.*, **27**, 259–272.

57. van Lenthe, J.H. and Balint-Kurti, G.G. (1983) *J. Chem. Phys.*, **76**, 5699–5713.

58. (a) Coulson, C.A. and Fischer, I. (1949) *Philos. Mag.*, **40**, 306. (b) Mueller, C.R. and Eyring, H. (1951) *J. Chem. Phys.*, **19**, 1495.

59. Mo, Y., Lin, Z., Wu, W., and Zhang, Q. (1996) *J. Phys. Chem.*, **100**, 11569–11572.

60. Muller, P. (1994) *Pure Appl. Chem.*, **66**, 1077–1184.

61. Coulson, C.A. and Altmann, S.L. (1952) *Trans. Faraday Soc.*, **48**, 293–302.

62. Shaik, S.S., Hiberty, P.C., Ohanessian, G., and Lefour, J.M. (1988) *J. Phys. Chem.*, **92**, 5086–5094.

63. Dewar, M.J.S. and De Llano, C. (1969) *J. Am. Chem. Soc.*, **91**, 789–795.

64. Boys, S.F. (1950) *Proc. R. Soc. London, A*, **200**, 542–544.
65. Mulliken, R.S. (1965) *J. Chem. Phys.*, **43**, S2–S11.
66. Schleyer, P.v.R., Maerker, C., Dransfeld, A., Jiao, H., and van Eikema Hommes, N. (1996) *J. Am. Chem. Soc.*, **118**, 6317–6318.
67. Kutzelnigg, W. (1980) *Isr. J. Chem.*, **19**, 193–200.
68. Steinmann, S.N., Vogel, P., Mo, Y., and Corminboeuf, C. (2011) *Chem. Commun.*, **47**, 227–229.
69. Murcko, M.A., Castejon, H., and Wiberg, K.B. (1996) *J. Phys. Chem.*, **100**, 16162–16168.
70. Mo, Y. and Gao, J. (2007) *Acc. Chem. Res.*, **40**, 113–119.
71. (a) Pophristic, V. and Goodman, L. (2001) *Nature*, **411**, 565–568. (b) Schreiner, P.R. (2002) *Angew. Chem. Int. Ed.*, **41**, 3579–3581. (c) Bickelhaupt, F.M. and Baerends, E.J. (2003) *Angew. Chem. Int. Ed.*, **42**, 4183–4188. (d) Weinhold, F. (2003) *Angew. Chem. Int. Ed.*, **42**, 4188–4194.
72. (a) Pitzer, K.S. (1951) *Discuss. Faraday Soc.*, **10**, 66–73. (b) Pitzer, R.M. (1983) *Acc. Chem. Res.*, **16**, 207–210. (c) Smith, L.G. (1949) *J. Chem. Phys.*, **17**, 139–167. (d) Wilson, E.B. Jr., (1959) *Adv. Chem. Phys.*, **2**, 367–393. (e) Allen, L.C. (1968) *Chem. Phys. Lett.*, **2**, 597–601. (f) Lowe, J.P. (1970) *J. Am. Chem. Soc.*, **92**, 3799–3800. (g) Halpern, A.M. and Glendening, E.D. (2003) *J. Chem. Phys.*, **119**, 11186–11191. (h) Hirota, E., Saito, S., and Endo, Y. (1979) *J. Chem. Phys.*, **71**, 1183–1187.
73. (a) Payne, P.W. and Allen, L.C. (1977) in *Mordern Theoretical Chemistry*, vol. 4 (ed. H.F. Schaefer III,), Plenum Press, New York and London, pp. 29–108. (b) Bader, R.F.W., Cheeseman, J.R., Laidig, K.E., Wiberg, K.B., and Breneman, C. (1990) *J. Am. Chem. Soc.*, **112**, 6530–6536. (c) Liu, S. (2007) *J. Chem. Phys.*, **126**, 244103. (d) Liu, S. and Govind, N. (2008) *J. Phys. Chem. A*, **112**, 6690–6699.
74. (a) England, W. and Gordon, M.S. (1971) *J. Am. Chem. Soc.*, **93**, 4649–4657. (b) Epiotis, N.D., Cherry, W.R., Shaik, S., Yates, R.L., and Bernardi, F. (1977) *Topics in Current Chemistry: Structural Theory of Organic Chemistry*, vol. 70, Springer-Verlag, Berlin, Heidelberg, New York. (c) Brunck, T.K. and Weinhold, F. (1979) *J. Am. Chem. Soc.*, **101**, 1700–1709.
75. Mulliken, R.S. (1939) *J. Chem. Phys.*, **7**, 339–352.
76. Reed, A.E. and Weinhold, F. (1991) *Isr. J. Chem.*, **31**, 277–285.
77. Song, L., Lin, Y., Wu, W., Zhang, Q., and Mo, Y. (2005) *J. Phys. Chem. A*, **109**, 2310–2316.
78. Cramer, C.J. (1998) in *Encyclopedia of Computational Chemistry* (ed. P.v.R. Schelyer), John Wiley & Sons, Ltd, Berlin, pp. 1294–1298.
79. Fisher, L.S., McNeil, K., Butzen, J., and Holme, T.A. (2000) *J. Phys. Chem. B*, **104**, 3744–3751.
80. (a) Thorne, L.R., Suenram, R.D., and Lovas, F.J. (1983) *J. Chem. Phys.*, **78**, 167–171. (b) Trudel, S. and Gilson, D.F.R. (2003) *Inorg. Chem.*, **42**, 2814–2816. (c) Klooster, W.T., Koetzle, T.F., Siegbahn, P.E.M., Richardson, T.B., and Crabtree, R.H. (1999) *J. Am. Chem. Soc.*, **121**, 6337–6343.
81. (a) Dapprich, S. and Frenking, G. (1995) *J. Phys. Chem.*, **99**, 9352–9362. (b) Dill, J.D., Schleyer, P.v.R., and Pople, J.A. (1975) *J. Am. Chem. Soc.*, **97**, 3402–3409. (c) Eoeggen, I. (1992) *Chem. Phys.*, **162**, 271–284. (d) Holme, T.A. and Truong, T.N. (1993) *Chem. Phys. Lett.*, **215**, 53–57. (e) Jonas, V., Frenking, G., and Reetz, M.T. (1994) *J. Am. Chem. Soc.*, **116**, 8741–8753. (f) Peyerimhoff, S.D. and Buenker, R.J. (1968) *J. Chem. Phys.*, **49**, 312.(g) Umeyama, H. and Morokuma, K. (1976) *J. Am. Chem. Soc.*, **98**, 7208–7220. (h) Branchadell, V., Sbai, A., and Oliva, A. (1995) *J. Phys. Chem.*, **99**, 6472–6476. (i) Anane, H., Boutalib, A., Nebot-Gil, I., and Tomas, F. (1998) *J. Phys. Chem. A*, **102**, 7070–7073. (j) Bauschlicher, C.W. Jr., and Ricca, A. (1995) *Chem. Phys. Lett.*, **237**, 14–19. (k) Barrios, R., Skurski, P., Rak, J., and Gutowski, M. (2000) *J. Chem. Phys.*, **113**, 8961–8968. (l) Dillen, J. and Verhoeven, P. (2003) *J. Phys. Chem. A*, **107**, 2570–2577. (m) Horvath, V., Kovacs, A., and Hargittai, I.

(2003) *J. Phys. Chem. A*, **107**, 1197–1202. (n) Timoshkin, A.Y., Suvorov, A.V., Bettinger, H.F., and Schaefer, H.F. III, (1999) *J. Am. Chem. Soc.*, **121**, 5687–5699.

82. (a) Custelcean, R. and Jackson, J.E. (2001) *Chem. Rev.*, **101**, 1963–1980. (b) Richardson, T., de Gala, S., Crabtree, R.H., and Siegbahn, P.E.M. (1995) *J. Am. Chem. Soc.*, **117**, 12875–12876.

83. (a) Grimme, S. (2004) *J. Comput. Chem.*, **25**, 1463–1473. (b) Perdew, J.P., Burke, K., and Ernzerhof, M. (1996) *Phys. Rev. Lett.*, **77**, 3865–3868.

84. Boys, S.F. and Bernardi, F. (1970) *Mol. Phys.*, **19**, 553–566.

85. Bondi, A. (1964) *J. Phys. Chem.*, **68**, 441–451.

86. (a) Sironi, M., Genoni, A., Civera, M., Pieraccini, S., and Ghitti, M. (2007) *Theor. Chem. Acc.*, **117**, 685–698. (b) Genoni, A., Merz, K.M. Jr.,, and Sironi, M. (2008) *J. Chem. Phys.*, **129**, 054101.

87. (a) Mo, Y. (2010) *Nat. Chem.*, **2**, 666–671. (b) Jia, J.-F., Wu, H.-S., and Mo, Y. (2012) *J. Chem. Phys.*, **136**, 144315.

7
The Conceptual Density Functional Theory Perspective of Bonding

Frank De Proft, Paul W. Ayers, and Paul Geerlings

7.1
Introduction

In Pauling's *Nature of the Chemical Bond*, for decades *the* reference for every chemist interested in the intricacies of what a chemical bond really is, and which is commemorated in this volume, Pauling concentrates on two main types of chemical bonds: the covalent bond and the ionic bond [1]. His chapter on the hydrogen bond is left out of consideration in our text in view of its classification as a "weak interaction," and the one on the metallic bond is also left out here owing to its fundamentally different nature. In both types of bonds (electrostatic and covalent), electrons play a fundamental role. In the ionic bond, a full electron transfer occurs between both partners (at least formally), as the bond is formed by an electrostatic interaction between a cation and an anion. In the covalent bond, one or more electron pairs are shared between the two atoms. The main factors that describe the energetics of ionic bonds are simpler: the atomic ionization energy and the electron affinity, the experimental determination of the latter values at the time being in its infancy (here the lattice energy is left out for simplicity). In the covalent bond, an orbital description is adopted based on the valence bond approach, and bonding occurs by overlapping of hybridized orbitals on the two partners, a concept worked out by Pauling himself in the mid-thirties. The characteristics of these bonds are related to the charge distribution resulting from a difference in electronegativity between the two partners, for which Pauling himself constructed a scale, for a long time considered *the* scale. Note that the Mulliken scale introduced during the same period uses the combination of the abovementioned concepts, namely, ionization energy and electron affinity, as a measure for electronegativity [2]. In this highly simplified description of some main aspects and seminal concepts highlighted in Pauling's *magnum opus*, the reader will notice that the electron density is less prominent than it is in a modern perspective.

X-ray diffraction leading to *experimental* density plots of sufficient resolution fully entered the chemical literature much later on as described by Coppens [3, 4], enabling their systematic analysis using, for example, Bader's Quantum Theory of Atoms in Molecules (QTAIM) [5, 6]. However, note again that Pauling was a pioneer in the use of X-ray diffraction for the structure determination of crystals.

The Chemical Bond: Fundamental Aspects of Chemical Bonding, First Edition.
Edited by Gernot Frenking, Sason Shaik.

From the purely *theoretical* point of view, the fundamental role of the electron density as carrier of information in the quantum chemical description of atoms, molecules, and the solid state was not fully recognized even at the time of the third Edition of the *Nature of the Chemical Bond* in 1960 [1].

The rest of the development is known to the broader chemical community: in 1964, Hohenberg and Kohn proved that the electron density can indeed replace the wave function as carrier of information through their *energy as a functional of the density* existence theorem, putting the much simpler electron density at the forefront for studying the electronic properties of atoms and molecules [7], thus formally establishing the field of density functional theory (DFT) [8–12]. Bader's seminal contributions, highlighted by Popelier in this volume, insisted on electron density as the central quantity in the study of bonding. However, the question how to obtain these densities, thereby avoiding the cumbersome *ab initio* techniques for wave functions, remained. It is at least in principle possible, because the variational principle applied to the energy density functional, $E = E[\rho]$, leads to the ground-state density in the same way as the analogous variational principle for the wave function leads to the ground-state wave function. Direct optimization of the electron density turned out to be hard to achieve. A practical ansatz, still by far the most popular one, which reintroduces orbitals, was formulated by Kohn and Sham [13]. Their equations contain, however, an unknown term in the Hamiltonian of the resulting one-electron equations, which is the price to be paid for simplifying the Schrödinger equation by going from a $4N$-variable spin-resolved wave function to a three-variable density.

This breakthrough entered the (quantum) chemical community in full force in the early nineties when the Kohn–Sham equations were implemented in the wave function package *par excellence*, John Pople's *Gaussian* [14]. The DFT has flourished since then as the computational workhorse of relatively accurate calculations on medium- to large-sized systems with thousands of DFT-based papers being published every year.

The following question can be asked: Did DFT contribute to new insights into the chemical bond that wave function theory could not? We think that the answer is a twofold "yes."

While discussing this point, we would like to make a distinction between "bond" and "bonding." *Bond* refers to the final situation when both partners have completed their association, whereas *bonding* refers to the initial phase of the formation of a bond, that is, the bond-formation processes.

The first "yes" to the question above is an indirect one. From a practical point of view, DFT offers faster methods to obtain the density, which can then be studied in much more detail and for much larger systems, both in a chemical bond and on bond formation (bonding). In our contribution, we briefly focus on Berlin's approach [15] that was cast in a conceptual DFT [8, 10, 11, 16–22] version (*vide infra*) as an example of such a combined approach. The basic aspects of DFT highlighting the role of electron density a carrier of information and the way to obtain it are outlined in Section 7.2.

On the other hand there is a second "yes", coming from the so-called conceptual DFT where from the late seventies onward, Parr and coworkers developed a theory based on a perturbational ansatz of interacting partners through the $E = E[N,\nu]$ functional and where the electron density itself and the electronic chemical potential turn out to be the key ingredients [8, 10, 11, 16–22]. In this scheme, reactivity descriptors were developed systematically as energy (functional) derivatives with respect to N and/or $\nu(\mathbf{r})$, representing the response of an atom or a molecule to perturbations in the number of electrons and/or the potential that the electrons move in. From this perspective, chemical bonding is a special type of perturbation. We describe succinctly the main descriptors of this type in Section 7.3. These descriptors enable us to disentangle the driving force for chemical bonding, leading to insights into the characteristics of the bond formed:

- Electronegativities should be equalized (*i*) [23–27].
- Hard partners prefer to interact (globally and locally) with other hard partners. Similarly, soft partners prefer soft partners (*ii*) [28–32].

Once the bond has been formed, its charge distribution may be derived through an electronegativity equalization-based methodology derived from (*i*).

In principle (*i*), the electronegativity (difference) as put forward by Pauling is the principal actor in defining the charge distribution after bond formation has been realized, although modulated, however, by the hardness/softness of the interacting partners. The hardness/softness of the partners is related to their polarizabilities, to which Pauling attached lesser importance in the process of bond formation.

An example is given for each concept/principle, the final case being the nuclear Fukui function [33–40] where the link with Berlin's picture of the chemical bond [15], directly concentrating on the density, is made.

7.2 Basics of DFT: The Density as a Fundamental Carrier of Information and How to Obtain It

The accurate solution of the (nonrelativistic) time-independent Schrödinger equation, yielding the energy of the system E and the wave function Ψ

$$\hat{H}\Psi = E\Psi \tag{7.1}$$

is considered to be the ultimate road to (get an insight into) the electronic structure of atoms and molecules. In Eq. (7.1), $\hat{H}$ is the Hamiltonian for a system of M nuclei and N electrons, in the absence of any external field, and given by (in a.u.)

$$\hat{H} = -\frac{1}{2}\sum_{A=1}^{M}\nabla_A^2 - \frac{1}{2}\sum_{i=1}^{N}\nabla_i^2 - \sum_{i=1}^{N}\sum_{A=1}^{M}\frac{Z_A}{|\mathbf{r}_i - \mathbf{R}_A|} + \sum_{i=1}^{N}\sum_{j>i}^{N}\frac{1}{|\mathbf{r}_i - \mathbf{r}_j|} + \sum_{A=1}^{M}\sum_{B>A}^{M}\frac{Z_A Z_B}{|\mathbf{R}_A - \mathbf{R}_B|} \tag{7.2}$$

In this expression, the first two terms are the kinetic energy of the nuclei and electrons, respectively, the third term expresses the electron–nucleus attraction, the fourth term the electron–electron repulsion, and the fifth term the nucleus–nucleus repulsion. The potential due to the nuclei, also called the *external potential* $v(\mathbf{r})$, can thus be written as

$$v(\mathbf{r}) = -\sum_{A=1}^{M} \frac{Z_A}{|\mathbf{r} - \mathbf{R}_A|} \tag{7.3}$$

All properties of the chemical system can be obtained from this wave function. In the 1920s, Thomas and Fermi developed a theory where the electron density $\rho(\mathbf{r})$ of the system was used as the basic carrier of information instead of the wave function [41, 42], which was extended to include exchange by Dirac [43]. In the 1950s, Slater used this local Dirac expression to replace the nonlocal exchange operator in the Hartree–Fock theory, yielding the so-called X_α method [44–46].

The electron density of the system, $\rho(\mathbf{r})$, gives the probability of finding any of the N electrons in the system in an elementary volume $\mathrm{d}\mathbf{r}$ around $\mathbf{r}$, regardless of the position and spin of the other $N-1$ electrons. It is readily obtained from the wave function as

$$\rho(\mathbf{r}_1) = N \int \dots \int \left|\Psi\left(\mathbf{x}_1, \mathbf{x}_2, \dots \mathbf{x}_N\right)\right|^2 \mathrm{d}s_1 \mathrm{d}\mathbf{x}_2 \dots \mathrm{d}\mathbf{x}_N \tag{7.4}$$

where $\mathbf{x}_i$ denotes the combined position and spin coordinate of electron i, and s_1 the spin coordinate of electron 1.

As can be seen, the electron density yields the total number of electrons of the system on integration over all spaces

$$\int \rho(\mathbf{r})\mathrm{d}\mathbf{r} = N \tag{7.5}$$

In 1964, Hohenberg and Kohn formally proved that the electron density can indeed be used as the basic variable determining all atomic and molecular properties [7]. They showed that the electron density of the system, up to a trivial additive constant, uniquely determines the external potential $v(\mathbf{r})$ of the system given in Eq. (7.3).

As the electron density integrates to the number of electrons of the system [Eq. (7.5)] and $\rho(\mathbf{r})$ determines $v(\mathbf{r})$, the density fixes the Hamiltonian and the wave function of the system, and consequently all properties of the system. The energy of the system and the different parts of the (electronic) energy can thus be written as a functional of the electron density[1)]:

$$E[\rho(\mathbf{r})] = T[\rho(\mathbf{r})] + V_{\mathrm{ne}}[\rho(\mathbf{r})] + V_{\mathrm{ee}}[\rho(\mathbf{r})] \tag{7.6}$$

where $T[\rho(\mathbf{r})]$ is the kinetic energy functional, $V_{\mathrm{ne}}[\rho(\mathbf{r})]$ the nucleus–electron attraction functional, and $V_{\mathrm{ee}}[\rho(\mathbf{r})]$ the electron–electron repulsion functional. For $V_{\mathrm{ne}}[\rho(\mathbf{r})]$, the exact expression is known:

$$V_{\mathrm{ne}}[\rho] = \int \rho(\mathbf{r})v(\mathbf{r})\mathrm{d}\mathbf{r} \tag{7.7}$$

1) A functional is a rule for going from a function to a number; it can be denoted as $F[f]$, where $f(x)$ is a function.

The other two parts of the energy functional in Eq. (7.6) are largely unknown and are often combined into the so-called Hohenberg–Kohn functional

$$F_{\mathrm{HK}}[\rho(\mathbf{r})] = T[\rho(\mathbf{r})] + V_{\mathrm{ee}}[\rho(\mathbf{r})] \tag{7.8}$$

Now that the density has been recognized as a fundamental carrier of information, the next problem is "how to get it".

In a second theorem, Hohenberg and Kohn established a variational principle: for any trial density $\tilde{\rho}(\mathbf{r})$, such that $\int \tilde{\rho}(\mathbf{r})\mathrm{d}\mathbf{r} = N$ and $\tilde{\rho}(\mathbf{r}) \geq 0, \forall \mathbf{r}$, $E[\tilde{\rho}(\mathbf{r})] \geq E_0[\rho_0(\mathbf{r})]$, where E_0 is the ground-state energy with corresponding ground-state electron density $\rho_0(\mathbf{r})$. If $\tilde{\rho}(\mathbf{r}) = \rho_0(\mathbf{r})$, then $E[\tilde{\rho}(\mathbf{r})] = E_0$ [7].

One can now minimize the electronic energy $E[\rho(\mathbf{r})]$, given in Eq. (7.6), with respect to the electron density $\rho(\mathbf{r})$, subject to the constraint that during minimization, the electron density should at all times integrate to the total number of electrons N of the system. Attaching a Lagrange multiplier μ to the normalization constraint yields the following equation:

$$\frac{\delta}{\delta\rho(\mathbf{r})}\left[E\left[\rho(\mathbf{r})\right] - \mu\left(\int \rho(\mathbf{r})\mathrm{d}\mathbf{r} - N\right)\right] = 0 \tag{7.9}$$

or

$$\frac{\delta E}{\delta\rho(\mathbf{r})} - \mu = 0 \tag{7.10}$$

Using Eq. (7.6) and (7.7), Eq. (7.9) can be rewritten as

$$\nu(r) + \frac{\delta(T[\rho(\mathbf{r})] + V_{\mathrm{ee}}[\rho(\mathbf{r})])}{\delta\rho(\mathbf{r})} - \mu = 0 \tag{7.11}$$

which finally becomes

$$\nu(r) + \frac{\delta F_{\mathrm{HK}}}{\delta\rho(\mathbf{r})} = \mu \tag{7.12}$$

This expression is often called the *DFT analogue* of the Schrödinger equation. If F_{HK} were known exactly, this would give the exact solution for the ground-state density and associated energy. Unfortunately, large parts of F_{HK} are unknown and have to be guessed. V_{ee} can be written as an exactly known classical contribution to the electron–electron repulsion $J[\rho(\mathbf{r})]$ and a residual unknown nonclassical part E_{ncl}:

$$V_{\mathrm{ee}}[\rho] = \frac{1}{2}\iint \frac{\rho(\mathbf{r})\rho(\mathbf{r}')}{|\mathbf{r} - \mathbf{r}'|}\mathrm{d}\mathbf{r}\mathrm{d}\mathbf{r}' + E_{\mathrm{ncl}}[\rho] = J[\rho] + E_{\mathrm{ncl}}[\rho] \tag{7.13}$$

However, a similar separation for the kinetic energy functional is not possible; Kohn and Sham circumvented this problem by introducing orbitals again [13]. They considered a noninteracting model system with the same electron density as the exact, fully interacting system. For the noninteracting reference system, the kinetic energy T_{s} can be computed exactly as

$$T_{\mathrm{s}}[\rho] = \sum_{i=1}^{N}\left\langle \phi_i \left| -\frac{1}{2}\nabla^2 \right| \phi_i \right\rangle \tag{7.14}$$

where ϕ_i are the so-called Kohn–Sham orbitals. The electron density is given exactly as

$$\rho(\mathbf{r}) = \sum_{i=1}^{N} |\phi_i(\mathbf{r})|^2 \tag{7.15}$$

The final Kohn–Sham energy density functional can thus be expressed as

$$E_{KS}[\rho(\mathbf{r})] = T_s[\rho] + (T[\rho] - T_s[\rho]) + V_{ne}[\rho] + J[\rho] + (V_{ee}[\rho] - J[\rho]) \tag{7.16}$$

One defines the exchange–correlation energy functional as

$$E_{XC}[\rho] = (T[\rho] - T_s[\rho]) + (V_{ee}[\rho] - J[\rho]) \tag{7.17}$$

so that the Kohn–Sham energy expression becomes

$$E_{KS}[\rho] = T_s[\rho] + J[\rho] + E_{XC}[\rho] + V_{ne}[\rho] \tag{7.18}$$

This energy expression would lead to an exact method if the exact expression of $E_{XC}[\rho]$ were to be known. In the past 25 years, accurate approximations derived from the first principle considerations, using parameterizations, or combinations of both, for the exchange–correlation energy functional have become available; hence, many atomic and molecular properties can now be computed very accurately. A number of problems persist, including the accurate description of reaction barriers, multiplet energies of transition metals, the accurate description of excited state properties, and dispersion interactions. A recent review by Yang gives a detailed account and perspective on the field of DFT and scrutinizes the many challenges the theory still continues to face [47].

7.3 Conceptual DFT: A Perturbative Approach to Chemical Reactivity and the Process of Bond Formation

7.3.1 Basics: Global and Local Response Functions

A central observation in DFT is that energy changes from one ground state to another are associated with electron density changes between the states, $E = E[\rho]$, or, equivalently, changes in the number of electrons N, the external potential v, or both, $E = E[N,v]$ [48]. As the electron density of the system uniquely determines both the number of electrons through Eq. (7.5) and the external potential v, through the first Hohenberg–Kohn theorem, changes of $E[\rho]$ from one ground state to another, for example, in a chemical reaction are equivalent to changes in N and v, that is, $E = E[N,v]$.

This is the key ingredient in the so-called perturbative perspective on chemical reactivity, which is the central point of discussion in this section [19]. Changes between ground states could, for example, be associated with the initial stages of a chemical reaction when reagents approach each other or with the initial stages of the formation of a chemical *bond*, that is, chemical *bonding*.

This approach can be considered as complementary to another DFT-based approach to chemical bonding using the atomic and molecular Kohn–Sham molecular orbitals. Using these orbitals to describe the chemical bonding can indeed be considered to be highly relevant because these orbitals are directly related to the exact electron density of the system through Eq. (7.15). In addition, a framework can be introduced for the decomposition of the bond energy in Pauli repulsion, electrostatic and orbital interactions, as outlined in detail in the chapter by Frenking and Bickelhaupt. Electrostatic and orbital interaction contributions can additionally be linked to hardness- and softness-related quantities as outlined in this section. These models are complementary to other approaches, such as valence bond theory (Chapter 7) or natural bond orbitals (NBOs, Chapter 7), that use a so-called mechanistic picture of bonding, enabling the study of structure, bonding, and reactivity from the properties of the isolated fragments or reactancts.

Consider a chemical system A and an approaching molecule B in the initial stages of a chemical reaction or of a bond-formation process. The approach of molecule B will result in a perturbation in the number of electrons of A, N_A, and/or the external potential $v_A(\mathbf{r})$. The corresponding initial energy change of A, E_A, can be expressed using a Taylor series expansion around the initial number of electrons N_A^0 and initial external potential $v_A^0(\mathbf{r})$

$$\begin{aligned}
E_A\left[N_A^0 + \Delta N_A, v_A^0(\mathbf{r}) + \Delta v_A(\mathbf{r})\right] &= E_A\left[N_A^0, v_A^0(\mathbf{r})\right] \\
&+ \left(\frac{\partial E_A}{\partial N_A}\right)_{\substack{N_A = N_A^0 \\ v_A(\mathbf{r}) = v_A^0(\mathbf{r})}} \Delta N_A + \int \left[\frac{\delta E_A}{\delta v_A(\mathbf{r})}\right]_{\substack{N_A = N_A^0 \\ v_A(\mathbf{r}) = v_A^0(\mathbf{r})}} \Delta v_A(\mathbf{r})\mathrm{d}\mathbf{r} \\
&+ \frac{1}{2}\left(\frac{\partial^2 E_A}{\partial N_A^2}\right)_{\substack{N_A = N_A^0 \\ v_A(\mathbf{r}) = v_A^0(\mathbf{r})}} \Delta N_A^2 + \int \left[\frac{\partial \delta E_A}{\partial N_A \delta v_A(\mathbf{r})}\right]_{\substack{N_A = N_A^0 \\ v_A(\mathbf{r}) = v_A^0(\mathbf{r})}} \Delta N_A \Delta v_A(\mathbf{r})\mathrm{d}\mathbf{r} \\
&+ \frac{1}{2}\iint \left[\frac{\delta^2 E_A}{\delta v_A(\mathbf{r})\,\delta v_A(\mathbf{r}')}\right]_{\substack{N_A = N_A^0 \\ v_A(\mathbf{r}) = v_A^0(\mathbf{r})}} \Delta v_A(\mathbf{r})\Delta v_A(\mathbf{r}')\mathrm{d}\mathbf{r}\mathrm{d}\mathbf{r}' \\
&+ \ldots
\end{aligned} \tag{7.19}$$

An essential observation that was made was that many of the derivatives that emerged in this Taylor series expansion (response functions) can be identified with interesting chemical properties. Most of these properties were only known qualitatively in chemistry, and the mathematical framework outlined provides the first, accurate and sharp definitions of these quantities, allowing them to be computed from the first principles. We now present a brief overview of the different response functions that emerge from this Taylor series expansion and discuss how they can be computed practically. It should be noted that other response functions of reactivity indices have been introduced without considering this energy expansion. Some of these will also be briefly mentioned in this section. This area of research within DFT gained considerable interest starting in the late 1970s and continues to attract much attention.

7.3.1.1 Global Response Functions

In the first part, we treat the global (i.e., nonlocal, not depending on the position in space) reactivity indices that emerge from Eq. (7.19). The first-order derivative of E_A with respect to N_A is called the *electronic chemical potential* μ; this quantity was shown to correspond to the Lagrange multiplier introduced in Eq. (7.9) of the DFT energy minimization problem

$$\mu_A = \left(\frac{\partial E_A}{\partial N_A}\right)_{\substack{N_A = N_A^0 \\ v_A(\mathbf{r}) = v_A^0(\mathbf{r})}} \tag{7.20}$$

Parr *et al.* showed that this quantity could be identified with the negative of the electronegativity of the system, χ_A [49], which is a well-known and very important quantity in the description of the chemical bond [50]:

$$\mu_A = -\chi_A \tag{7.21}$$

in line with the work of Ickowski and Margrave [51], and the earlier work of Pritchard and Sumner [52]. It should be mentioned that a number of contributions have suggested to treat electronegativity and chemical potential as two distinct properties (for recent detailed accounts, see e.g., Refs. [53, 54] and references therein).

Electronegativity was introduced by Pauling (cf. Section 7.1) as the *power of an atom in a molecule to attract electrons to itself*. It thus constitutes a very important quantity in describing the polarity of covalent chemical bonds between atoms of different types [1].

As can be seen, the chemical potential (or minus the electronegativity) is defined as a derivative with respect to the number of electrons, which basically requires that the number of electrons of the system be treated as a continuous variable. In DFT, the treatment of systems with fractional number of electrons is extremely important and continues to be a topic of great interest; for a detailed analysis, we refer the reader to Refs. [55–58]. The basic working equation for computation of the electronegativity is given as

$$\chi_A = \frac{I_A + A_A}{2} \tag{7.22}$$

where I_A and A_A are the vertical ionization energy and electron affinity of A, respectively. This rationalizes the Mulliken electronegativity scale that was also mentioned in Section 7.1 [2]. For an N–electron system, these quantities can be computed as

$$I_A = E_A(N-1) - E_A(N) \tag{7.23}$$

$$A_A = E_A(N) - E_A(N+1) \tag{7.24}$$

where $E_A(N+1)$, $E_A(N)$, and $E_A(N-1)$ are the energies of the N+1, N, and N−1 electron systems, computed at the geometry of the N−electron system. In many cases, a Koopmans'-type of approximation [59] is used to compute the electronegativity, using the frontier molecular orbital energies (either computed at Hartree–Fock and Kohn–Sham level) as

$$\chi_A = -\frac{\varepsilon_{\mathrm{HOMO}} + \varepsilon_{\mathrm{LUMO}}}{2} \tag{7.25}$$

that is, the average of the HOMO and LUMO orbital energy. It has been shown that for exchange–correlation functionals averaging over the so-called integer discontinuity, this constitutes a reasonable approximation for this global quantity [60].

It should be mentioned that early approaches and the approach of Mulliken refer to the energies of valence-state atoms and ions, whereas in the present approach, use is made of ground-state energies. More recent approaches have also successfully used valence-state quantities, as witnessed, for example, in Ref. [61].

The chemical hardness of the system is defined as the second derivative of the energy with respect to the number of electrons:

$$\eta_A = \left(\frac{\partial^2 E_A}{\partial N_A^2}\right)_{\substack{N_A = N_A^0 \\ v_A(\mathbf{r}) = v_A^0(\mathbf{r})}} \tag{7.26}$$

for which the following working equation can be derived:

$$\eta_A = I_A - A_A \tag{7.27}$$

sometimes preceded by a factor of 1/2. Equation (7.24) was introduced in 1983 by Parr and Pearson as the absolute hardness [62]. The inverse of the (global) chemical hardness of the system is the (global) softness [63]:

$$S_A = \frac{1}{\eta_A} \tag{7.28}$$

This quantity has been shown to be related to the polarizability of the system [64–68].

Hardness and softness had been introduced already during the 1960s by Pearson in order to rationalize trends in complexation energies between Lewis acids and Lewis bases [28–32]. This resulted in the so-called hard and soft acids and bases (HSAB) principle, states that hard acids prefer to *bind* to hard bases and soft acids prefer to *bind* to soft bases. As such, this principle plays a fundamental role in the description of *bond formation* or *bonding*, and will be discussed in greater detail later on. Also, the concept of hardness has also been invoked to investigate molecular aromaticity [69–71] through the principle of maximum hardness [72–77]. This principle, also formulated by Pearson, states that molecules arrange themselves to be as hard as possible. Although not universally valid, this principle has been applied in the study of molecular stability and chemical reactivity. (e.g., see Ref. [78]).

Again, using a Koopmans' type of approximation [59], the chemical hardness can be expressed as

$$\eta_A = \varepsilon_{LUMO} - \varepsilon_{HOMO} \tag{7.29}$$

that is, the HOMO–LUMO gap. This method of evaluating the hardness has been shown to lead to a large underestimation, and a correction scheme was proposed by Tozer and De Proft [79].

It should be noted that when the chemical potential, chemical hardness, and softness are evaluated for gas-phase molecules, one often encounters the problem of anion metastability, resulting in negative values for the electron affinity. Different approaches exist to circumvent this problem; we skip their details in this contribution, and refer the reader to Refs. [79, 80]. It has also been recently discussed whether or not to use the negative electron affinities in the evaluation of the chemical potential and the chemical hardness [80].

Table 7.1, recently composed and compiled by us, provides the best (recent) values (to our knowledge) for the ionization potential and electron affinity of atoms, along with the associated values for chemical hardness and electronic chemical potential [80]. These values might be of use for discussing the covalent bond through the notion of electronic chemical potential, and (cf. Section 7.1) the ionic bond whose energetics is governed by I and A of the interacting partners.

The derivative of the chemical hardness with respect to the number of electrons is called hyperhardness; [81] this and higher order derivatives with respect to the number of electrons are usually not taken into account when studying bond-formation processes [21, 81–84]. For a detailed account on the third- and higher order derivatives, the reader is referred to Ref. [21].

A global reactivity descriptor that combines the chemical potential and chemical hardness is the so-called electrophilicity index ω_A, introduced by Parr *et al.* as [85–87]

$$\omega_A = \frac{\mu_A^2}{2\eta_A} \tag{7.30}$$

7.3.1.2 Local Response Functions

The first local response function emerging in expression (7.19) is the derivative of the energy with respect to the external potential at constant number of electrons. From the first-order perturbation theory, one can easily show that this quantity corresponds to the electron density, which is the central quantity in DFT

$$\rho_A(\mathbf{r}) = \left[\frac{\delta E_A}{\delta v_A(\mathbf{r})}\right]_{\substack{N_A = N_A^0 \\ v_A(\mathbf{r}) = v_A^0(\mathbf{r})}} \tag{7.31}$$

The first-order mixed derivative in Eq. (7.19) is the so-called Fukui function, introduced by Parr and Yang [88, 89]

$$f_A(\mathbf{r}) = \left[\frac{\partial\delta E_A}{\partial N_A \delta v_A(\mathbf{r})}\right]_{\substack{N_A = N_A^0 \\ v_A(\mathbf{r}) = v_A^0(\mathbf{r})}} \tag{7.32}$$

Table 7.1 Ionization energies, electron affinities, chemical hardnesses, and chemical potentials of the atoms in the Periodic Table.

Atom	I (eV)	A (eV)	η (eV)	μ (eV)
H	13.598	0.754	12.84	−7.18
He	24.587	Not stable	24.59	–
Li	5.392	0.618	4.77	−3.01
Be	9.323	Not stable	9.323	−4.66
B	8.298	0.280	8.02	−4.29
C	11.260	1.262	10.00	−6.26
N	14.534	−0.07	14.61	−7.23
N	14.534	−0.1809	14.71	−7.18
N	14.534	−1.0612	15.60	−6.74
O	13.6181	1.461	12.16	−7.54
F	17.423	3.401	14.02	−10.41
Ne	21.565	Not stable	21.57	–
Na	5.139	0.548	4.59	−2.84
Mg	7.646	Not stable	7.65	−3.82
Al	5.986	0.433	5.55	−3.21
Si	8.152	1.390	6.76	−4.77
P	10.487	0.747	9.74	−5.62
S	10.360	2.077	8.28	−6.22
Cl	12.968	3.613	9.36	−8.29
Ar	15.760	Not stable	15.76	–
K	4.341	0.501	3.84	−2.42
Ca	6.113	0.02455	6.09	−3.07
Sc	6.561	0.188	6.37	−3.37
Ti	6.828	0.079	6.75	−3.45
V	6.746	0.525	6.22	−3.64
Cr	6.767	0.666	6.10	−3.72
Mn	7.434	−0.498	7.93	−3.47
Fe	7.902	0.151	7.75	−4.03
Co	7.881	0.662	7.22	−4.27
Ni	7.640	1.156	6.48	−4.40
Cu	7.726	1.235	6.49	−4.48
Ga	5.999	0.43	5.57	−3.21
Ge	7.899	1.232712	6.67	−4.57
As	9.789	0.814	8.98	−5.30
Se	9.752	2.02067	7.73	−5.89
Br	11.814	3.363588	8.45	−7.59
Kr	14.000	Not stable	14.00	–
Rb	4.177	0.48592	3.69	−2.33
Sr	5.695	0.05206	5.64	−2.87
Y	6.217	0.307	5.91	−3.26
Zr	6.634	0.426	6.21	−3.53
Nb	6.759	0.893	5.87	−3.83
Mo	7.092	0.748	6.34	−3.92

(continued overleaf)

Table 7.1 (continued)

Atom	I (eV)	A (eV)	η (eV)	μ (eV)
Tc	7.280	0.55	6.73	−3.92
Tc	7.280	0.636	6.64	−3.96
Ru	7.361	1.05	6.31	−4.21
Rh	7.459	1.137	6.32	−4.30
Pd	8.337	0.562	7.78	−4.45
Ag	7.576	1.302	6.27	−4.44
In	5.786	0.3	5.49	−3.04
Sn	7.344	1.112	6.23	−4.23
Sb	8.608	1.046	7.56	−4.83
Te	9.010	1.971	7.04	−5.49
I	10.451	3.059	7.39	−6.76
Xe	12.130	Not stable	12.13	–
Cs	3.894	0.471	3.42	−2.18
Ba	5.211664	0.14462	5.07	−2.68
La	5.577	0.47	5.11	−3.02
Ce	5.539	0.955	4.58	−3.25
Ce	5.539	0.65	4.89	−3.09
Pr	5.473	0.962	4.51	−3.22
Eu	5.670	0.864	4.81	−3.27
Tm	6.184	1.029	5.16	−3.61
Yb	6.254	−0.02	6.27	−3.12
Lu	5.426	0.34	5.09	−2.88
Hf	6.825	Not stable	6.83	–
Ta	7.550	0.322	7.23	−3.94
W	7.864	0.815	7.05	−4.34
Re	7.834	0.15	7.68	−3.99
Os	8.438	1.10	7.34	−4.77
Ir	8.967	1.564	7.40	−5.27
Pt	8.959	2.128	6.83	−5.54
Au	9.226	2.309	6.92	−5.77
Hg	10.438	0.42	10.02	−5.43
Tl	6.108	0.20	5.91	−3.15
Pb	7.417	0.364	7.05	−3.89
Bi	7.286	0.942	6.34	−4.11
Po	8.414	1.90	6.51	−5.16
At		2.80	−2.80	−1.40
Rn	10.749	Not stable	10.75	–
Fr	4.0727	0.46	3.61	−2.27
Ac	5.17	0.35	4.82	−2.76
Uuo	–	0.056	–	–
Ubu	–	0.57	–	–

Compilation taken from Ref. [80]; for details, see references in this paper.
Reproduced by permission of the PCCP Owner Societies.

This function that can be seen as an extension to frontier molecular orbital reactivity theory is of utmost importance in describing the regioselectivity of chemical reactions and covalent bond-formation processes.

Using a Maxwell relation, the Fukui function can also be expressed as

$$f_{\mathrm{A}}(\mathbf{r}) = \left[\frac{\delta \mu_{\mathrm{A}}}{\delta v_{\mathrm{A}}(\mathbf{r})}\right]_{\substack{N_{\mathrm{A}} = N_{\mathrm{A}}^0 \\ v_{\mathrm{A}}(\mathbf{r}) = v_{\mathrm{A}}^0(\mathbf{r})}} = \left(\frac{\partial \rho_{\mathrm{A}}(\mathbf{r})}{\partial N}\right)_{\substack{N_{\mathrm{A}} = N_{\mathrm{A}}^0 \\ v_{\mathrm{A}}(\mathbf{r}) = v_{\mathrm{A}}^0(\mathbf{r})}} \tag{7.33}$$

Owing to the discontinuity of the electron density with respect to the number of electrons, the left and right side derivatives are different and describe the system's propensity toward an electrophilic or nucleophilic attack, respectively,

$$f_{\mathrm{A}}^{-}(\mathbf{r}) = \left(\frac{\partial \rho_{\mathrm{A}}(\mathbf{r})}{\partial N}\right)^{-}_{\substack{N_{\mathrm{A}} = N_{\mathrm{A}}^0 \\ v_{\mathrm{A}}(\mathbf{r}) = v_{\mathrm{A}}^0(\mathbf{r})}} = \rho_{\mathrm{A},N}(\mathbf{r}) - \rho_{\mathrm{A},N-1}(\mathbf{r}) \tag{7.34}$$

$$f_{\mathrm{A}}^{+}(\mathbf{r}) = \left(\frac{\partial \rho_{\mathrm{A}}(\mathbf{r})}{\partial N}\right)^{+}_{\substack{N_{\mathrm{A}} = N_{\mathrm{A}}^0 \\ v_{\mathrm{A}}(\mathbf{r}) = v_{\mathrm{A}}^0(\mathbf{r})}} = \rho_{\mathrm{A},N+1}(\mathbf{r}) - \rho_{\mathrm{A},N}(\mathbf{r}) \tag{7.35}$$

where $\rho_{\mathrm{A},N+1}(\mathbf{r})$, $\rho_{\mathrm{A},N}(\mathbf{r})$, and $\rho_{\mathrm{A},N-1}(\mathbf{r})$ are the electron densities of the N+1, N, and N−1 electron systems, respectively, all computed at the geometry of the N−electron system.

The Fukui function for the attack of a neutral (radical) reagent, denoted by $f_{\mathrm{A}}^{0}(\mathbf{r})$, has been introduced as the average of $f_{\mathrm{A}}^{+}(\mathbf{r})$ and $f_{\mathrm{A}}^{-}(\mathbf{r})$

$$f_{\mathrm{A}}^{0}(\mathbf{r}) = \frac{f_{\mathrm{A}}^{+}(\mathbf{r}) + f_{\mathrm{A}}^{-}(\mathbf{r})}{2} \tag{7.36}$$

It should be pointed out that these expressions are exact if the exact electron densities are used. As approximate electron densities are used to compute these quantities in the majority of cases, the abovementioned finite difference expressions become approximate. As can be seen, $f_{\mathrm{A}}^{-}(\mathbf{r})$ quantifies the change of the electron density when the total number of electrons of the system is reduced by one. Areas with large values of $f_{\mathrm{A}}^{-}(\mathbf{r})$ can thus be identified with regions where an electrophilic attack is more likely to occur (consequently, these regions can be categorized as nucleophilic). Analogously, $f_{\mathrm{A}}^{+}(\mathbf{r})$ gives the change of the electron density when the total number of electrons is increased; therefore, areas with large values of this function are more prone to a nucleophilic attack (and can be called *electrophilic regions*). As such, the electronic Fukui function is an important, often decisive, player in deciphering the initial stage of covalent bonding.

It should be noted that evaluation of $f_{\mathrm{A}}^{+}(\mathbf{r})$ for gas-phase molecules using Eq. (7.35) might suffer from problems associated with metastable anions. For a detailed discussion of these problems, we again refer the reader to Ref. [90].

Different contributions have been devoted to the analytic calculation of the Fukui function through Eqs. (7.34) and (7.35). Recently, an approach has been published affording the analytic evaluation of the Fukui function within the Kohn–Sham

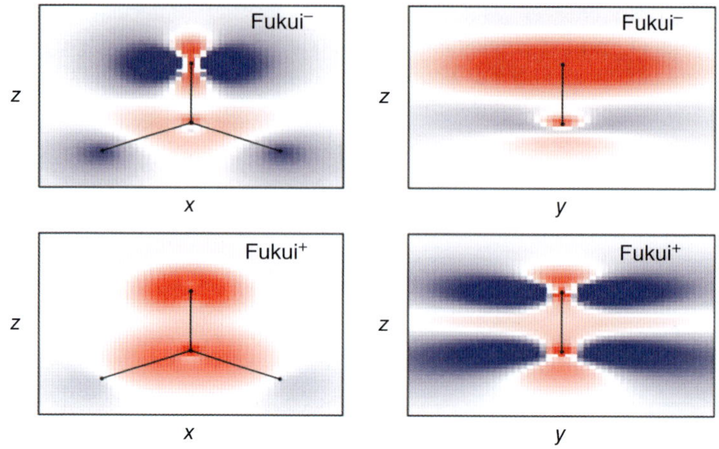

Figure 7.1 Plots of the Fukui functions $f_A^-(\mathbf{r})$ (Fukui⁻) and $f_A^+(\mathbf{r})$ (Fukui⁺) for H_2CO both in the molecular plane and perpendicular to this plane (Reprinted with permission from Ref. [91]. Copyright 2012, AIP Publishing LLC).

approach, of course within the approximation adopted for the exchange–correlation functional [91].

In Figure 7.1, contour plots of the Fukui functions $f_A^-(\mathbf{r})$ and $f_A^+(\mathbf{r})$, computed analytically for a representative simple organic molecule, formaldehyde H_2CO, are given. As can be seen, the contours of $f_A^-(\mathbf{r})$ are the most extended contours around the oxygen atom of the molecule, thus predicting that bonding with an electrophile will predominantly occur on this atom. The contours of $f_A^+(\mathbf{r})$ on the other hand are the most extended contours around the carbon atom, completely in line with the fact that bonding with a nucleophile occurs on this atom.

Note that these exact expressions can be approximated by the corresponding frontier molecular orbital density, that is,

$$f_A^-(\mathbf{r}) \approx |\phi_{A,HOMO}|^2 \tag{7.37}$$

$$f_A^+(\mathbf{r}) \approx |\phi_{A,LUMO}|^2 \tag{7.38}$$

where $\phi_{A,HOMO}$ and $\phi_{A,LUMO}$ are the HOMO and LUMO orbitals of the system. The correction term to the frontier–molecular–orbital approximation, is precisely the relaxation of the occupied orbitals on ionization, and was already put forward by Yang *et al.* in the early 1980s [92].

Often, it is interesting to also consider values of these reactivity indices condensed to atoms in molecules [93]. These estimates can be obtained using population analysis method of one's choice as

$$f_{A,k}^- = p_{A,k}(N) - p_{A,k}(N-1) \tag{7.39}$$

$$f_{A,k}^+ = p_{A,k}(N+1) - p_{A,k}(N) \tag{7.40}$$

where $p_{A,k}(N+1)$, $p_{A,k}(N)$, and $p_{A,k}(N-1)$ are the atomic population on atom k in the N+1, N, and N−1 electron systems, respectively, all considered at the geometry of the N−electron system. It should be mentioned that these equations result from the so-called response of the molecular fragment approach; that is, the electron densities of N, N+1, or N−1 systems are partitioned separately, and the Fukui function is computed subsequently as the differences of atomic populations. In another approach, denoted as the fragment-of-the-molecular response, the Fukui function is computed first, then partitioned over atomic regions. The two approaches usually yield the same conclusions about regioselectivity [94].

When using approximations (7.37) and (7.38), it is evident that the Fukui function will be strictly positive in all regions of space. When using the finite difference approximations (7.34) and (7.35), however, the Fukui function is also positive in most regions of space, although this function can exhibit negative values, for which theoretical arguments have been given [95–99]. When considering this observation, the dilemma occurs that apparently no analogy of the orbital phase in conceptual DFT is available, a property that plays a very important role in the rationalization of the Woodward–Hoffmann rules governing bonding in pericyclic reactions [100–103]. As no reactivity theory can be considered as complete without explaining these seminal bonding rules, this aspect was scrutinized by the present authors. An important step forward in this aspect was the use of the initial hardness response, which is the response of the hardness along the initial stages of the reaction coordinate of the pericyclic reaction [104]. A more intuitive approach has used the so-called dual descriptor [105, 106], introduced as the derivative of the Fukui function with respect to the number of electrons:

$$f_A^{(2)}(\mathbf{r}) = \left[\frac{\delta \eta_A}{\delta v_A(\mathbf{r})}\right]_{\substack{N_A = N_A^0 \\ v_A(\mathbf{r}) = v_A^0(\mathbf{r})}} = \left(\frac{\partial f_A(\mathbf{r})}{\partial N}\right)_{\substack{N_A = N_A^0 \\ v_A(\mathbf{r}) = v_A^0(\mathbf{r})}} = \left(\frac{\partial^2 \rho_A(\mathbf{r})}{\partial N^2}\right)_{\substack{N_A = N_A^0 \\ v_A(\mathbf{r}) = v_A^0(\mathbf{r})}} \tag{7.41}$$

Using a finite difference approximation, this quantity can be computed as the difference between the Fukui functions for a nucleophilic and an electrophilic attack

$$f_A^{(2)}(\mathbf{r}) \approx f_A^{+}(\mathbf{r}) - f_A^{-}(\mathbf{r}) \tag{7.42}$$

This quantity has been shown to provide a "one-shot" picture of the chemical reactivity of the molecule; positive regions indicate areas where the nucleophilic attack is more probable, whereas negative areas correspond to regions with a higher probability for an electrophilic attack [105, 106].

When considering the Woodward–Hoffmann rules for pericyclic reactions, an intuitive reactivity rule was adopted, stating that bonding, predicting an allowed mode of the reaction, will occur when the nucleophilic regions of the molecules (i.e., regions with $f_A^{(2)}(\mathbf{r}) < 0$) are aligned with electrophilic regions ($f_A^{(2)}(\mathbf{r}) > 0$). The driving force for the formation of mostly nonpolar bonds is thus retrieved via conceptual DFT [107].

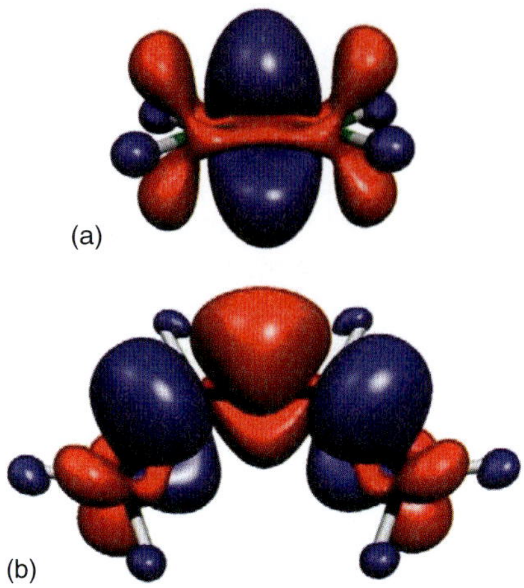

Figure 7.2 Dual descriptor plots of ethylene (a) and 1,3-butadiene (b). Red regions correspond to areas where $f_A^{(2)}(\mathbf{r}) > 0$, whereas blue regions correspond to areas of $f_A^{(2)}(\mathbf{r}) < 0$. (Reprinted with permission from Ref. [107]. Copyright 2007 John Wiley and Sons.)

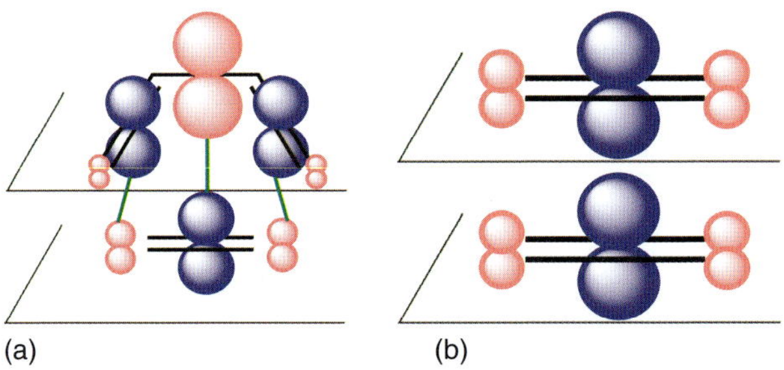

Figure 7.3 Interactions in (a) 1,3-butadiene–ethylene, corresponding to a Woodward–Hoffmann allowed mode (green line) and (b) ethylene–ethylene, corresponding to a Woodward–Hoffmann forbidden mode. Pink regions correspond to areas in which $f_A^{(2)}(\mathbf{r}) > 0$, whereas blue regions correspond to areas of $f_A^{(2)}(\mathbf{r}) < 0$. (Reprinted with permission from Ref. [107]. Copyright 2007 John Wiley and Sons.)

In Figure 7.2, the dual descriptors for 1,3-butadiene and ethylene have been plotted. It can be seen from Figure 7.3a that a [4+2] suprafacial/suprafacial addition of butadiene to ethane corresponds to a favorable matching of the two reagents (i.e., matching of positive regions of one reagent with the negative ones of the other and vice versa); this indeed corresponds to a Woodward–Hoffmann allowed reaction. The [2+2] addition (Figure 7.3b) reaction between two ethene molecules clearly shows "repulsive" interactions (blue/blue and red/red interactions), confirming the Woodward–Hoffmann forbidden character of this interaction mode.

Other pericyclic reactions were also treated within this framework [108–111].

In many studies, the Fukui function has been proven to be a very interesting tool to study the intramolecular reactivity, predicting regioselectivity of an attacking molecule. In order to investigate intermolecular reactivity sequences (e.g., as is done within the framework of the HSAB principle), one needs to consider the so-called local softness s of the molecule, introduced as [112]

$$s_A(\mathbf{r}) = \left[\frac{\partial \rho_A(\mathbf{r})}{\partial \mu_A}\right]_{\substack{N_A = N_A^0 \\ v_A(\mathbf{r}) = v_A^0(\mathbf{r})}} = \left(\frac{\partial N}{\partial \mu_A}\right)_{\substack{N_A = N_A^0 \\ v_A(\mathbf{r}) = v_A^0(\mathbf{r})}} \left(\frac{\partial \rho_A(\mathbf{r})}{\partial N}\right)_{\substack{N_A = N_A^0 \\ v_A(\mathbf{r}) = v_A^0(\mathbf{r})}} = S f_A(\mathbf{r}) \tag{7.43}$$

This quantity can indeed be computed by multiplying the global softness with the Fukui function; this quantity results from the distribution of the global softness over the different regions of the molecule using the Fukui function. It identifies regions of favorable orbital-controlled bond-formation processes (i.e., soft–soft interactions) [113, 114].

It should be pointed out that this quantity naturally emerges when considering the so-called open-system description of the chemical reactivity problem, which will be discussed in greater detail in the later sections in this chapter.

The definition of a local counterpart of the global hardness has been less straightforward and many, even very recent, contributions have been devoted to this problem. For recent accounts on this matter, we refer to the references listed in [115–119]. Often, the molecular electrostatic potential $V(\mathbf{r})$ [120, 121], given in Eq. (7.44), is used as a measure of the local hardness of the molecule to probe regions of favorable charge-controlled bond-formation processes (i.e., hard–hard interactions) [122, 123]. This quantity gives the interaction energy of the molecule with a unit positive charge in the absence of any polarization or geometry relaxation effects:

$$V(\mathbf{r}) = \sum_A \frac{Z_A}{|\mathbf{r} - \mathbf{R}_A|} - \int \frac{\rho(\mathbf{r}')}{|\mathbf{r} - \mathbf{r}'|} d\mathbf{r}' \tag{7.44}$$

7.3.1.3 **Nonlocal Response Functions: the Linear Response Kernel**

The linear response kernel (also called the linear response function or the polarizability kernel) is introduced as the second derivative of the energy with respect to the external potential [124]:

$$\omega_A(\mathbf{r}, \mathbf{r}') = \left[\frac{\delta^2 E_A}{\delta v_A(\mathbf{r}) \delta v_A(\mathbf{r}')}\right]_{\substack{N_A = N_A^0 \\ v_A(\mathbf{r}) = v_A^0(\mathbf{r})}} = \left[\frac{\delta \rho_A(\mathbf{r})}{\delta v_A(\mathbf{r}')}\right]_{\substack{N_A = N_A^0 \\ v_A(\mathbf{r}) = v_A^0(\mathbf{r})}} \tag{7.45}$$

As can be seen, it expresses the response of the electron density to a perturbation in the external potential of the system. This quantity is symmetric, that is

$$\omega_A(\mathbf{r}, \mathbf{r}') = \omega_A(\mathbf{r}', \mathbf{r}) \tag{7.46}$$

and the integral over space of this quantity is zero

$$\int \omega_A(\mathbf{r}, \mathbf{r}') d\mathbf{r} = \int \omega_A(\mathbf{r}, \mathbf{r}') d\mathbf{r}' = 0 \tag{7.47}$$

Several papers have been devoted to the investigation of theoretical and formal aspects associated to this quantity (e.g., refer to [125]), but it has been only recently found that its chemical interpretation was addressed through explicit calculations of the kernel. An important quantity in this aspect is the so-called atom condensed linear response matrix that is composed of elements obtained by integration of the position variables $\mathbf{r}$ and $\mathbf{r}'$ over the atomic volumes V_A and V_B of atoms A and B, respectively [126–128],

$$\omega_{AB} = \int_{V_A}\int_{V_B} \omega(\mathbf{r},\mathbf{r}')\mathrm{d}\mathbf{r}\mathrm{d}\mathbf{r}' \tag{7.48}$$

A limited number of approaches, including the analytical evaluation within the framework of self-consistent perturbed Kohn–Sham theory, have been proposed recently [91].

An approximate expression for the calculation of the linear response kernel can be obtained from the first-order perturbation theory, which yields the following form for methods involving a single Slater determinant wave function (typically the HF or KS DFT approaches) [129]

$$\omega_s(\mathbf{r},\mathbf{r}') = 2\sum_{\sigma}\sum_{i=1}^{N_\sigma}\sum_{a=N_\sigma}^{\infty} \frac{\phi_i^{\sigma *}(\mathbf{r})\phi_a^{\sigma}(\mathbf{r})\phi_a^{\sigma *}(\mathbf{r}')\phi_i^{\sigma}(\mathbf{r}')}{\varepsilon_i^{\sigma} - \varepsilon_a^{\sigma}} \tag{7.49}$$

where the index i covers all the occupied molecular orbitals $\phi_i^{\sigma}(\mathbf{r})$ and a covers all unoccupied molecular orbitals $\phi_a^{\sigma}(\mathbf{r})$. ε_i^{σ} and ε_a^{σ} are the corresponding orbital energies.

For a closed-shell system, this equation can be simplified to

$$\omega_s(\mathbf{r},\mathbf{r}') = 4\sum_{i=1}^{N_0/2}\sum_{a=(N_0/2)+1}^{\infty} \frac{\phi_i^{*}(\mathbf{r})\phi_a(\mathbf{r})\phi_a^{*}(\mathbf{r}')\phi_i(\mathbf{r}')}{\varepsilon_i - \varepsilon_a} \tag{7.50}$$

It was shown that the values of the condensed linear response function can be used as a measure of electron delocalization; as such, this quantity can be linked to the concept of aromaticity [130–132]. If the inductive effect is dominant, the values of the atom-condensed linear response functions decay exponentially with the increasing distance between the atoms. If mesomeric effects are present, high values of the linear response functions remain present even at large distance; an illustration of this can be seen in Figure 7.4, in which the difference in the transmission of the inductive and mesomeric effects can be well appreciated. In this figure, values of the condensed linear response function, which are essentially elements of the matrix equation (7.48), are plotted [126–128].

In Figure 7.4, the linear response elements ω_{OC_i} are plotted; these are the elements between O and the different C atoms in the saturated molecule 1-hexanol versus the unsaturated molecule hexa-1,3,5-trien-1-ol. As can be seen, in the case of the alkane derivative, the response of a carbon atom's density on an external potential perturbation at the oxygen decreases monotonically with the distance of this carbon to O. The triene case, however, yields a totally different pattern: values alternately go up and down. This pattern has been linked with the different

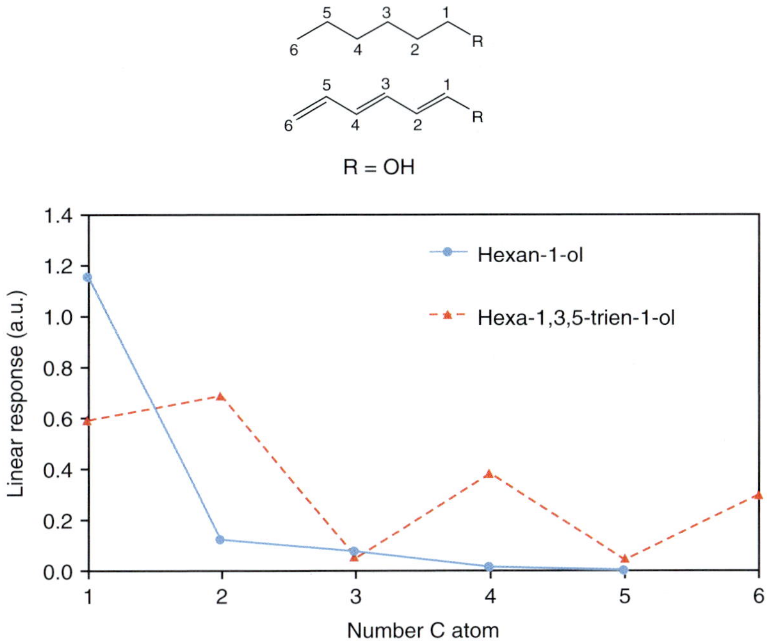

Figure 7.4 Linear response elements ω_{OC_i} between oxygen and the various carbon atoms in hexan-1-ol and hexa-1,3,5-trien-1-ol. The contributions of the various hydrogen atoms are summed into the values of the carbon atoms they are attached to (Reprinted (adapted) with permission from Ref. [126]. Copyright 2010 American Chemical Society).

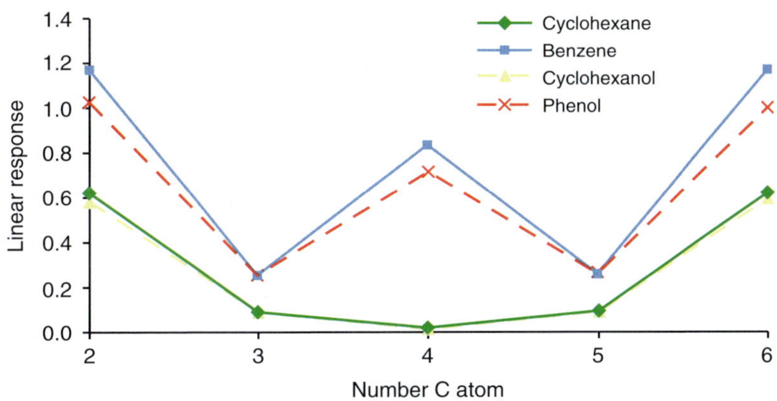

Figure 7.5 Linear response elements ω_{CC_i} between a reference carbon and the other carbon atoms in cyclohexane and benzene, and cyclohexanol and phenol. (Reprinted from Ref. [127], Copyright 2010, with permission from Elsevier.)

mesomeric structures that can be written for this compound. It can thus be clearly seen that the transmission of the external potential perturbation through saturated and unsaturated systems is fundamentally different. In Figure 7.5, a similar plot is presented, now for cyclohexane and benzene (and also for the similar case of a substituted ring, i.e., cyclohexanol and phenol). Again, the linear response function matrix elements between a reference carbon and the other ring carbons decrease monotonically in the case of the saturated compound, whereas they have shown an alternating pattern in the case of the unsaturated compound. The magnitude of this effect in the latter compounds has been put forward as a measure of degree of aromaticity of the ring.

Overall, the linear response function, nowadays available via different computational schemes, offers an unbiased approach to evaluate the response of atoms in molecules to perturbations, or, to put it alternatively, to investigate the propagation of a perturbation through bonds.

7.3.2 Combined use of DFT-Based Reactivity Indices and Principles in the Study of Chemical Bonding

In the previous section, it was shown that a series of reactivity indices emerge that can provide interesting information about the initial stages of bond formation. In this section, we elaborate on this and present several examples combining these reactivity indices with a series of reactivity and stability principles, in order to describe the onset of bonding or the bonding process itself.

7.3.2.1 Principle of Electronegativity Equalization

During 1950s, Sanderson put forward the so-called stability ratio of an atom as a measure of the electronegativity of the atom. The stability ratio of an atom A, $\chi_{S,A}$, is defined as [26, 27]

$$\chi_{S,A} = \frac{\overline{\rho}_A}{\overline{\rho}_0} \tag{7.51}$$

where $\overline{\rho}_A$ is the average electron density of A, computed as

$$\overline{\rho}_A = \frac{Z_A}{\frac{4}{3}\pi r_A^3} \tag{7.52}$$

with Z_A the nuclear charge (i.e., the number of electrons of the system if it concerns a neutral atom) and r_A the radius of A.

$\overline{\rho}_0$ is the average electron density of a hypothetical noble gas atom with the same number of electrons as A.

The use of this quantity to probe the electronegativity of an atom can be understood as follows. Atoms with a low stability ratio (and low average electron density) are less able to bind their electrons and will have lesser tendency to bind additional electrons on bond formation; as such, they have low electronegativities. On the other hand, atoms with a high stability ratio strongly bind their electrons and can easily bind extra electrons on bond formation; this is suggestive of a high electronegativity. Sanderson also formulated the principle of electronegativity equalization, which states that when two or more atoms form a molecule, their electronegativities will equalize to an intermediate value; he postulated that this intermediate value was the geometric mean of the atomic Sanderson electronegativities $\chi_{S,i}$; for a system containing n atoms, this equalized electronegativity becomes

$$\chi_S = (\chi_{S,A}\chi_{S,B}\chi_{S,C}\cdots\chi_{S,n})^{\frac{1}{n}} \tag{7.53}$$

Using this principle, Sanderson also designed a method to obtain partial atomic charges on the atoms after bond formation. Assuming that the molecule NaF has a ionicity of 90% (subsequently this value was adjusted to 80%), it was empirically derived that the change in Sanderson electronegativity on addition or subtraction of an entire electron for a given atom type A was given by

$$\Delta S_A = 1.56\, S_A^{\frac{1}{2}} \tag{7.54}$$

As a result, the partial charge δ_A on the atom A in the molecule on bond formation can be expressed as

$$\delta_A = \frac{S - S_A}{1.56\,\Delta S_A} \tag{7.55}$$

Theoretical justification for the electronegativity equalization principle followed in the subsequent years; a seminal paper in this context was contributed by Parr, Donnelly, Levy, and Palke, identifying also the chemical potential, the Lagrange multiplier introduced in the DFT energy minimization problem with the negative of the electronegativity. Sanderson subsequently used the atomic charges obtained through Eq. (7.55) to obtain and rationalize trends in bond energies.

Within DFT, a simple and elegant derivation directly links the bond energy with difference in electronegativity [8]. Consider the bond formation between two atoms A and B. Considering only a change in the number of electrons N on bond formation (and neglecting a change in the external potential), the change in the chemical potential of both atoms can be expressed to first order as

$$\mu_A = \mu_A^0 + \left(\frac{\partial \mu_A}{\partial N_A}\right)_{v_A} \Delta N_A = \mu_A^0 + \eta_A^0 \Delta N_A \tag{7.56}$$

$$\mu_B = \mu_B^0 + \left(\frac{\partial \mu_B}{\partial N_B}\right)_{v_B} \Delta N_B = \mu_B^0 + \eta_B^0 \Delta N_B \tag{7.57}$$

In these expressions, μ_A^0 and μ_B^0 denote the chemical potential of the isolated atoms A and B, respectively; η_A^0 and η_B^0 are the isolated atom hardnesses. ΔN_A and ΔN_B are the charge transfers to A and B, respectively. On formation of a bond between A and B, their electronegativities (or chemical potentials) will equalize, that is,

$$\mu_A = \mu_B \tag{7.58}$$

Because, obviously

$$\Delta N_A = -\Delta N_B = \Delta N \tag{7.59}$$

one finds that

$$\Delta N = \frac{\mu_B^0 - \mu_A^0}{\eta_A^0 + \eta_B^0} \tag{7.60}$$

or, because $\mu = -\chi$

$$\Delta N = \frac{\chi_A^0 - \chi_B^0}{\eta_A^0 + \eta_B^0} \tag{7.61}$$

As can be seen, the charge transfer on bond formation will be larger if the electronegativity difference between the two atoms is larger; it can also be seen that the hardness modulates charge transfer.

The corresponding energy changes to first order can be computed as

$$E_A = E_A^0 + \left(\frac{\partial E_A}{\partial N_A}\right)_{v_A} \Delta N_A = E_A^0 + \mu_A^0 \Delta N_A \tag{7.62}$$

$$E_B = E_B^0 + \left(\frac{\partial E_B}{\partial N_B}\right)_{v_B} \Delta N_B = E_B^0 + \mu_B^0 \Delta N_B \tag{7.63}$$

Using Eq. (7.62), the changes in energy of A and B on bond formation become

$$\Delta E_A = E_A - E_A^0 = \mu_A^0 \Delta N \tag{7.64}$$

$$\Delta E_B = E_B - E_B^0 = -\mu_B^0 \Delta N \tag{7.65}$$

and, using Eq. (7.61), the total energy change can finally be obtained as

$$\Delta E = \Delta E_A + \Delta E_B = -\frac{(\mu_B^0 - \mu_A^0)^2}{\eta_A^0 + \eta_B^0} = -\frac{(\chi_A^0 - \chi_B^0)^2}{\eta_A^0 + \eta_B^0} \tag{7.66}$$

It can be seen that the energy change is always negative and is proportional to the square of the difference in electronegativity between atoms A and B. Also, this equation partially provides a proof for the HSAB principle; the combination of two soft species on bond formation (i.e., to atoms with a small hardness) leads to the largest stabilization.

Equation (7.66) can also be linked to Pauling's extra ionic resonance energy Δ, playing a fundamental role in *the* electronegativity scale; [1] this quantity is related to the energy difference between a bond between different atoms A and B and the energy of a covalent bond between A and A and B and B:

$$\Delta = D(A - B) - \frac{1}{2}[D(A - A) + D(B - B)] \tag{7.67}$$

where Δ corresponds to the bond energy. Assuming that

$$E_A^0 \approx \frac{1}{2} D(A - A) \tag{7.68}$$

$$E_B^0 \approx \frac{1}{2} D(B - B) \tag{7.69}$$

and

$$E_A + E_B \approx D(A - B) \tag{7.70}$$

it can be seen that

$$\Delta E \approx \Delta \tag{7.71}$$

Pauling defined the quantity Δ' as

$$\Delta' = D(\text{A}-\text{B}) - [D(\text{A}-\text{A})D(\text{B}-\text{B})]^{\frac{1}{2}} \tag{7.72}$$

which now considers the geometric mean of D(A–A) and D(B–B). This quantity can then be related to the difference in Pauling electronegativities χ_A^P and χ_B^P of atoms A and B

$$\Delta' = 30\,(\text{kcal}\,\text{mol}^{-1})(\chi_\text{A}^P - \chi_\text{B}^P)^2 \tag{7.73}$$

which can be compared with Eq. (7.66); as can be seen, in Eq. (7.66), the energy change is also dependent on the sum of the hardnesses of the two atoms involved, whereas in Eq. (7.73), only a constant factor is in front of the square of the electronegativity difference.

Equations (7.64) and (7.65) only considered the energy change to first order in the number of electrons of the bond-forming atoms.

Expressions (7.56) and (7.57) can also be rewritten to include the atomic charges on the atoms; in terms of the electronegativities, these become

$$\chi_\text{A} = \chi_\text{A}^0 + \eta_\text{A}^0 q_\text{A} \tag{7.74}$$

$$\chi_\text{B} = \chi_\text{B}^0 + \eta_\text{B}^0 q_\text{B} \tag{7.75}$$

These expressions were also proposed by Huheey [133–137]. Using the principle of electronegativity equalization and the fact that the sum of the atomic charges should be equal to the total charge of the molecule, this provides a method to obtain atomic charges in an economical way. However, using the equations in this form, leads to atomic charges that do not depend on the connectivity between atoms because the external potential change on molecule formation was neglected. A famous extension was provided by Mortier *et al.* using the electronegativity equalization method (EEM) that employs a simple electrostatic model to represent the dependence on the external potential. The resulting expression for the electronegativity of an atom in a molecule is [138–142]

$$\chi_\text{A} = \chi_\text{A}^* + \eta_\text{A}^* q_\text{A} + \sum_{\text{B}\neq\text{A}} \frac{q_\text{B}}{R_\text{AB}} \tag{7.76}$$

which, combined with the condition that

$$\sum_\text{A} q_\text{A} = q_\text{tot} \tag{7.77}$$

where q_tot is the total molecular charge, can be used to obtain a more refined charge distribution. χ_A^* and η_A^* are now effective electronegativities and hardnesses that were obtained by calibration of the method against Mulliken atomic charges, initially obtained at the Hartree–Fock STO-3G level of theory. Subsequently, a large number of approaches have elaborated on the electronegativity equalization principle (or chemical potential equalization principle), many of which aiming at the calculation of atomic charges at a very low computational cost, for use in force-field calculations [143].

We finally return to the energy expression given in Eq. (7.66). More accurate energy expression can be obtained by writing expression (7.19) including now Δv perturbations for both A and B and considering again that $\Delta N_A = -\Delta N_B = \Delta N$ [8, 144]

$$\begin{aligned}\Delta E = E_A + E_B - E_A^0 - E_B^0 &= (\mu_A - \mu_B)\Delta N + (\eta_A + \eta_B)\Delta N^2 \\ &+ \int \rho_A(\mathbf{r})\Delta v_A(\mathbf{r})d\mathbf{r} + \int \rho_B(\mathbf{r})\Delta v_B(\mathbf{r})d\mathbf{r} \\ &+ \Delta N\left[\int f_A(\mathbf{r})\,\Delta v_A(\mathbf{r})d\mathbf{r} - \int f_B(\mathbf{r})\Delta v_B(\mathbf{r})d\mathbf{r}\right] \\ &+ \frac{1}{2}\int \omega_A(\mathbf{r},\mathbf{r}')\Delta v_A(\mathbf{r})\Delta v_A(\mathbf{r}')d\mathbf{r}d\mathbf{r}' \\ &+ \frac{1}{2}\int \omega_B(\mathbf{r},\mathbf{r}')\Delta v_B(\mathbf{r})\Delta v_B(\mathbf{r}')d\mathbf{r}d\mathbf{r}' \end{aligned} \tag{7.78}$$

Minimizing the energy change with respect to the charge transfer yields

$$\Delta N = \frac{-(\mu_A - \mu_B) - \left[\int f_A(\mathbf{r})\,\Delta v_A(\mathbf{r})d\mathbf{r} - \int f_B(\mathbf{r})\Delta v_B(\mathbf{r})d\mathbf{r}\right]}{2(\eta_A + \eta_B)} \tag{7.79}$$

which can then be substituted back in Eq. (7.78) to give

$$\begin{aligned}\Delta E = &-\frac{\left((\mu_A - \mu_B) + \left[\int f_A(\mathbf{r})\,\Delta v_A(\mathbf{r})d\mathbf{r} - \int f_B(\mathbf{r})\Delta v_B(\mathbf{r})d\mathbf{r}\right]\right)^2}{2(\eta_A + \eta_B)} \\ &+ \int \rho_A(\mathbf{r})\Delta v_A(\mathbf{r})d\mathbf{r} + \int \rho_B(\mathbf{r})\Delta v_B(\mathbf{r})d\mathbf{r} \\ &+ \frac{1}{2}\int \omega_A(\mathbf{r},\mathbf{r}')\Delta v_A(\mathbf{r})\Delta v_A(\mathbf{r}')d\mathbf{r}d\mathbf{r}' \\ &+ \frac{1}{2}\int \omega_B(\mathbf{r},\mathbf{r}')\Delta v_B(\mathbf{r})\Delta v_B(\mathbf{r}')d\mathbf{r}d\mathbf{r}' \end{aligned} \tag{7.80}$$

As can be seen, the interaction energy on the *initial stages of bond formation* naturally gives rise to three terms. The first term, corresponding to the charge-transfer contribution to the bond formation, gives the covalent contribution to bonding; it plays an important role in orbital-controlled interactions. The second term (combined with the contribution for the nuclear–nuclear repulsion) is the electrostatic contribution, which is important for charge-controlled interactions. The third term finally includes the effect of polarization.

7.3.2.2 **Hard and Soft Acids and Bases Principle**

The HSAB principle was formulated by Pearson in the 1960s on the basis of a large collection of experimental data; at that time, no sharp definition of hardness and softness was available [28–31]. The mathematical definition of these concepts within DFT afforded opportunities to prove this principle [145, 146]. Recent insights into the validity of the HSAB principle have been provided by Ayers by considering, for example, exchange reactions of Lewis acids and bases [147–149].

In order to prove the principle, one turns to a so-called open-system picture to describe the system, due to the fact that the number of electrons can fluctuate but the chemical potential of the system is well defined (this is motivated by the open nature of atoms, functional groups, or other subunits within a molecular system); in this picture, one uses changes in the chemical potential, rather than changes in the number of electrons, to control the change-state of a molecule, while keeping the number of electrons fixed [48].

The state function of the open-system picture is the so-called grand potential $\Omega[\mu, v]$, which is readily obtained through the Legendre transform of the energy [48]

$$\Omega[\mu, v] = E[N, v] - \left(\frac{\partial E}{\partial N}\right)_v N = E[N, v] - \mu N \tag{7.81}$$

The change in the grand potential from one ground state to another, to first order, can be expressed as

$$\Delta\Omega = \left(\frac{\partial\Omega}{\partial\mu}\right)_v \Delta\mu + \int \left[\frac{\delta\Omega}{\delta \upsilon(\mathbf{r})}\right]_\mu \Delta\upsilon(\mathbf{r})d\mathbf{r} \tag{7.82}$$

where

$$\left(\frac{\partial\Omega}{\partial\mu}\right)_v = -N \tag{7.83}$$

and

$$\left[\frac{\delta\Omega}{\delta \upsilon(\mathbf{r})}\right]_\mu = \rho(\mathbf{r}) \tag{7.84}$$

Considering the change to first order in the number of electrons and electron density on variation of μ and v yields

$$\Delta N = \left(\frac{\partial N}{\partial\mu}\right)_v \Delta\mu + \int \left[\frac{\delta N}{\delta \upsilon(\mathbf{r})}\right]_\mu \Delta\upsilon(\mathbf{r})d\mathbf{r} \tag{7.85}$$

In this expression, the first derivative with respect to μ, a global quantity, is the global softness of the system (cf. Eq. (7.43))

$$\left(\frac{\partial N}{\partial\mu}\right)_v = \left[\left(\frac{\partial\mu}{\partial N}\right)_v\right]^{-1} = \eta^{-1} = S = -\left(\frac{\partial^2\Omega}{\partial\mu^2}\right)_v \tag{7.86}$$

The first derivative with respect to v, a local descriptor, is the aforementioned local softness of the system,

$$-\left[\frac{\delta N}{\delta \upsilon(\mathbf{r})}\right]_\mu = \left(\frac{\partial\rho(\mathbf{r})}{\partial\mu}\right)_N = s(\mathbf{r}) = \left(\frac{\partial\delta\Omega}{\partial\mu\delta \upsilon(\mathbf{r})}\right) \tag{7.87}$$

It is also interesting to consider the change in the electron density on changes in the chemical and external potential

$$\Delta\rho(\mathbf{r}) = \left(\frac{\partial\rho(\mathbf{r})}{\partial\mu}\right)_v \Delta\mu + \int \left[\frac{\delta\rho(\mathbf{r})}{\delta \upsilon(\mathbf{r}')}\right]_\mu \Delta\upsilon(\mathbf{r}')d\mathbf{r}' \tag{7.88}$$

Again, the first response function in this equation is the local softness. The second quantity is the so-called softness kernel $s(\mathbf{r}, \mathbf{r}')$ [124]

$$s(\mathbf{r}, \mathbf{r}') \equiv \left[\frac{\delta \rho(\mathbf{r})}{\delta v(\mathbf{r}')}\right]_{\mu} = \left[\frac{\delta^2 \Omega}{\delta v(\mathbf{r})\, \delta v(\mathbf{r}')}\right]_{\mu} \tag{7.89}$$

which is related to the linear response function by the Berkowitz–Parr relationship [124]

$$\omega(\mathbf{r}, \mathbf{r}') = -s(\mathbf{r}, \mathbf{r}') + \frac{s(\mathbf{r})s(\mathbf{r}')}{S} \tag{7.90}$$

The softness kernel integrates to the local softness of the molecule

$$\int s(\mathbf{r}, \mathbf{r}')\mathrm{d}\mathbf{r}' = s(\mathbf{r}') \tag{7.91}$$

The inverse of the softness kernel

$$\int s(\mathbf{r}, \mathbf{r}')\eta(\mathbf{r}', \mathbf{r}'')\mathrm{d}\mathbf{r}' = \delta(\mathbf{r} - \mathbf{r}'') \tag{7.92}$$

is the hardness kernel

$$\eta(\mathbf{r}, \mathbf{r}') \equiv \left[\frac{\delta v(\mathbf{r})}{\delta \rho(\mathbf{r}')}\right]_{\mu} = \frac{\delta^2 F_{\mathrm{HK}}}{\delta \rho(\mathbf{r})\delta \rho(\mathbf{r}')} \tag{7.93}$$

where F_{HK} is the Hohenberg–Kohn functional.

As stated above, the open-picture description provides an ideal framework to provide support for the HSAB principle [150–154]. Gazquez *et al.* [150, 151] derived a working equation expressing the energy change on bond formation between atoms A and B as a function of the change in the chemical and external potential

$$\Delta E_{\mathrm{AB}} = \Delta E_{\mathrm{AB},\nu} + \Delta E_{\mathrm{AB},\mu} \tag{7.94}$$

with

$$\Delta E_{\mathrm{AB},\nu} \approx -\frac{1}{2}\frac{(\mu_{\mathrm{A}} - \mu_{\mathrm{B}})^2}{S_{\mathrm{A}} + S_{\mathrm{B}}} S_{\mathrm{A}} S_{\mathrm{B}} \left(= -\frac{1}{2}\frac{(\mu_{\mathrm{A}} - \mu_{\mathrm{B}})^2}{\eta_{\mathrm{A}} + \eta_{\mathrm{B}}} \right) \tag{7.95}$$

and

$$\Delta E_{\mathrm{AB},\mu} \approx -\frac{1}{2}\frac{\lambda}{S_{\mathrm{A}} + S_{\mathrm{B}}} \tag{7.96}$$

The first term $\Delta E_{\mathrm{AB},\nu}$ expresses the stabilization of the system due to the equalization of the chemical potentials at constant external potential. The second term is a rearrangement term at constant chemical potential, which involves a constant λ, that considers that effective number of valence electrons involved in the interaction between A and B. These equations have also been extended to the local level as to include the local softness of the interacting atoms in the bond-formation process [150–154].

In the final part of this chapter, we briefly discuss two applications of the DFT-based reactivity concepts to so-called dual reactivity behavior; that is, a molecule with a hard and a soft center where a shift from orbital to charge control in the onset of bonding can occur.

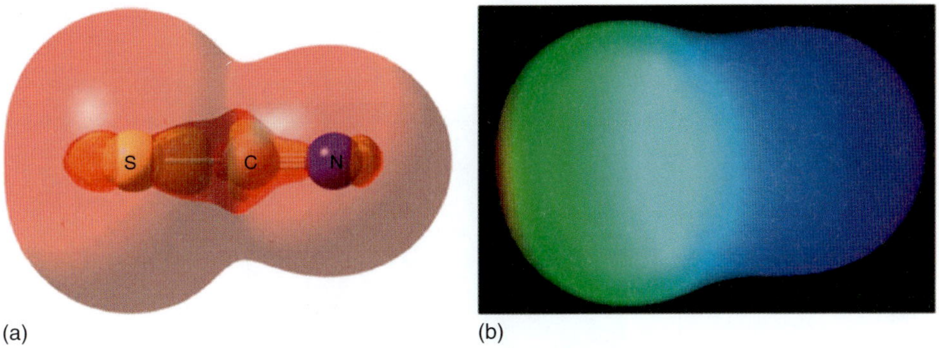

Figure 7.6 Reactivity indices for the ambident thiocyanate ion SCN^-: (a) isosurface plot of the Fukui function for an electrophilic attack and (b) electrostatic potential plotted on the van der Waals surface. The blue color corresponds to a region with a negative potential and the red color with a (small) positive value of this quantity (reproduced with kind permission from Springer Science and Business Media from Ref. [159]).

As a first example, we consider the thiocyanate ion SCN^-, a so-called ambident nucleophile [155], which has been the subject of many studies within conceptual DFT [156–162].

As can be seen from Figure 7.6, the electrostatic potential turns out to be the most negative around the nitrogen atom identifying this atom as the hard side of the molecule, and thus predicting that this is the preferred site for the charge-controlled addition of a generic hard electrophile. The Fukui function on the other hand is the most extended function around the sulfur atom, thus predicting that this is the soft side of the molecule and the preferred site for the orbital-controlled addition of a generic soft electrophile.

Considering Eq. (7.19) to first order (including the nuclear repulsion terms), Ayers *et al.* have derived a so-called all-purpose reactivity indicator Ξ that enables to probe the dual reactivity behavior and to quantify the shift in the reactivity (site-selectivity) depending on the nature of the electrophile (hard or soft), which can be used for reactions that are neither charge- nor orbital-controlled [160, 161]. In the case of an electrophilic attack, that is, when $\Delta N \leq 0$ (transfer of electrons from the nucleophile to the attacking electrophile), this index for the attack on a site α of the nucleophile can be expressed as

$$\Xi_\alpha^\kappa = (\kappa + 1)q_\alpha^0 - \Delta N(\kappa - 1)f_\alpha^- \tag{7.97}$$

where q_α^0 is the charge on atom α in the nucleophile and f_α^- the value of the condensed Fukui function for electrophilic attack on α; the parameter κ contains information about the nature of the attacking electrophile and is proportional to the sum of the charge on the interacting site of the electrophile $q^0_{\text{electrophile}}$ and the amount of charge that is transferred to this site ΔN multiplied with the value of the Fukui function $f^+_{\text{electrophile}}$ on this site

$$\kappa \sim q^0_{\text{electrophile}} + \Delta N f^+_{\text{electrophile}} \tag{7.98}$$

Table 7.2 "Reactivity transition table" for the electrophilic attack on the thiocyanate ion SCN^-.

ΔN\k	1	0.8	0.6	0.4	0.2	0	−0.2	−0.4	−0.6	−0.8	−1
−1	−1.59843	−1.50432	−1.41878	−1.43619	−1.45359	−1.471	−1.48841	−1.50581	−1.52322	−1.54063	−1.55803
−0.9	−1.59843	−1.49775	−1.39706	−1.38945	−1.39127	−1.3931	−1.39493	−1.39675	−1.39858	−1.4004	−1.40223
−0.8	−1.59843	−1.49117	−1.38392	−1.34271	−1.32895	−1.3152	−1.30144	−1.28769	−1.27394	−1.26018	−1.24643
−0.7	−1.59843	−1.4846	−1.37077	−1.29596	−1.26663	−1.2373	−1.20796	−1.17863	−1.14929	−1.11996	−1.09062
−0.6	−1.59843	−1.47803	−1.35762	−1.24922	−1.20431	−1.15939	−1.11448	−1.06956	−1.02465	−0.97974	−0.93482
−0.5	−1.59843	−1.47145	−1.34448	−1.2175	−1.14199	−1.108149	−1.021	−0.9605	−0.90001	−0.83951	−0.77902
−0.4	−1.59843	−1.46488	−1.33133	−1.19778	−1.07967	−1.00359	−0.92752	−0.85144	−0.77536	−0.69929	−0.62321
−0.3	−1.59843	−1.45831	−1.31818	−1.17806	−1.03794	−0.92569	−0.83403	−0.74238	−0.65072	−0.55907	−0.46741
−0.2	−1.59843	−1.45174	−1.30504	−1.15834	−1.01164	−0.86495	−0.74055	−0.63332	−0.52608	−0.41884	−0.31161
−0.1	−1.59843	−1.44516	−1.29189	−1.13862	−0.98535	−0.83208	−0.67881	−0.52554	−0.40144	−0.27862	−0.1558
0	−1.59843	−1.43859	−1.27875	−1.1189	−0.95906	−0.79922	−0.63937	−0.47953	−0.31969	−0.15984	
Nitrogen						Sulfur					

Reproduced with permission from Ref. [158].
Reproduced by permission of the PCCP Owner Societies.

It turns out that the value of κ is between -1 and $+1$ in many cases. As can be seen, the case of $\kappa = -1$ represents a reaction that is essentially electron-transfer (or orbital) controlled (i.e., in the case of soft reagents), whereas the case of $\kappa = +1$ corresponds to a charge-controlled process (hard reagents). This reactivity index thus has been computed for these ambident nucleophiles [158].

In Table 7.2, we present a so-called reactivity transition table of the electrophilic attack on SCN^-; for different choices of κ and ΔN, the value of Ξ_α^κ for the most reactive atom is listed (most negative value), and this atom is identified by color-coding the cell of the table. This table clearly illustrates that the orbital-controlled process occurs preferentially on S, whereas the charge-controlled attack occurs on N; it also well captures the change in the bonding interaction from N to S on changing the nature of the electrophile.

As a second and final example, we briefly treat the dual reactivity of 2,6-dichloropyridine [162]. A plot of both the dual descriptor (to probe reactivity toward soft reagents), and the electrostatic potential is given in Figure 7.7. As can be seen, these reactivity indices predict that a soft electrophile will attach to C3 and C5 (because the dual descriptor is negative on these atoms), whereas a hard electrophile will attach to the nitrogen atom. As the dual descriptor is positive around C4, a nucleophilic attack is predicted on this site. All these predictions are in agreement with experimental data [162].

7.3.2.3 Berlin's Approach in a Conceptual DFT Context: the Nuclear Fukui Function

As stated in Section 7.1, the central quantity of DFT itself, the electron density, has also been used numerous times in order to describe the properties of chemical bonds, as in the quantum chemical topology approach [5, 6], reviewed by Popelier in this volume, or in the electron localization function ELF [163, 164], as reviewed by Savin, Silvi and Grin.

An interesting approach using the electrostatic picture of chemical bonding using the electron density was initiated by Berlin [15, 165] which we now combine here with the conceptual DFT in the nuclear Fukui function [33–40]. Using the Hellmann–Feynman theorem [166, 167], the electrostatic force on an atom α in a molecule $\mathbf{F}_\alpha$ due to the electrons and the other electrons in the molecule can be written as

$$\mathbf{F}_\alpha = -\nabla_\alpha E = \int \frac{Z_\alpha(\mathbf{r}-\mathbf{R}_\alpha)\rho(\mathbf{r})}{|\mathbf{r}-\mathbf{R}_\alpha|^3}\mathrm{d}\mathbf{r} - \sum_{\beta\neq\alpha} \frac{Z_\beta Z_\alpha(\mathbf{R}_\alpha-\mathbf{R}_\beta)}{|\mathbf{R}_\beta-\mathbf{R}_\alpha|^3} \tag{7.99}$$

Now introducing the binding function F_B as the virial of the electrostatic forces that are necessary to hold all the nuclei in the molecule fixed

$$F_\mathrm{B} = -\sum_\alpha \mathbf{R}_\alpha \cdot \mathbf{F}_\alpha = \int \rho(\mathbf{r}) \sum_\alpha \frac{Z_\alpha(\mathbf{R}_\alpha - \mathbf{r})\cdot\mathbf{R}_\alpha}{|\mathbf{r}-\mathbf{R}_\alpha|^3}\mathrm{d}\mathbf{r} - \sum_\alpha\sum_{\beta\neq\alpha} \frac{Z_\beta Z_\alpha}{|\mathbf{R}_\alpha-\mathbf{R}_\beta|} \tag{7.100}$$

This function can be considered as measuring the binding effect in the molecule. One usually puts the geometric center of the molecule at the origin of the Cartesian coordinate system. Consequently, F_B will be related to the work that must be done

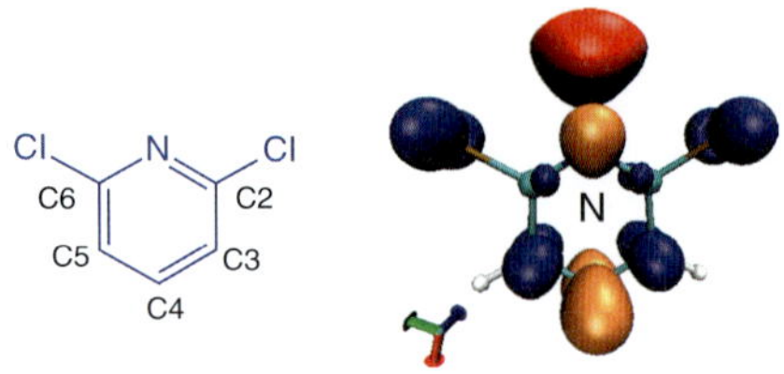

Figure 7.7 Dual descriptor and electrostatic potential of 2,6-dichloro-pyridine. The blue isosurface corresponds to $f^{(2)}(r) = -0.004$ a.u.; the orange isosurface corresponds to $f^{(2)}(r) = 0.004$ a.u. The red isosurface corresponds to a value of −0.035 a.u. for the electrostatic potential. (Reprinted (adapted) with permission from Ref. [162]. Copyright 2009 American Chemical Society.)

to symmetrically scale the coordinates of the initial molecule, that is, all bonds in the molecule will stretch or shrink to the same scale.

Using the approach of Berlin, the binding function can be rewritten as

$$F_{\mathrm{B}} = \int \rho(\mathbf{r}) f_v(\mathbf{r}) \mathrm{d}\mathbf{r} - \sum_{\alpha} \sum_{\beta \neq \alpha} \frac{Z_\beta Z_\alpha}{|\mathbf{R}_\alpha - \mathbf{R}_\beta|} \tag{7.101}$$

where Berlin's function, $f_v(\mathbf{r})$, is defined as

$$f_v(\mathbf{r}) = \sum_{\alpha} \frac{Z_\alpha (\mathbf{R}_\alpha - \mathbf{r}) \cdot \mathbf{R}_\alpha}{|\mathbf{r} - \mathbf{R}_\alpha|^3} \tag{7.102}$$

This function can now be used to separate the molecular space into binding and antibinding regions; the binding function can be rewritten as

$$F_{\mathrm{B}} = \int_{f_v > 0} \rho(\mathbf{r}) f_v(\mathbf{r}) \mathrm{d}\mathbf{r} + \int_{f_v < 0} \rho(\mathbf{r}) f_v(\mathbf{r}) \mathrm{d}\mathbf{r} - \sum_{\alpha} \sum_{\beta \neq \alpha} \frac{Z_\beta Z_\alpha}{|\mathbf{R}_\alpha - \mathbf{R}_\beta|} \tag{7.103}$$

As $\rho(\mathbf{r}) \geq 0$, a pileup of electron density in regions with a positive $f_v(\mathbf{r})$ has a positive contribution to the binding function and tends to shrink the molecule; conversely, accumulation of density in regions of negative $f_v(\mathbf{r})$ will weaken the overall binding.

The force considered hitherto is the force on the nuclei and is often referred to as Feynman force. It should be distinguished from the Ehrenfest force that has also been used to investigate chemical bonding, which is directly related to the electronic stress tensor and finds its place in QTAIM [5, 168–170].

Within the context of conceptual DFT, the following response function has been defined [33]

$$\mathbf{\Phi}_\alpha = \left(\frac{\partial \mathbf{F}_\alpha}{\partial N} \right)_v \tag{7.104}$$

which is the change of the force acting on nucleus α on changing the number of electrons. This so-called "nuclear Fukui function" can be expressed as the force due to the electronic Fukui function

$$\mathbf{\Phi}_\alpha = \int \frac{Z_\alpha (\mathbf{r} - \mathbf{R}_\alpha) f(\mathbf{r})}{|\mathbf{r} - \mathbf{R}_\alpha|^3} \mathrm{d}\mathbf{r} \tag{7.105}$$

as can be seen when deriving the right side of Eq. (7.101) with respect to N.

Note that this quantity is a second-order response function. Considering that $\mathbf{F}_\alpha = -\left(\frac{\partial E}{\partial \mathbf{R}_\alpha}\right)_N$, so that

$$\mathbf{\Phi}_\alpha = \left(\frac{\partial^2 E}{\partial \mathbf{R}_\alpha \partial N}\right)_\nu \tag{7.106}$$

where variations in $\mathbf{R}_\alpha$ are a particular choice of generating variations in $\nu(\mathbf{r})$.

The change in binding function $\mathrm{d}F_\mathrm{B}$ can then be written as

$$\mathrm{d}F_\mathrm{B} = \left(\frac{\partial F_\mathrm{B}}{\partial N}\right)_\nu \mathrm{d}N \tag{7.107}$$

with

$$\left(\frac{\partial F_\mathrm{B}}{\partial N}\right)_\nu = \int f(\mathbf{r}) f_\nu(\mathbf{r}) \mathrm{d}\mathbf{r} \tag{7.108}$$

so that

$$\mathrm{d}F_\mathrm{B} = \int f(\mathbf{r}) f_\nu(\mathbf{r}) \mathrm{d}\mathbf{r} \mathrm{d}N = -\sum_\alpha \mathbf{R}_\alpha \cdot \mathbf{\Phi}_\alpha \mathrm{d}N \tag{7.109}$$

As such, this quantity can be used to probe the changes in the forces on the atoms induced by electron transfer to or from the molecule; it can thus be invoked to study bond-formation or -breaking processes induced by changing the number of electrons. We illustrate this in the following by considering the simple case of the H_2O molecule [38]. As an example, we consider the case of ionization of H_2O. In an orbital picture, the electron is taken away from the nonbonding orbital perpendicular to the molecular plane. In Figure 7.8, the nuclear Fukui function vectors $\mathbf{\Phi}_\alpha^-$ (multiplied by $\Delta N = -1$) are given superposed on a $f(\mathbf{r})f_\nu(\mathbf{r})$ plot. Owing

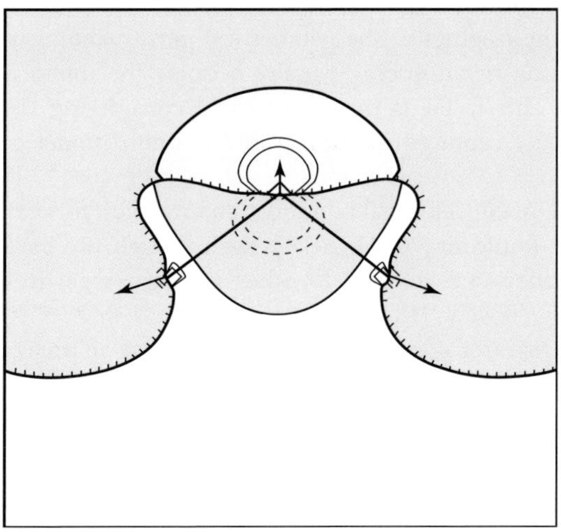

Figure 7.8 $f(\mathbf{r})f_\nu(\mathbf{r})$ values (shaded parts are the regions with negative value and the binding/antibinding contour is hashed) and values of the $\mathbf{\Phi}_\alpha^-$ atomic vectors multiplied by ΔN for the H_2O molecule in the molecular plane. (Reprinted with permission from Ref. [38]. Copyright 2001, AIP Publishing LLC.)

to a delicate interplay of bonding and antibonding regions the resulting nuclear Fukui function vectors tell us that, on ionization, the OH bond will stretch, and the bond angle is enlarged as found experimentally.

7.4 Conclusions

The fundamental descriptor of DFT, the electron density, plays a central role in many approaches to study bonding. Conceptual DFT augments the formal framework and computational procedures of DFT with chemical quantities that describe the initial stages of chemical reactions and, in particular, bond-formation processes. The chemical quantities it introduces are response functions that express how the system's energy changes when the number of electrons (electron-transfer), the external potential (electrostatic interactions, changes in molecular geometry), or both change. These response functions, in turn, allow conceptual DFT to provide a framework for rationalizing many chemical principles (such as Sanderson's electronegativity equalization principle and Pearson's HSAB principle), and even introducing new principles (most famously, the maximum hardness principle). The electronegativity equalization principle and HSAB principle can be invoked to study energy changes related to bond formation in a (semi)quantitative way. The basic idea – the core of the perturbative perspective introduced by conceptual DFT – is that the reactivity is associated with perturbations in the molecular fragments and that prospective reactions with favorable (energy-lowering or low-barrier interactions) interactions are favored. It should be stressed that this basically describes the reactivity at the onset of the reaction and that along the reaction path, next to the interaction between the fragments, the geometrical perturbation can become highly important, resulting in an energy penalty because the interacting molecules become strained. This factor is important when considering the chemical reactivity and is explicitly captured in the activation strain model of Bickelhaupt *et al.* [171]

In this review, we have focused on conventional bonding. (an extension to weak interactions can, for example, be found in [76, 172, 173]) Even though, we have neglected many aspects, and readers are referred to other recent perspectives [159, 174] and more comprehensive reviews [18–22] for more details.

Summarizing, conceptual DFT features a density-centered perspective on molecular bonding and reactivity, providing a box of tools (response functions and quantities derived from them) that elucidate both the nature of the chemical bond and the chemical driving forces associated with bond formation and cleavage.

Acknowledgments

FDP and PG wish to acknowledge the Vrije Universiteit Brussel (VUB) and Research Foundation Flanders (FWO) for continuous support of their research group. They also wish to acknowledge past and present group members for interesting discussions on many different aspects of conceptual DFT. PWA thanks NSERC for funding.

References

1. Pauling, L. (1960) *The Nature of the Chemical Bond*, 3rd edn, Cornell University Press, Ithaca, NY.
2. Mulliken, R.S. (1934) *J. Chem. Phys.*, **2**, 782.
3. Coppens, P. (2005) *Angew. Chem. Int. Ed.*, **44**, 6810.
4. Koritsanszky, T.S. and Coppens, P. (2001) *Chem. Rev.*, **101**, 1583.
5. Bader, R.F.W. (1990) *Atoms in Molecules: A Quantum Theory*, Clarendon, Oxford.
6. Popelier, P.L.A. (2000) *Atoms in Molecules: An Introduction*, Pearson, Harlow.
7. Hohenberg, P. and Kohn, W. (1964) *Phys. Rev. B.*, **136**, 864.
8. Parr, R.G. and Yang, W. (1989) *Density Functional Theory of Atoms and Molecules*, Oxford University Press, New York.
9. Dreizler, R.M. and Gross, E.K.U. (1990) *Density Functional Theory*, Springer-Verlag, Berlin, Heidelberg, New York.
10. Parr, R.G. and Yang, W. (1995) *Ann. Rev. Phys. Chem.*, **46**, 710.
11. Kohn, W., Becke, A.D., and Parr, R.G. (1996) *J. Phys. Chem.*, **100**, 978.
12. Koch, W. and Holthausen, M. (2001) *A Chemist's Guide to Density Functional Theory*, 2nd edn, Wiley-VCH Verlag GmbH, Weinheim.
13. Kohn, W. and Sham, L. (1965) *J. Phys. Rev. A*, **140**, 1133.
14. Pople, J.A. *et al.* (2010) GAUSSIAN 09, Gaussian Inc., Wallingford, CT, 2010 and previous versions.
15. Berlin, T. (1951) *J. Chem. Phys.*, **19**, 208.
16. Chermette, H. (1999) *J. Comput. Chem.*, **20**, 129.
17. Geerlings, P., De Proft, F., and Langenaeker, W. (1999) *Adv. Quantum Chem.*, **33**, 303.
18. Geerlings, P., De Proft, F., and Langenaeker, W. (2003) *Chem. Rev.*, **103**, 1793.
19. Ayers, P.W., Anderson, J.S.M., and Bartolotti, L.J. (2005) *Int. J. Quantum Chem.*, **101**, 520.
20. Gázquez, J.L. (2008) *J. Mex. Chem. Soc.*, **52**, 3.
21. Geerlings, P. and De Proft, F. (2008) *Phys. Chem. Chem. Phys.*, **10**, 3028.
22. Liu, S.B. (2009) *Acta Phys.-Chem. China*, **25**, 590.
23. Sanderson, R.T. (1952) *Science*, **116**, 41.
24. Sanderson, R.T. (1952) *J. Chem. Educ.*, **29**, 539.
25. Sanderson, R.T. (1955) *Science*, **121**, 207.
26. Sanderson, R.T. (1976) *Chemical Bonds and Bond Energy*, Academic Press, New York.
27. Sanderson, R.T. (1983) *Polar Covalence*, Academic Press, New York.
28. Pearson, R.G. (1963) *J. Am. Chem. Soc.*, **85**, 3533.
29. Pearson, R.G. (1966) *Science*, **151**, 172.
30. Pearson, R.G. (1973) *Hard and Soft Acids and Bases*, Dowden, Hutchinson & Ross, Stroudenbury, PA.
31. Pearson, R.G. (1993) in *Chemical Hardness*, Structure and Bonding, vol. 80 (ed. K.D. Sen), Springer-Verlag, Berlin, p. 1.
32. Pearson, R.G. (1997) *Chemical Hardness*, Wiley-VCH Verlag GmbH, Weinheim.
33. Cohen, M.H., Ganduglia-Pirovano, M.V., and Kudrnovsky, J. (1994) *J. Chem. Phys.*, **101**, 8988.
34. Cohen, M.H., Ganduglia-Pirovano, M.V., and Kudrnovsky, J. (1995) *J. Chem. Phys.*, **103**, 3543.
35. Baekelandt, B.G. (1996) *J. Chem. Phys.*, **105**, 4664.
36. De Proft, F., Liu, S.B., and Geerlings, P. (1998) *J. Chem. Phys.*, **108**, 7549.

37. Balawender, R. and Geerlings, P. (2001) *J. Chem. Phys.*, **114**, 682.
38. Balawender, R., De Proft, F., and Geerlings, P. (2001) *J. Chem. Phys.*, **114**, 4441.
39. Cardenas, C., Lamsabhi, A.M., and Fuentealba, P. (2006) *Chem. Phys.*, **322**, 303.
40. Cardenas, C., Chamorro, E., Galvan, M., and Fuentealba, P. (2007) *Int. J. Quantum Chem.*, **107**, 807.
41. Thomas, L.H. (1927) *Proc. Cambridge Philos. Soc.*, **23**, 542.
42. Fermi, E. (1928) *Z. Phys.*, **48**, 73.
43. Dirac, P.A.M. (1930) *Proc. Cambridge Philos. Soc.*, **26**, 376.
44. Slater, J.C. (1951) *Phys. Rev.*, **81**, 385.
45. Slater, J.C. (1972) *Adv. Quantum Chem.*, **6**, 1.
46. Slater, J.C. (1974) *The Self Consistent Field for Molecules and Solids*, McGraw Hill, New York.
47. Cohen, A.J., Mori-Sánchez, P., and Yang, W. (2011) *Chem. Rev.*, **112**, 289.
48. (a) Nalewajski, R.F. and Parr, R.G. (1982) *J. Chem. Phys.*, **77**, 399. (b) Nalewajski, R.F. and Korchowiec, J. (1997) *Charge Sensitivity Approach to Electronic Structure and Chemical Reactivity*, World Scientific.
49. Parr, R.G., Donnelly, R.A., Levy, M., and Palke, W.E. (1978) *J. Chem. Phys.*, **68**, 3801.
50. For a detailed account of different scales of electronegativity, see e.g. Mullay, J. in *Electronegativity, Structure and Bonding*, Vol. 66, Sen, K.D.; Jørgenson, C. K. Ed.; Springer-Verlag, Berlin, Heidelberg, 1987, p. 1.
51. Iczkowski, R.P. and Margrave, J.L. (1961) *J. Am. Chem. Soc.*, **83**, 3547.
52. Pritchard, H.O. and Sumner, F.H. (1956) *Proc. R. Soc. A (London)*, **235**, 136.
53. Politzer, P., Shields, Z.P.-I., Bulat, F.A., and Murray, J.S. (2011) *J. Chem. Theory Comput.*, **7**, 377.
54. Datta, D., Shee, N.K., and von Szentpály, L. (2013) *J. Phys. Chem. A*, **117**, 200.
55. Perdew, J.P., Parr, R.G., Levy, M., and Balduz, J.L. Jr., (1982) *Phys. Rev. Lett.*, **49**, 1691.
56. Zhang, Y.K. and Yang, W. (2000) *Theor. Chem. Acc.*, **103**, 346.
57. Yang, W., Zhang, Y.K., and Ayers, P.W. (2000) *Phys. Rev. Lett.*, **84**, 5172.
58. Ayers, P.W. (2008) *J. Math. Chem.*, **43**, 285.
59. Koopmans, T. (1934) *Physica*, **1**, 104.
60. Perdew, J.P. and Levy, M. (1983) *Phys. Rev. Lett.*, **51**, 1884.
61. Glasser, L. and von Szentpály, L. (2006) *J. Am. Chem. Soc.*, **128**, 12314.
62. Parr, R.G. and Pearson, R.G. (1983) *J. Am. Chem. Soc.*, **105**, 7512.
63. Yang, W. and Parr, R.G. (1985) *Proc. Natl. Acad. Sci. U.S.A.*, **82**, 6723.
64. Politzer, P. (1987) *J. Chem. Phys.*, **86**, 1072.
65. Vela, A. and Gazquez, J.L. (1990) *J. Am. Chem. Soc.*, **112**, 1490.
66. Ghanty, T.K. and Ghosh, S.K. (1993) *J. Phys. Chem.*, **97**, 4951.
67. Hati, S. and Datta, D. (1994) *J. Phys. Chem.*, **98**, 10451.
68. Simon-Manso, Y. and Fuentealba, P.J. (1998) *Phys. Chem. A*, **102**, 2029.
69. Zhou, Z., Parr, R.G., and Garst, J.F. (1988) *Tetrahedron Lett.*, **29**, 4843.
70. Zhou, Z. and Parr, R.G. (1989) *J. Am. Chem. Soc.*, **111**, 7371.
71. De Proft, F. and Geerlings, P. (2004) *Phys. Chem. Chem. Phys.*, **6**, 242.
72. Pearson, R.G. (1987) *J. Chem. Educ.*, **64**, 561.
73. Parr, R.G. and Chattaraj, P.K. (1991) *J. Am. Chem. Soc.*, **113**, 1854.
74. Pearson, R.G. and Palke, W.E. (1992) *J. Phys. Chem.*, **96**, 3283.
75. Pearson, R.G. (1993) *Acc. Chem. Res.*, **26**, 250.
76. Pearson, R.G. (1999) *J. Chem. Ed.*, **76**, 267.
77. Ayers, P.W. and Parr, R.G. (2000) *J. Am. Chem. Soc.*, **122**, 2010.
78. Torrent-Sucarrat, M., Luis, J.M., Duran, M., and Sola, M. (2002) *J. Chem. Phys.*, **117**, 10561.
79. Tozer, D.J. and De Proft, F. (2005) *J. Phys. Chem. A*, **109**, 8923.
80. Cardenas, C., Ayers, P.W., De Proft, F., Tozer, D.J., and Geerlings, P. (2011) *Phys. Chem. Chem. Phys.*, **13**, 2285.
81. Fuentealba, P. and Parr, R.G. (1991) *J. Chem. Phys.*, **94**, 5559.

82. Senet, P. (1996) *J. Chem. Phys.*, **105**, 6471.
83. Ayers, P.W. and Parr, R.G. (2008) *J. Chem. Phys.*, **129**, 054111.
84. Cardenas, C., Echegaray, E., Chakraborty, D., Anderson, J.S.M., and Ayers, P.W. (2009) *J. Chem. Phys.*, **130**, 244105.
85. Parr, R.G., Von Szentpaly, L., and Liu, S.B. (1999) *J. Am. Chem. Soc.*, **121**, 1922.
86. Chattaraj, P.K., Maiti, B., and Sarkar, U. (2003) *J. Phys. Chem. A*, **107**, 4973.
87. Chattaraj, P.K., Sarkar, U., and Roy, D.R. (2006) *Chem. Rev.*, **106**, 2065.
88. Parr, R.G. and Yang, W. (1984) *J. Am. Chem. Soc.*, **106**, 4049.
89. For a perspective seeAyers, P.W. and Levy, M. (2000) *Theor. Chem. Acc.*, **103**, 353.
90. Tozer, D.J. and De Proft, F. (2007) *J. Chem. Phys.*, **127**, 034108.
91. Yang, W., Cohen, A.J., De Proft, F., and Geerlings, P. (2012) *J. Chem. Phys.*, **136**, 144110.
92. Yang, W., Parr, R.G., and Pucci, R. (1984) *J. Chem. Phys.*, **81**, 2862.
93. Yang, W. and Mortier, W.J. (1986) *J. Am. Chem. Soc.*, **108**, 5708.
94. Bultinck, P., Fias, S., Van Alsenoy, C., Ayers, P.W., and Carbó-Dorca, R. (2007) *J. Chem. Phys.*, **127**, 034102.
95. Roy, R.K., Pal, S., and Hirao, K. (1999) *J. Chem. Phys.*, **110**, 8236.
96. Roy, R.K., Pal, S., and Hirao, K. (2000) *J. Chem. Phys.*, **113**, 1372.
97. Bultinck, P., Carbó-Dorca, R., and Langenaeker, W. (2003) *J. Chem. Phys.*, **118**, 4349.
98. Bultinck, P. and Carbó-Dorca, R. (2003) *J. Math. Chem.*, **34**, 67.
99. Melin, J., Ayers, P.W., and Ortiz, J.V. (2007) *J. Phys. Chem. A*, **111**, 10017.
100. Hoffmann, R. and Woodward, R.B. (1965) *J. Am. Chem. Soc.*, **87**, 395.
101. Hoffmann, R. and Woodward, R.B. (1965) *J. Am. Chem. Soc.*, **87**, 2046.
102. Hoffmann, R. and Woodward, R.B. (1965) *J. Am. Chem. Soc.*, **87**, 2511.
103. Woodward, R.B. and Hoffmann, R. (1970) *The Conservation of Orbital Symmetry*, Academic Press, New York.
104. De Proft, F., Ayers, P.W., Fias, S., and Geerlings, P. (2006) *J. Chem. Phys.*, **125**, 214101.
105. Morell, C., Grand, A., and Toro-Labbé, A. (2005) *J. Phys. Chem. A*, **109**, 205.
106. Morell, C., Grand, A., and Toro-Labbé, A. (2006) *Chem. Phys. Lett.*, **425**, 342.
107. Ayers, P.W., Morell, C., De Proft, F., and Geerlings, P. (2007) *Chem. Eur. J.*, **13**, 8240.
108. De Proft, F., Chattaraj, P.K., Ayers, P.W., Torrent-Sucarrat, M., Elango, M., Subramanian, V., Giri, S., and Geerlings, P. (2008) *J. Chem. Theory Comput.*, **4**, 595.
109. Sablon, N., De Proft, F., and Geerlings, P. (2009) *Croat. Chem. Acta*, **82**, 157.
110. Jaque, P., Correa, J.V., De Proft, F., Toro- Labbé, A., and Geerlings, P. (2010) *Can. J. Chem.*, **88**, 858.
111. Geerlings, P., Ayers, P.W., Toro-Labbé, A., Chattaraj, P.K., and De Proft, F. (2012) *Acc. Chem. Res.*, **55**, 683.
112. Lee, C., Yang, W., and Parr, R.G. (1988) *J. Mol. Struct. (THEOCHEM)*, **163**, 305.
113. Li, Y. and Evans, J.N.S. (1995) *J. Am. Chem. Soc.*, **117**, 7756.
114. Chattaraj, P.K. (2001) *J. Phys. Chem. A*, **105**, 511.
115. Chattaraj, P.K., Roy, D.R., Geerlings, P., and Torrent-Sucarrat, M. (2007) *Theor. Chem. Acc.*, **118**, 923.
116. Ayers, P.W. and Parr, R.G. (2008) *J. Chem. Phys.*, **128**, 184108.
117. Cuevas-Saavedra, R., Rabi, N., and Ayers, P.W. (2011) *Phys. Chem. Chem. Phys.*, **13**, 19594.
118. Gal, T., Geerlings, P., De Proft, F., and Torrent-Sucarrat, M. (2011) *Phys. Chem. Chem. Phys.*, **13**, 15003.
119. Gal, T. (2012) *Theor. Chem. Acc.*, **131**, 1223.
120. Bonaccorsi, R., Scrocco, E., and Tomasi, J. (1970) *J. Chem. Phys.*, **52**, 5270.
121. *Molecular Electrostatic Potentials – Concepts and Applications, Theoretical and Computational Chemistry*, Vol. 3, Murray, J. S., Sen, K. D., Eds. Elsevier, Amsterdam, 1996.

122. Berkowitz, M., Ghosh, S.K., and Parr, R.G. (1985) *J. Am. Chem. Soc.*, **107**, 6811.
123. Langenaeker, W., De Proft, F., and Geerlings, P. (1995) *J. Phys. Chem.*, **99**, 6424.
124. Berkowitz, M. and Parr, R.G. (1988) *J. Chem. Phys.*, **88**, 2554.
125. Liu, S., Li, T., and Ayers, P.W. (2009) *J. Chem. Phys.*, **131**, 114106.
126. Sablon, N., De Proft, F., and Geerlings, P. (2010) *J. Phys. Chem. Lett.*, **1**, 1228.
127. Sablon, N., De Proft, F., and Geerlings, P. (2010) *Chem. Phys. Lett.*, **498**, 192.
128. Sablon, N., De Proft, F., Ayers, P.W., and Geerlings, P. (2010) *J. Chem. Theory Comput.*, **6**, 3671.
129. Ayers, P.W. (2007) *Faraday Discuss.*, **135**, 161.
130. Sablon, N., De Proft, F., Sola, M., and Geerlings, P. (2012) *Phys. Chem. Chem. Phys.*, **14**, 3960.
131. Fias, S., Geerlings, P., Ayers, P.W., and De Proft, F. (2013) *Phys. Chem. Chem. Phys.*, **15**, 2882.
132. Fias, S., Boisdenghien, Z., Stuyver, T., Audiffred, M., Merino, G., Geerlings, P., and De Proft, F. (2013) *J. Phys. Chem. A*, **117**, 3556.
133. Huheey, J.E. (1983) *Inorganic Chemistry*, 3rd edn, Harper and Row, New York.
134. Politzer, P., Huheey, J.E., Murray, J.S., and Grodzicki, M. (1992) *J. Mol. Struct. (THEOCHEM)*, **259**, 99.
135. Huheey, J.E. (1965) *J. Phys. Chem.*, **69**, 3284.
136. Huheey, J.E. and Watts, J.C. (1971) *Inorg. Chem.*, **10**, 1553.
137. Huheey, J.E. (1971) *J. Org. Chem.*, **36**, 204.
138. Mortier, W.J. (1987) in *Electronegativity*, Structure and Bonding, vol. 66 (ed. K.D. Sen), Springer-Verlag, Berlin, p. 125.
139. Mortier, W.J., Van Genechten, K., and Gasteiger, J. (1985) *J. Am. Chem. Soc.*, **107**, 829.
140. Mortier, W.J., Ghosh, S.K., and Shankar, S. (1986) *J. Am. Chem. Soc.*, **108**, 4315.
141. Van Genechten, K.A., Mortier, W.J., and Geerlings, P. (1986) *Chem. Commun.*, 1278.
142. Van Genechten, K.A., Mortier, W.J., and Geerlings, P. (1987) *J. Chem. Phys.*, **86**, 5063.
143. For a recent method, see e.g.Verstraelen, T., Van Speybroeck, V., and Waroquier, M. (2009) *J. Chem. Phys.*, **131**, 044127.
144. Berkowitz, M. (1987) *J. Am. Chem. Soc.*, **109**, 4823.
145. Chattaraj, P.K. and Parr, R.G. In *Chemical Hardness*, Structure and Bonding, Vol. 80 Sen, K. D., Ed.; Springer-Verlag: Berlin, 1993, p. 11.
146. Chattaraj, P.K., Lee, H., and Parr, R.G. (1991) *J. Am. Chem. Soc.*, **113**, 1855.
147. Ayers, P.W., Parr, R.G., and Pearson, R.G. (2006) *J. Chem. Phys.*, **124**, 194107.
148. Chattaraj, P.K., Ayers, P.W., and Melin, J. (2007) *Phys. Chem. Chem. Phys.*, **9**, 3853.
149. Ayers, P.W. and Cárdenas, C. (2013) *J. Chem. Phys.*, **138**, 181106.
150. Mendez, F. and Gazquez, J.L. (1994) *J. Am. Chem. Soc.*, **116**, 9298.
151. Gazquez, J.L. and Mendez, F. (1994) *J. Phys. Chem.*, **98**, 4591.
152. Gazquez, J.L. (1997) *J. Phys. Chem. A*, **101**, 4657.
153. Geerlings, P. and De Proft, F. (2000) *Int. J. Quantum Chem.*, **80**, 227.
154. Damoun, S., Van de Woude, G., Mendez, F., and Geerlings, P. (1997) *J. Phys. Chem.*, **101**, 886.
155. Klopman, G. (1974) in *Chemical Reactivity and Reaction Paths* (ed. G. Klopman), John Wiley & Sons, Inc., New York. p. 1974.
156. Langenaeker, W., De Proft, F., and Geerlings, P. (1996) *J. Mol. Struct. (THEOCHEM)*, **362**, 175.
157. De Proft, F., Martin, J.M.L., and Geerlings, P. (1996) *Chem. Phys. Lett.*, **256**, 400.
158. Anderson, J.S.M. and Ayers, P.W. (2007) *Phys. Chem. Chem. Phys.*, **9**, 2371.
159. Johnson, P.A., Bartolotti, L.J., Ayers, P.W., Fievez, T. and Geerlings, P. in *Modern Charge-Density Analysis*, Gatti, C.; Macchi, P. Eds., Springer, 2010, p. 715.

160. Anderson, J.S.M., Melin, J., and Ayers, P.W. (2007) *J. Chem. Theory Comput.*, **3**, 358.

161. Anderson, J.S.M., Melin, J., and Ayers, P.W. (2007) *J. Chem. Theory Comput.*, **3**, 375.

162. Cardenas, C., Rabi, N., Ayers, P.W., Morell, C., Jaramillo, P., and Fuentealba, P. (2009) *J. Phys. Chem. A*, **113**, 8660.

163. Becke, A.D. and Edgecombe, K.E. (1990) *J. Chem. Phys.*, **92**, 5397.

164. Silvi, B. and Savin, A. (1994) *Nature*, **371**, 683.

165. Wang, X. and Peng, Z. (1993) *Int. J. Quantum Chem.*, **47**, 393.

166. Hellmann, H. (1937) *Einführung in die Quantenchemie*, F. Deutick, Leipzig.

167. Feynman, R.P. (1939) *Phys. Rev.*, **56**, 340.

168. Bader, R.F.W. and Austen, M.A. (1997) *J. Chem. Phys.*, **107**, 4271.

169. Hernandez-Trujillo, J., Cortes-Guzman, F., Fang, D.-C., and Bader, R.F.W. (2007) *Faraday Discuss.*, **135**, 79.

170. Ayers, P.W. and Jenkins, S. (2009) *J. Chem. Phys.*, 130–154104.

171. van Zeist, W.-J. and Bickelhaupt, F.M. (2010) *Org. Biomol. Chem.*, **8**, 3118.

172. Ozen, A., Aviyente, V., De Proft, F., and Geerlings, P. (2006) *J. Phys. Chem. A*, **110**, 5860.

173. Geerlings, P. (2009) in *Chemical Reactivity Theory – A Density Functional View* (ed. P.K. Chattaraj), CRC Press, Taylor & Francis Group, Boca Raton, FL, p. 395.

174. Chattaraj, P.K. (ed.) (2009) *Chemical Reactivity Theory – A Density Functional View*, CRC Press, Taylor & Francis Group, Boca Raton, FL.

8
The QTAIM Perspective of Chemical Bonding

Paul Lode Albert Popelier

8.1
Introduction

In this chapter the "Quantum Theory of Atoms in Molecules (QTAIM)" is discussed in the spirit of this book, which combines an educational style with an awareness of current scientific boundaries, while avoiding too many equations in the main text. Because QTAIM has been reviewed many times before it has become harder to add value to what has already been written. Still, the current text seeks to achieve added value, by an alternative angle of exposition of QTAIM, by including both a historic narrative as well as a pointer to future research.

Before starting to explain QTAIM we mention background texts that readers may want to consult. An authoritative source on QTAIM is a book [1] written by Richard Bader, who inspired and oversaw the development of QTAIM since its inception. Apart from a didactic monograph [2] on QTAIM there are also a number of accounts, reviews, and edited books on the subject. An early review [3] on so-called molecular *virial fragments* was written in 1975. This account covered only one aspect of QTAIM, namely that of the virial fragment. We will discuss this concept in detail later in the text but for now it is best to think of a virial fragment as "the atom" in QTAIM. Obviously this is an important cornerstone of QTAIM, and was the early driving force for QTAIM's development. It is important to realize, already now, that the definition of an atom (within the framework of QTAIM) is energetic in nature. It demands that not only should atomic energy exist but it should also be well defined. This 1975 account did not introduce bonds nor did it define them; in fact, it shied away from the general idea of a bond, as implied by its title *Molecular Fragments or Chemical Bonds?*. Here Bader suggested that, in the search to understand the properties of a total system in terms of its parts, one can choose atoms rather than bonds as fundamental parts. Later, in a more mature version of QTAIM, this "either or" view was not sustained.

Ten years later, in 1985, a second account [4] was published in the same journal by the same author. This account presented QTAIM in its entirety, including

The Chemical Bond: Fundamental Aspects of Chemical Bonding, First Edition.
Edited by Gernot Frenking, Sason Shaik.

a treatment of bonding through the topology of the electron density. What we mean by topology will become clear later in this chapter. In his 1985 account, Bader was careful enough to write that *It is the atom and its properties that are defined by quantum mechanics. The bond paths (BPs) and the structure they define just mirror and summarize in a convenient way what the atoms are doing*. The 1985 account would effectively form the basis for Bader's book on QTAIM, which was published 5 years later. This book appeared more or less simultaneously with a long review [5], presenting QTAIM as deduced from first principles. Ten years later a more introductory and didactic book [2] appeared, followed by two reports soon after [6, 7] listing the by then hundreds of applications of QTAIM in fields ranging from mineralogy to biochemistry, together with developments, both methodological and algorithmic. In a philosophical review [8] of QTAIM as a quantum mechanical basis of conceptual chemistry, Bader listed all atomic theorems [9] derived to date. Two years later, in 2007, an extensive compilation of QTAIM contributions appeared in a book [10] edited by Matta and Boyd. This book's subtitle *From Solid State to DNA and Drug Design* reaffirmed the enormous breadth of application that QTAIM had acquired by the mid 2000s. In another personal account entitled *Everyman's Derivation of the Theory of Atoms in Molecules* Bader presented QTAIM as an extension of quantum mechanics to subspaces. Finally, in his swansong publication [11], posthumously published in *Foundations of Chemistry*, Bader emphasized the bounded, space-filling, and nonoverlapping feature of QTAIM atoms, a topic discussed [12] at great length seven years earlier, and in the context of short-range intermolecular interactions.

Three final remarks are in order. First, the references cited in this Introduction are not exhaustive but merely provide background material to make up for the brevity of the QTAIM exposition in this chapter. These references were ordered chronologically, in line with the historical character of the exposition below. Second, another chapter in this book, written by Scherer *et al.* (Chapter 9), assumes background knowledge of QTAIM in their account on chemical bonding from the point of view of high-resolution crystallography. Third and finally, the way QTAIM partitions the electron density, through the concept of the gradient path (GP), has been adopted by other approaches such as the topological analysis [13] of the electron Localization function (ELF) [14]. It makes sense to bundle the topological analysis of the electron density, its Laplacian (which are both part of QTAIM), ELF, and many other quantum mechanical functions into one approach called *Quantum Chemical Topology* (*QCT*). This name has been justified in detail over the last decade in various sources, such as footnote 19 in Ref. [15], Section 2 of Ref. [12], the Appendix of Ref. [16] and the Introduction of Ref. [17]. More details on QCT can be found in Box 8.1 of this chapter. Box 8.1 lists the various quantum mechanical functions that have been investigated via the language of topology and dynamical systems, and thereby outlines what QCT is.

Box 8.1 Quantum Chemical Topology (QCT)

There is a growing body of work that uses the language of topology to extract chemical information from modern *ab initio* wave functions. The important point is that such topological analyses use three-dimensional quantum mechanical functions other than the electron density. A well-documented example of a function that features as a chapter in this book is the ELF. In 1994, bonds were for the first time classified [13] by introducing the topology of ELF. Central to that work was again the concept of the GP. A GP is a trajectory of steepest ascent through a given three-dimensional (3D) function. The full collection of GPs, which draws itself in a molecule or cluster of molecules, is called the *gradient vector field*. This field was first studied as it operated on the electron density. This is how QTAIM started: the gradient vector field of the electron density partitioned a molecule into atoms (that have a well-defined kinetic energy). However, many 3D quantum mechanical functions, other than the electron density, also have maxima that attract GPs. Therefore one can find basins (i.e., subspaces containing GPs terminating at an attractor) in these other 3D functions. The identity of the partitioned function will determine which type of chemical information one recovers. It is important that all activity and results that use the central idea of a gradient vector field partitioning are collectively referred to under one single name. A sensible name has already been proposed [15] in 2003, which is QCT.

A non-exhaustive list of 3D topologically partitioned quantum mechanical functions includes:

- Electron density $\rho(\mathbf{r})$ (started with Ref. [20]).
- The Laplacian of ρ, $\nabla^2\rho(\mathbf{r})$ (started with Refs [68, 69] and full topology first explored in Refs [70–72]).
- (Bare) nuclear potential $V_{nuc}(\mathbf{r})$ (early start with Ref. [73] but elaborate and self-contained study [17]).
- ELF [14] (started with Ref. [13] and reviewed in Ref. [74]).
- Electrostatic potential [75] (started with thorough studies [76, 77] and continued with Ref. [78], applied in the area of chemical reactions [79], Lewis acidity [80] or electron diffraction study [81]).
- Virial field (or trace of the Schrödinger stress tensor) (topology explored in Ref. [57].).
- Magnetically induced molecular current distributions (started with [82]).
- Intracule density (started with Ref. [83], which reveals correlation cages).
- Ehrenfest force field (topology first investigated [84] in 2012).
- Energy partitioning (beyond the kinetic energy and atom virial theorem) (Coulomb potential energy partitioning started with [37] and culminated into the theory of IQA [24], leading to energetic underpinning for the topological expression of chemical bonding [85]).

Alternative methods of interpretative quantum chemistry [86–92] do not share the central concept of the gradient vector field. This crucial difference draws together the topological analysis of the various functions under the heading of QCT, which is thus distinct from non-QCT methods.

8.2 Birth of QTAIM: the Quantum Atom

The birth of QTAIM was marked by work that, with hindsight, could be called a *false start* actually. Two publications appeared in 1971, one [18] in *Chemical Physics Letters* and one [19] in *Journal of the American Chemical Society*, where Bader and coworkers explained their "natural partitioning." It is interesting to pause here and highlight what these publications convey because this reveals the driving force behind QTAIM and what it is essentially. The opening sentence of the *JACS* paper states that "*There is a history of attempts (...) to partition the total electronic charge between the nuclei in the system. The prime reason for proposing any scheme which assigns some number of electrons to each nucleus in a molecule is to provide a measure of the charge transfer* ..." Let us make clear here, and keep in mind for later in this chapter, that a population analysis indeed provides a measure of charge transfer, nothing more and nothing less.

In the abstract of the 1971 *JACS* paper the authors term their method a "natural partitioning", "... *as it is suggested by the nature of the charge distribution itself; the point along the internuclear axis at which the charge density attains its minimum value between a pair of bonded nuclei defines the position of the partitioning surface.*" They immediately illustrated their partitioning recipe on the *linear* molecule FC≡N. In this system a first partitioning plane was positioned at the point on the molecular axis with the lowest electron density value between the F nucleus and the C nucleus. The second partitioning plane was positioned in the same manner but now in between the C nucleus and the N nucleus. Figure 8.1 of the 1971 *JACS* paper shows two panels, which we do not reproduce here. In the top panel there is a one-dimensional profile of the electron density along the molecular axis, partitioned in the way described just earlier. The bottom panel shows these partitioning planes again but now cutting through a two-dimensional representation consisting of contour lines of constant electron density. Instead of reproducing their original figure we show a partitioned mountain landscape in Figure 8.1. The reason for this example from macroscopic reality is that it reminds us how the brain would separate two (mountain) peaks that are mathematically similar to the peaks appearing in a 1D profile of the electron density.

Mountain landscapes also appear in the context of chemical reactivity where saddle points or paths of steepest ascent need to be illustrated, often in connection with a new configuration sampling algorithm. Here the metaphor of the mountain landscape simply fulfills the role of emphasizing how "the human eye" naturally partitions an object into two *nonoverlapping* parts. No one would readily think

Figure 8.1 A system of two mountains partitioned into a left mountain (L) and a right mountain (R) by a red line going through the point of lowest altitude between the two peaks.

of mountain L as an object that overlaps with mountain R. There is a sharp boundary between the two objects, and few would argue with this assertion. In macroscopic reality there are many examples of sharp boundaries separating two entities, ranging from borders between countries to cell walls. Yet, when it comes to atoms in a molecule, a sharply delineated atom does not (yet) constitute the mainstream view.

We now return to the question why the natural partitioning proposed in 1971 was a false start. The concomitant 1971 Letter [18] applied the "natural partitioning" method to another linear molecule: carbon dioxide or O=C=O. An important new concept was invoked, which is transferability. In the broadest sense, this transferability gauges the extent to which a molecule can be constructed from atoms or molecular fragments. The Letter first looked at transferability in terms of "bond properties," which were restricted to bonded and nonbonded electron populations. It is easy to imagine that a different type of partitioning plane, this time through a nucleus, can separate a bonded population from a nonbonded one. The blue line in Figure 8.2 marks an example of such a plane, going through O here. Left of this plane is situated the nonbonded electron density, which is segregated from the bonded density appearing to the right of this blue plane.

Envisage that one wants to first obtain the *total* population of the left oxygen in O=C=O. For that purpose the electron density is integrated from infinity (at the left) up to the partitioning plane (red line in Figure 8.2) at the minimum point between the left O nucleus and the C nucleus next to it. Both partitionings (red and blue line) may appear cavalier in the light of the mature theory that QTAIM later became. However, the 1971 Letter already gives the impression that, in the long run, "not all is well" with such a red partitioning. Understanding the reason for this is important to comprehend the essence of QTAIM and what it strives to achieve.

Noting the relative insensitivity of populations to changes in bonding (e.g., going from CO to CO_2) the authors directed their attention to the transferability of a fragment's *energy*. They focused on the kinetic energy only, working with two

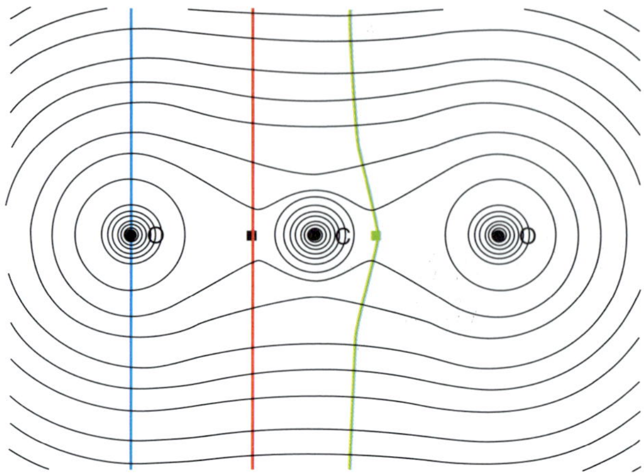

Figure 8.2 Partitioning of the linear molecule carbon dioxide (CO_2) superimposed on a set of contours of constant electron density. The red line marks the "false start" or "natural partitioning" between O and C, while the green curve marks the correct QTAIM boundary between C and O. The blue line marks a way to separate nonbonded (left) and bonded (right) electron density, as suggested in the 1971 Letter (see main text).

alternative kinetic energy densities, denoted $K(\mathbf{r})$ and $G(\mathbf{r})$ (see Box 8.2). They showed that when $K(\mathbf{r})$ and $G(\mathbf{r})$ are integrated over a subspace bounded by the red line then one obtains very different numerical values. In order for these two values to be identical, the 1971 Letter correctly concluded that the subspace, denoted by Ω, had to be such that the Laplacian of the electron density, $\nabla^2\rho$, vanishes when integrated over the subspace's volume (see Box 8.2),

$$\int_{\Omega} \mathrm{d}V \nabla^2 \rho(\mathbf{r}) = 0 \tag{8.1}$$

However, the 1971 Letter failed to see this fact through to its final conclusion and thereby failed to draw the green boundary in Figure 8.2. This had to wait until the next paper [20] appeared in 1972, containing a consistent treatment. This 1972 paper can be considered as actually heralding the birth of QTAIM. Here, for the first time, the characteristic shape of a molecular fragment (i.e., atom) with a well-defined kinetic energy was born. This paper illustrated this molecular fragment by means of simple lithium-containing diatomics, both neutral and positively charged. Figure 8.2 shows this characteristic shape by means of the green curve. When reflected to the left side of the molecule, this green curve provides a complete partitioning of CO_2 into three atomic subspaces with a nonplanar shape. We could call such a subspace a kinetic atom, because it has a well-defined kinetic energy. However, one could give this subspace the more general name *quantum atom*, which has been used elsewhere in the literature (e.g., Chapter 7 in Ref. [2]. and in Refs [21–24].). This name is appropriate because much work of the Bader group during the rest of the 1970s aimed at strengthening the link between this quantum atom and quantum mechanics itself. The need for strengthening this link follows

Box 8.2 A Molecular Fragment with a Well-defined Kinetic Energy

The purpose of this box is to show how to define a molecular fragment that has a well-defined kinetic energy. The starting point is having a definition of *local* kinetic energy, which is the kinetic energy at a particular point *per unit volume*. This quantity is thus a kinetic energy *density*, which when integrated over a volume, gives the kinetic energy of the electrons in that volume. If this volume corresponds to that of a molecular fragment then one recovers the kinetic energy of that fragment.

A potential problem immediately arises because there is a quantum mechanical ambiguity [93] in defining this kinetic energy density. Although there are an infinite number of expressions for the kinetic energy density it is sufficient to choose only two possible expressions to develop the argument of this box. Two alternative definitions, previously used [94] by Bader and coworkers in this argument, are given by Eqs. (B2.1) and (B2.2),

$$K(\mathbf{r}) = -\frac{1}{4}N\int d\tau'[\psi^*\nabla^2\psi + \psi\nabla^2\psi^*] \tag{B2.1}$$

$$G(\mathbf{r}) = \frac{1}{2}N\int d\tau'\nabla\psi^* \cdot \nabla\psi \tag{B2.2}$$

where N is the total number of electrons in the system, ψ the system's N-electron wave function, and $\int d\tau'$ the integration over all electrons except one. This integration reduces the $3N$ dimensional character of the integrand to a three-dimensional function. To simplify matters spin is not considered here.

In a way, $G(\mathbf{r})$ resembles classical kinetic energy in that it is always positive. It is easy to show that these two kinetic energy densities are linked via the Laplacian of the electron density, $\nabla^2\rho$,

$$K(\mathbf{r}) = G(\mathbf{r}) - \frac{1}{4}\nabla^2\rho(\mathbf{r}) \tag{B2.3}$$

When integrated over whole space the Laplacian vanishes, or

$$\int_{\text{whole space}} dV\,\nabla^2\rho(\mathbf{r}) = 0 \tag{B2.4}$$

As a result, integrating both sides of Eq. (B2.3) over whole space, yields

$$K(\text{molecule}) = G(\text{molecule}) = T(\text{molecule}) \tag{B2.5}$$

where T expresses *the* kinetic energy regardless of whether it was calculated from $K(\mathbf{r})$ or $G(\mathbf{r})$. Because a single molecule in the gas phase occupies the whole space, one indeed recovers the kinetic energy of the molecule by integration over the whole space. This energy is well-defined in that it is unique: both $K(\mathbf{r})$ and $G(\mathbf{r})$ give the same answer.

The main question is now if this unique result can also be obtained for a molecular fragment. Let us consider the subspace of an *arbitrary* molecular fragment, denoted by ⊗. For such an arbitrary subspace one finds that

$$\int_{\otimes} dV\,\nabla^2\rho(\mathbf{r}) \neq 0 \tag{B2.6}$$

Hence, one will not recover a unique kinetic energy for such a fragment, or

$$K(\otimes) \neq G(\otimes) \tag{B2.7}$$

However, there are special subspaces Ω for which

$$\int_{\Omega} dV\,\nabla^2\rho(\mathbf{r}) = 0 \tag{B2.8}$$

such that it makes sense to speak of a unique and hence well-defined kinetic energy $T(\Omega)$ associated with those subspaces,

$$K(\Omega) = G(\Omega) = T(\Omega) \tag{B2.9}$$

from a concern raised by Bader and Beddall themselves at the end of their 1972 "QTAIM birth" paper [20]. They wrote that *While the concept of the energy of a fragment appears to be at variance with the quantum mechanical definition of the energy as the expectation value of the Hamiltonian operator averaged over all space, we are led to this concept by our observations.* As is perhaps always the case with mature and fruitful theories (such as QTAIM) they have their roots in observation and practicality. As another example, one could quote Feynman path integrals, which were used successfully long before it was shown they were mathematically allowed. The Dirac delta function is another example.

Returning to QTAIM, Srebrenik joined the Bader group in the mid 1970s and started the development of subspace quantum mechanics, a topic also pursued by others at that time. When the expectation value of an operator is evaluated in *full space* quantum mechanics, one obtains a real value when the operator is Hermitian. However, the evaluation of an expectation value over a *subspace* generally yields a complex number. This problem can be solved by taking the mean of the (non-real) expectation value and its complex conjugate, thereby eliminating the purely imaginary part. More details can be found in Ref. [25], Chapter 7 of Ref. [2] and Ref. [1].

In the next section we introduce the language of dynamical systems to reinterpret the quantum atom from a topological view.

8.3
The Topological Atom: is it also a Quantum Atom?

In order to make some new concepts concrete right away we work with a simple molecule, HFC=O, which is still general enough. The fact that this molecule is planar eliminates some unnecessary complexity at the outset. Figure 8.3 shows how the electron density varies within HFC=O. The electron density determining

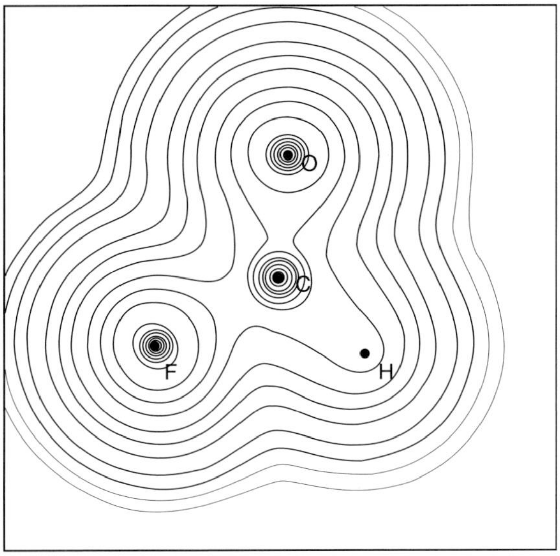

Figure 8.3 Electron density contour plot of HFC=O in the plane of symmetry. The molecule has been geometry optimized at the MP2/cc-pVDZ level of theory. The carbon is placed at the origin and the bold square box marks the −5 au and +5 au horizontal and vertical boundaries of the plot. The electron density values of the contour lines are 1×10^n, 2×10^n, 4×10^n, and 8×10^n au where n starts at −3 and increases with unity increments.

the contour lines (roughly) doubles its value in going from one contour line to the next. From the (almost) linear spacing of the contour lines one deduces that the electron density increases exponentially, reaching a local peak at each of the four nuclei.

These peaks dominate the electron density landscape to such an extent that one may have the impression that the positions of the nuclei are all that the electron density indicates about the molecule. This information is what a routine crystallographic study gathers and this may be all one needs to know if one is only interested in the geometry of a newly synthesized molecule, for example. However, the field of high-resolution crystallography [26] has pushed the interpretation of the electron density further. A natural way of eliminating the dominance of the molecular electron density near the nuclei is subtracting from it another electron density with large peaks. The latter electron density corresponds to that of a superposition of isolated atoms that have not been chemically hybridized or polarized. The resulting difference density is called the *deformation (electron) density*. In principle, it reveals the chemical features that the molecular electron density contains but which are drowned by the peaks at the nuclei. Deformation densities constitute a useful idea at first sight, and have indeed been used much in the 1980s, but open issues surrounding the exact nature of the isolated atoms (i.e., which quantum mechanical state or which part of their electron density, valence or core) can spoil the interpretation of deformation densities. The current section will show that there is no need for a reference electron density (i.e., a superposition of isolated atoms). Indeed, one can use the *internal* difference within a single molecular electron density to extract chemical information from it. The key concept to achieve this is the gradient vector, as explained later, but first we focus on obtaining topological atoms in HFC=O.

Let us look again at the constant electron density contour lines of Figure 8.3, and focus on the outer line, which corresponds to $\rho = 0.001$ au. Because this contour can be considered as the practical edge of the molecule [27] it encompasses the whole molecule. If the value of the constant electron density is now increased the rather "blobby" and shapeless outer contour line becomes increasingly shaped. Higher electron density contour lines start protruding toward the region between the bonded nuclei. Like a corset, the contour lines start increasingly following the shape of the individual atoms that form the molecule. Figure 8.4 zooms in on the essential characteristics of the patterns emerging from the contour lines that enclose individual atoms or groups of atoms. Starting with the lowest electron density value, $\rho = 0.25$ au, the first contour line we encounter encloses the whole molecule. In fact, this electron density value is the highest possible value yielding a contour line that still encloses the whole molecule; any higher value of ρ will enclose parts of the molecule. More precisely, a contour line corresponding to $\rho > 0.25$ au will enclose the fluorine atom on its own. This atom now becomes a

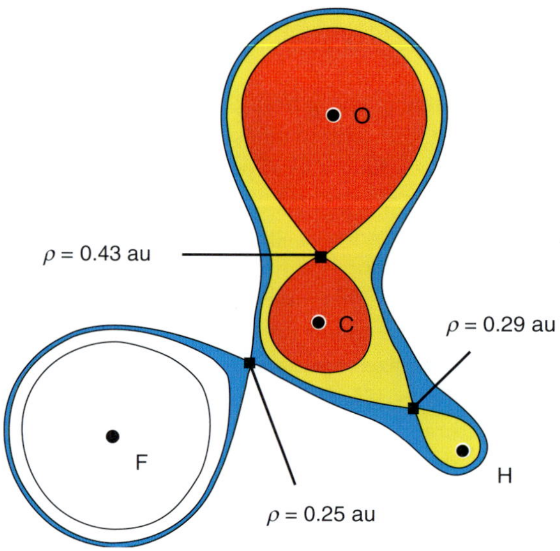

Figure 8.4 Diagram illustrating how HFC=O falls apart into four atoms, by following three disconnection events, at three different electron density contour values. First, when $\rho = 0.25$ au, F becomes disconnected from HC=O. The blue zone marks contour lines with $\rho > 0.25$ au but still < 0.29 au, at which point the H atom splits off from C=O, when $\rho = 0.29$ au. The yellow zone marks contour lines with $\rho > 0.29$ au but still < 0.43 au, at which point the C and O atoms also become disconnected. Finally, contour lines with $\rho > 0.43$ au enclose all atoms as completely separated entities.

separate object that is disconnected from the rest of the molecule. The full meaning of the special point in Figure 8.4, marked by the little square where $\rho = 0.25$ au, will be explained later.

A little black square in Figure 8.4 is reminiscent of the point of lowest altitude between two mountains, where the red line in Figure 8.1 partitions the overall landscape in two separate mountains. Of course, in the light of the discussion on defining atomic kinetic energy it would be uninformed to still draw such a line through each of the black squares in order to partition the molecule into atoms. Instead, the equivalent of the green boundary in Figure 8.2 needs to be found, which is shown in Figure 8.5. In this figure three bold black curves mark the edges of the four topological atoms in HFC=O. The little black squares appear in the center of each of these three bold curves marking the sharp boundaries of the atoms. These curves are actually intersections between the plotting plane and surfaces in three dimensions called *interatomic surfaces.*

There are five comments to make on Figure 8.5. First, one can prove again that the Laplacian of the electron density, $\nabla^2\rho$, vanishes when integrated over the subspace's volume of each of the four atoms. Therefore, the atoms displayed in Figure 8.5 are quantum atoms. In the current section we establish why we can also refer to the atoms in Figure 8.5 as topological atoms. It should be clear already that all topological atoms are quantum atoms but not all possible quantum atoms are topological atoms [21]. Second, the topological atoms in Figure 8.5 do not overlap. This nonoverlapping picture may appear alien to many (quantum) chemists. A prevailing picture appears to be one of fuzzy atoms that penetrate each other. Although this image is possibly inspired by the cloudlike nature of an electron

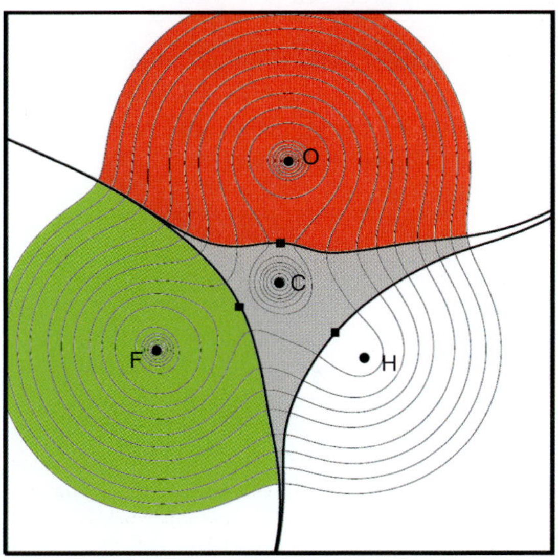

Figure 8.5 The molecule HFC=O is partitioned into four topological atoms: F (green), C (gray), O (red), and H (white).

density, one should keep in mind that topological atoms also emerge from that same electron density. Third, topological atoms do not leave any gaps in between. This space-filling feature of topological atoms is also at variance with a more mainstream picture where some regions of space are not allocated to any atom. This is especially common in molecular complexes held together by hydrogen bonds and/or van der Waals interactions. For example, a ligand that docks into an enzyme's active site is typically thought of as a molecule of finite size, even if it finds itself in a condensed matter situation, which is the case when inside the active site. In this picture, therefore, there are regions of space, between the ligand and the enzyme, that are unassigned to any molecule and thus any atom. Similarly, water molecules in the liquid state do not leave gaps between them: all of space belongs to one water molecule or another. This is a result of each water molecule being completely bounded by interatomic surfaces. Fourth, the fluorine atom distorts the carbon atom more than the hydrogen atom. QTAIM provides a picture of atoms reminding us of the macroscopic world where objects can be distorted to whichever degree of malleability they possess. Fifth and finally, there is "no direct contact" between the oxygen and the fluorine, that is, topologically they do not touch. In fact, they are kept apart by a very thin wedge of space belonging to the carbon atom, although it is not visible in this plot. Similarly, there is no direct contact between the oxygen and the hydrogen, which are again separated by a wedge of carbon that is now more clearly visible than in the previous case. The same is true for fluorine and hydrogen. In any of these three ligand· · ·ligand pairs (F· · ·H, O· · ·H, and O· · ·F), direct contact would mean that there is a little black square at the center of the interatomic surface separating the atoms of interest. If the nuclear configuration is sufficiently changed, such little black squares can be created.

Figure 8.6 reveals the three-dimensional shape of the three interatomic surfaces, in two different types of graphical representation and in a slightly different orientation (Figure 8.6a,b). In order to obtain a deeper insight into the nature of a topological atom we need to introduce the fundamental concept of the GP.

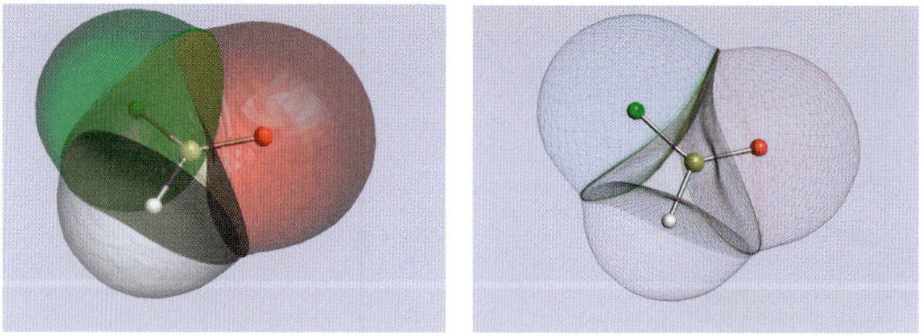

(a) (b)

Figure 8.6 Three-dimensional representation of the topological atoms in FHC=O, using the same colors as in Figure 8.5. (a) Semitransparent solid surfaces and (b) wireframe surfaces. The small purple spheres correspond to the black squares in Figure 8.5.

A *GP* is a path of steepest ascent through a multidimensional function, here one depending on three spatial coordinates. In the part of QTAIM that we discuss first, this function is the electron density $\rho(\mathbf{r})$. In the second part of QTAIM, this function will be the Laplacian of the electron density. In general, this function can still be another quantum mechanical function of interest (see Box 8.1). The GP is the cornerstone of QCT, which was briefly mentioned in the Introduction. Returning to the GP, it "wastes no time" in reaching a local maximum. Indeed, the GP pierces through surfaces of constant electron density such that it reaches a local maximum in the quickest possible way. The direction a GP takes is orthogonal to a very small (i.e., infinitesimal) plane, locally representing a surface of constant ρ value. Because the gradient vector is also orthogonal to such a plane one deduces that a GP is in fact a succession of gradient vectors. Imagine the gradient vector of the electron density, denoted $\nabla\rho$, being evaluated at a given point. This vector points in the direction of steepest ascent. Follow the gradient vector $\nabla\rho$ over a very short (i.e., infinitesimal) stretch and reevaluate $\nabla\rho$ at its endpoint. Continue this process iteratively. The resulting succession of very short gradient vectors can be regarded as a smooth curve. This curve is the GP.

Earlier we stated that a deformation density introduces a reference electron density. QTAIM eliminates the need for this reference density. Indeed, the gradient, as a differential, operates on a *single* electron density. This differential does not represent a difference between two densities but effectively constitutes an *internal* difference, within the same density, which is the molecular electron density ρ. This is an example of how QTAIM obeys Occam's razor or the *principle of minimalism.*

Figure 8.7 illustrates a multitude of GPs in HFC=O. The majority of the GPs originate at infinity and travel through space until they reach a local maximum, which practically coincides with a nuclear position. A collection of such types of GPs forms a spider web-like pattern, marking the subspace of each topological atom. Because the GPs "draw themselves," the topological atom also draws itself. The space it occupies is naturally carved out by the minimal prescription of tracing GPs, wherever they lead us. The resulting atomic shapes can be complicated but the prescription that generated them is minimal. By loose analogy, Newton's second law is also minimal but it also generates many complicated patterns of motion (e.g., of celestial bodies in the solar system).

It is clear from Figure 8.7 that the web-like patterns of GPs forming the topological atom are bounded by the interatomic surfaces. In other words, some GPs graze along an interatomic surface, sometimes indiscernibly closely. These grazing GPs move toward the center of an interatomic surface, marked by the little black square, but then, as they approach this center, they suddenly bend and terminate at a nucleus. This behavior suggests that the interatomic surface itself consists of GPs that do not terminate at a nucleus but at a mysterious little black square. In that sense, the trajectories of GPs constituting an interatomic surface can be considered as a limiting case.

It is time to zoom in on what these little black squares are, and thereby reveal a more complete picture of topological relationships.

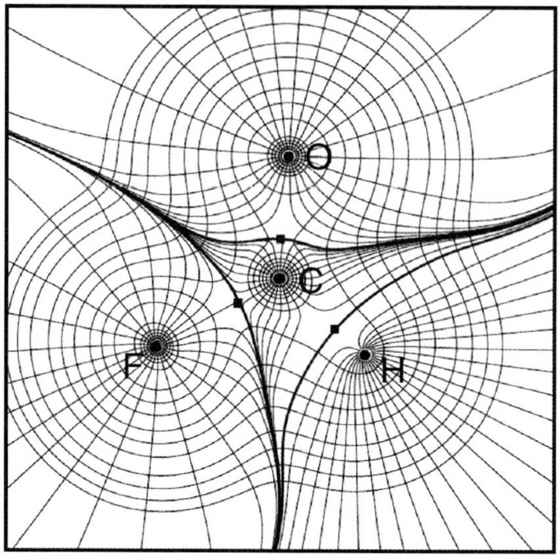

Figure 8.7 Gradient paths traversing the subspace of each topological atom in HFC=O. The interatomic surfaces shown in Figure 8.5 are repeated here in bold.

8.4
The Bond Critical Point and the Bond Path

Let us start with the GP, which was the central concept of the last section. We know that a GP is oriented: it has a beginning and an end. We also know that a great many GPs can come together at one point, practically coinciding with a nuclear position. We also know from Box 8.3 that, if at a given point $\nabla\rho \neq \mathbf{0}$, then there is only one GP containing this point. Thus, by the logical construction *modus tollens*, we can state that if a given point belongs to more than one GP then, at that point, it is not true that $\nabla\rho \neq \mathbf{0}$. In other words, a point where more than one GP meets has the property that $\nabla\rho = \mathbf{0}$. Such a special point is called a *critical point* (CP).

Box 8.3 The Gradient Path and Natural Coordinates

The gradient vector is all that is needed to map out the potentially complicated internal structure of a molecule in terms of the topological atoms that make up the molecule. A GP can be seen as succession of very short gradient vectors, in the simplest possible description. Thus, the tangent to such a GP, in a given point, is a gradient vector. This statement can be rewritten as a system of three ordinary differential equations,

$$\frac{d\mathbf{r}}{d\ell} = \frac{\nabla\rho}{|\nabla\rho|} \qquad \text{(B8.1)}$$

where $\mathbf{r}$ is a position vector describing the path and ℓ is the path length. The simplest way to solve this system is to use the Euler method, which is essentially tracing the paths by following the gradient vector over a very short stretch and reevaluating it at every new end point. Although this method may suffice for visual purposes it is not that accurate. The Runge–Kutta method (and its more modern variants such as the Cash–Karp method) is more appropriate and so is a predictor-correct method. Unpublished research showed that the more modern Bulirsch–Stoer method is not performing better than the Runge–Kutta or Cash–Karp method. Analytical expressions for the solutions of Eq. (B2.1) have been investigated before [95]. Note that one can trace only one GP through a point where the gradient of the electron density does not vanish. Slightly rephrased, if at a given point $\nabla\rho \neq \mathbf{0}$, then there is only one GP containing this point, or in other words, going through this point.

Let us introduce a path coordinate s, which is related to the actual path length ℓ via the equation $ds = d\ell/|\nabla\rho|$. One can then rewrite Eq. (B8.1) in a slightly simpler way,

$$\frac{d\mathbf{r}(s)}{ds} = \nabla\rho \tag{B8.2}$$

For a given initial point $\mathbf{r}_0(s=0)$ (where $\nabla\rho \neq \mathbf{0}$), a unique GP follows from solving Eq. (B8.2). At the origin of the GP $s = -\infty$ while at its terminus $s = +\infty$. The parameter s can be regarded as a coordinate describing the path over its full length. We know from the main text in Section 8.4 (Figure 8.8) that a GP always originates at a CP and terminates at another CP. If the distance between these two CPs is finite then the GP's length will also be finite. This means that a finite curve in Cartesian space is described by a path coordinate that spans the full range of $[-\infty, +\infty]$. To complete the set of coordinates that describe the topological atom to three coordinates we need two more coordinates. Imagine a sphere of radius β, centered at the nucleus within a given topological atom. The radius β is such that the sphere lies completely within the volume of the atom. Any point on the surface of the sphere can be uniquely described by θ and φ, the familiar angular spherical coordinates. Equation (B8.3) makes clear

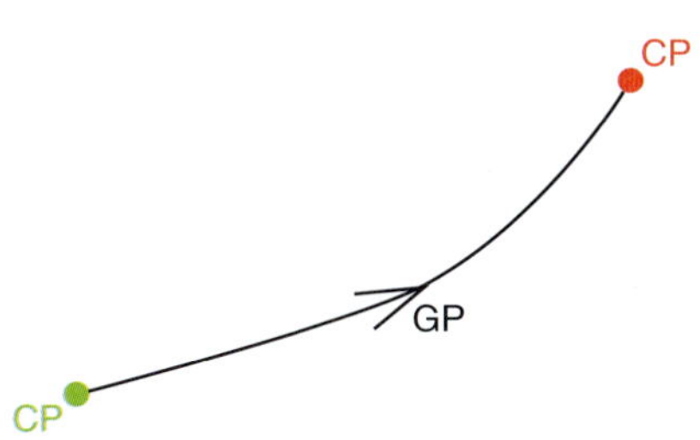

Figure 8.8 Each gradient path (GP) has a direction, originates at a critical point (CP) and terminates in a CP.

how an initial point $\mathbf{r}_0$ (x_0, y_0, z_0) is described as the GP is traced in the forward (ascending) and backward (descending) direction,

$$x_0 = \beta \sin\theta \cos\varphi$$
$$y_0 = \beta \sin\theta \sin\varphi$$
$$z_0 = \beta \cos\theta \qquad \text{(B8.3)}$$

Now we have three coordinates that each cover their full range, that is, $s \in [-\infty, +\infty]$, $\theta \in [0, \pi]$ and $\varphi \in [0, 2\pi]$. These are called *natural coordinates*. The interesting point is that they describe the topological atom "from within," as if it was a full (and unbounded) space. This "complete world" is of course not complete in Cartesian space, where its description in terms of the x, y, z coordinates clearly shows areas where the atom is absent. The coordinate mapping (s, θ, φ) → (x, y, z) could be interesting from the point of view of a mathematical branch called *differential geometry*. Such an analysis has already been carried out [96] on interatomic surfaces, which are described by their own two natural coordinates.

Because a gradient vector ceases to exist at a CP, this point is not associated with a direction pointing at a higher electron density value. Thus, the CP can be considered as the end point of a GP: as the path arrives at a CP there is no gradient vector to guide it beyond this point. Now, one can ask if a CP can also serve as the starting point of a GP. The answer is yes and again, this situation arises in a straightforward manner. If a GP starts in a given point then there cannot be a GP trajectory preceding it. This is exactly the case if $\nabla\rho = 0$ at that point, and that confirms that the GP started at a CP. Figure 8.8 summarizes the explanation above, introducing GP as a convenient acronym for gradient path and CP for critical point.

The most common type of GP is the one that originates at infinity and terminates at a maximum in ρ, which for all practical purposes coincides with a nucleus. There are many examples of this type of GP in Figure 8.7, which collectively form the *atomic basin*, that is, the subspace carved out inside a larger quantum system, forming the topological atom. From this picture one deduces that there is an ocean of CPs at infinity. The maximum in ρ at the nucleus is a maximum in three dimensions. In other words, in whichever of the three mutually orthogonal directions one approaches the maximum, ρ will increase.

Each interatomic surface in Figure 8.7 also consists of a bundle of GPs, sharing with the GPs forming the atomic basin, the fact that they originate at infinity. However, the GPs of the interatomic surface terminate at a so-called *bond critical point* (*BCP*), which is the proper name for the mysterious black squares in Figure 8.7 (and some earlier figures). Locally, the interatomic surface may be regarded as a plane containing the BCP. GPs travel a very short stretch in this plane, on their way to the BCP, which acts a local maximum in this plane. The BCP is thus a maximum in two dimensions. In this case these two dimensions describe the molecular plane. Is the BCP a maximum or a minimum in the third dimension, which is characterized by the direction orthogonal to the plane?

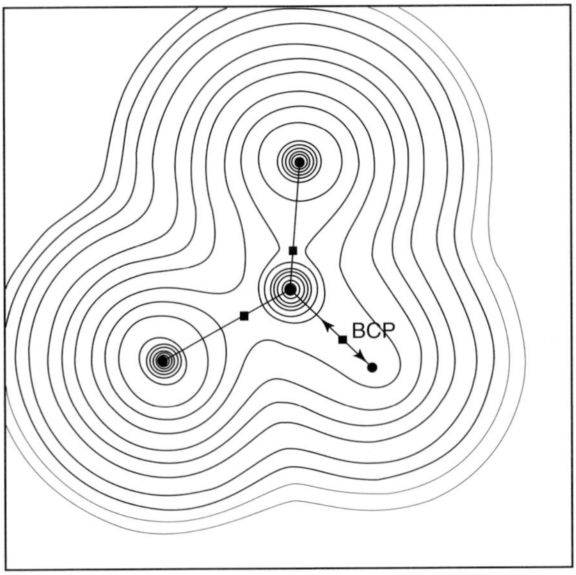

Figure 8.9 Examples of bond critical points (BCP), which are the three black squares in HFC=O, one of which is explicitly marked by the acronym BCP, and lies in between C and H. One gradient path originates at this BCP and terminates at C, while another gradient path originates at the same BCP but terminates at H. Together they form the bond path linking C and H.

Figure 8.9 answers this question by focusing on one BCP, lying in between C and H. The behavior of the third dimension is marked by two arrows, which show the direction of steepest ascent. It is clear that the BCP is a minimum in this third dimension. The BCP is thus a saddle point, or a CP with mixed local behavior in terms of the directions in which it is a minimum or a maximum. In three-dimensional space there are four types of CPs in this sense: a maximum, a BCP, a minimum and a second type of saddle point that we will discuss later. In general, there are $(n+1)$ types of CPs in an n-dimensional space. Box 8.4 provides more details on CP classification and local curvature. This Box also explains the various types of GPs, as determined by the type of CP they link.

Box 8.4 Classification of Critical Points

In two dimensions there can be only three types of CP: a maximum, a minimum or a saddle point. A saddle point is a CP that acts both as a minimum and a maximum depending on the direction one scans in. In three dimensions there are four types of CP: a maximum, a minimum and two types of saddle point. Any CP is typified by the local curvature of the function in which the CP appears. For our purpose, this function is the electron density, denoted by ρ, and the CP is a point where the gradient vanishes, or $\nabla\rho = 0$. The statements made here, about the number of CPs in two or three dimensions, are only true if this curvature is not zero, that is, the function is *not* flat at the CP. This curvature is quantified by the eigenvalues of the Hessian of the 3D function of interest. The Hessian is a matrix of partial second derivatives, each matrix entry H_{ij} being defined by

$$H_{ij} = \frac{\partial^2 f}{\partial q_i \partial q_j} \quad \text{where } i \text{ and } j = 1, 2, \text{or } 3 \tag{B4.1}$$

where q_1, q_2, and q_3 stands for *x*, *y*, and z, respectively, if the Hessian is constructed in three dimensions. A second derivative naturally measures curvature but the individual Hessian matrix elements cannot be analyzed directly as gauges of local curvature. The reason is that the numerical values of the entries in the Hessian matrix themselves depend on the choice of the coordinate system with respect to which the coordinates *x*, *y*, and *z* are expressed. However, the Hessian's eigenvalues are independent of this choice and thus solve this problem.

To each eigenvalue corresponds an eigenvector that points out the direction in which this eigenvalue determines the curvature. In order to illustrate this statement, let us take one type of CP, the BCP, because it is the most relevant for this chapter. The figure given here shows the three eigenvectors $\mathbf{u}_1$, $\mathbf{u}_2$, and $\mathbf{u}_3$, associated with a BCP. The latter eigenvector points in the direction in which the CP is a minimum. Accordingly, the corresponding eigenvalue λ_3 is positive. The figure shows how the electron density, starting at nucleus A, decreases along the BP and reaches its lowest value at the BCP, to then increase again to reach a local maximum value at the nucleus B.

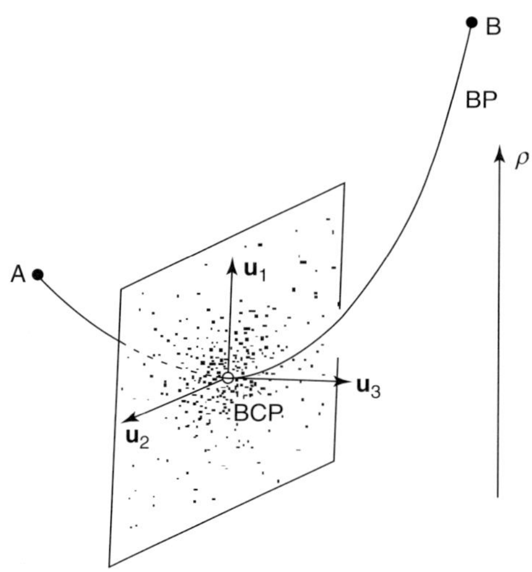

The figure also shows how eigenvectors $\mathbf{u}_1$ and $\mathbf{u}_2$ mark the directions in which the BCP is a maximum, as amplified by the density of points in the plane spanned by these eigenvectors. As expected, the corresponding eigenvalues are negative, or $\lambda_1 < \lambda_2 < 0$.

The other type of saddle point is called *a ring critical point* (RCP). It occurs in the center of a benzene ring, for example. Clearly, the electron density is a minimum at the RCP with respect to any displacement away from it in the molecular plane. Hence, there are two positive Hessian eigenvalues. In the direction orthogonal to the molecular plane the RCP acts as a maximum. Hence, the remaining eigenvalue is negative.

A compact notation, denoted (r, s), is often used to refer to any of the four types of CPs. The *rank* (r) of a CP refers to the number of nonzero eigenvalues and the *signature* (s) to the sum of the signs of the eigenvalues. More precisely, the latter means that a negative eigenvalue is associated with "minus one" (−1), and a positive eigenvalue with "plus one" (+1). The minus/plus ones can then be added. CPs of rank 3 are the most common, that is, those points with strictly nonzero Hessian eigenvalues. In this notation, the BCP is written as (3, −1) because it has three nonzero eigenvalues and because the "sum of eigenvalue signs" is $(-1)+(-1)+1=-1$. Another useful number is the so-called *index*, which is the number of positive eigenvalues of the Hessian. In summary, the four types of rank-3 CPs in three dimensions are denoted (3, −3), (3, −1), (3, +1), and (3, +3) corresponding to the index values 0, 1, 2, and 3.

Figure 8.8 made clear that GPs always link two CPs. This fact can be used to characterize all possible GPs. This was achieved [72], for the first time, in 2003. All details can be found in this paper but the final result is reproduced here in Table B4.1, which lists all possible types of GP in the electron density. This classification is universal, beyond the electron density, and is valid for other 3D scalar fields, such as the Laplacian of the electron density. The key to understanding how many GPs can originate or terminate at a CP is determined by the eigenvalues and associated eigenvectors of the Hessian evaluated at the CP. A negative eigenvalue corresponds to a direction in which the CP is a maximum and hence acts as a sink (or terminus) for GPs. Conversely, a positive eigenvalue corresponds to a direction in which the CP is a minimum. In this direction the CP acts as a source (or origin) of GPs.

As an example, let us look at the second row in Table B4.1, which shows the case of GPs in an interatomic surface (the name for this manifold if one works with the electron density). The notation for such a manifold is a [3, 1], or $[I_{source}, I_{sink}]$ in general, where the source and sink CPs are characterized by their index. Here, the first CP at the origin acts a source in all three directions and therefore does not restrict the dimensionality of the manifold that connects this CP and the CP at the terminus (or sink). It is clear that the dimensionality of the connecting manifold is determined by the CP at the terminus. Because the terminal CP can receive GPs in two directions the dimensionality of the connecting manifold is two. Obviously, the number of directions the terminal CP can receive GPs from, determines the dimensionality of the first three GP types in the Table, that is, [3, 2], [3, 1], and [3, 0]. In general, the dimensionality of a manifold connecting two CPs is the dimensionality of the source CP or that of the sink CP, whichever one is lowest. Formally, $\dim_{manifold}=\min(\dim_{source}, \dim_{sink})$.

Table B4.1 Survey of all nine possible types[a] of gradient path (GP) in the electron density (or any 3D scalar field).

GP type[c]	$(r, s)_{origin} \rightarrow (r, s)_{terminus}$	Dimensionality[b] Origin	Terminus	Manifold[d]	Example (name is for ρ only)
[3, 2]	(3, +3) → (3, +1)	3	1	1	Ring line
[3, 1]	(3, +3) → (3, −1)	3	2	2	Interatomic surface
[3, 0]	(3, +3) → (3, −3)	3	3	3	Atomic basin
[2, 2]	(3, +1) → (3, +1)	2	1	1	
[2, 1]	(3, +1) → (3, −1)	2	2	2	
[2, 0]	(3, +1) → (3, −3)	2	3	2	Ring surface
[1,2]	(3, −1) → (3, +1)	1	1	1	
[1, 1]	(3, −1) → (3, −1)	1	2	1	GP in ρ in a conflict structure
[1, 0]	(3, −1) → (3, −3)	1	3	1	Bond path

[a]Only critical points (CPs) of rank 3 are considered.
[b]For a CP at the origin this is the number of directions in which the CP can "send" GPs, while for a CP at the terminus this is the number of directions in which the CP can "receive" GPs. For the manifold this is the dimension of the connecting topological object, that is, curve, surface or basin, corresponding to one, two or three dimensions, respectively.
[c]Each gradient path is characterized by two critical points, an origin and a terminus. The critical points are denoted by their indices, I_{origin} and $I_{terminus}$, respectively, and the gradient path as $[I_{origin}, I_{terminus}]$.
[d]This is the set of gradient paths that connect two given critical points.

Figure 8.9 shows a GP that originates at the one explicitly marked BCP and that terminates at the carbon nucleus. The other GP, originating at the same BCP, terminates at the hydrogen nucleus. Together they form the so-called *BP* linking C and H. This name was first proposed [28] by the Bader group in 1977, *many years after the work on the topological atom was published.* This 1977 article immediately spotted the chemically intuitive character of the *BP*, when it is curved, for example, as in the case of ring strain. Figure 8.10 shows the simple example of cyclopropane, in which the CC *BP*s bulge outward. This deviation from a straight line connecting the nuclei can be quantified by straightforward geometrical measures.

At this juncture, a moment of reflection serves to obtain a better understanding of the overall philosophy and development of QTAIM. In his 1975 Accounts, Bader wrote that *While one cannot deny the important role played by the bond concept, we propose, as an alternative, a return to what is essentially the "atoms in molecules" approach to chemistry. Specifically, one seeks to understand or predict the properties of a total system in terms of the properties of its parts. Rather than bonds, we choose as our*

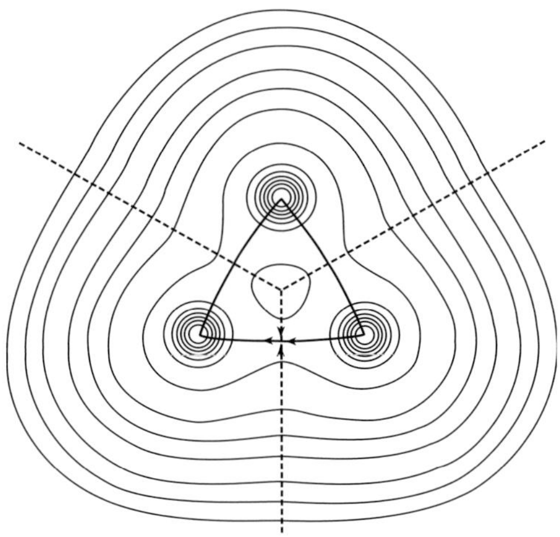

Figure 8.10 Contour plot of the electron density in the plane of the carbon nuclei in cyclopropane. The interatomic surfaces are marked by dashed line and the bond paths by solid lines. The little arrows mark the direction of the gradient paths in the vicinity of the bottom CC bond critical point. (Reproduced from Ref. [28].)

fundamental parts mononuclear fragments of the system with boundaries defined in real space.

About 2 years after this statement was published, an alternative view was proposed by the Bader group, this time very much in line with the traditional picture of a chemical bonding. The Bader group had *observed* that the topology of the electron density offered an interpretation of a bond that associated it with the *BP*, a topological object coexisting with topological atoms. So, whereas QTAIM started off as an energy partitioning scheme and viewed bonding as a result of atomic energies, the theory was now tempted to reinterpret bonding guided by the presence of BCPs and *BP*s. This dual approach, both resulting from the full topology of the electron density, should not be seen as a dilemma. Instead, both pictures should ideally be complementary aspects of the same thing: a chemical bond. In fact, in the last part of this chapter we spend time on exactly this complementarity. A question of essence is the following: *can we link the QTAIM energy partitioning with BPs?* We will show in Section 8.5 that this is indeed possible as demonstrated by a seminal paper published [29] in 2007, entitled *BPs as Privileged Exchange Channels.*

First, however, we look at how evaluating quantities at the BCP can offer an insight into the type of bond that it is a signature of. Box 8.5 provides details on an important function called the *Laplacian* of the electron density, $\nabla^2\rho$, which has received much attention in QTAIM. The function $L(\mathbf{r}) = -\nabla^2\rho(\mathbf{r})$ often appears in the literature when bonds are classified along the lines discussed in Box 8.5. This is because it is more intuitive to think of a positive $L(\mathbf{r})$ value as one associated with a concentration of electron density, while depletion is naturally reminiscent of a negative value. The sign of $L(\mathbf{r})$ at the BCP can be used as an indicator to characterize a bond,

Box 8.5 The Laplacian of the Electron Density

The Laplacian of the electron density, $\nabla^2\rho$, is the second quantum mechanical function ever to be analyzed topologically, after ρ itself. However, the first paper on the Laplacian (in a QTAIM context), which was finally published [97] in 1984, did not (yet) explore its topology. Instead, this work plotted $\nabla^2\rho$ its value for a variety of systems including covalent diatomics, hydrogen-bonded complexes, ionic systems, and van der Waals (noble gas) complexes. This work interpreted the sign of $\nabla^2\rho$ in order to make a connection with the energetics of atomic interactions. A variant of Eq. (B2.3), listed below as Eq. (B5.1), made this connection possible,

$$2G(\mathbf{r}) + V(\mathbf{r}) = \frac{1}{4}\nabla^2\rho(\mathbf{r}) \tag{B5.1}$$

where $G(\mathbf{r})$ is the kinetic energy density (defined in Eq. (B2.2)) and $V(\mathbf{r})$ is the electronic potential energy density. Note that this is a *local* relationship; when integrated over the volume of a topological atom Eq. (B5.1) leads to the atomic virial theorem, due to the fact that the Laplacian then vanishes (see Eq. (B2.8)).

In regions of a (molecular) system where $\nabla^2\rho < 0$, the electron density is locally concentrated. This interpretation of concentration is a general property of the Laplacian in a (3D) scalar field, disconnected from quantum mechanics. This interpretation can be seen as an extension of the second derivative in one dimension. Indeed, the Laplacian is basically a second derivative, and it essentially quantifies a local curvature. It is well known that, in any one dimension, a negative curvature corresponds to a local maximum. Therefore, a negative Laplacian corresponds a local concentration, even if this does not happen at a stationary point in the Laplacian (where $\nabla(\nabla^2\rho) = \mathbf{0}$). From Eq. (B5.1) one can deduce that the potential energy density then dominates if $\nabla^2\rho < 0$, because $G(\mathbf{r})$ is always positive and can therefore never contribute to an overall negative sign. When $\nabla^2\rho < 0$ is found in the internuclear region then one classifies the interatomic interaction as *shared*. Covalent and polar bonds fall in that class, and the interaction between atoms in such bonds is caused by a contraction of the electron density toward the interaction line linking the two nuclei. We imagine this situation at the BCP. This local state of affairs can be understood semiquantitatively by invoking the fact that $\nabla^2\rho = \lambda_1 + \lambda_2 + \lambda_3$, where the three terms are eigenvalues of the Hessian, as discussed in Box 8.4. The Laplacian can only be negative if the two manifestly negative eigenvalues, λ_1 and λ_2, dominate (in magnitude) the one positive eigenvalue, λ_3. This, in turn, means that the electron density is concentrated toward the BCP.

The opposite situation arises when $\nabla^2\rho > 0$, in which case the electron density is locally depleted. If we imagine this situation to arise again at a BCP then this time the positive eigenvalue λ_3 dominates. This means that the electron density is rapidly increasing away from the BCP. This situation arises in so-called *closed-shell interactions* such as ionic bonds, hydrogen bonds and van der Waals complexes. By virtue of Eq. (B5.1), the kinetic energy density is now in charge of the mechanics of the interaction.

At the time the 1984 paper [97] appeared the local energy density $H(\mathbf{r})$ was proposed [98] to complete the description of bonding along the lines of the earlier discussion. This function is defined in Eq. (B5.2),

$$H(\mathbf{r}) = G(\mathbf{r}) + V(\mathbf{r}) \tag{B5.2}$$

The driver to introduce this function was the observation that for about 100 covalent bonds (including F–F in F_2) $H(\mathbf{r}_b)$ is always negative. The opposite is true for ionic bonds, hydrogen bonds, and van der Waals bonds, thereby turning $H(\mathbf{r}_b)$ into a convenient discriminator between covalent and non-covalent bonds.

The second paper on the Laplacian [68] this type looked the topology of this function, starting with its CPs. The numbers and locations of the bonded and nonbonded concentrations of charge in the valence shell of a bonded atom, as determined by the maxima (CPs) of $L(\mathbf{r}) = -\nabla^2\rho(\mathbf{r})$, were found to be in general agreement with Gillespie's valence shell electron pair repulsion (VSEPR) model [99, 100]. It should be noted that this interpretation is based on observations only, and actually at variance with the conclusions based on an analysis of the Fermi hole [101] by Bader and Stephens in 1975. They found that regions that maximize Fermi correlation to yield localized groupings of electrons correspond to atomic cores, and not, in general, to localized pairs of bonded and nonbonded electrons as anticipated on the basis of the Lewis model.

In 1985, the concept of an *atomic graph* was proposed [102] in the context of reactivity. The idea of an atomic graph is loosely analogous to that of a molecular graph. Both represent a connectivity scheme of CPs, completely determined by their topological links, as embodied by GPs. Discussing this in detail is beyond the scope of this chapter but it is worth mentioning though that atomic graphs were introduced [103] in 1992, in the context of complementarity in chemistry and molecular recognition. Later, more complete insight into the Laplacian's topology (in terms of CPs and their connectivity only) was published [70, 104] and how this topology changes in response to a change in molecular geometry. A thorough study on a multitude of atomic graphs covering most common functional groups of organic chemistry revealed 16 highly transferable atomic graphs [105]. This work offers a physicalization of chemical graph theory, introducing building blocks rooted in quantum mechanics, through properties of the orbital-invariant electron density. In 2003, Laplacian basins were shown for the first time [106] thanks to the construction of a novel algorithm (based on the "octal tree") that also enabled integration the electron density over their volume. This led to the first ever study [71] using the full topology of the Laplacian, which scrutinized it as a physical basis for the VSEPR model. This study confirmed, by computed volume of Laplacian basins, that a nonbonding domain is indeed larger than a bonding one, in full accord with the VSEPR model. Second, these volume quantities for the first time being available also showed that a multiple-bond domain is indeed larger than a single-bond one, again in agreement with VSEPR.

However, the calculations did not corroborate the effect of the electronegativity of the central atom or ligand on the volume of bonding domains. More worryingly, this study yielded unexpectedly small electronic populations (by integration of ρ over Laplacian basins), nearer to one than to two, for non-hydrogen cores and bonding domains, while nonbonding domains could have populations much larger than two.

although it is safer to use it in combination with other indicators, the simplest being the value of ρ at the BCP, denoted by $\rho(\mathbf{r}_b)$. In a typical ionic bond, which is a *closed-shell interaction*, $\rho(\mathbf{r}_b)$ is typically of the order of 0.01 au whereas for a *shared interaction* it is typically about an order of magnitude larger, that is, 0.1 au. The need to characterize a bond by more than just $L(\mathbf{r}_b)$ is clear from the existence of bonds such as C≡O or F_2, which are characterized as *intermediate interactions*. This is because it has mixed features: a high $\rho(\mathbf{r}_b)$ value (normally found when $L(\mathbf{r}_b) > 0$) but $L(\mathbf{r}_b) < 0$. Box 8.5 introduces the measure $H(\mathbf{r}_b)$, which circumvents the appearance of intermediate interactions by always assigning a negative value to $H(\mathbf{r}_b)$ for any recognized covalent bond, including that in F_2. When complemented with the extra indicator $G(\mathbf{r}_b)$, $\rho(\mathbf{r}_b)$ atomic interaction can be classified in a way more sophisticated than in the 1980s. One such proposal came in 1998 from the world of experiment, in particular, high-resolution X-ray crystallography, where bonds in a transition metal dimer were classified [30] by topological indicators. Table 8.1 describes the main features of several types of interaction including between light (L) elements (e.g., H, C, B) and heavy (H) elements (e.g., As, Co). Note that this table takes into account the position of the BCP with respect to the behavior of $L(\mathbf{r})$ in terms of sign, sign change, or extrema. One of the conclusions of this extensive crystallographic study on $Co_2(CO)_6(AsPh_3)_2$ is that there is a covalent Co–Co bond, in spite of the lack of accumulation of the deformation

Table 8.1 Summary of topological indicators and features that characterize the atomic interaction.

	$\rho(r_b)$	Position of r_b wrt L(r) along the BP	$L(r_b)$	$G(r_b)/\rho(r_b)$	$H(r_b)$
Open-shell (covalent bonds)	Large	Close to a minimum	>0	<1	<0
Intermediate interactions (polar shared bonds, e.g., CO)	Large	Close to a nodal surface	Arbitrary	≥1	<0
Closed-shell	Small	Inside a flat region	<0	≥1	>0
Shared (e.g., Co–Co)	Small	Close to a maximum	~0	<1	<0
Donor–acceptor (e.g., Co–As)	Small	Close to a nodal surface	<0	~1	<0

density map. Note that all values in Table 8.1 were obtained experimentally but the kinetic energy density was indirectly obtained via a semiempirical method [31] that estimates $G(\mathbf{r}_b)$ from a function of ρ and its Laplacian.

An otherwise popular indicator that does not feature in Table 8.1 is the so-called *ellipticity* ε, which is defined as $\lambda_1/\lambda_2 - 1$. Note that the ellipticity is not bounded from above but, because $\lambda_1 < \lambda_2 < 0$, its minimum value is zero. This corresponds to a cylindrically symmetric electron density, which is perfectly reached at the carbon–carbon BCP in ethyne. The BCP of the central CC bond in butane has an ellipticity of 0.01, which is almost zero. The ellipticity reaches 0.3 at the C=C BCP in ethene, and is interpreted as a measure of the π character. Any CC BCP in benzene has the lower value of $\varepsilon = 0.18$, which is in agreement with the expected reduction in the π character compared to the "pure" double bond in ethene. Remarkably, the substantial ellipticity of 0.4 at a CC BCP in cyclopropane (Figure 8.10) is reminiscent of the widely recognized double bond character of (strained) three-membered carbon rings, in terms of their reactive behavior (e.g., rates of solvolysis).

Finally, we point out that the topological descriptors were introduced [32] in the area of molecular similarity through the concept of BCP space. In its original form, BCP space represented BCPs as points in a 3D space, spanned by $\rho(\mathbf{r}_b)$, $-L(\mathbf{r}_b)$ and $\varepsilon(\mathbf{r}_b)$. This compact description sufficed, compared to descriptions from whole electron densities, in providing descriptors for linear free energy relationships such as the one proposed by Hammett on acidity. Further research showed that the BCP space proved a reliable source of compact quantum mechanical descriptors for QSARs (quantitative structure–activity relationship) in a medicinal, ecological, physical organic (pK_a prediction), or toxicological context, culminating in a method called *Quantum Topological Molecular Similarity* (*QTMS*) [33].

8.5 Energy Partitioning Revisited

An early important result in the development of QTAIM was the establishment that an atom in a molecule has its own atomic virial theorem. This means that, for a single atom, there is a relation between the kinetic energy of this (topological) atom and its potential energy. This relationship can be obtained by integrating Eq. (B5.1) over the volume of a topological atom Ω, or

$$\int_{\Omega} d\mathbf{r}[2G(\mathbf{r}) + V(\mathbf{r})] = \frac{1}{4}\int_{\Omega} d\mathbf{r}\nabla^2\rho(\mathbf{r}) = 0 = 2T(\Omega) + V(\Omega) \tag{8.2}$$

where we emphasize again that topological atoms are remarkable subspaces in that they have a well-defined kinetic energy, denoted $T(\Omega)$. Thanks to this relationship there was no need to calculate $V(\Omega)$ in the early days of QTAIM. Calculating $T(\Omega)$ was sufficient, which required a 3D volume integration [34] over the volume of Ω. The literature on the integration algorithms over topological subspaces is fairly large and has been briefly reviewed in a recent paper [35] on fully analytical integration (over a spherical core inside a topological atom), a body of work to which one can add a most recent contribution [36]. In spite of the technical challenges of such 3D integration over topological atoms, they are computationally less demanding than the 6D integrations required to calculate $V(\Omega)$ (independently from the atomic virial theorem). The existence of an atomic virial theorem enables one to calculate only $T(\Omega)$ and then deduce $V(\Omega)$ from it as $-2T(\Omega)$. As a result, the total atomic energy $E(\Omega) = T(\Omega) + V(\Omega) = -T(\Omega)$. One can the sum the total atomic energies in a molecule and obtain the total molecular energy. Although this is convenient, the latter statement is true only if the forces on the nuclei all vanish, that is, at a stationary energy point, typically an energy minimum. A calculation of $V(\Omega)$, independent of the atomic virial theorem, would free QTAIM energy partitioning from being confined to equilibrium geometries only.

The latter was achieved [37] for the first time in 2001, in a paper that calculated the Coulomb interaction energy defined in Eq. (8.3),

$$E_{\text{Coul}}^{\text{AB}} = \int_{\Omega_{\text{A}}} d\mathbf{r}_1 \int_{\Omega_{\text{B}}} d\mathbf{r}_2 \frac{\rho_{\text{tot}}(\mathbf{r}_1)\rho_{\text{tot}}(\mathbf{r}_2)}{r_{12}} \tag{8.3}$$

where the *total* charge density, $\rho_{\text{tot}}(\mathbf{r})$, is the sum of the electron density (i.e., electronic) $\rho(\mathbf{r})$ and the nuclear charge density, and r_{12} is the distance between two infinitesimal pieces of charge density. This work was further developed completing the calculation of non-Coulomb interaction energies [38, 39]. However, in 2004, an efficient algorithm [40] made possible a full and systematic analysis of molecular energy in terms of intra-atomic and interatomic (pairwise) energy contributions, dubbed *Interacting Quantum Atoms* (IQA) [24]. All electron–electron interactions are determined by the diagonal of the second order reduced density matrix, $\rho_2(\mathbf{r}_1,\mathbf{r}_2)$. The electron–electron interaction energy between two topological atoms Ω_{A} and Ω_{B} can then be written as in Eq. (8.4),

$$V_{\text{ee}}^{\text{AB}} = \int_{\Omega_{\text{A}}} d\mathbf{r}_1 \int_{\Omega_{\text{B}}} d\mathbf{r}_2 \frac{\rho_2(\mathbf{r}_1, \mathbf{r}_2)}{r_{12}} \tag{8.4}$$

Interestingly, the electron–electron interaction can always be split into a classical (Coulombic), and a nonclassical (quantum mechanical) exchange-correlation (xc) part, as follows,

$$\rho_2(\mathbf{r}_1, \mathbf{r}_2) = \rho_1(\mathbf{r}_1)\rho_1(\mathbf{r}_2) + \rho_{\text{xc}}(\mathbf{r}_1, \mathbf{r}_2) \tag{8.5}$$

where the subscript "1" can be omitted if no emphasis is needed on the fact $\rho(\mathbf{r})$ is in fact the diagonal of the *first*-order reduced density matrix. Substituting Eq. (8.5) into Eq. (8.4) leads to

$$V_{\text{ee}}^{\text{AB}} = V_{\text{cl}}^{\text{AB}} + V_{\text{xc}}^{\text{AB}} \tag{8.6}$$

where $V_{\text{cl}}^{\text{AB}}$ is the purely electronic part of $E_{\text{Coul}}^{\text{AB}}$, that is, without any contributions of nuclear charge. Stimulatingly, this part measures the ionic-like contributions, while the second term $V_{\text{xc}}^{\text{AB}}$ measures the covalent-like contribution to a given atom–atom interaction (no matter how far apart the atoms are). This term is then of course defined as,

$$V_{\text{xc}}^{\text{AB}} = \int_{\Omega_{\text{A}}} d\mathbf{r}_1 \int_{\Omega_{\text{B}}} d\mathbf{r}_2 \frac{\rho_{\text{xc}}(\mathbf{r}_1, \mathbf{r}_2)}{r_{12}} \tag{8.7}$$

This genuine quantum mechanical index is key to enduring progress in the "holy grail" question of how (if at all) to extract chemical bonds from wave functions. It should be emphasized that this index is an *energetic* term, and therefore enables one to analyse bonding directly in terms of energy, which is the ultimate arbiter of chemical stability, even of chemical fragments. Second, the index V_{xc}^{AB} does not invoke a reference state, and hence any (risky) decision as to what that state might be. *An analysis based on V_{xc}^{AB} benefits from the minimalism discussed earlier.*

We now discuss an observation of paramount importance [29], made in 2007, which revealed a remarkable link between the presence of a BCP and V_{xc}^{AB}. The importance of this observation lies in *directly relating the existence of the topological object that decides on a bond to atomically partitioned covalent energy.* How this works will be explained using the example of water. We consider the minimum-energy dissociative adiabatic pathway (maintaining C_{2v} symmetry) that leads to the formation of an oxygen atom and a hydrogen molecule, H_2O ($^1A_1$) → O (^1D) + H_2 ($^1\Sigma_g{}^+$). A sequence of gradient vector fields for relevant snapshots along this pathway has already been published [41] in 1981, in the context of the QTAIM theory of structural change, which invokes catastrophe theory. We are interested in following this pathway in the reverse direction, that is, to form a water molecule by H_2 approaching an oxygen atom. This formation is controlled by a single (reaction) parameter *d*, which is the distance between the oxygen nucleus and the midpoint of H_2. Figure 8.11 plots this parameter as the *x*-axis against a *y*-axis of V_{xc}^{AB} energies. Starting at the right-hand side of this plot, H_2 is an essentially unperturbed molecule. Here, H_2 is a BCP expressing the fact that the two hydrogen atoms are bonded. In this regime V_{xc}^{HH} is markedly lower than V_{xc}^{OH} and this is why the BCP appears between H and H. As the control parameter *d* decreases, suddenly the V_{xc}^{OH} energy becomes lower than that of its competitor V_{xc}^{HH}. This happens near

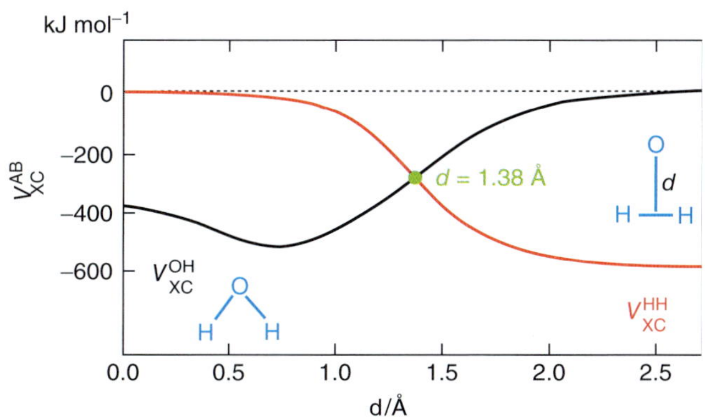

Figure 8.11 Competition between V_{xc}^{HH} and V_{xc}^{OH} in the formation of ground state water (left) from H_2 and an oxygen atom (right). Moving from the right to the left, the control parameter *d* decreases, and an intersection is hit at *d* = 1.38 Å, beyond which V_{xc}^{OH} becomes lower than V_{xc}^{HH}. Around this intersection, the O–H BCP is formed and the H–H BCP point is destroyed (ignoring the complication of a very short-lived ring critical point near this transition point).

$d = 1.38$ Å, at which point, the O–H BCP is formed and the H–H BCP is destroyed. Matters are actually slightly more complicated because the evolution of the gradient vector field passes through a very short-lived ring structure between $d = 1.391$ Å, and $d = 1.389$ Å. Note that in the ring structure both O–H BCPs and H–H BCPs coexist. The same energy-topology of the ρ relationship has been established both in the HCN to CNH isomerization and in a process in which a fluoride anion moves around LiF in a circle, passing from FLiF to LiFF.

The general conclusion of the 2007 paper [29] was confirmed very recently, in 2013, by Tognetti and Joubert [42]. They thoroughly analyzed admittedly approximate values for V_{xc}^{AB} (ignoring electron correlation and for the first time employing density functional theory (DFT), all calculations being performed in the monodeterminantal Kohn–Sham *ansatz* using B3LYP). They computed what they call E_{AB}^{x} values for a large number of intramolecular O···X interactions (where X is an O, S, or halogen atom) occurring in a set of 36 molecules, which also contained C=C bonds. The authors address many natural questions about the nature of the BCPs, for example, if they would always appear between any two nuclei. This misleading impression is often had when a one-dimensional interpretation of a minimum lying between two maxima (see Figure 8.1 and the mountain landscape) is wrongly extrapolated to three dimensions. One should not forget that for a BCP to exist, all three components of the gradient of the electron density must vanish, not just one component along a line connecting the two nuclei. A second type of misconception arises when one presumes that a BCP is merely a reflection of internuclear distance. Again, this is naïve: a BCP is not a docile signature of internuclear distance. There are surprizing cases where two O···O distances, identical up to three significant figures, appear in two different molecules where one displays a BCP in between these two oxygens and the other molecule does not. In fact, when surveying the many cases studied in this paper it becomes clear that a BCP, by its presence or absence, is a highly informative yet compact signature of the exchange energy between two given atoms. For the first time, numerical criteria are proposed. The authors conclude that the existence of BCPs indeed depends on the competition of what the 2007 paper calls "exchange channels." The authors go further and state that *the controversies about the existence or not of BCPs (…) are not sufficient (…) to undermine QTAIM's foundations.* Given these most interesting and intriguing observations there is a need for more studies of this ilk.

A natural and important question is then: can QTAIM, through patterns in V_{xc}^{AB} values, extract a Lewis diagram from a given molecular wave function? A very recent study, published [43] in 2013, set out to exactly answer this question, and the answer is yes, so far. In that paper the interatomic exchange-correlation energy V_{xc}^{AB} was investigated exhaustively for all atom–atom interactions in 31 small covalent molecules (including ions) and three van der Waals complexes. For the first time, clear clusters were revealed in the values of V_{xc}^{AB}, clusters separated by almost an order of magnitude (depending on the system at hand). This quantitative information, justified by a precise and minimal physical picture of topological energy partitioning, underpins the idea of a molecular graph. Such a full analysis

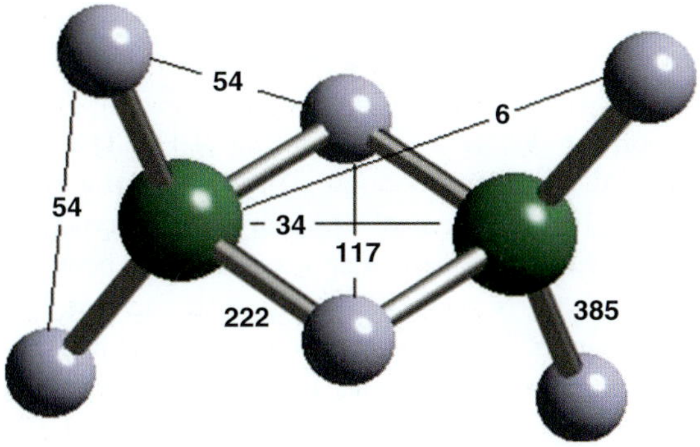

Figure 8.12 A ball-and-stick diagram of diborane, B_2H_6, expanded with numerical values of $-V_{xc}^{AB}$ in kJ mol^{-1}.

of V_{xc}^{AB}, for all atom–atom interactions in a system, reveals where to draw the lines in a Lewis diagram.

A useful example illustrating this success is that of diborane, B_2H_6, which at one time was controversial in terms of its Lewis structure. Figure 8.12 shows a ball-and-stick diagram of B_2H_6, endowed with $-V_{xc}^{AB}$ values (in kJ mol^{-1} and for HF/6-311G(d,p) wave functions). The two largest values (in terms of absolute values) are 385 and 222 kJ mol^{-1}, corresponding to the covalent bonds BH_{term} and BH_{bridge}, respectively. They provide the "sticks" of the full molecular graph because it can be completely built from just these two types of B–H bonds. The next strongest interaction is between the bridging hydrogens (117 kJ mol^{-1}). This relatively large value is compatible with diborane forming a H–H BCP upon small geometrical distortions, rather than forming a B–B BCP (note that the B–B interaction is more than three times weaker than the H–H interaction). Some one versed in molecular theory may find this fact surprising.

In summary, V_{xc}^{AB} displays a hierarchy of values, expressing both coarse-grained and fine-grained information about the covalent electronic structure of a molecule, without using molecular orbitals. The quantity V_{xc}^{AB} is (still) expensive to compute and numerical errors of a kilojoule per mole are common (due to the quadrature of large energy values). However, the fundamental patterns that V_{xc}^{AB} reveals are beyond this numerical error. The quantity V_{xc}^{AB} urgently needs to be computed in larger molecules and in molecular complexes. The most important feature of a V_{xc}^{AB} analysis is that it does *not* impose *a priori* chemical views (e.g., steric repulsion) onto the chemical bonding puzzle or controversy at hand, and that it does *not* invoke a reference state or arbitrary parameters either. Unfortunately, these two desirable features are absent in some recent studies [44–48] that propose interpretations of chemical bonding without using V_{xc}^{AB}. Many of these studies were vigorously refuted by Bader himself [49] and we will not repeat the arguments and his counterarguments here. However, we briefly discuss a case study that fully represents this fundamentally important debate on whether a particular atom–atom interaction is a bond or not, namely that of H···H interactions in aromatic hydrocarbons. We chose this example for three reasons: this case study is prototypical, it is the oldest controversy of this kind, and a preliminary V_{xc}^{AB} analysis of this type of interaction exists.

The full narrative, up to 2004, on the debate on the meaning of H···H BCPs (excluding the now widely accepted dihydrogen bond [50, 51]) is given in a book chapter (Section 4.3 of Ref. [12])). Salient element of this story will be repeated here. Between 1990 and 1992, a flurry of seven papers appeared on *BP*s between unexpected atoms (H and H, O and O, N and O), ending with the last one [52] on ortho-fluoro-substituted biphenyls. During the course of publishing this flurry, the authors changed their mind about whether the *BP*s they observed were indicative of bonds rather than "steric interactions." The fourth paper in the sequence [53] investigated kekulene, a highly symmetric polyaromatic consisting of a hexagon of 12 fused benzene rings. Remarkable highly curved *BP*s linking the six inner hydrogens were dismissed under the banner that they were *unable to attach any physical significance to these bonds.* In the sixth paper [54], focusing on "hydrogen–hydrogen" nonbonding interactions, the authors, now more confident, stated that *the term BP should be reserved for the interaction lines describing ordinary strong bonds.* Here they varied the central torsion angle φ ($C_2C_1C_{1'}C_{2'}$) in biphenyl to study its effect on the local topology of the interaction between the hydrogen bonded to C_2 and the hydrogen bonded to $C_{2'}$. The corresponding H–H distance acts as a control variable determining if a BCP has appeared between these two hydrogen atoms. A BCP appears in the planar transition state ($\varphi = 0°$), but not in the local energy minimum, when $\varphi = 45°$. Eventually, in the seventh and last paper, the authors settle their conundrum in favor of an interpretation that projects traditional chemical intuition onto their set of topological observations. They propose that *when the distance between two atoms is smaller than their contact interatomic separation, a (...) BCP appears, indicating a nonbonding repulsive interaction.* The authors claim this is a rigorous definition of *sterically crowded molecules,* superior to that obtained from van der Waals radii. Then, in 1995, an eighth and final paper [55] appeared on the matter, further elaborating steric crowding, now in perhalogenated hydrocarbons.

In a single author paper [56] Bader completely rejected the notion of steric repulsion as allegedly expressed by a BCP. He argued that *the question is not how the final geometry is attained in some mental process involving passage over a repulsive barrier, a situation that is in fact common to most chemical changes, but rather how the mechanics determines the final distribution of charge.* In technical terms, Bader based his rebuttal on the claim that the virial field and the electron density are homeomorphic [57]. This homeomorphism is not mathematically proven and only observed for only 15 molecules at that time. The second issue is that the homeomorphism is not perfect because it has several exceptions. For example, amongst others, the two fields behave differently in Li_2 and also in B_2H_6, where the virial graph shows a path linking the two borons, which is absent in the electron density. Because of the imperfection of the homeomorphism, and because the virial field is judged [56] to have the ultimate authority to decide on bonds, one may wonder why the electron density is still consulted on matters of bonding, in the first place. Nevertheless, the Bader group used exactly this angle of attack of the virial graph to present [58] a hydrogen–hydrogen BCP as a mark of *stabilizing* interaction in molecules and crystals. This type of bonding accounts for the existence of solid hydrogen, and as it falls in the class of "van der Waals" interactions, no different in kind, for example, from the intermolecular Cl–Cl bonding in solid chlorine [59]. The authors point out that these H–H interactions are ubiquitous and that they should not be confused with dihydrogen bonds (see above). According to their approach there is no steric repulsion between the *ortho*-hydrogens in biphenyl; rather the resultant H–H bonding contributes a stabilizing effect to the molecule's energy.

The debate continued when, in 2006, a paper [47] appeared that literally stated in its title that *Hydrogen-Hydrogen Bonding in Planar Biphenyl, Predicted by Atoms-in-Molecules Theory, Does Not Exist*. Of course, one could not be clearer than that, which is why Bader immediately wrote a rebuttal paper [60] in the same year, with an equally clear title stating *Pauli Repulsions Exist Only in the Eye of the Beholder*. The essence of Bader's rebuttal is indeed that the arguments presented in the attacking paper are based on an arbitrary partitioning of the energy into contributions from physically unrealizable states of the system. Bader invokes a statement [61] from Morokuma, one of the first architects of energy decomposition analyses, his own ostentatiously referred to as Energy Decomposition Analysis (EDA). Morokuma himself wrote that *There is no unique choice for the intermediate wave functions, and they do not correspond to reality (e.g., "not all" satisfy the Pauli principle!*. In the many EDA schemes that followed, several variants of such imagined reference states appeared. The IQA method, which was explained above, is of course also an energy decomposition analysis, triggered by the pioneering calculation [37] of the Coulomb energy between two topological atoms (see Eq. (8.3)). However, following the principle of minimalism, IQA does not invoke any reference state at all, let alone one that does not exist in Nature.

The seminal 2007 paper [29] on *BPs as Privileged Exchange Channels*, also joined the debate on controversial H–H BCPs, providing new data, unlike Bader in his 2006 paper. Five systems were investigated, R–H$\cdots$H–R (where R = H, Li, or CH_3), *cis*-butadiene and biphenyl itself, all at Hartree–Fock level, which has proven to be sufficient for the debate at hand. The distance between the two hydrogen atoms was taken as a control parameter *d*, generating a number of geometries, and against which the V_{xc}^{HH} energies could be plotted. The profiles of all five systems turned out to be remarkably similar. Hence, this observed universality puts the H$\cdots$H interaction in biphenyl on a par with the well-accepted attractive interaction that H$\cdots$H embodies in the R–H$\cdots$H–R systems. In summary, the H$\cdots$H interactions in biphenyl are not destabilizing (i.e., repulsive) but stabilizing.

One should note that the origin of the rotation barrier in biphenyl was not determined in the 2007 paper because this would require a full IQA analysis, which was lacking at the time. Indeed, many atomic self-energy and atom–atom interaction energies change substantially between molecular configurations and all have to be captured. The most important point to remember is that chemical behavior emerges as a *net* result (i.e., sum) of many different large and often opposing energy terms. At no point does IQA ascribe any chemical behavior (i.e., rotation barrier) to the interaction between just two atoms. Such ascription (i.e., isolation of fragment behavior and projection onto the whole) often happens in Chemistry (for a case study criticizing the secondary interaction hypothesis [62]). We note that IQA does not fall into the trap of singling out highly local interactions and then projecting the overall behavior onto them. In summary, we can say that the *overall* repulsion barrier in biphenyl is *not* due to a H· · ·H BCP representing a locally repulsive interaction.

Finally, a brief discussion of another attacking paper [44], written by a number of German authors in 2009, is in order here. This case study on phenanthrene studied experimentally (using Raman spectroscopy), gives the impression of settling the matter on how to interpret a H· · ·H BCP (this time appearing in the "bay region" of the molecule). This paper was vigorously disputed [49] by Bader, again in a single-author paper, and again immediately, in the same year. The core of his counterattack amounts to a perfectly valid point on the interpretation of the "interaction constant" (better called the *coupling force constant*) *k*, but unfortunately it takes up to only one column of text. This main and powerful argument might better have displaced much of the philosophy and "back-to-physics" outbursts that wrapped this argument in the rest of the paper.

Moreover, it should be emphasized that the German authors invoked the "architecture" of the MM3 force field in order to separate molecular potential energy into "chemically meaningful parts." Of course, this is the wrong way around. Indeed, popular force fields such as AMBER (or MM3 for that matter) have little authority, if any, when it comes to partitioning energy in a physically rigorous manner [63, 64]. The right way around is to construct a rigorous force field from an actual quantum mechanical energy partitioning scheme. Pivotally, however, the German authors used the arbitrary energy partitioning that underpins MM3 to support their main argument of the repulsive nature of H· · ·H interactions. There is another and, quite frankly, better route however. If one is happy with the main idea of topological partitioning, which is that of QCT, then one can trust IQA and proceed with it. The minimal quantity V_{xc}^{AB} that IQA offers provides a route to cut the vicious circle of interpreting bonding with schemes that impose bonding. Is it not safer to calculate a minimal and physically well-defined quantity and observe the bonding pattern it reveals rather than perpetuate primitive chemical intuition, by deciding *a priori* which interactions are bonds and which are not?

8.6 Conclusion

A relatively underexposed paper published in 1972 marked the birth of a completely novel way of thinking about how to partition a quantum mechanical system such as a molecule. This paved the way to what is often called the *Quantum Theory of Atoms in Molecule* (QTAIM). After an incubation period, the key vision in this 1972 paper developed into a methodology that fully embraced the language of dynamical systems as a vehicle to extract chemical information from wave functions. The idea of using a (gradient) vector field to partition full space into subspaces was initially applied to the electron density, leading to the concept of topological atoms. Stating that these are the atoms of chemistry is perhaps based on a leap of faith, but then one guided by beauty and minimalism.

The idea of "vector field partitioning" remains curiously confined to the field of Chemistry. Instead, one would expect this idea to be more generally applied in completely different knowledge fields, such as sociology or meteorology, given its universality. Still, within chemistry, this partitioning method has been applied to quantum functions other than the electron density, leading to the more general and accurate name "Quantum Chemical Topology (QCT)." Moreover, in the area of high-resolution X-ray crystallography, QCT is now mainstream.

An area of future QCT research is that of intermolecular interactions. Understanding and quantifying these interactions is crucial for the future of Chemistry, which increasingly develops as a science of molecular assembly. This is where the concept of the chemical bond needs to be scrutinized, widened, or even (re)introduced, beyond the limitation of hydrogen bonds or the corset of covalent interactions. QCT provides an appealing potential to tackle the question of detection, classification, and quantification of intermolecular interactions. The reason is that understanding and defining bonding patterns, in the widest sense, must ultimately be based on the atomic partitioning of energy. Topological atoms offer a successful route to guide and execute this partitioning; let us not forget that they deliver well-defined atomic kinetic energies. Moreover, they offer a visual picture that will appeal to those willing to open their mind, and see chemistry as a ballet of open and malleable boxes, with fluctuating shapes of amazing variety.

Acknowledgment

Mark Griffiths is thanked for preparing Figure 8.6 using in-house software [65–67].

References

1. Bader, R.F.W. (1990) *Atoms in Molecules. A Quantum Theory*, Oxford University Press, Oxford.
2. Popelier, P.L.A. (2000) *Atoms in Molecules. An Introduction*, Pearson Education, London.
3. Bader, R.F.W. (1975) Molecular fragments or chemical bonds? *Acc. Chem. Res.*, **8**, 34–40.
4. Bader, R.F.W. (1985) Atoms in molecules. *Acc. Chem. Res.*, **18**, 9–15.
5. Bader, R.F.W. (1991) A quantum-theory of molecular-structure and its applications. *Chem. Rev.*, **91**, 893–928.
6. Popelier, P.L.A., Aicken, F.M., and O'Brien, S.E. (2000) *Chemical Modelling: Applications and Theory*, Royal Socity of Chemistry.
7. Popelier, P.L.A. and Smith, P.J. (2002) in *Chemical Modelling: Applications and Theory*, Vol. 2: Royal Society of Chemistry Specialist Periodical Report, Chapter 8 (ed. A. Hinchliffe), Royal

Society of Chemistry, Cambridge, pp. 391–448.
8. Bader, R.F.W. (2005) The quantum mechanical basis of conceptual chemistry. *Monatsh. Chem.*, **136**, 819–854.
9. Bader, R.F.W. and Popelier, P.L.A. (1993) Atomic theorems. *Int. J. Quantum Chem.*, **45**, 189–207.
10. Matta, C.F. and Boyd, R.J. (2007) *The Quantum Theory of Atoms in Molecules. From Solid State to DNA and Drug Design*, Wiley-VCH Verlag GmbH, Weinheim.
11. Bader, R.F.W. and Matta, C.F. (2013) Atoms in molecules as non-overlapping, bounded, space-filling open quantum systems. *Found. Chem.*, **15**, 253–276. doi: 10.1007/s10698-012-9153-1
12. Popelier, P.L.A. (2005) in *Structure and Bonding. Intermolecular Forces and Clusters*, vol. 115 (ed. D.J. Wales), Springer, Heidelberg, pp. 1–56.
13. Silvi, B. and Savin, A. (1994) Classification of chemical bonds based on topological analysis of electron localization functions. *Nature (London)*, **371**, 683–686.
14. Becke, A.D. and Edgecombe, K.E. (1990) A simple measure of electron localization in atomic and molecular systems. *J. Chem. Phys.*, **92**, 5397–5403.
15. Popelier, P.L.A. and Aicken, F.M. (2003) Atomic properties of amino acids: computed atom types as a guide for future force field design. *ChemPhysChem*, **4**, 824–829.
16. Devereux, M., Popelier, P.L.A., and McLay, I.M. (2009) Towards an ab initio fragment database for bio-isosterism: dependence of qct properties on level of theory, conformation and chemical environment. *J. Comput. Chem.*, **30**, 1300–1318.
17. Popelier, P.L.A. and Brémond, É.A.G. (2009) Geometrically faithful homeomorphisms between the electron density and the bare nuclear potential. *Int. J. Quantum Chem.*, **109**, 2542–2553.
18. Bader, R.F.W. and Beddall, P.M. (1971) The spatial partitioning and transferability of molecular energies. *Chem. Phys. Lett.*, **8**, 29–36.
19. Bader, R.F.W., Beddall, P.M., and Cade, P.E. (1971) Partitioning and characterization of molecular charge disitributions. *J. Am. Chem. Soc.*, **93**, 3095–3107.
20. Bader, R.F.W. and Beddall, P.M. (1972) Virial field relationship for molecular charge distributions and the spatial partitioning of molecular properties. *J. Chem. Phys.*, **56**, 3320–3329.
21. Nasertayoob, P. and Shahbazian, S. (2008) Revisiting the foundations of Quantum Theory of Atoms in Molecules (QTAIM):the variational procedure and the zero-flux conditions. *Int. J. Quantum Chem.*, **108**, 1477–1484.
22. Delle Site, L. (2002) Bader's interatomic surfaces are unique. *Theor. Chem. Acc.*, **107**, 378–380.
23. Bader, R.F.W. (2010) Definition of molecular structure: by choice or by appeal to observation? *J. Phys. Chem. A*, **114**, 7431–7444.
24. Blanco, M.A., Pendas, A.M., and Francisco, E. (2005) Interacting quantum atoms: a correlated energy decomposition scheme based on the quantum theory of atoms in molecules. *J. Chem. Theor. Comp.*, **1**, 1096–1109.
25. Srebrenik, S. and Bader, R.F.W. (1975) Towards the development of the quantum mechanics of a subspace. *J. Chem. Phys.*, **63**, 3945–3961.
26. Koritsanszky, T.S. and Coppens, P. (2001) Chemical applications of x-ray charge-density analysis. *Chem. Rev.*, 1583–1627.
27. Bader, R.F.W., Carroll, M.T., Cheeseman, J.R., and Chang, C. (1987) Properties of atoms in molecules – atomic volumes. *J. Am. Chem. Soc.*, **109**, 7968–7979.
28. Runtz, G.R., Bader, R.F.W., and Messer, R.R. (1977) Definition of bond paths and bond directions in terms of the molecular charge distribution. *Can. J. Chem.*, **55**, 3040–3045.
29. Martm-Pendás, A., Francisco, E., Blanco, M.A., and Gatti, C. (2007) Bond paths as privileged exchange channels. *Chem. Eur. J.*, **13**, 9362–9371.
30. Macchi, P., Proserpio, D.M., and Sironi, A. (1998) Experimental electron

density in a transition metal dimer: metal- metal and metal-ligand bonds. *J. Am. Chem. Soc.*, **120**, 13429–13435.

31. Abramov, Y.A. (1997) On the possibility of kinetic energy density evaluation from the experimental electron density distribution. *Acta Crystallogr.*, **A53**, 264–272.
32. Popelier, P.L.A. (1999) Quantum molecular similarity. 1. BCP space. *J. Phys. Chem. A*, **103**, 2883–2890.
33. Popelier, P.L.A. (2010) in *Quantum Biochemistry: Electronic Structure and Biological Activity*, vol. Part IV (ed. C.F. Matta), Wiley-VCH Verlag GmbH, Weinheim, pp. 669–692.
34. Biegler-Koenig, F.W., Bader, R.F.W., and Tang, T.H. (1982) Calculation of the average properties of atoms in molecules. 2. *J. Comput. Chem.*, **3**, 317–328.
35. Popelier, P.L.A. (2011) Fully analytical integration over the 3D volume bounded by the β sphere in topological atoms. *J. Phys. Chem. A*, **115**, 13169–13179.
36. Polestshuk, P.M. (2013) Accurate integration over atomic regions bounded by zero-flux surfaces. *J. Comput. Chem.*, **34**, 206–219.
37. Popelier, P.L.A. and Kosov, D.S. (2001) Atom-atom partitioning of intramolecular and intermolecular Coulomb energy. *J. Chem. Phys.*, **114**, 6539–6547.
38. Rafat, M. and Popelier, P.L.A. (2007) Atom-atom partitioning of total (super)molecular energy: the hidden terms of classical force fields. *J. Comput. Chem.*, **28**, 292–301.
39. Rafat, M. and Popelier, P.L.A. (2007) in *Quantum Theory of Atoms in Molecules*, vol. 5 (eds C.F. Matta and R.J. Boyd), Wiley-VCH Verlag GmbH, Weinheim, pp. 121–140.
40. Martin-Pendas, A., Blanco, M.A., and Fransisco, E. (2003) Two-electron integrations in the quantum theory of atoms in molecules. *J. Chem. Phys.*, **120**, 4581–4592.
41. Bader, R.F.W., Nguyen-Dang, T.T., and Tal, Y. (1981) A topological theory of molecular-structure. *Rep. Prog. Phys.*, **44**, 893–948.
42. Tognetti, V. and Joubert, L. (2013) On the physical role of exchange in the formation of an intramolecular bond path between two electronegative atoms. *J. Chem. Phys.*, **138**, 024102.
43. Garcia-Revilla, M., Francisco, E., Popelier, P.L.A., and Martin-Pendas, A.M. (2013) Domain-averaged exchange correlation energies as a physical underpinning for chemical graphs. *ChemPhysChem*, **14**, 1211–1218.
44. Grimme, S., Mück-Lichtenfeld, C., Erker, G., Kehr, G., Wang, H., Beckers, H., and Willner, H. (2009) When do interacting atoms form a chemical bond? Spectroscopic measurements and theoretical analyses of dideuteriophenanthrene. *Angew. Chem. Int. Ed.*, **48**, 2592–2595.
45. Cerpa, E., Krapp, A., Vela, A., and Merino, G. (2008) The implications of symmetry of the external potential on bond paths. *Chem. Eur. J.*, **14**, 10232–10234.
46. Cerpa, E., Krapp, A., Flores-Moreno, R., Donald, K.J., and Merino, G. (2009) Influence of endohedral confinement on the electronic interaction between He atoms: A $He_2@C_{20}H_{20}$ case study. *Chem. Eur. J.*, **15**, 1985–1990.
47. Poater, J., Sola, M., and Bickelhaupt, F.M. (2006) Hydrogen–Hydrogen bonding in planar biphenyl, predicted by atoms-in-molecules theory, does not exist. *Chem. Eur. J.*, **12**, 2889–2895.
48. Haaland, A., Shorokhov, D.J., and Tverdova, N.V. (2004) Topological analysis of electron densities: is the presence of an atomic interaction line in an equilibrium geometry a sufficient condition for the existence of a chemical bond? *Chem. Eur. J.*, **10**, 4416–4421.
49. Bader, R.F.W. (2009) Bond paths are not chemical bonds. *J. Phys. Chem. A*, **113**, 10391–10396.
50. Richardson, T.B., de Gala, S., Crabtree, R.H., and Siegbahn, P.E.M. (1995) Unconventional hydrogen bonds: intermolecular B-H.cntdot..cntdot..cntdot.H-N interactions. *J. Am. Chem. Soc.*, **117**, 12875–12876.

51. Popelier, P.L.A. (1998) Characterization of a dihydrogen bond on the basis of the electron density. *J. Phys. Chem. A*, **102**, 1873–1878.
52. Cioslowski, J. and Mixon, S.T. (1992) Topological properties of electron density in search of steric interactions in molecules: electronic structure calculations in ortho-substituted biphenyls. *J. Am. Chem. Soc.*, **114**, 4382–4387.
53. Cioslowski, J., O'Connor, P.B., and Fleischmann, E.D. (1991) Is superbenzene superaromatic? *J. Am. Chem. Soc.*, **113**, 1086–1089.
54. Cioslowski, J. and Mixon, S.T. (1992) Universality among topological properties of electron density associated with the hydrogen-hydrogen nonbonding interactions. *Can. J. Chem.*, **70**, 443–449.
55. Cioslowski, J., Edgington, L., and Stefanov, B.B. (1995) Steric overcrowding in perhalogenated cyclohexanes, dodecahedranes, and [60]fulleranes. *J. Am. Chem. Soc.*, **117**, 10381–10384.
56. Bader, R.F.W. (1998) A bond path: a universal indicator of bonded interactions. *J. Phys. Chem. A*, **102**, 7314–7323.
57. Keith, T.A., Bader, R.F.W., and Aray, Y. (1996) Structural homeomorphism between the electron density and the virial field. *Int. J. Quantum Chem.*, **57**, 183–198.
58. Matta, C.F., Hernandez-Trujillo, J., Tang, T.-H., and Bader, R.F.W. (2003) Hydrogen-hydrogen bonding: a stabilizing interaction in molecules and crystals. *Chem. Eur. J.*, **9**, 1940–1951.
59. Tsirelson, V.G., Zou, P.F., Tang, T.H., and Bader, R.F.W. (1995) Topological definition of crystal-structure – determination of the bonded interactions in solid molecular chlorine. *Acta Crystallogr. A*, **51**, 143–153.
60. Bader, R.F.W. (2006) Pauli repulsions exist only in the eye of the beholder. *Chem. Eur. J.*, **12**, 2896–2901.
61. Morokuma, K. (1971) Molecular orbital studies of hydrogen bonds. III C=O … H-O hydrogen bond in $H_2CO \ldots H_2O$ and $H_2CO \ldots 2H_2O$. *J. Chem. Phys.*, **55**, 1236–1244.
62. Popelier, P.L.A. and Joubert, L. (2002) The elusive atomic rationale for DNA base pair stability. *J. Am. Chem. Soc.*, **124**, 8725–8729.
63. Popelier, P.L.A. (2012) in *Modern Charge-Density Analysis*, vol. 14 (eds C. Gatti and P. Macchi), Springer, pp. 505–526.
64. Popelier, P.L.A. (2012) Quantum chemical topology: knowledgeable atoms in peptides. *AIP Conf. Proc.*, **1456**, 261–268.
65. Popelier, P.L.A. (1996) MORPHY, a program for an automated "atoms in molecules" analysis. *Comput. Phys. Commun.*, **93**, 212–240.
66. Rafat, M., Devereux, M., and Popelier, P.L.A. (2005) Rendering of quantum topological atoms and bonds. *J. Mol. Graphisc Modell.*, **24**, 111–120.
67. Rafat, M. and Popelier, P.L.A. (2007) Visualisation and integration of quantum topological atoms by spatial discretisation into finite elements. *J. Comput. Chem.*, **28**, 2602–2617.
68. Bader, R.F.W., MacDougall, P.J., and Lau, C.D.H. (1984) Bonded and nonbonded charge concentrations and their relation to molecular-geometry and reactivity. *J. Am. Chem. Soc.*, **106**, 1594–1605.
69. Bader, R.F.W., Gillespie, R.J., and MacDougall, P.J. (1988) A physical basis for the VSEPR model of molecular geometry. *J. Am. Chem. Soc.*, **110**, 7329–7336.
70. Popelier, P.L.A. (2000) On the full topology of the laplacian of the electron density. *Coord. Chem. Rev.*, **197**, 169–189.
71. Malcolm, N.O.J. and Popelier, P.L.A. (2003) The full topology of the Laplacian of the electron density: scrutinising a physical basis for the VSEPR model. *Faraday Discuss.*, **124**, 353–363.
72. Malcolm, N.O.J. and Popelier, P.L.A. (2003) An improved algorithm to locate critical points in a 3D scalar field as implemented in the program MORPHY. *J. Comput. Chem.*, **24**, 437–442.
73. Tal, Y., Bader, R.F.W., and Erkku, J. (1980) Structural homeomorphism between the electron density and

the nuclear potential of a molecular system. *Phys. Rev. A*, **21**, 1–11.

74. Polo, V., Andres, J., Berski, S., Domingo, L.R., and Silvi, B. (2008) Understanding reaction mechanisms in organic chemistry from catastrophe theory applied to the electron localization function topology. *J. Phys. Chem. A*, **112**, 7128–7136.
75. Naray-Szabo, G. and Ferenczy, G.G. (1995) Molecular electrostatics. *Chem. Rev.*, **95**, 829–847.
76. Gadre, S.R., Kulkarni, S.A., and Shrivastava, I.H. (1992) Molecular electrostatic potentials: a topographical study. *J. Chem. Phys.*, **96**, 5253–5261.
77. Gadre, S.R. and Shrivastava, I.H. (1991) Shapes and sizes of molecular aniosn via topographical analysis of electrostatic potential. *J. Chem. Phys.*, **94**, 4384–4391.
78. Balanarayan, P. and Gadre, S.R. (2003) Topography of molecular scalar fields. I. Algorithm and Poincare-Hopf relation. *J. Chem. Phys.*, **119**, 5037–5043.
79. Balanarayan, P., Kavathekar, R., and Gadre, S.R. (2007) Electrostatic potential topography for exploring electronic reorganizations in 1,3 dipolar cycloadditions. *J. Phys. Chem. A*, **111**, 2733–2738.
80. Aray, Y., Rodriguez, J., Coll, S., Rodrı́guez-Arias, E.N., and Vega, D. (2005) Nature of the lewis acid sites on molybdenum and ruthenium sulfides: an electrostatic potential study. *J. Phys. Chem. B*, **109**, 23564–23570.
81. Tsirelson, V.G., Avilov, A.S., Lepeshov, G.G., Kulygin, A.K., Stahn, J., Pietsch, U., and Spence, J.C.H. (2001) Quantitative analysis of the electrostatic potential in rock-salt crystals using accurate electron diffraction data. *J. Phys. Chem. B*, **105**, 5068–5074.
82. Keith, T.A. and Bader, R.F.W. (1993) Topological analysis of magnetically induced molecular current distributions. *J. Chem. Phys.*, **99**, 3669–3682.
83. Cioslowski, J. and Liu, G.H. (1999) Topology of electron-electron interactions in atoms and molecules. II. The correlation cage. *J. Chem. Phys.*, **110**, 1882–1887.
84. Pendas, A.M. and Hernandez-Trujillo, J. (2012) The Ehrenfest force field: topology and consequences for the definition of an atom in a molecule. *J. Chem. Phys.*, **137**, 134101.
85. Pendas, A.M., Francisco, E., Blanco, M.A., and Gatti, C. (2007) Bond paths as privileged exchange channels. *Chem. A Eur. J.*, **13**, 9362–9371.
86. Webster, B. (1990) *Chemical Bonding Theory*, Blackwell, Oxford.
87. Reed, A.E., Curtiss, L.A., and Weinhold, F. (1988) Intermolecular interactions from a natural bond orbital, donor-acceptor viewpoint. *Chem. Rev.*, **88**, 899–926.
88. Stone, A.J. (1981) Distributed multipole analysis, or how to describe a molecular charge-distribution. *Chem. Phys. Lett.*, **83**, 233–239.
89. Mulliken, R.S. (1955) Electronic population analysis on LCAO-MO molecular wave functions. I. *J. Chem. Phys.*, **23**, 1833–1840.
90. Pearson, R.G. (2007) Applying the concepts of density functional theory to simple systems. *Int. J. Quantum Chem.*, **108**, 821–826.
91. Kovacs, A., Esterhuysen, C., and Frenking, G. (2005) The nature of the chemical bond revisited: an energy-partitioning analysis of nonpolar bonds. *Chem. Eur. J.*, **11**, 1813–1825.
92. McWeeny, R. (1992) *Methods of Molecular Quantum Mechanics*, 2nd edn, Academic Press, SanDiego, CA.
93. Cohen, L. (1978) Local kinetic energy in quantum mechanics. *J. Chem. Phys.*, **70**, 788–799.
94. Bader, R.F.W. and Preston, H.J.T. (1969) The kinetic energy of molecular charge distributions and molecular stability. *Int. J. Quantum Chem.*, **3**, 327–347.
95. Popelier, P.L.A. (1994) An analytical expression for interatomic surfaces in the theory of atoms in molecules. *Theor. Chim. Acta*, **87**, 465–476.
96. Popelier, P.L.A. (1996) On the differential geometry of interatomic surfaces. *Can. J. Chem.*, **74**, 829–838.
97. Bader, R.F.W. and Essen, H. (1984) The characterization of atomic interactions. *J. Chem. Phys.*, **80**, 1943–1960.

98. Cremer, D. and Kraka, E. (1984) Chemical bonds without bonding electron density - does the difference electron-density analysis suffice for a description of the chemical bond? *Angew. Chem.*, **23**, 627–628.
99. Gillespie, R.J. (1972) *Molecular Geometry*, Van Nostrand Reinhold, London.
100. Gillespie, R.J. (1998) The physical basis of the VSEPR model. *Struct. Chem.*, **9**, 73–76.
101. Bader, R.F.W. and Stephens, M.E. (1975) Spatial localization of the electronic pair and number distributions in molecules. *J. Am. Chem. Soc.*, **97**, 7391–7399.
102. Bader, R.F.W. and MacDougall, P.J. (1985) Toward a theory of chemical-reactivity based on the charge-density. *J. Am. Chem. Soc.*, **107**, 6788–6795.
103. Bader, R.F.W., Popelier, P.L.A., and Chang, C. (1992) Similarity and complementarity in chemistry. *J. Mol. Struct. (THEOCHEM)*, **255**, 145–171.
104. Malcolm, N.O.J. and Popelier, P.L.A. (2001) On the full topology of the Laplacian of the electron density II: umbrella inversion of the ammonia molecule. *J. Phys. Chem. A*, **105**, 7638–7645.
105. Popelier, P.L.A., Burke, J., and Malcolm, N.O.J. (2003) Functional groups expressed as graphs extracted from the laplacian of the electron density. *Int. J. Quantum Chem.*, **92**, 326–336.
106. Malcolm, N.O.J. and Popelier, P.L.A. (2003) An algorithm to delineate and integrate topological basins in a three-dimensional quantum mechanical density function. *J. Comput. Chem.*, **24**, 1276–1282.

9
The Experimental Density Perspective of Chemical Bonding

Wolfgang Scherer, Andreas Fischer, and Georg Eickerling

9.1
Introduction

The experimental feasibility of measuring and exploring charge density distributions of molecules and solids was first proposed in 1921 by Bragg [1]. In the light of the enormous experimental complexity and instrumental demands of charge density studies, it is quite surprising that this proposal came up 8 years after the pioneering X-ray diffraction experiments of Friedrich *et al.* [2]. At that time, the occurrence of *atomic regularity in solids* was proved, but the mapping of electron density distributions was clearly beyond the scope of imagination. However, during this pioneering period Bragg "observed" the symmetry-forbidden (222) reflection of diamond by an ambitious X-ray single-crystal study, which should be absent in case of atoms displaying merely spherical-symmetric charge density distributions. Accordingly, Bragg concluded that the charge density distribution of atoms in solids and molecules is distorted as a consequence of chemical bond formation: "*It is necessary, therefore, to suppose that the attachment of one atom to the next is due to some directed property, and that the carbon atom has four such special directions: as indeed the tetra-valency of the atom might suggest*" [1]. In this respect, Bragg provided the first experimental support for Debye's prediction in 1915 that "*it should be possible to determine the arrangement of the electrons in atoms*"[1)] [3]. However, a few years later, theoretical structure factor calculations by Ewald and Hönl [4] and the discovery of *Umweganregungen* (multiple scattering) in diamond by Renninger [5] questioned Bragg's observation of a pronounced (222) reflection. Indeed, recent charge density studies of diamond [6] using high-resolution powder diffraction data of a third-generation synchrotron source [7] reveal that the intensity of the (222) reflection in diamond was overestimated by Bragg by about one order of magnitude – most likely due to multiple scattering (Figure 9.1). Thus, the example of diamond demonstrates the challenge to experimentally determine precise charge density distributions. However, owing to the tremendous achievements in the

1) Translated from German: "[...] auf diesem Wege sollte es dann gelingen, die besondere Anordnung der Elektronen im Atom experimentell festzustellen."

The Chemical Bond: Fundamental Aspects of Chemical Bonding, First Edition.
Edited by Gernot Frenking, Sason Shaik.

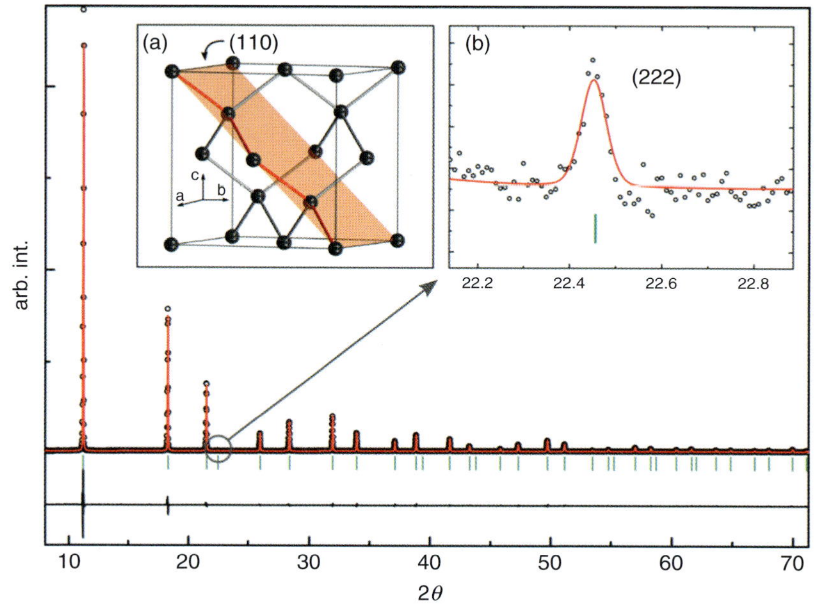

Figure 9.1 High-resolution powder diffraction data of diamond determined at Spring-8 [7, 9] and fitted by simultaneous Rietveld and *extended Hansen–Coppens* (EHC) multipolar refinements. Observed (black circles), calculated (solid red line), and difference pattern (solid black line, bottom). Inset (a) shows the diamond crystal lattice and the highlighted (110) plane containing a zigzag chain of carbon atoms. This is our reference plane in all subsequent contour plots of diamond. Inset (b) shows the accuracy of the fit of the (222) reflection which is "symmetry forbidden" in the case of the standard independent atoms model (IAM). The nonzero Bragg intensity of the weakly pronounced (222) reflection provides a direct and model-free measure of the asphericity of the charge density deformation at the carbon atoms due to chemical bonding. (Adapted from Ref. [6].)

recent developments of high-flux X-ray sources and in the design of highly efficient and precise area detectors, charge density determinations of organic and inorganic compounds can be routinely performed in many cases and can be considered an analytical technique. Indeed, "charge densities come of age" as pointed out by Coppens in 2005 [8].

In combination with Bader's *Quantum Theory of Atoms in Molecules* (QTAIM) [10], charge density analyses can be directly employed to retrieve physical properties of molecules and solids to investigate chemical bonding scenarios in detail [11–13]. In contrast to the charge density – a physically observable property – the direct observation of wave functions is prevented by their complex nature. However, it remains to be seen whether "constrained wave functions" [14–18], which are restrained by experimental structure factor amplitudes, might provide a more direct linkage between theoretical and experimental electronic structure analyses in future. In combination with spin-polarized neutron diffraction experiments [19] or resonant X-ray scattering [20], spin density distributions are also experimentally accessible, which might provide complex electronic information, for example, on relativistic effects, spin and electron correlation, and the admixture of excited states. The most recent developments in this field were pioneered by Deutsch *et al.* [21], who could demonstrate that spin-dependent electron density distributions can be determined to very high precision by joint multipolar refinements of X-ray and polarized neutron diffraction data. Hence, the precision of experimentally derived spin, charge, and momentum density distributions might furnish suitable reference maps for theoretical calculations and eventually inspire the development of more advanced density functional theory (DFT) functionals for the study of highly correlated electronic systems in extended solids. The present frontier of research in experimental charge density studies also encloses experiments performed at subatomic resolution, at short time scales and/or aiming at excited states [6, 22, 23]. The possibility of obtaining experimental charge density distributions of matter in electronically excited states (e.g., in photochemically active systems) or to study the electronic structure of molecular crystals or solids at a femtosecond timescale has been demonstrated recently [24, 25].

9.2 Asphericity Shifts and the Breakdown of the Standard X-ray Model

As outlined in the subsequent section, a true deconvolution of thermal smearing and chemical bonding effects is the prerequisite to retrieve reliable structural parameters and thermal displacement parameters of molecules and solids at high precision. This is illustrated in Figure 9.2, where the C–C bond distances of several molecules, complexes, and solids **1–5** are depicted. These systems are compared with our reference models for a C–C single bond (diamond), double bond (ethylene), and triple bond (acetylene). Note that the distance parameters of our reference systems are model independent, as they rely only on the precision of (i) the lattice parameter determination (diamond) or (ii) the accuracy in the determination of IR/Raman frequencies (ethylene, acetylene) [26, 27]. In the case of the more complex systems **1–5**, no model-free structural parameters are accessible. Furthermore, the geometrical parameters of the molecular systems are affected by crystal packing effects, which do not allow an unbiased geometrical comparison between the molecules in condensed phase (experiment) and their theoretical models assuming isolated noninteracting molecules. The effects of crystal packing on the geometrical parameters [28], the electron density distribution [29–31] and physical properties [32, 33] of molecules in the condensed phase have been very recently reviewed by Spackman [34] and are therefore not further discussed in this contribution. In the case of standard X-ray diffraction experiments, the situation becomes further complicated by systematic errors owing to the so-called asphericity shifts [35–38].

The origin of these asphericity shifts can be illustrated by inspecting the C–C bond distances in $[Ag(\eta^2\text{-}C_2H_2)]^+[A']$ **5**, a model system for a silver acetylene complex displaying a weakly coordinating anion ($A' = [Al(OC(CH_3)(CF_3)_2)_4]^-$) [44, 45]. The analysis of X-ray diffraction data using the standard *independent*

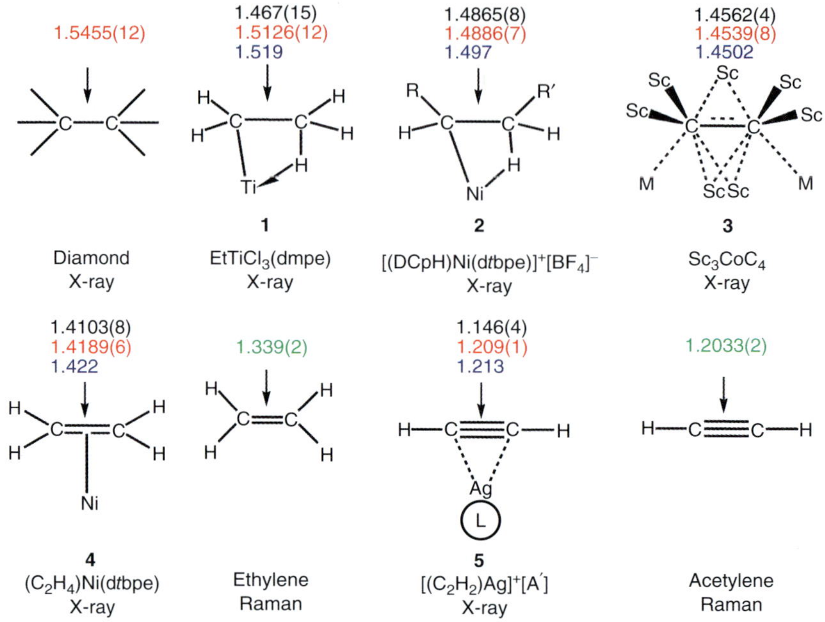

Figure 9.2 Comparison of C–C bond distances obtained from the IAM model (black values), the HC multipolar model (red values) and DFT calculations (blue values) for benchmarks **1** [13, 39], **2** [40], **3** [41, 42], **4** [43], and **5** [44, 45]. Model-free C–C bond lengths are specified for diamond [6], ethylene [27] and acetylene [26] (green values).

atom model (IAM) yields an unreasonably short C–C bond distance of 1.146(4) Å for the coordinated acetylene moiety, which is significantly shorter than the value of the free ligand (1.2033(2) Å)[2)] [26]. On the other hand, DFT calculations predict the elongation of the C–C triple bond upon coordination of the acetylene moiety in **5** (1.213 Å). This artificial discrepancy between experimental and theoretical C–C bond lengths represents an extreme case of an asphericity shift. In the case of our benchmark **5**, the underlying phenomenon can be explained by the pronounced density accumulation in the bonding domain of covalent C–C triple bonds. This interaction density cannot be accounted for by the standard X-ray model of independent (pseudo) atoms (IAM) displaying spherically averaged density distributions. Hence, the resulting residual density is (partially) compensated for in the least-squares refinements by an artificial shortening of the C–C distance (and an erroneous modeling of the thermal displacement parameters). This artificial shorting of the C–C bonds is, however, prevented by standard multipolar refinements that fully account for the density accumulation between the carbon atoms and thus yield a C–C bond length of 1.209(1) Å in fine agreement with the theoretical model (1.213 Å) – if we assume that crystal packing plays a minor role in **5**.

2) The corresponding value of acetylene (C–C = 1.193(6) Å) in the condensed phase (neutron diffraction study at 15 K; Ref. [139]) has unfortunately not been precisely determined owing to the low resolution of the experimental data ($\sin\theta/\lambda \leq 0.74$ Å^{-1}).

The presence of asphericity shifts is a well-known phenomenon that was first identified and explained in terms of a charge density model in the 1950s [35, 36]. These studies were, however, based on a direct comparison of IAM refinement results with those obtained from alternative structural determination techniques (gas electron diffraction, neutron diffraction, microwave or IR/Raman spectroscopy) – long before the advent of reliable computational methods [35–38]. Hence, asphericity shifts provided – besides the presence of the (222) reflection in diamond (see preceding text) – another univocal experimental evidence for the aspherical nature of atoms in molecules and solids. Apparently, more sophisticated models that are capable of modeling aspherical density distributions of atoms due to chemical bonding are needed. In the next section we outline that multipolar models such as the standard *Hansen–Coppens* (HC) model are highly suitable to eliminate or to minimize asphericity shifts. The performance of the experimental HC model in deriving reliable structural parameters is already documented at this stage, and can be seen in Figure 9.2. The multiple bond character of the C–C bonds increases systematically in **1–5** and the C–C bond distances are systematically larger in the HC model with respect to the IAM model, owing to the proper description of aspherical electron density. These examples clearly underpin the necessity to establish comparisons between theoretical and experimental structural data exclusively on the basis of multipolar refinements when the experimental model relies only on X-ray diffraction data. Finally, we note that in the case of the transition metal carbide Sc_3CoC_4, the differences between the IAM and the HC model become rather negligible [41, 42]. This appears to be a general phenomenon and is caused by the reduced thermal motion of the atoms in close-packed and rigid solid state compounds, allowing a more reliable refinement of bond lengths unbiased by thermal smearing.

9.3
Precision of Charge Density Distributions in Experimental and Theoretical Studies

In addition to spectroscopic measurements, the experimental charge density as a physical observable property provides an ideal benchmark to probe the precision of theoretical methods and experimental approaches in real space. Therefore, it is not surprising that the comparison of topological features of charge density distributions became a routine technique to identify systematic experimental errors (e.g., extinction, absorption, crystal decay, and imperfections) or to trace deficiencies in the level of approximation used in theory (e.g., frozen-core approximation, treatment of electronic correlation, and relativistic effects). Furthermore, the static charge density allows a direct comparison between theory and experiment in solids, as it can be extracted from the experimental structure factors, F_{exp}, via multipolar refinements using, for example, the standard HC model (Boxes 9.1–9.3). The charge density is also readily available from model wave functions (WFN) based on theoretical methods.

Box 9.1 The Spherical Atom Kappa Formalism

Already in 1977, a multipolar formalism was suggested by Stewart [46] to model the electron density distribution of crystalline matter. This approach relies on a decomposition of the total charge density into pseudo atoms. Nowadays, most experimental charge density studies are based on the related HC multipolar model [47], which provides an efficient and robust frozen-core algorithm to extract charge density distributions $\rho(\mathbf{r})$ from experimental X-ray diffraction data. In the following, the basic concept of this standard and popular multipolar model is briefly introduced. We first focus on the spherical-atom kappa formalism [48, 49], which provides perhaps the simplest Ansatz to overcome the severe limitations of the IAM. This formalism accounts for the charge transfer between atoms in solids or molecules in condensed phase by the following definition of pseudo atoms:

$$\rho_{\text{at}}(r) = \rho_{\text{core}}(r) + P_{\text{v}}\kappa^3\rho_{\text{valence}}(\kappa, r) \tag{9.1}$$

While the spherical core density (first term of Eq. (9.1)) is kept frozen, the spherical valence density (second term of Eq. (9.1)) is used to account for the charge transfer between atoms by its occupation factor P_{v}. The radial expansion/contraction of each pseudo atom as a consequence of the charge transfer is modeled by the κ-parameter in Eq. (9.1). The κ-formalism therefore simply relies on the original Slater concept [50]. Accordingly, the screening parameter κ fine-tunes the ζ-exponents of the radial part of the Slater-type orbitals (STOs) of the individual pseudo atoms (Eq. (9.2)).

$$R(\kappa, r) = Nr^{(n-1)}e^{-\kappa\xi r} \tag{9.2}$$

A value of $\kappa > 1$ thus indicates a radial contraction, while a value of $\kappa < 1$ is signaling a radial expansion of the atomic valence electron density distribution with respect to the corresponding neutral atom in its ground state (N is a normalization factor, while n represents the main quantum number in Eq. (9.2)). In recent databank files of multipolar models, the unperturbed core density distribution (first term in Eq. (9.1)) is usually taken from atomic ground state wave functions of the respective atoms. In the case of the popular Volkov–Macchi (VM) bank file as implemented in the XD program [51], the core and valence density terms are represented by the same quadruple-ζ STOs as implemented in the Amsterdam density functional (ADF) program [52–54]; hence, comparisons between theoretical and experimental charge density distributions can be performed using the same STO expansion in both cases [51].

The general suitability of the HC model to extract electron density distributions from experimental data has been documented in numerous case studies; see, for example, Refs. [13, 17, 55–62]. However, a more sophisticated evaluation of the performance of the standard multipolar model can be obtained by comparing density distributions obtained (i) directly from wave functions (e.g., $\rho(\mathbf{r})_{\text{LAPW}}$ based on the linearized augmented planewave (LAPW) method) and (ii) multipolar refinements using the corresponding theoretical structure factors, F_{LAPW}, (e.g., $\rho(\mathbf{r})_{\mathrm{F,LAPW}}$). In the following examples, we used periodic full-potential LAPW DFT calculations to compute the LAPW wave function and the corresponding theoretical structure factors F_{LAPW} for diamond.

Box 9.2 The Multipolar Expansion of the Electron Density

The κ-model (as introduced in Box 9.1) still does not account for the deformation of the valence density due to chemical bonding. This is, however, accomplished by the more advanced HC-multipole model, which expands the total electron density distribution also in terms of pseudo atoms, $\rho_{\mathrm{at}}(\mathbf{r})$, but extends Eq. (9.1), by a third term that models the aspherical contributions to the valence density (deformation density):

$$\rho_{\mathrm{at}}(\mathbf{r}) = P_{\mathrm{c}}\rho_{\mathrm{core}}(r) + P_{\mathrm{v}}\kappa^{3}\rho_{\mathrm{valence}}(\kappa, r) + \sum_{l=0}^{l_{\max}}(\kappa_l')^3 R_l(\kappa_l', r)\sum_{m=0}^{l} P_{lm\pm} d_{lm\pm}(\theta, \phi) \quad (9.3)$$

The valence deformation density is expanded in density functions consisting of a radial part $R_l(\kappa_l', r)$ and the density-normalized spherical harmonic functions $d_{lm}(\theta,\phi)$. In contrast to the core and spherical valence density, the radial functions $R_l(\kappa_l', r)$ are taken as nodeless single-ζ functions of the type

$$R_l(\kappa_l', r) = \frac{\xi^{n_l+3}}{(n_l+2)!}(\kappa_l')^3(\kappa_l' r)^{n_l} e^{-\kappa_l' \xi r} \quad (9.4)$$

by introducing a second, l-dependent radial scaling factor κ_l' for the deformation density and the coefficients n_l that have to fulfill the condition $n_l \geq l$ [46]. The ζ-exponents of these functions are often simply based on the energy-optimized single-ζ functions derived by Clementi and Raimondi, which need to be multiplied by a factor of 2 in order to be used as exponents for the density functions $R_l(\kappa_l', r)$ [63].

The core and valence occupation factors, P_{c} and P_{v} in Eq. (9.3), account for the electronic configuration of the individual pseudo atoms and the charge transfer between them in the same manner as already defined in the κ-formalism (Box 9.1).

Finally, the renormalized real spherical harmonics in Eq. (9.3) are derived from the associated Legendre functions $L_{lm}(\cos\theta)$ as

$$d_{lm+} = N_{lm} L_{lm}(\cos\theta) \cdot \cos m\phi \quad (9.5)$$

$$d_{lm-} = N_{lm} L_{lm}(\cos\theta) \cdot \sin m\phi \quad (9.6)$$

and renormalized according to

$$\int |d_{lm}| d\Omega = 2 \quad \text{for } l > 0 \quad (9.7)$$

$$\int |d_{lm}| d\Omega = 1 \quad \text{for } l = 0 \tag{9.8}$$

This approach adopts the spherical harmonics for the representation of the deformation electron density distribution relative to the spherical valence density contribution $\rho_{\text{valence}}(\mathbf{r})$. The term d_{00} introduces a spherically symmetric contribution to the deformation density term and might correlate strongly with the spherical valence density (second term in Eq. (9.3)). Indeed, this term is often omitted when modeling experimental diffraction data, but may be used to introduce additional radial flexibility into the model.

The occupation parameters $P_{lm\pm}$ can be adopted to the experimental data by least-squares refinements and represent a weighting of the individual density-normalized spherical harmonic functions $d_{lm}(\theta, \phi)$, which describe the aspherical features of the valence deformation density as exemplified in Box 9.3. As outlined in Box 9.4, the standard HC model can be further expanded in case of first row elements to include effects similar to the contraction/expansion of the inner (i.e., nonvalence) shell by including an additional radial scaling parameter κ_{core} and the core population P_c into the list of refinable parameters (Eq. (9.9)). The latter model will be denoted as the extended Hansen–Coppens (EHC) model [6] in the following.

$$\rho_{\text{at}}(\mathbf{r}) = P_c \kappa_{\text{core}}^3 \rho_{\text{core}}(\kappa_{\text{core}}, r) + P_v \kappa^3 \rho_{\text{valence}}(\kappa, r) + \sum_{l=0}^{l_{\max}} (\kappa')^3 R_l(\kappa', r) \sum_{m=0}^{l} P_{lm\pm} d_{lm\pm}(\theta, \phi) \tag{9.9}$$

These calculations were performed using the Elk program [64] and the Perdew–Burke–Ernzerhof (PBE) generalized gradient approximation (GGA) functional [65]. A direct comparison of the $\rho(\mathbf{r})_{\text{LAPW}}$ and $\rho(\mathbf{r})_{\text{F,LAPW}}$ maps should therefore reveal potential deficiencies of the multipolar model and of the data resolution as the same theoretical methodology has been employed to compute the LAPW wave function and the F_{LAPW} structure factors.

Box 9.3 Modeling the Deformation Density – a Practical Example

In this example, the valence deformation density of a hypothetical carbon atom displays an aspherical valence density, which has been polarized along the z-axis by bond formation with a neighboring atom. Such polarization is modeled in our simplistic example by populating the following density-normalized spherical harmonic functions (denoted as "multipoles" in the following): d_{10} with $P_{10} = +0.3$ and d_{20} with $P_{20} = +0.3$ (Figure a). The linear combination of the dipole (d_{10}) and the quadrupole (d_{20}) yields a composed density function ($0.3d_{10} + 0.3d_{20}$), which displays a pronounced (i) dipolar polarization along and (ii) a quadrupolar polarization parallel and perpendicular to the *z*-direction. Adding these deformation terms to the spherical valence density (Figure b) formed by the four valence electrons of our hypothetical carbon atom, a nonspherical deformation of the electron density results. According to the definition of the employed multipole model, a positive value of $P_{lm\pm}$ leads to a shift of the electron density from the spatial regions of the negative lobes of the $d_{lm\pm}$ multipoles into the regions of the positive lobes and vice versa. Note that the spherical valence electron density $\rho_{\text{valence}}(\mathbf{r})$ is positive per definition, as it is obtained by the squared STOs employed. Adding the resulting deformation density to the undistorted atomic density of the carbon 1s electrons ($\rho_{\text{core}}(\mathbf{r})$) yields the total, nonspherical electron density of our hypothetical carbon pseudo atom $\rho_{\text{at}}(\mathbf{r})$.

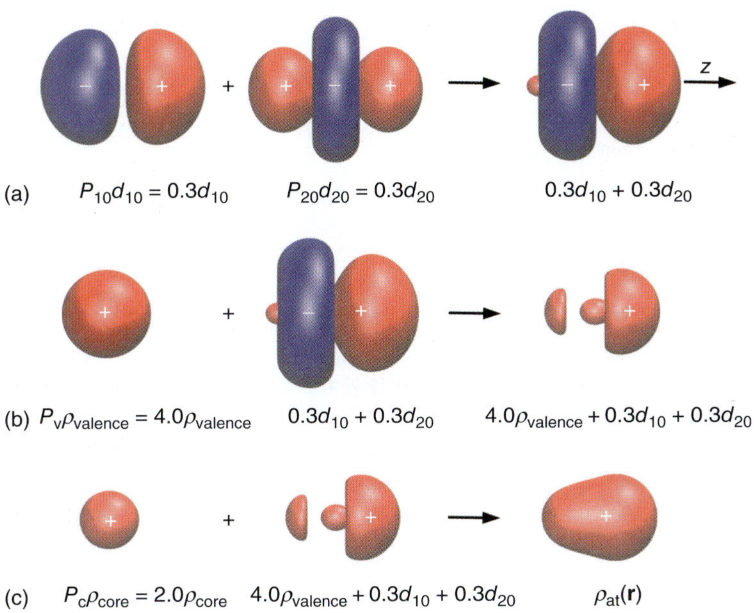

Figure a–c: Sequence of isosurface plots demonstrating the reconstruction of the aspherical charge density features of a hypothetical carbon atom displaying a polarized valence charge density along the *z*-axis. (a) Linear combination of the dipole d_{10} and the quadrupol d_{20} (blue and red color coding relates to positive/negative signs of the lobes of the individual multipoles, respectively, isosurface values ± 0.01 Bohr^{-3}); (b) deformation of the spherical valence electron density $\rho_{\text{valence}}(\mathbf{r})$ (isosurface value 0.2 e Bohr^{-3}) by a superposition with the composite multipoles d_{10} and d_{20}; and (c) reconstruction of $\rho_{\text{at}}(\mathbf{r})$ by adding the resulting (positive) deformation density to the undistorted core density $\rho_{\text{core}}(\mathbf{r})$.

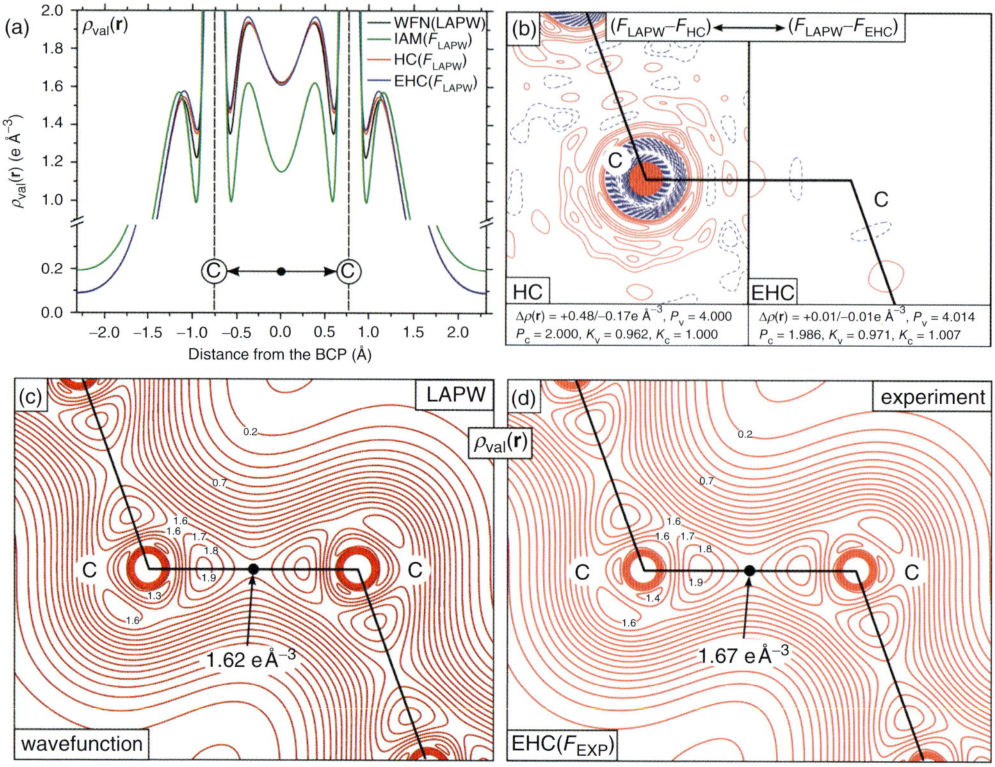

Figure 9.3 (a) Comparison of the theoretical valence density distribution along the C–C bond in diamond as derived (i) directly from the LAPW DFT wave function and by refinements of theoretical structure factors (F_{LAPW}) using (ii) the standard IAM X-ray model, (iii) the standard HC multipolar model, and (iv) the more flexible extended HC (EHC) model [6]; (b) Residual density maps, $\Delta\rho(\mathbf{r})$ ($F_{LAPW} - F_{HC}$), in the (110) lattice plane of diamond revealing the limited flexibility of the HC multipolar model in fitting the theoretical structure factors of diamond in the core region of the carbon atoms; contour step widths 0.01 e Å^{-3}; positive/negative contour values are marked by solid/broken lines. 2D valence density distributions of diamond in the (110) lattice plane based (c) on the LAPW wave function and (d) on the EHC model employing experimental structure factors (F_{exp}); contour step width 0.1 e Å^{-3}. Both multipolar models employ scattering factors of a carbon atom in its prepared valence ^{5}S state [66, 67].

This is illustrated in Figure 9.3a showing the valence density, $\rho(\mathbf{r})_{val}$, in diamond along one C–C single bond as derived (i) directly from the wave function (WFN LAPW), and alternatively by refinements of theoretical structure factors, F_{LAPW}, using (ii) the standard IAM X-ray model, (iii) the standard HC multipolar model, and (iv) a more flexible EHC model.

Figure 9.3a reveals the dramatic failure of the standard IAM in fitting the C–C bonding density as it is based on noninteracting atoms and thus ignores chemical bonding effects. Unfortunately, the IAM still represents the workhorse of structural determinations using X-ray diffraction techniques. Therefore, it has been employed in more than 99.9% of all published X-ray diffraction studies despite its inherent systematic deficiencies, which might cause significant errors in the determination of atomic coordinates (e.g., because of asphericity shifts) [35–38] and atomic displacement (thermal) parameters. Indeed, Figure 9.3a demonstrates the much better performance of the HC model with respect to the IAM in describing the valence density in the C–C bonding regime. We note, however, already at this stage that the "frozen-core" approximation of the HC model prevents a similar successful density description in the core and cusp regions of atoms. But this limitation can be overcome using the more flexible EHC model, which also accounts for density deformation in the core regions of atoms (Boxes 9.2 and 9.4). The increased flexibility of this model is best demonstrated by inspection of the residual density maps, $\Delta\rho(\mathbf{r})$, in the (110) lattice plane of diamond (Figure 9.3b). These clearly document the excellent performance of the EHC multipolar model to precisely recover the density distribution also in the core and cusp regions of the carbon atoms of diamond. Hence, the discrepancies between the theoretical structure factors and the ones fitted by the EHC model ($F_{\text{LAPW}} - F_{\text{EHC}}$) become small enough to yield rather featureless residual density maps with $\Delta\rho(\mathbf{r}) = \pm 0.01\,\text{e}\,\text{Å}^{-3}$ (at a resolution of $\sin\theta/\lambda \leq 2.0\,\text{Å}^{-1}$). Furthermore, Figure 9.3c,d shows the excellent agreement of the 2D valence density maps directly obtained from the LAPW wave function and from the experimental F_{exp} structure factors by employing the EHC model.

Another way to test the performance of the HC and the EHC model relies on the comparison of experimental and theoretical charge density maps using the QTAIM method. Such analysis is suitable to reveal even subtle differences by comparing the electronic structures of different theoretical models and allows an identification of potential deficiencies of experimental and/or theoretical approaches. Such a case study is shown in Figure 9.4, where various experimental and theoretical contour maps of the negative Laplacian of the electron density, $L(\mathbf{r}) = -\nabla^2\rho(\mathbf{r})$, in the (110) lattice plane of diamond are depicted. In the following discussion, we use the experimental EHC model (Figure 9.4g) and a theoretical model employing a flexible LAPW wave function (Figure 9.4f) as our experimental and theoretical references, respectively.

We first inspect the topology of the total electron density as obtained from the experimental IAM, HC, and EHC models (Figure 9.4a,d,g) to quantify the rudimental density distribution as obtained by the interaction-free IAM. Indeed, the density at the bond critical point (BCP) of the covalent C–C bond in diamond is dramatically underestimated in comparison with the HC model ($\rho(\mathbf{r}) = 1.15$ vs $1.68\,\text{e}\,\text{Å}^{-3}$, respectively). The IAM also causes an artificial lack of charge concentration (CC) in the C–C bonding domain, which is reflected by a positive sign and small value of the Laplacian at the BCP: $\nabla^2\rho(\mathbf{r}) = 1.92$ versus $-14.9\,\text{e}\,\text{Å}^{-5}$, for the IAM and HC model, respectively (Figure 9.4a,d). Further increasing the flexibility of the standard HC model, that is, by employing the EHC model, lowers

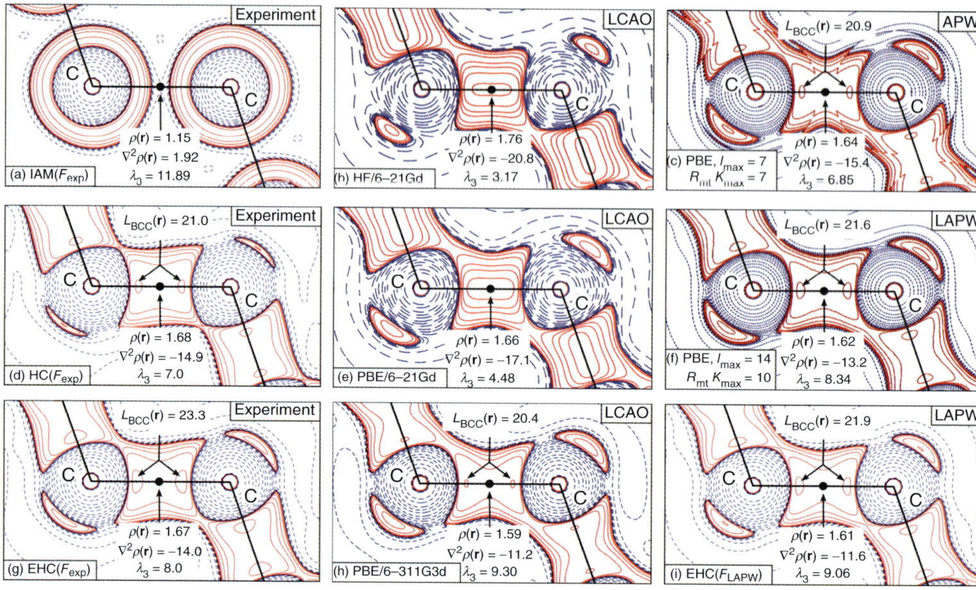

Figure 9.4 Experimental and theoretical $L(\mathbf{r}) = -\nabla^2\rho(\mathbf{r})$ maps, in the (110) lattice plane of diamond; negative (blue, dashed) and positive (red, solid) contour lines were drawn at $\pm 2.0\times 10^n$, $\pm 4.0\times 10^n$, $\pm 8.0\times 10^n$ e Å^{-5} with $n = \pm 3, \pm 2, \pm 1, 0$. The first column shows the systematic improvement of salient experimental topological parameters at the C–C bond critical point (BCP) and of the bonded charge concentration in the valence shell of the carbon atom. The $\rho(\mathbf{r})$ values are specified in e Å^{-3}, whereas $\nabla^2\rho(\mathbf{r})$, $L(\mathbf{r})$, and λ_3 are given in e Å^{-5}; λ_3 reports the positive curvature of the density at the C–C BCP. The second and third columns show the Laplacian maps based on periodic LCAO calculations using the Crystal06 code [68] and the periodic full-potential (L)APW DFT method. (Figure 9.4d,g were adapted from Ref. [6].)

$\rho(\mathbf{r}_{\mathrm{BCP}})$ only slightly to our experimental reference value[3] of 1.67 e Å^{-3}. Accordingly, the differences between both models are mainly manifested in the core and cusp density distribution, where the frozen-core approach of the HC model causes severe residual densities, as outlined earlier (Figure 9.3b).

The electron density distributions in diamond obtained from different *ab initio* approaches employing varying levels of approximation and different basis sets also reveal a systematic trend if the influence of (i) the model Hamiltonian and that of (ii) the basis-set flexibility on the topology of $\rho(\mathbf{r})$ is analyzed independently. In comparison with our experimental EHC reference system (Figure 9.4g) a periodic Hartree–Fock (HF)/6-21G(d) calculation seems to *overestimate* $\rho(\mathbf{r}_{\mathrm{BCP}})$ significantly (1.76 e Å^{-3}) (Figure 9.4b). Also, the topology of $L(\mathbf{r})$ appears to be unreasonable, as at this level of approximation no maximum of the valence shell charge concentration

3) We note that in the original publication [6] an even lower value of 1.63 eÅ^{-3} was found. This value turned out to be underestimated by a programming error in the property module of the JANA2006 program. This has been corrected by the authors of the program in the meantime.

(VSCC) is found along the C–C bond. The presence of such a bonded charge concentration (BCC) is a characteristic indicator of covalent bonding between *sp*-block elements. This obvious deficiency of the HF/6-21G(d) model appears to be a consequence of the rather limited basis-set size employed [69, 70]. Similar conclusions were derived from a systematic study on the basis-set choice in periodic HF calculations on molecular crystals by Spackman *et al.* [70]. Hence, the $L(\mathbf{r})$ topology does not improve significantly by taking electronic correlations via DFT calculation (PBE/6-21G(d) into account. However, this approach clearly *reduces* $\rho(\mathbf{r}_{BCP})$ to 1.66 e Å^{-3}, which is already close to our experimental reference value, 1.67 e Å^{-3}, which naturally incorporates electronic correlation and relativistic effects (especially for heavier elements). The reduction of $\rho(\mathbf{r}_{BCP})$ owing to the inclusion of correlation effects is a well-known phenomenon and has been studied in detail for a large variety of molecular benchmark systems by employing various model Hamiltonians; see, for example, the pioneering work by Gatti, McDougall, and Bader in 1988 and subsequent studies [71–74]. These studies document that HF/DFT tend to overestimate/underestimate $\rho(\mathbf{r}_{BCP})$ with respect to MP2 and CCSD (coupled-cluster with singles and doubles) calculations, respectively.[4] Increasing the size of the basis set to, for example, 6-311G(3d) (Figure 9.4h), PBE calculations further *reduce* $\rho(\mathbf{r}_{BCP})$ to 1.59 e Å^{-3}, which again demonstrates the underestimation of $\rho(\mathbf{r}_{BCP})$ by DFT for reasonable basis-set sizes with respect to our experimental reference value. The latter observation also holds true for (L)APW basis sets, as demonstrated in Figure 9.4c,f. Employing a standard APW basis ($l_{max} = 7/R_{mt}k_{max} = 7$)[5] $\rho(\mathbf{r}_{BCP})$ is found to be 1.64 e Å^{-3}, which is close to our experimental reference value but biased by limited basis-set flexibility.[6] Hence, increasing the flexibility of our model (LAPW with $l_{max} = 14/R_{mt}k_{max} = 10$) further reduces $\rho(\mathbf{r}_{BCP})$ to 1.62 e Å^{-3}.

We finally tested whether the EHC model is flexible enough to recover the highly flexible PBE/LAPW model density (Figure 9.4f) from the respective theoretical structure factors. Apparently, the reconstructed density employing the EHC model (Figure 9.4i), does not differ significantly from the one of the original PBE/LAPW wave function (Figure 9.4f). This proves the capability of the EHC multipolar approach to model even highly flexible electron density distributions. Accordingly, the development of post-HC models (such as the EHC model) represents a vivid and fruitful research topic in the charge density community to date and

4) See also related studies of correlation effects on integrated properties derived from the electron density distributions, for example, Refs. [140, 141].

5) Here, l_{max} specifies the maximum angular momentum used for the expansion of the core potential and the density inside the muffin-tin spheres defined by the radius R_{mt}. The product $R_{mt}k_{max}$ thus represents a dimensionless value defining the size of the plane wave basis set employed outside the spheres, where the value of k_{max} defines the largest reciprocal wave vector included.

6) We note that the discontinuities observed at the muffin-tin boundary in $L(\mathbf{r})$ in Figure 9.4c are due to the fact that within the (L)APW ansatz the magnitude (and slope) of the density, but not its curvature, are matched at the sphere boundary. The effect of such discontinuities can be significantly reduced by increasing l_{max} of the density expansion inside the MT spheres (see Figure 9.4f).

Table 9.1 Topological parameters at the C–C BCP of ethane obtained from calculations employing various model Hamiltonians and basis sets, $\rho(\mathbf{r}_{BCP})$ and $L(\mathbf{r}_{BCP})$ are given in e Å^{-3} and e Å^{-5}, respectively.

	$\rho(\mathbf{r}_{BCP})$	$L(\mathbf{r}_{BCP})$
6-21G(d)		
HF	1.76	21.9
DFT (PBE)	1.68	18.4
MP2	1.71	19.7
CISD	1.72	20.2
6-31G(d)		
HF	1.71	16.0
DFT (PBE)	1.62	12.6
MP2	1.65	13.8
CISD	1.67	14.4
6-311G(3d)		
HF	1.69	15.3
DFT (PBE)	1.58	11.6
MP2	1.61	12.6
CISD	1.63	13.4

parallels/supports the tremendous recent achievements in acquisition of high-quality and high-resolution data (see, for example, Refs. [29, 75–86] and references quoted therein).

Owing to the difficulties of systematically increasing the level of approximation in periodic calculations on diamond, we may exemplify these results by studying the molecular model benchmark system for a covalent C–C single bond: ethane. Comparing $\rho(\mathbf{r}_{BCP})$ at the C–C BCP of ethane with the respective values obtained for diamond at the same level of approximation, we observe the same trends (compare topological values in Table 9.1 to Figure 9.4b,e,h): (i) increasing the basis-set size *reduces* $\rho(\mathbf{r}_{BCP})$; (ii) HF *overestimates* $\rho(\mathbf{r}_{BCP})$; (iii) DFT *underestimates* $\rho(\mathbf{r}_{BCP})$ with respect to the MP2 and CI/CC Hamiltonians; and (iv) MP2 and truncated CI/CC results do not show a consistent trend even though in Table 9.1 CISD is consistently giving larger values compared to MP2 [74]. Finally, we note that the CI/CC $\rho(\mathbf{r}_{BCP})$ values in ethane are all centered (1.63–1.72 e Å^{-3}) around the experimental values $\rho(\mathbf{r}_{BCP})$ of the HC and EHC model (1.68/1.67 e Å^{-3}) in diamond.

9.4
Core Density Deformations Induced by Chemical Bonding

Usually, the electron density distribution of the energetically low-lying inner shells, that is, the "core density" of an atom is considered to be rather inert and unaffected by chemical bonding. Indeed, numerous computational charge density studies employing effective core potentials (ECPs) or the frozen-core approximation still rely on the general validity of this simple assumption. However, the high resolution and precision of modern diffraction experiments questions this simple approximation. Indeed, we will demonstrate that the distortion of the core density of the carbon atoms in diamond can be experimentally resolved and is inherently connected with the formation of covalent C–C bonds. In the case of high-resolution studies (d-spacings smaller than 0.35 Å) these core density deformations can therefore no longer be ignored and need to be accounted for by the flexible EHC model (see preceding text), which employs a κ- and multipolar formalism for each individual atomic shell [6]. We have already documented the necessity of this approach in Figure 9.3b, where the theoretical residual density maps $\Delta\rho(\mathbf{r})$ in the (110) lattice plane of diamond clearly reveal the limited flexibility of the HC multipolar model to describe the charge density of atoms in the core region. Indeed, the standard HC model yields large positive and negative residual densities (−0.17 to 0.48 e Å^{-3} at a resolution of $d \geq 0.25$ Å) mainly in the core region of the carbon atoms documenting the breakdown of the frozen-core approximation even in the case of a light atom (Figure 9.3b). Smooth and featureless residual density maps can therefore only be obtained by considering (i) a *charge transfer* between core (core population: $P_c = 1.986$) and valence density ($P_v = 4.014$) and (ii) the inherently connected *core contraction* due to the depletion of the core density (Figure 9.3b) [6]. In the EHC model of diamond (Figure 9.3b), the carbon's core contraction is simply described by a radial core contraction/expansion parameters ($\kappa_c = 1.007$), which scales the exponent ($\exp[-(\kappa_c \zeta r]$) of the Slater-type core orbitals (core-STOs) (see Eq. (9.9), Boxes 9.2 and 9.4) [6]. Accordingly, the κ_c value larger of unity suggests a contraction of the carbon's core density in line with its decreased core population [48, 49]. This results in a significant depletion of the density at the carbon's cusp: $\rho_{cusp,diamond,EHC} = 857$ e Å^{-3} (Figure 9.5 and Box 9.4) in comparison with the cusp density of a carbon atom in its ^3P ground state, $\rho_{cusp,C(^3P)} = 865$ e Å^{-3}. Vice versa,

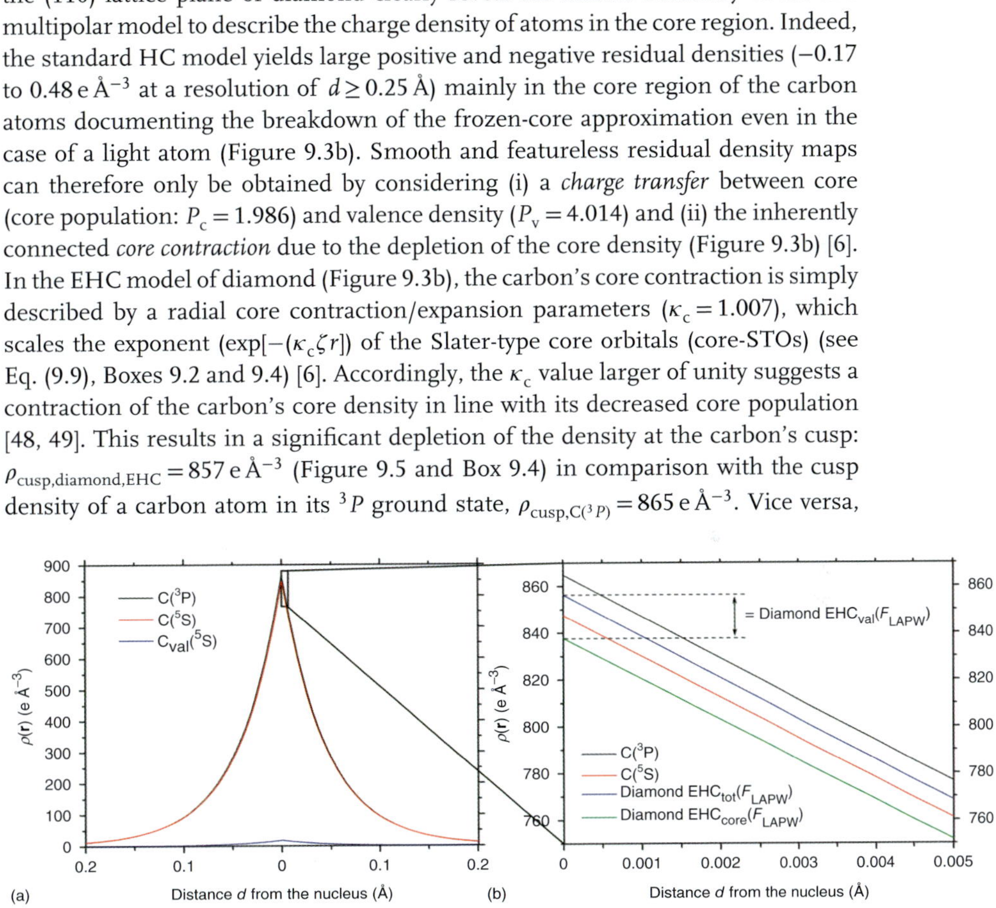

Figure 9.5 The black/red solid lines are obtained from atomic calculations [66, 67] and depict the radial density distribution, $\rho(r)$, of a reference carbon atom in its electronic ground state, C(^{3}P), and prepared valence state, C(^{5}S), respectively. (a) The tiny valence density contribution to the core region is illustrated by a blue line. (b) The core and total density contributions of the carbon atom in the EHC diamond model are shown by solid green and blue lines, respectively.

the increase of the valence density population parameter causes an expansion of the valence shell, which is accounted for by a radial valence contraction/expansion parameter ($\kappa_v = 0.971$) smaller than unity [6]. This scenario appears to be a natural consequence of the involvement of carbon 2s states in the C–C bond formation [6]. This contribution of the 2s atomic orbital (AO) to the core density is relevant as witnessed by its local density maximum at the cusp (Figure 9.5a). Accumulation of the bonding density in the C–C bond is thus inherently linked with a subtle core depletion. It is interesting also to point out that an even lower cusp density (compared with the situation in diamond) is found for a carbon atom in its "prepared valence state" (^{5}S; sp^3 hybridization; Figure 9.5). Hence, the ratio of the chemically induced mixing of the "2s/2p" states appears to control the cusp density in diamond [6].

Box 9.4 The Carbon's Cusp Density in Diamond Derived by the EHC Model

The contribution of the valence 2s/2p AOs to the density at the cusp of a carbon atom in diamond can be analytically expressed by the EHC model (see Eq. (9.9), Box 9.2) via Eq. (9.10)

$$\rho(r) = P_c\kappa_c^3\rho_{1s}(\kappa_c, r) + P_v\kappa_v^3[\rho_{2s}(\kappa_v, r) + \rho_{2p}(\kappa_v, r)] \quad (9.10)$$

where the P_c represents the population parameter of the core density, ρ_{1s}, and P_v is the population parameter of the valence density, $\rho_{2s} + \rho_{2p}$, while κ_c and κ_v are the radial contraction/expansion parameters of the core and valence shell, respectively.

Note that the core and valence density contribution are simply obtained from the squared AOs based on atomic calculations of the carbon atom in its ground state [66, 67]. Owing to their angular nodal structure, basis functions with $l > 0$ cannot contribute to the electron density at the cusp and therefore only the radial part of multiple-ζ Slater-type atomic basis functions $b(n, r)$ needs to be considered, that is,

$$b(n, r) = \sum_i c_i N_i r^{(n-1)} e^{-\kappa\zeta_i r}$$

Furthermore, the term $r^{(n-1)}$ prevents contribution to the cusp density ($r = 0$) from s-type basis functions with $n > 1$. Accordingly, Eq. (9.10) can be simply written as

$$\rho(0) = P_c\kappa_c^3\left[\sum_i c_i N_i\right]^2 + P_v\kappa_v^3\left[\sum_j c_j N_j\right]^2 \quad \text{with } N_{i,j} = \sqrt{\frac{\zeta_{i,j}^3}{\pi}} \quad (9.11)$$

where c_i are the coefficients of the STO basis set employed containing i and j basis functions with principle quantum number n equal to 1 for the core and valence orbitals, while the N_i are the corresponding normalization factors.

In our EHC model (Figures 9.3d and 9.4g), we employed the Su–Coppens–Macchi (SCM) [66, 67] wave function of the free carbon atom in its ^{5}S prepared valence state, which contains two s-type basis functions with $n = 1$ for the 1s and 2s orbitals:

$$b_{1s}(1,0) = 0.363018\sqrt{\frac{8.568164^3}{\pi}} + 0.442181\sqrt{\frac{4.885374^3}{\pi}}$$

$$b_{2s}(1,0) = 0.075015\sqrt{\frac{8.493421^3}{\pi}} - 0.456424\sqrt{\frac{4.862298^3}{\pi}}$$

From this, we obtain the following expression, which reveals the entanglement of the core and valence density at the cusp

$$\rho(0) = 61.3175\, P_c\kappa_c^3 + 2.93555\left(P_v\frac{x}{4}\right)\kappa_v^3 \quad (9.12)$$

where x depends on the selected valence configuration $s^x p^{4-x}$

In our case (EHC model of diamond using the F_{LAPW} structure factors) $\rho(0) = 127.05$ au $= 857\,\text{e}\,\text{Å}^{-3}$; ($P_c = 1.986$, $P_v = 4.014$, $x = 1$, $\kappa_c = 1.007$ and $\kappa_v = 0.971$) in agreement with Figure 9.5.

At this stage of our analysis theoretical structure factors from the LAPW DFT calculations were used to rule out any bias from experimental data artifacts. The use of these multipolar EHC parameters to fit the experimental structure factors is therefore also affecting the experimental valence density, as documented in Figure 9.4g. In comparison with the "frozen-core" HC model (Figure 9.4d), the density in the EHC model (Figure 9.4g) decreases slightly at the C–C BCP, while the $L(\mathbf{r})$ value is increased relative to the HC model at the BCCs in the carbon valence shell (Figure 9.4g). In the subsequent section, we therefore outline that core deformation phenomena need to be considered to achieve a complete deconvolution of thermal motion and chemical bonding effects in experimental studies.

9.5 How Strongly Is the Static Electron Density Distribution Biased by Thermal Motion?

The correlation between thermal smearing and chemical deformation of the electron density distribution usually hinders the determination of precise structural and thermal displacement parameters using the standard IAM X-ray model. This severe limitation can be systematically overcome by increasing the data resolution and the flexibility of the multipolar model. The successful deconvolution of thermal smearing and electron density deformation due to bonding effects is documented in Figure 9.6 for our benchmark system diamond. Owing to the high point-group symmetry of the carbon atoms, their time-averaged thermal motion can be simply described with a single isotropic Debye–Waller factor, B_{iso}, and is not affected by significant anharmonic motion [87]. Furthermore, the large thermal conductivity (2000 $\text{W}\cdot\text{K}^{-1}\cdot\text{m}^{-1}$ at 300 K) of diamond, which is connected with a rather large Debye temperature of 1946(20) °C, warrants that the B_{iso} value remains rather constant over a wide temperature range [88, 89]. Accordingly, the phononic

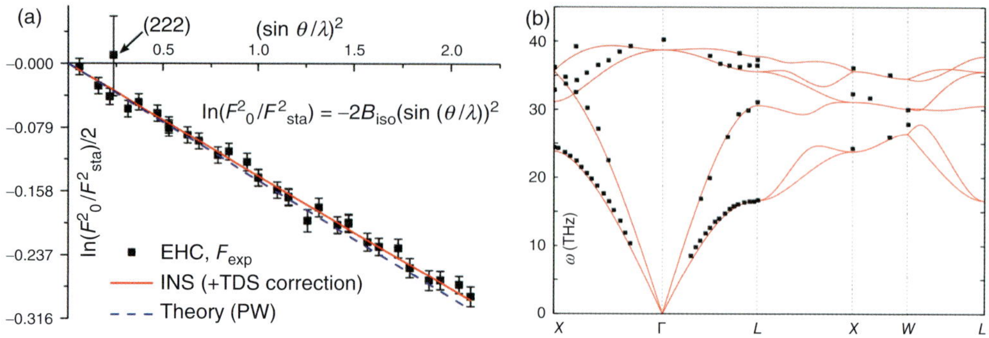

Figure 9.6 (a) Wilson plot $\ln(F^2{}_o/F^2{}_{sta})/2$ for the Spring-8 diamond powder diffraction data [7] using the EHC model at 300 K [6] and yielding a $B_{iso,EHC}$ value of 0.139(4) Å^2. Systematically absent reflections (except for the (222)-reflection) were excluded from the plot and the fit. Error bars from combined multipolar and Rietveld refinements [6]. The red solid line is based on the experimental $B_{iso,INS,TDS}$ value of 0.139 Å^2 derived from inelastic neutron scattering. The blue broken line originates from a calculated B_{iso} value of 0.146 based on phonon dispersion curves derived from plane wave (PW) DFT calculations for diamond.[7] (b) Experimental (squares) and theoretical (lines) phonon dispersion curves derived from PW DFT calculations for diamond along the k-path X (1/2,0,1/2) → Γ (0,0,0) → L (1/2, 1/2, 1/2) → X → W (1/2, 1/4, 3/4)→ L. (experimental data taken from Ref. [90].)

behavior of diamond provides an ideal benchmark to compare the performance of experimental and theoretical methods by refining/computing the respective B_{iso} value at a certain temperature. We note that experimental and theoretical phonon dispersion curves based on inelastic neutron scattering (INS) studies and LAPW DFT calculations agree well, while only the optical modes appear to be slightly underestimated by theory (Figure 9.6). Therefore, we also find a good agreement for the B values ($B_{iso,theor} = 0.146$ Å^2; $B_{iso,INS} = 0.149$ Å^2), which can be derived from the theoretical and observed phonon frequency spectrum [90–92]. We note that the experimental value can be corrected for thermal diffuse scattering, as proposed by Stewart [92], yielding $B_{iso,INS,TDS} = 0.139$ Å^2, which is basically the same value as obtained from the EHC charge density refinements ($B_{iso,EHC} = 0.139(4)$ Å^2; Figure 9.6a). This clearly documents the successful deconvolution of thermally and chemically smeared deformation densities by the EHC model. In this respect, the EHC model outperforms the standard HC model, which underestimates the B-values ($B_{iso,HC} = 0.127(4)$ Å^2) by 9% [6].

9.6
Relativistic Effects on the Topology of Electron Density

The extension of the HC model into the inner electronic shells of atoms imposes new requirements with respect to the theoretical models employed for comparative

7) The force calculations were done employing the PBE [65] functional within the VASP program package [142–145]. Phonon dispersion relations were calculated employing PHONOPY [146].

studies. If the above-mentioned density analyses are to be extended from light elements such as carbon to compounds containing heavy main group or even transition metal elements, a proper treatment of relativistic effects becomes important. In general, this can be achieved by two different approaches, namely, (i) the use of all-electron (AE) basis sets in combination with (approximate) relativistic Hamiltonians or (ii) the use of ECPs that have been derived from fully relativistic atomic calculations.

Even though the first approach is obviously the preferred ansatz, in case of large systems containing many heavy elements, fully relativistic four-component *ab initio* calculations employing reasonable basis-set sizes are not feasible in many cases. However, with respect to a subsequent topological analysis of the electron density within the QTAIM, the use of ECPs imposes further obstacles. Most importantly, the overall qualitative topology of the electron density is changed by removing the local density maxima, that is, the attractors from $\rho(\mathbf{r})$ at the nuclear positions. As demonstrated, for example, for the molecular model system $MoOCl_4$, depending on the number of atomic shells included into the ECP, these basis sets may result in a failure to locate the correct number of BCPs [93–95]. If integrated atomic properties such as atomic moments or charges are to be investigated, the "valence" electron densities obtained from ECP calculations have therefore to be augmented by the missing core electron density distributions. Such an augmentation has been used, for example, by Vyboishchikov *et al.*, Cioslowski and Piskorz, and Tiana *et al.* [95–97] or more recently by Keith and Frisch [98]. Figure 9.7 depicts a comparison of the Laplacian of the electron density in the molecular plane of the model complex $[Ag(C_2H_2)]^+$ as obtained from (augmented) ECP, AE calculations as well as from an experimental study on the silver complex $[Ag(C_2H_2)A']$ **5** (Figure 9.2) [45]. As seen from this comparison, the missing core shells in the ECP calculations have direct consequences for the absolute values of the VSCCs. In the AE case (DKH2 Hamiltonian with basis sets of triple-ζ quality [100]), the contributions of the inner atomic shells lead to a dampening of the VSCC of the $n=4$ shell [101] so that the local maximum CC1 is, for example, found to be at $340\,e\,Å^{-5}$ (Figure 9.7b) [45]. This dampening effect is missing in the ECP calculation, which results in a larger radial extension of the VSCC and, more importantly, a significant overestimation of its absolute value, that is, CC1 values at $1106\,e\,Å^{-5}$ (Figure 9.7a). The augmentation of the ECP density with the core density, as implemented in GAUSSIAN09 [102], resolves this artifact incorrectly, as it causes a significant overestimate of the dampening. This seems to be caused by an incorrect radial scaling of the core–shells added to the ECP valence density (Figure 9.7c) so that CC1 values at merely $48\,e\,Å^{-5}$. Even more severe artifacts become evident in case of the augmented ECP densities when the inner shells of the transition metal atoms are inspected. Even though the $n=1$ and the $n=2$ shell can be resolved in $L(\mathbf{r})$ of the augmented densities, the third shell is still not present as a region with $L(\mathbf{r}) > 0$, in contrast to the findings from the AE calculations (Figure 9.7b,c). Taking into account the good agreement between our experimental reference system **5** (Figure 9.7d) and the AE calculation ($L(\mathbf{r})_{CC1} = 365/340\,e\,Å^{-5}$, respectively), it is obvious that the augmented ECP calculations are not providing sufficient accuracy

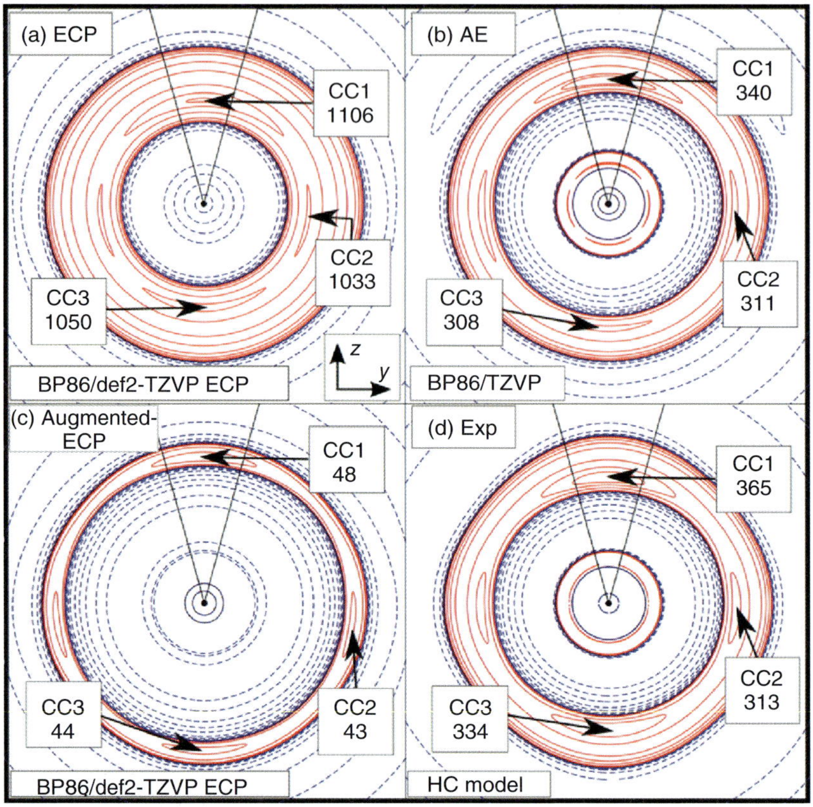

Figure 9.7 Negative Laplacian, $L(\mathbf{r}) = -\nabla^2\rho(\mathbf{r})$, maps in the molecular plane of the model complex $[Ag(C_2H_2)]^+$ as obtained from (a) ECP, (b) DKH2 all electron, (c) ECP calculations with subsequent addition of the core density, and (d) an experimental charge density study of the complex $[Ag(C_2H_2)\ Al(OC(CH_3)(CF_3)_2)_4]$ (**5**). Contours are drawn at ± 2, ± 4, $\pm 8 \times 10^n$ e Å^{-5}, $n = -1$, 0, 1, 2, 3, 4; positive values in red, negative values in blue, extra contours at (a) 1025 and 1100, (b) 300, 320, 67300 and 68000, (d) 285, 335, 54485 eÅ^{-5}, $L(\mathbf{r})$ values for the CCs are specified in eÅ^{-5}.

to meet the requirements needed for a direct comparison of the calculated and experimental fine structure of $L(\mathbf{r})$ in the VSCC.

Regardless of the accuracy of the augmented electron densities, in order to study core contraction/expansion or even core polarization phenomena, an AE approach and therefore the use of appropriate (quasi-)relativistic Hamiltonians is mandatory. The indirect treatment of relativistic effects by the use of effective core potentials derived from calculations on free atoms is insufficient to recover polarizations of core–shells of the atoms induced by chemical bonding. The salient question, which approximate relativistic model Hamiltonian is able to recover the relativistic effects on the topology of the electron density, has been therefore addressed by Eickerling *et al.* [103]. Within this study, four-component Dirac–Hartree–Fock (DHF) calculations on the model compounds $M(C_2H_2)$ (M = Ni, Pd, Pt) employing

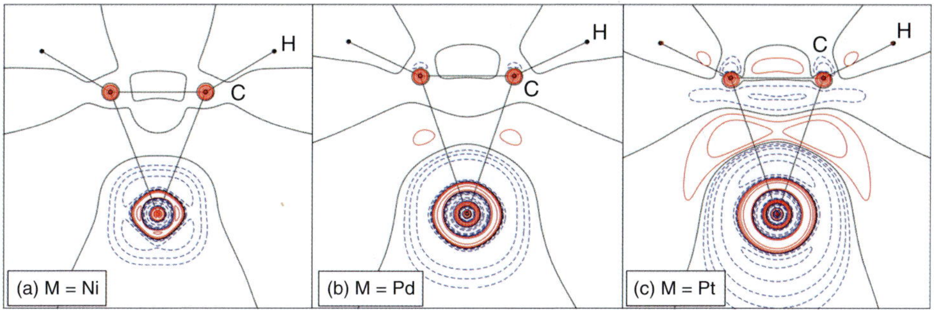

Figure 9.8 Isocontour representations of the difference densities $\Delta\rho(\mathbf{r}) = \rho(\mathbf{r})_{4\text{comp}} - \rho(\mathbf{r})_{\text{nonrel}}$ as obtained from four-component DHF and nonrelativistic HF calculations on $[M(C_2H_2)]$ with M = Ni, Pd, and Pt. Contours are drawn at ±2, 4, 8×10^n ($n = -2, -1, 0, 1, 2, 3$) $e\,Å^{-3}$, positive values in red, negative values in blue. For details on the calculations and the basis sets employed. (See Ref. [103].)

a basis set close to the HF limit were used as a reference to compare the performance of several approximate relativistic model Hamiltonians. As shown in Figure 9.8, the relativistic effects on the topology of the electron density are not restricted to the core region where scalar-relativistic effects are expected to be large, but especially for M = Pt they also extend into the bonding region between the metal atom and the ligand and even into the C–C bonding region. At the Pt–C BCP, for example, the density differs by 6% (1.00 vs 0.94 $e\,Å^{-3}$ for the four-component and the nonrelativistic calculation, respectively), while the Laplacian $L(\mathbf{r})$ at the same critical point (CP) differs even by 90% (−3.09 vs −5.88 $e\,Å^{-5}$ for the four-component and the nonrelativistic calculation, respectively) [103]. However, these effects can be faithfully recovered by employing the zeroth-order regular approximation (ZORA) ($\rho(\mathbf{r}_{\text{BCP}}) = 1.00\,e\,Å^{-3}$, $L(\mathbf{r}_{\text{BCP}}) = -3.29\,e\,Å^{-5}$) or the DKH2 Hamiltonian ($\rho(\mathbf{r}_{\text{BCP}}) = 1.00\,e\,Å^{-3}$, $L(\mathbf{r}_{\text{BCP}}) = -3.19\,e\,Å^{-5}$) so that the computational demand of the according AE calculation on such systems can be kept within reasonable limits.

This can also be demonstrated by comparison of the magnitude of the individual CCs in the model complex $[Ag(C_2H_2)]^+$. Neglecting relativistic effects, the value of CC1 is overestimated in a HF calculation employing the basis set specified in [45] by ~3% compared to the four-component DHF result (354 vs 344 $e\,Å^{-5}$), while application of the DKH2 Hamiltonian reduces this deviation significantly (345 $e\,Å^{-5}$). We point out that another interesting trend can be observed for the magnitude of the CCs: neglecting electron correlation effects by employing the HF ansatz overestimates the value of $L(\mathbf{r})$. In a previous study, CC1 in $[Ag(C_2H_2)]^+$ was found to be 322 and 323 $e\,Å^{-5}$ for BP86 ZORA- and four-component calculations, respectively [45]. This result is similar to the findings for C–C BCPs in diamond (Figure 9.4) and ethane (Table 9.1), where the neglect of electronic correlation overestimates the density at the BCP, while DFT causes an underestimation. We may summarize this section by stating that relativistic effects on the topology of the electron density can be observed in the core and valence region of transition metal atoms and that these can be accurately described by approximate relativistic Hamiltonians. Even though these effects, together with correlation effects, seem to reflect systematic trends within the calculated results, a comparison with reliable experimental studies is needed. This warrants further exploration by ultrahigh-resolution experimental studies and also stresses the necessity to increase the flexibility of the multipolar model as discussed in the previous sections.

9.7 The Topology of the Laplacian and the MO Picture – Two Sides of the Same Coin?

In the previous section we illustrated that topological analysis of the negative Laplacian, $L(\mathbf{r}) = -\nabla^2\rho(\mathbf{r})$, of the total electron density provides a powerful tool to identify even subtle electronic features such as the fine structure of the VSCC in metals (Figure 9.7). Already in the early 1950s, it became evident that "the Laplacian of a scalar field is a very important property, for it determines where the field is locally *concentrated* and *depleted*" [10]. Indeed, it has been demonstrated by Bader *et al.* that the sign of $L(\mathbf{r})$ indicates where the charge density is locally concentrated ($L(\mathbf{r}) > 0$) or locally depleted ($L(\mathbf{r}) < 0$) [104, 105]. The $L(\mathbf{r})$ function can, in addition, be employed to resolve the shell structure of atoms [106, 107], even though the fourth, fifth, and sixth shell for elements of periods 4–6, respectively, is not revealed as positive maxima in $L(\mathbf{r})$[8] [108–110]. Bader *et al.* therefore suggested as a convention that the outermost shell of CC of an atom (i.e., the fourth shell of CC of the silver atom in **5**) represents its effective VSCC [94, 111]. Owing to the presence of the acetylene and aluminate ligands in **5**, the VSCC of the silver atom displays, however, also a fine structure, the so-called *atomic graph*, which is characterized by four zones of charge concentration (denoted "CC" in Figure 9.7d) and charge depletion (CD). The depletion zones in the approximately planar-coordinated Ag d^{10} complex **5** are situated between the four CCs and are due to the depletion of the d_{yz} orbital as a consequence of the subtle $Ag(d_{yz}) \rightarrow \pi^*(C_2H_2)$ back donation in the framework of the Dewar–Chatt–Duncanson (DCD) model [45, 112, 113]. For a definition of the local coordinate system, see Figure 9.7a. Hence, the presence of the *four charge depletion zones* in the VSCC and the presence of the $Ag(d_{yz}) \rightarrow \pi^*(C_2H_2)$ *back donation in the molecular orbital (MO) picture* basically represent two sides of the same coin. The chemical consequences of the ligand-induced polarization of the VSCC are, however, better demonstrated by the β-agostic d^8 species $[DCpHNi(dt\text{-}bpe)]^+[BF_4]^-$ (dt-bpe = $t\text{-}Bu_2PCH_2CH_2Pt\text{-}Bu_2$) (**2**) [40] (Figures 9.2 and 9.9).

The alkyl backbone in the agostic species **2** is characterized by an acute $\angle Ni{-}C_\alpha{-}C_\beta$ valence angle and one of the most activated $C{-}H_\beta$ bonds ($C{-}H_\beta = 1.20(1)$ [1.205] Å) observed in a molecular species (in its electronic ground

8) Furthermore, four component calculations on spherical symmetric atoms (Au) by Kohout *et al.* showed that the 5th and 6th shell of elements of period 6 might not be resolved in the second derivative of the radial density (see Refs. [101, 147–149]).

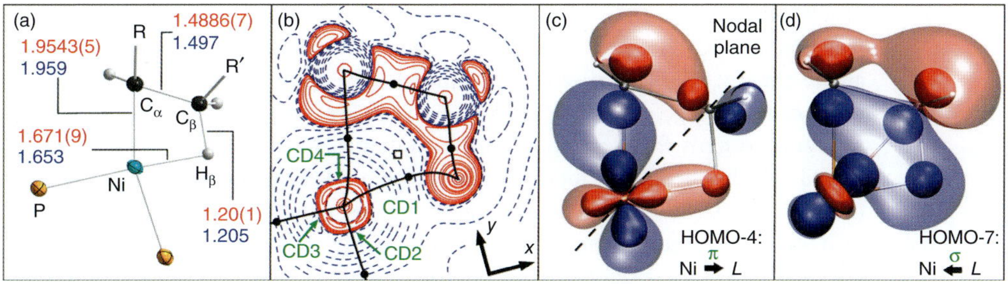

Figure 9.9 (a, b) Experimental geometry and $L(\mathbf{r})$ contour map in the agostic $NiC_\alpha C_\beta H_\beta$ moiety of **2**. Bond paths are drawn as black solid lines; the BCPs and the RCP by filled circles and open squares, respectively; positive (solid) and negative (dashed) contour lines are drawn at 0, $\pm 2.0 \times 10^n$, $\pm 4.0 \times 10^n$, $\pm 8.0 \times 10^n$ e Å^{-5} with $n = \pm 3, \pm 2, \pm 1, 0$; one contour level deleted (800 e Å^{-5}); extra level at 30 and 1100 e Å^{-5}. (c, d) Multicenter molecular orbitals in the model system $[EtNi]^+$ (isodensity map at 0.05 au) establishing the Ni → L π back donation and Ni ← L σ donation (L = alkyl unit). The inner lobes are schematic ones to clarify the salient atomic orbital contributions. (Adopted from Ref. [40]).

state). The formation of CD1–CD4 is a direct consequence of the Ni → ligand (L) π back-donation (highest occupied molecular orbital (HOMO)-4; Figure 9.9c) causing a significant depopulation of the $Ni(d_{x^2-y^2})$ orbital relative to the other d-orbitals. This is in line with the *relatively* small $P(d_{x^2-y^2})$ population of only 1.62(2) electrons obtained from HC multipolar refinements [114]. Note that the local x, y coordinate axes (Figure 9.9b) are located in the molecular plane spanned by the β-agostic moiety ($Ni–C_\alpha–C_\beta–H_\beta$). As the four CD zones (denoted CD1-4 in Figure 9.9b) in the charge density picture are directly connected with the depletion of the metal $d_{x^2-y^2}$ orbital in the MO picture, the angle between these CD zones is constrained and dictates the position of all ligand atoms in a key and lock scenario. However, the electronic situation is more complex and the agostic interaction is further established by additional Ni ← L π (HOMO-6)[9] and σ-donation (HOMO-7; Figure 9.9d). Accordingly, the bonding in β-agostic late transition metal complexes can be considered in terms of an extended DCD model and is related to the scenario in the silver acetylene complex **5** (Figure 9.7) or the $[M(C_2H_2)]$ model systems with M = Ni, Pd, and Pt (Figure 9.8). Note that the presence of the σ-donation component is also directly reflected in the charge density picture as this interaction is directly responsible for the unusually large density accumulation inside the cyclic β-agostic (Ni, C_α, C_β, H_β) moiety. The competition of the σ-donation and π-back-donation component in the MO picture therefore controls the amount of electron density accumulation inside the (Ni, C_α, C_β, H_β) cycle. This competition of both DCD components is the electronic origin

9) The additional and weak Ni ← *L* π donation (which complements the σ-donation and π-back-donation component in the classical DCD model for olefine complexes) simply reflects the increased functionality of β-agostic alkyl ligands by involving the β-H atom in the ligand to metal bonding.

of a M···C–H bond catastrophe scenario, which causes the annihilation of the β-agostic M···C–H bond path [13, 39, 115, 116] when the density accumulation at the (M, C_α, C_β, H_β) ring critical point (RCP) approaches the corresponding value of the charge density at the M···C–H BCP. Even in **2**, representing a true reference system for a predominantly covalent β-agostic M···H–C interaction, the M···H bond path is topologically instable. Indeed, the density at the RCP (0.533(7) [0.507] e Å^{-3}; calculated value in brackets) and the Ni···H BCP (0.553(4) [0.569] e Å^{-3}) of the agostic moiety are not significantly different. Hence, a DCD scenario, where the σ-donation dominates the π back-donation component (e.g., in case of agostic d^0 complexes), is usually reflected within the charge density picture by a merging of the M···H_β BCP and the RCP of the (M, C_α, C_β, H_β)-agostic moiety, a topological scenario leading to the fission of the agostic M···H bond path. In general, it could therefore be shown that the weakening or lack of a X–H···M (M = transition metal; X = Si, C) bond path in β-agostic alkyls or in σ-silane complexes [117, 118] is a natural consequence of these competing interactions and highlights the delocalized nature of the bonding situation in agostic species and σ-silane moieties.

A related DCD scenario has been identified for the change of bond path topology (metallacyclopropane type vs T-shaped) in transition metal olefin or acetylene model complexes. The electronic situation of these transition metal complexes has been pioneered by Frenking and coworkers [119, 120] via combined MO and theoretical charge density analyses. Also, in a subsequent experimental charge density study of the $Ni(COD)_2$ complex (COD = 1,5 cyclooctadiene) the various bond path topologies were directly explained in the terminology of the DCD model [121]. This linkage between the MO and charge density picture could then be extended by establishing a direct connection between the fine structure of the Laplacian in the VSCC at the metal and the d-orbitals involved in the individual σ-donation and π-back-donation DCD components [43, 61]. The competition between both DCD components has a strong impact on the bond path topology as can be illustrated by a series of experimental charge density studies (Figure 9.10). In this series of compounds, the Ni-olefin complex $(C_2H_4)Ni(dt\text{-bpe})$ **4** represents

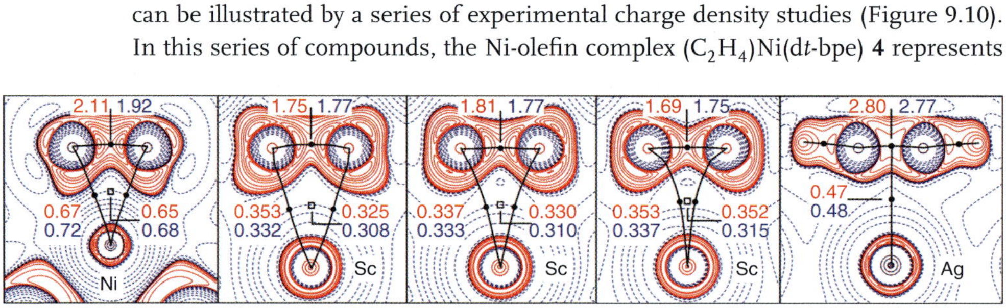

$(C_2H_4)Ni(dt\text{-bpe})$ Sc_3FeC_4 Sc_3CoC_4 Sc_3NiC_4 $(C_2H_2)AgA'$

Figure 9.10 Experimental $L(\mathbf{r})$ maps showing the bond path topologies in the MC_2 moieties of **4**, **5** and the transition metal carbides Sc_3MC_4 (M = Fe (**6**), M = Co (**3**), M = Ni (**7**)). For a specification of the contour levels and symbols, see Figures 9.7 and 9.9. The $\rho(\mathbf{r})$ values at the RCP, the C–C and Sc–C BCPs are specified in e Å^{-3}; experimental/theoretical values in red/blue color, respectively.

our benchmark characterized by pronounced covalent M–C interactions and thus the predominance of the Ni → $\pi^*(C_2H_4)$ back-donation component. This electronic scenario is therefore reflected in the charge density picture by a metalla-cylcopropane-type bond path topology for the M-C-C moiety. The other extreme case is marked by the silver acetylene complex $[Ag(C_2H_2)A']$ **5** displaying a predominant $Ag(d_{yz}) \leftarrow \sigma\ (C_2H_2)$ donation yielding a T-shaped bond path scenario. The transition metal carbides Sc_3MC_4 (M = Fe (**6**), M = Co (**3**), M = Ni (**7**)) thus represent intermediate cases of the bond path scenarios displayed by **5** and **4** [122].

Hence, in our molecular benchmark systems **4**, **5** and the solids **6**, **3**, and **7**, a direct linkage between the MO model and the $L(\mathbf{r})$ topology emerges.

9.8 Elusive Charge Density Phenomena: Nonnuclear Attractors

The determination of charge density distributions of metals has always been a challenge for experimentalists, even in the case of seemingly simple cases such as beryllium metal. The delocalization of the itinerant electrons in the conduction band led Larsen and Hansen in 1984 to conclude that "*it is questionable whether the scale factor of diffraction data from a light metallic substance can be determined very reliably by high-order refinement even using data of very high sinθ/λ cut-off values*" [123]. However, as outlined in the previous sections, the determination of precise charge density maps appears to be rather a matter of the flexibility of the multipolar model employed and of a suitable resolution of the diffraction experiment. Indeed, we outline in the case of the benchmark system beryllium that a simultaneous core and valence density refinement recovers even elusive density features such as nonnuclear attractors (NNAs), in excellent agreement with state-of-the-art theoretical calculations. We note that the theoretical prediction of these elusive density features in metals by Besnainou *et al.* has stimulated the perhaps most controversial debate in the charge density community on the observability of these weakly pronounced charge density features [124]. Indeed, an unequivocal experimental proof of the existence of NNAs is difficult owing to the minute density accumulation at these (3, −3) CPs. Thus, the pioneering work of Sakata [125] and Iversen [126, 127], who first traced NNAs by means of experimental studies, is still a matter of intense debate as outlined in the following.

Figure 9.11 illustrates the hexagonal close-packing scenario of beryllium and the location of salient CPs. A QTAIM analysis based on the experimental and theoretical charge densities (Table 9.2) spots an NNA inside the trigonal bipyramidal voids and identifies a cage critical point (CCP) – as a local density minimum – in the octahedral voids. The density values of the NNA and the CCP, which represent (3, −3) and (3, +3) CP points, respectively, differ in theory and experiment by $\Delta\rho(\mathbf{r}_c)_{theor} = 0.084$ and $\Delta\rho(\mathbf{r}_c)_{exp} = 0.092\ e\ Å^{-3}$. These significant but rather small density differences stress the remarkable flatness of the charge density distribution in this itinerant electronic system. The NNA itself is linked in a trigonal-bipyramidal manner by five bond paths and five BCPs (3 × BCP1; 2 × BCP2) with its neighboring

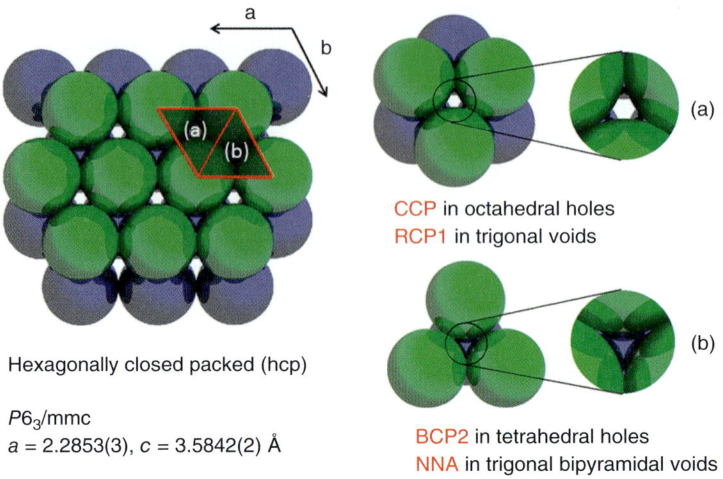

Figure 9.11 (a, b) Hexagonal close packing of beryllium metal and location of salient critical points: BCP2, RCP1 and CCP denote bond, ring, and cage critical points. The notation NNA represents a so-called nonnuclear attractor which constitutes a local maximum in the total charge density distribution in interatomic regions. (Unit cell constants from Ref. [123].)

beryllium atoms, which are acting as nuclear attractors. The RCP1 in the same hexagonal (001) plane ($z = 0.25$) are located above and below the octahedral CCP formed by six Be atoms. They are thus located in trigonal voids formed by three neighboring beryllium atoms (Figures 9.11 and 9.12).

Figure 9.12b shows the valence density map formed by four neighboring beryllium atoms in the (001) plane ($z = 0.25$) of beryllium (highlighted in red in Figure 9.11a). This planar moiety thus comprises the NNA, three BCPs ($3 \times$ BCP1), RCP1, and the four Be atoms as nuclear attractors. Note the presence of two chemically different voids, which are denoted (a) and (b) in Figure 9.11. Accordingly, point (a) is located in a trigonal void marking the position of the RCP1, while point (b) resides in a trigonal-bipyramidal void and defines the position of the NNA. We first note the surprisingly shallow density region around the NNA. The experimental density at the NNA and BCP1 differs only at the fifth decimal place on a cubic angstrom scale, which is clearly outside the range of the experimental precision of this study. Also, in the case of the LAPW density, the discrimination of the NNA and BCP1 is at the limit of the numerical precision and both density values ($(\rho(\mathbf{r})_{BCP1} = 0.283\,e\,Å^{-3})$ and $(\rho(\mathbf{r})_{NNA} = 0.284\,e\,Å^{-3})$ differ by <0.4% (Table 9.2, Figure 9.12). Hence, it remains questionable whether the theoretical and experimental studies can provide an unequivocal proof of the presence of elusive phenomena such as NNAs.

It is therefore not surprising that the location of the NNA, BCP1, RCP1, and CCP in the (001) plane ($z = 0.25$) of beryllium has always been a matter of controversial debate in earlier experimental studies, which relied on maximum-entropy analyses [126–128], the standard HC model [126], or the constrained wave function

Table 9.2 Coordinates and topological parameters $\rho(\mathbf{r}_c)$ (e Å^{-3}), $\nabla^2\rho(\mathbf{r}_c)$ (e Å^{-5}) at selected critical points in beryllium metal.

Critical point	LAPW wave function[a]		Coordinates	EHC (F_{Exp})		Coordinates
	$\rho(\mathbf{r}_c)$	$\nabla^2\rho(\mathbf{r}_c)$		$\rho(\mathbf{r}_c)$	$\nabla^2\rho(\mathbf{r}_c)$	
Be	239.6	—	1/3, 2/3, 1/4	240.2	—	1/3, 2/3, 1/4
NNA	0.284	−0.16	2/3, 1/3, 1/4	0.292[b]	−0.16	2/3, 1/3, 1/4
BCP1	0.283	−0.27	0.54981, 0.45019, 1/4	0.292[b]	−0.17	0.64978, 0.35022, 1/4
BCP2	0.277	−0.12	2/3, 1/3, 0.07391	0.282	−0.21	2/3, 1/3, 0.0743
RCP1	0.213	0.41	0, 0, 1/4	0.209	0.59	0, 0, 1/4
RCP2	0.273	−0.20	1/2, 1/2, 1/2	0.270	0.02	1/2, 1/2, 1/2
CCP	0.200	0.57	0, 0, 0	0.200	0.49	0, 0, 0

[a]The periodic full-potential LAPW DFT calculations employed the Elk program [64] and the PBE GGA-functional [65]. The experimental results are based on structure factors measured by Larsen and Hansen [123].
[b]The experimental density at the NNA and BCP1 differs at the fifth decimal place (Å^3 scale), which is outside the range of the experimental precision. Also, in the case of the LAPW density, the discrimination of the NNA and BCP1 is at the limit of the numerical precision.

method [129]. However, none of the previous approaches yielded a topology in full agreement with the results obtained by state-of-the-art LAPW calculations, even though all experimental models employed the same X-ray diffraction data of Larsen and Hansen [123].

However, the topological features derived from the flexible EHC model and the LAPW calculations (Table 9.2) provide at least a consistent topological scenario and support the findings of Pendás *et al.* [130], who well demonstrated that the presence of NNAs might just be a natural consequence in the bond formation step of atoms at decreasing bond distances. This is especially true in cases where bonding partners displaying at least one nonconvex region in the radial atomic density (e.g., atoms with $Z = 3$–6, 16–32, and 45–92) are involved. We exemplify this behavior in Figure 9.13, in which the superposition of the radial atomic densities of two Be atoms obtained by numerical calculations at varying distances is shown.

For the limiting cases of long (>1.9 Å) or short interatomic distances (<1.7 Å) local minima are observed at the midpoint between the atoms in line with the presence of a BCP. However, for the interatomic distances in the range from 1.7 to 1.9 Å, a local maximum at the midpoint between the atoms is formed, signaling the presence of an NNA (red spheres in Figure 9.13). The NNA in these simple models would thus result from the overlap of the convex density regions of the Be atoms at ~1 Å distance from the nucleus (blue arrows in Figure 9.13a). These convex density regions are clearly revealed as local maxima in the negative Laplacian, $L(\mathbf{r})$ (green arrows in Figure 9.13b). In case of beryllium, these extrema denote where the VSCC is at a maximum and are thus a consequence of the atomic shell structure of beryllium. It is therefore reasonable to use the $L(\mathbf{r})$ distributions of paired atoms to estimate the critical distance range where NNAs can occur (if chemical

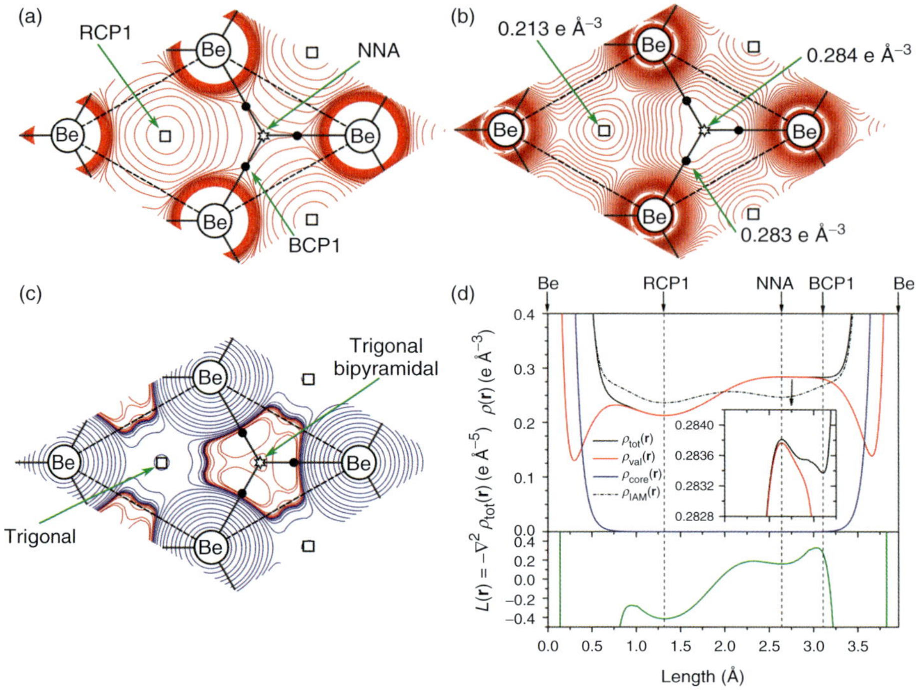

Figure 9.12 (a) Theoretical $\rho(\mathbf{r})$ maps of beryllium in the hexagonal (001) plane at $z = 0.25$ based on the LAPW calculation as specified in Table 9.2. Contour levels from 0.0 to 0.6 e Å^{-3} in steps of 0.01 e Å^{-3} and one additional line at 0.283 e Å^{-3}. (b) $\rho_{val}(\mathbf{r})$ from 0.0 to 0.3 e Å^{-3} in steps of 0.005 e Å^{-3}. (c) $L(\mathbf{r})$ map, negative (blue, solid) and positive (red, solid) contour lines were drawn at $\pm 2.0 \times 10^n$, $\pm 4.0 \times 10^n$, $\pm 8.0 \times 10^n$ e Å^{-5} with $n = \pm 3, \pm 2, \pm 1, 0$. One extra line was drawn at $+0.17$ e Å^{-5}. Bond paths are displayed as black solid lines. Black stars mark nonnuclear attractors (NNAs), black dots mark BCPs, and squares mark RCPs. (d) 1D profiles in the (001) plane at $z = 0.25$ along the [110] direction.

bonding effects are not considered in the first approximation). In that case, the NNA formation between atom pairs is a consequence of the superposition of the two local VSCC maxima, which in between a critical distance range merge into a single feature (Figure 9.13b). We note that the shortest Be–Be bond distances in metallic Beryllium (2.23 Å for atoms in different hexagonal layers (i.e., $z = 0.25$ and $z = 0.75$) and 2.29 Å inside one hexagonal close-packed layer) are outside the above-mentioned stability range of NNAs. Hence, the presence of an NNA suggests that density contributions from chemical bonding cannot be ignored. Furthermore, in the case of metallic beryllium, the situation is more complex than in the diatomic case, as five atoms are simultaneously involved in the formation of the NNA in the trigonal bipyramidal voids (Be(eq)-NNA = 1.32 Å for the equatorial positions in the (001) plane of the hexagonal layers; Be(ax)-NNA = 1.79 Å between axial Be positions above and below the trigonal-bipyramidal voids; Figure 9.11). Indeed, the superposition of atomic densities of beryllium atoms assuming the experimental

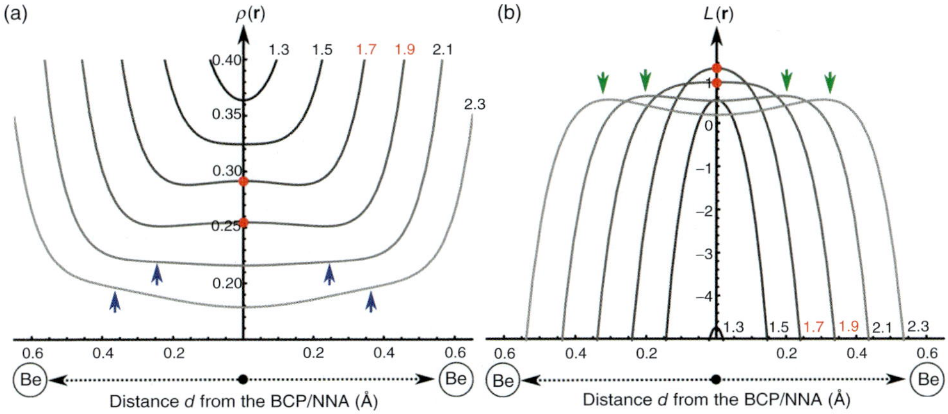

Figure 9.13 One-dimensional profiles obtained by the superposition of (a) the electron density (e Å^{-3}) and (b) the Laplacian of the electron density (e Å^{-5}) of two noninteracting Be atoms at varying distances from $d = 1.3–2.3$ Å as obtained from fully numerical four-component Dirac–Fock calculations [131] (*x*-axis centered at the midpoint between the atoms; distances in Å). The formation of an NNA at the midpoint between the atoms (red spheres) for distances between 1.5 and 1.9 Å is indicated by a local maximum in both, the electron density and its Laplacian. The position of the convex density regions in (a) and of the VSCC in (b) are indicated by blue and green arrows, respectively.

geometry of metallic beryllium does not yield any NNA formation. The NNA formation in the trigonal bipyramidal voids can also not be enforced by an artificial shrinkage of the unit cell.[10]

As outlined, the presence of NNAs in beryllium metal cannot be simply explained by the presence of convex regions in the radial atomic density. Hence, an electron–electron mediated coupling of the ground state to low-lying excited states, as suggested by Bersuker *et al.* [132], might still be a prerequisite for the formation of NNAs. Another proposal has been made by Madsen *et al.* [133]. These authors suggested that NNAs might be caused by a relatively small extension of the core–shell densities. This brings us to another controversial topic that has been extensively addressed by theoretical and experimental charge density studies: do the core–shells of atoms in metals expand as a consequence of metallic bonding and the presence of itinerant electrons?

In the following, we therefore critically discuss the experimental evidence for a potential core deformation of beryllium. We note that this controversial debate has been initiated by Stewart in 1977 [134] and it also marks one of the first attempts to experimentally determine the total charge density using a multipolar

10) NNA formation is, however, observed at different positions. Indeed, shrinkage of the Be–Be distances between atoms of adjacent hexagonal layers below 2.00 Å triggers the NNA formation in the interatomic region. Upon reduction of the Be–Be distances below 1.90 Å, these features disappear but new ones in the tetrahedral voids show up when the Be–Be distance is further lowered to 1.34 Å. Below that distance margin, NNA formation switches to the octahedral voids. Below 1.11 Å all NNAs vanish.

formalism. However, data deficiencies due to severe extinction effects prevented a precise reconstruction of the static density of beryllium metal at that time. However, for the first time the idea was born that the core density of a metal might be distorted as a consequence of metallic bonding. This fundamental issue gave rise to systematic experimental and theoretical studies providing pros and cons for a potential core expansion or contraction in beryllium. Dovesi suggested in 1982 [135, 136], on the basis of a Compton profile analysis based on periodic LCAO calculations, that the kinetic energy of the core 1s electrons in beryllium is lowered by about 1 eV relative to a free beryllium atom. Assuming a simple particle in the box model, the lowering of the kinetic energy could be connected with an enlarged box size – and thus with an enlargement of the core [137]. In this simple model, the enlarged box size would coincide with a reduction of the slope of the wave function and thus cause a lowering of the kinetic energy and also of the total energy. This argument has been already employed by Ruedenberg in 1962 [138] to account for the valence density deformations of bonded atoms when he concluded: "*By virtue of the uncertainty principle, it is to be expected that this smoothening of the density, inherent in the interference effect, is associated with a lowering of the kinetic energy. It is related in character to the lowering of the kinetic energy of free electrons when the containing box is increased in size*". However, this model is too simplistic and cannot be applied in case of atomic cores as it ignores the decrease of the absolute value of the potential energy during a hypothetical core expansion. Indeed, this is due to the virial theorem which relates the kinetic and the potential energy as $V = -2T$. Furthermore, the Hilbert-space partitioning of core (1s) and valence (2s) states cannot be employed in real space to analyze core contraction/expansion phenomena, because the 2s valence AO of beryllium also contributes to the density at the cusp, as outlined in the case of the carbon atom in diamond earlier. Hence, studies of this phenomenon should not be based on energy partitioning schemes but rather on real space methods such as the analysis of the topology of the electron density distribution. Indeed, inspection of the radial density distributions of free beryllium atoms in the ^{1}S ground state and in Be metal show a minute charge depletion in the proximity of the cusp in the latter case on the basis of periodic LAPW DFT calculations as well as in the experimental EHC refinements: $\rho_{\mathrm{cusp,Be(^1S)}} = 242.8\,\mathrm{e\,Å^{-3}}$; $\rho_{\mathrm{cusp,Be,LAPW}} = 239.6\,\mathrm{e\,Å^{-3}}$; $\rho_{\mathrm{cusp,Be,EHC,theor}} = 239.0\,\mathrm{e\,Å^{-3}}$; $\rho_{\mathrm{cusp,Be,EHC,exp}} = 240.2\,\mathrm{e\,Å^{-3}}$. Accordingly, the charge transfer between the core and valence density in the EHC model is also minute (only 0.006 e when the theoretical LAPW structure factors were employed). Therefore, the core and valence population parameters were fixed in our experimental EHC model ($P_c = 2.0$ and $P_v = 2.0$), while the respective radial screening parameters for core ($\kappa_c = 0.996(1)$) and valence ($\kappa_v = 1.004(18)$) were optimized.[11] The value of the core κ-parameter based on the experimental structure factors is close to

11) VM bank [51], ξ_{2s}=0.956 Bohr^{-1}, Extinction Type II [150]: $\rho_{iso} = 0.12(7)$, $U_{11} = 0.00629(3)$, $U_{33} = 0.00553(3)$, $\kappa_v = 1.004(18)$, $\kappa'_v = 1.04(18)$, $\kappa_c = 0.996(1)$, $P_{20} = -0.003(24)$, $P_{33-} = -0.048(22)$, $P_{40} = 0.031(25)$, $P_{53-} = 0.008(25)$, $P_{60} = 0.054(38)$, $P_{66+} = -0.018(36)$, $R_1 = 0.25\%$, $wR_1 = 0.46\%$, weighting scheme $1/\sigma^2(F_o)$, $\Delta\rho = +0.03/-0.02\,\mathrm{e\,Å^{-3}}$, scale = 1.019(2), $n_l = (2,2,2,3,4,5,6)$ for $0 \leq l \leq 6$; local coordinate system: $z = [0,0,1]$, $y = [1,2,0]$.

unity in the EHC multipolar model and therefore does not support any significant core expansion. In contrast, the above results rather suggests a core depletion in connection with a core contraction. Such a core contraction would mainly originate from the contributions of the 2s states to the core region. As the 2s orbital contributes via its inner radial node directly to the core–shell density, the core depletion appears to be a natural consequence of the shifting of the 2s density into the interatomic regions and the subsequent sp^x hybridization due to the $s \rightarrow p$ promotion of the valence electrons.

We therefore conclude that the core–shell in beryllium appears to be minutely contracted as a consequence of the metallic bonding. Hence, the presence of NNAs in beryllium can neither be explained by an extension of the core–shell densities as suggested by Madsen *et al.* (see subsequent text), nor by a simple superposition of atoms displaying nonconvex regions in their radial atomic density as proposed by Pendás *et al.* [130]. Hence, the presence of NNAs as a highly elusive charge density phenomenon appears to be triggered by chemical bonding effects in extended solids.

References

1. Bragg, W.H. (1921) *Proc. Phys. Soc. London*, **33**, 304–311.
2. Friedrich, W., Knipping, P., and Laue, M. (1913) *Ann. Phys.*, **346**, 971–988.
3. Debye, P. (1915) *Ann. Phys.*, **351**, 809–823.
4. Ewald, P.P. and Hönl, H. (1936) *Ann. Phys.*, **4**, 281–308.
5. Renninger, M. (1937) *Z. Phys.*, **106**, 141–176.
6. Fischer, A., Tiana, D., Scherer, W., Batke, K., Eickerling, G., Svendsen, H., Bindzus, N., and Iversen, B.B. (2011) *J. Phys. Chem. A*, **115**, 13061–13071.
7. Nishibori, E., Sunaoshi, E., Yoshida, A., Aoyagi, S., Kato, K., Takata, M., and Sakata, M. (2006) *Acta Crystallogr. A*, **63**, 43–52.
8. Coppens, P. (2005) *Angew. Chem. Int. Ed.*, **44**, 6810–6811.
9. Svendsen, H., Overgaard, J., Busselez, R., Arnaud, B., Rabiller, P., Kurita, A., Nishibori, E., Sakata, M., Takata, M., and Iversen, B.B. (2010) *Acta Crystallogr. A*, **66**, 458–469.
10. Bader, R.F.W. (1994) *Atoms in Molecules: A Quantum Theory (International Series of Monographs on Chemistry)*, Oxford University Press, Oxford.
11. Matta, C.F. and Boyd, R.J. (2007) *The Quantum Theory of Atoms in Molecules*, Wiley-VCH, Weinheim.
12. Gatti, C. and Macchi, P. (2012) *Modern Charge-Density Analysis*, Springer, Dordrecht, New York.
13. Scherer, W., Herz, V., and Hauf, C. (2012) in *Electron Density and Chemical Bonding I* (ed. D. Stalke), Springer, Berlin, Heidelberg, pp. 159–207.
14. Jayatilaka, D. and Grimwood, D.J. (2001) *Acta Crystallogr. A*, **57**, 76–86.
15. Grimwood, D.J. and Jayatilaka, D. (2001) *Acta Crystallogr. A*, **57**, 87–100.
16. Grimwood, D.J., Bytheway, I., and Jayatilaka, D. (2003) *J. Comput. Chem.*, **24**, 470–483.
17. Farrugia, L.J. and Macchi, P. (2012) in *Electron Density and Chemical Bonding I* (ed. D. Stalke), Springer, Berlin, Heidelberg, pp. 127–158.
18. Jayatilaka, D. (2012) in *Modern Charge-Density Analysis* (eds C. Gatti and P. Macchi), Springer, Dordrecht, pp. 213–257.
19. Shirane, G. (2002) *Neutron Scattering with a Triple-Axis Spectrometer: Basic Techniques*, Cambridge University Press, New York.
20. Lovesey, S.W. and Collins, S.P. (1996) *X-Ray Scattering and Absorption by*

Magnetic Materials, Clarendon Press, Oxford.

21. Deutsch, M., Claiser, N., Pillet, S., Chumakov, Y., Becker, P., Gillet, J.-M., Gillon, B., Lecomte, C., and Souhassou, M. (2012) *Acta Crystallogr. A*, **68**, 675–686.
22. Coppens, P. (2011) *J. Phys. Chem. Lett.*, **2**, 616–621.
23. Elsaesser, T. and Woerner, M. (2010) *Acta Crystallogr. A*, **66**, 168–178.
24. Legrand, V., Pillet, S., Souhassou, M., Lugan, N., and Lecomte, C. (2006) *J. Am. Chem. Soc.*, **128**, 13921–13931.
25. Glover, T.E., Fritz, D.M., Cammarata, M., Allison, T.K., Coh, S., Feldkamp, J.M., Lemke, H., Zhu, D., Feng, Y., Coffee, R.N., Fuchs, M., Ghimire, S., Chen, J., Shwartz, S., Reis, D.A., Harris, S.E., and Hastings, J.B. (2012) *Nature*, **488**, 603–608.
26. Fast, H. and Welsh, H.L. (1972) *J. Mol. Spectrosc.*, **41**, 203–221.
27. Dowling, J.M. and Stoicheff, B.P. (1959) *Can. J. Phys.*, **37**, 703–721.
28. Popelier, P., Lenstra, A.T.H., Alsenoy, C.V., and Geise, H.J. (1991) *Struct. Chem.*, **2**, 3–9.
29. Spackman, M.A., Byrom, P.G., Alfredsson, M., and Hermansson, K. (1999) *Acta Crystallogr. A*, **55**, 30–47.
30. Dovesi, R., Causa, M., Orlando, R., Roetti, C., and Saunders, V.R. (1990) *J. Chem. Phys.*, **92**, 7402–7411.
31. Coppens, P., Dam, J., Harkema, S., Feil, D., Feld, R., Lehmann, M.S., Goddard, R., Krüger, C., Hellner, E., Johansen, H., Larsen, F.K., Koetzle, T.F., McMullan, R.K., Maslen, E.N., and Stevens, E.D. (1984) *Acta Crystallogr. A*, **40**, 184–195.
32. Spackman, M.A., Munshi, P., and Dittrich, B. (2007) *ChemPhysChem*, **8**, 2051–2063.
33. Koritsanszky, T.S. and Coppens, P. (2001) *Chem. Rev.*, **101**, 1583–1628.
34. Spackman, M.A. (2013) *Phys. Scr.*, **87**, 048103.
35. Cochran, W. (1956) *Acta Crystallogr.*, **9**, 924–928.
36. Tomiie, Y. (1958) *J. Phys. Soc. Jpn.*, **13**, 1030–1037.
37. Chieux, P. and Dachs, H. (1978) *Neutron Diffraction*, Springer-Verlag, Berlin.
38. Coppens, P. (1997) *X-ray Charge Densities and Chemical Bonding*, Oxford University Press, Oxford.
39. Scherer, W. and McGrady, G.S. (2004) *Angew. Chem. Int. Ed.*, **43**, 1782–1806.
40. Scherer, W., Herz, V., Brück, A., Hauf, C., Reiner, F., Altmannshofer, S., Leusser, D., and Stalke, D. (2011) *Angew. Chem. Int. Ed.*, **50**, 2845–2849.
41. Scherer, W., Hauf, C., Presnitz, M., Scheidt, E.-W., Eickerling, G., Eyert, V., Hoffmann, R.-D., Rodewald, U.C., Hammerschmidt, A., Vogt, C., and Pöttgen, R. (2010) *Angew. Chem. Int. Ed.*, **49**, 1578–1582.
42. Eickerling, G., Hauf, C., Scheidt, E.-W., Reichert, L., Schneider, C., Muñoz, A., Lopez-Moreno, S., Romero, A.H., Procher, F., André, G., Pöttgen, R., and Scherer, W. (2013) *Z. Allg. Anorg. Chem.*, **639**, 1985–1995.
43. Scherer, W., Eickerling, G., Shorokhov, D., Gullo, E., McGrady, G.S., and Sirsch, P. (2006) *New J. Chem.*, **30**, 309–312.
44. Reisinger, A., Trapp, N., Krossing, I., Altmannshofer, S., Herz, V., Presnitz, M., and Scherer, W. (2007) *Angew. Chem. Int. Ed.*, **46**, 8295–8298.
45. Himmel, D., Trapp, N., Krossing, I., Altmannshofer, S., Herz, V., Eickerling, G., and Scherer, W. (2008) *Angew. Chem. Int. Ed.*, **47**, 7798–7801.
46. Stewart, R.F. (1977) *Isr. J. Chem.*, **16**, 124–131.
47. Hansen, N.K. and Coppens, P. (1978) *Acta Crystallogr. A*, **34**, 909–921.
48. Griffin, J.F. and Coppens, P. (1975) *J. Am. Chem. Soc.*, **97**, 3496–3505.
49. Coppens, P., Guru Row, T.N., Leung, P., Stevens, E.D., Becker, P.J., and Yang, Y.W. (1979) *Acta Crystallogr. A*, **35**, 63–72.
50. Slater, J.C. (1932) *Phys. Rev.*, **42**, 33–43.
51. Volkov, A., Macchi, P., Farrugia, L.J., Gatti, C., Mallinson, P., Richter, R., and Koritsanszky, T. (2006) A Computer Program for Multipole Refinement, Topological Analysis of Charge Densities, and Evaluation of Intermolecular Energies from Experimental or Theoretical Structure Factors. XD2006 (Version 5.42).

52. te Velde, G., Bickelhaupt, F.M., Baerends, E.J., Fonseca Guerra, C., van Gisbergen, S.J.A., Snijders, J.G., and Ziegler, T. (2001) *J. Comput. Chem.*, **22**, 931–967.
53. ADF2009.01, SCM, Theoretical Chemistry, Vrije Universiteit, Amsterdam, The Netherlands.
54. Fonseca Guerra, C., Snijders, J.G., te Velde, G., and Baerends, E.J. (1998) *Theor. Chem. Acc.*, **99**, 391–403.
55. Flierler, U. and Stalke, D. (2012) in *Electron Density and Chemical Bonding I* (ed. D. Stalke), Springer, Berlin, Heidelberg, pp. 75–99.
56. Overgaard, J., Grin, Y., Takata, M., and Iversen, B.B. (2012) in *Modern Charge-Density Analysis* (eds C. Gatti and P. Macchi), Springer, Dordrecht, pp. 469–504.
57. Spackman, M.A. (2012) in *Modern Charge-Density Analysis* (eds C. Gatti and P. Macchi), Springer, Dordrecht, pp. 553–572.
58. Overgaard, J. and Iversen, B.B. (2010) in *Electron Density and Chemical Bonding I* (ed. D. Stalke), Springer, Berlin, Heidelberg, pp. 53–74.
59. Flierler, U., Stalke, D., and Farrugia, L.J. (2012) in *Modern Charge-Density Analysis* (eds C. Gatti and P. Macchi), Springer, Dordrecht, pp. 435–467.
60. Jelsch, C., Domagała, S., Guillot, B., Liebschner, D., Fournier, B., Pichon-Pesme, V., and Lecomte, C. (2012) in *Modern Charge-Density Analysis* (eds C. Gatti and P. Macchi), Springer, Dordrecht, pp. 527–552.
61. Scherer, W., Eickerling, G., Hauf, C., Presnitz, M., Scheidt, E.-W., Eyert, V., and Pöttgen, R. (2012) in *Modern Charge-Density Analysis* (eds C. Gatti and P. Macchi), Springer, pp. 359–385.
62. Tsirelson, V.G. and Ozerov, R.P. (1996) *Electron Density and Bonding in Crystals: Principles, Theory and X-Ray Diffraction Experiments in Solid State Physics and Chemistry*, Taylor & Francis Group, New York.
63. Clementi, E. and Raimondi, D.L. (1963) *J. Chem. Phys.*, **38**, 2686–2689.
64. The Elk FP-LAPW Code (2012) Program ELK Version 1.4.18, *http://elk.sourceforge.net*.
65. Perdew, J.P., Burke, K., and Ernzerhof, M. (1996) *Phys. Rev. Lett.*, **77**, 3865–3868.
66. Su, Z. and Coppens, P. (1998) *Acta Crystallogr. A*, **54**, 646–652.
67. Macchi, P. and Coppens, P. (2001) *Acta Crystallogr. A*, **A57**, 656–662.
68. Dovesi, R., Saunders, V.R., Roetti, C., Orlando, R., Zicovich-Wilson, C.M., Pascale, F., Civalleri, B., Doll, K., Harrison, N., Bush, J., D'Arco, P., and Llunell, M. (2006) *CRYSTAL06 User's Manual*, Università di Torino, Torino.
69. Popelier, P.L.A., Burke, J., and Malcolm, N.O.J. (2003) *Int. J. Quantum Chem.*, **92**, 326–336.
70. Spackman, M.A. and Mitchell, A.S. (2001) *Phys. Chem. Chem. Phys.*, **3**, 1518–1523.
71. Gatti, C., MacDougall, P.J., and Bader, R.F.W. (1988) *J. Chem. Phys.*, **88**, 3792–3804.
72. Boyd, R.J. and Wang, L.-C. (1989) *J. Comput. Chem.*, **10**, 367–375.
73. Volkov, A., Gatti, C., Abramov, Y., and Coppens, P. (2000) *Acta Crystallogr. A*, **56**, 252–258.
74. Tognetti, V. and Joubert, L. (2011) *J. Phys. Chem. A*, **115**, 5505–5515.
75. Swaminathan, S., Craven, B.M., Spackman, M.A., and Stewart, R.F. (1984) *Acta Crystallogr. B*, **40**, 398–404.
76. Howard, S.T., Hursthouse, M.B., Lehmann, C.W., and Poyner, E.A. (1995) *Acta Crystallogr. B*, **51**, 328–337.
77. Spackman, M.A. and Byrom, P.G. (1996) *Acta Crystallogr. B*, **52**, 1023–1035.
78. Zavodnik, V., Stash, A., Tsirelson, V., De Vries, R., and Feil, D. (1999) *Acta Crystallogr. B*, **55**, 45–54.
79. Volkov, A., Abramov, Y., Coppens, P., and Gatti, C. (2000) *Acta Crystallogr. A*, **56**, 332–339.
80. Pillet, S., Souhassou, M., Lecomte, C., Schwarz, K., Blaha, P., Rerat, M., Lichanot, A., and Roversi, P. (2001) *Acta Crystallogr. A*, **57**, 290–303.
81. Volkov, A. and Coppens, P. (2001) *Acta Crystallogr. A*, **57**, 395–405.
82. Bytheway, I., Chandler, G.S., and Figgis, B.N. (2002) *Acta Crystallogr. A*, **58**, 451–459.

83. Lippmann, T., Blaha, P., Andersen, N.H., Poulsen, H.F., Wolf, T., Schneider, J.R., and Schwarz, K.-H. (2003) *Acta Crystallogr. A*, **59**, 437–451.
84. Coppens, P. and Volkov, A. (2004) *Acta Crystallogr. A*, **60**, 357–364.
85. Poulain-Paul, A., Nassour, A., Jelsch, C., Guillot, B., Kubicki, M., and Lecomte, C. (2012) *Acta Crystallogr. A*, **68**, 715–728.
86. Koritsanszky, T., Volkov, A., and Chodkiewicz, M. (2012) in *Electron Density and Chemical Bonding II* (ed. D. Stalke), Springer, Berlin, Heidelberg, pp. 1–25.
87. Spackman, M.A. (1991) *Acta Crystallogr. A*, **47**, 420–427.
88. Slack, G.A. (1973) *J. Phys. Chem. Solids*, **34**, 321–335.
89. Desnoyehs, J.E. and Morrison, J.A. (1958) *Philos. Mag.*, **3**, 42–48.
90. Warren, J.L., Yarnell, J.L., Dolling, G., and Cowley, R.A. (1967) *Phys. Rev.*, **158**, 805–808.
91. Blackman, M. and Fowler, R.H. (1937) *Math. Proc. Cambridge Philos. Soc.*, **33**, 380–384.
92. Stewart, R.F. (1973) *Acta Crystallogr. A*, **29**, 602–605.
93. Sierraalta, A. and Ruette, F. (1994) *J. Comput. Chem.*, **15**, 313–321.
94. Bader, R.F.W., Gillespie, R.J., and Martćn, F. (1998) *Chem. Phys. Lett.*, **290**, 488–494.
95. Tiana, D., Francisco, E., Blanco, M.A., and Pendás, A.M. (2009) *J. Phys. Chem. A*, **113**, 7963–7971.
96. Vyboishchikov, S.F., Sierraalta, A., and Frenking, G. (1997) *J. Comput. Chem.*, **18**, 416–429.
97. Cioslowski, J. and Piskorz, P. (1996) *Chem. Phys. Lett.*, **255**, 315–319.
98. Keith, T.A. and Frisch, M.J. (2011) *J. Phys. Chem. A*, **115**, 12879–12894.
99. Krapp, A. and Frenking, G. (2008) *Angew. Chem. Int. Ed.*, **47**, 7796–7797.
100. Dyall, K.G. (2007) *Theor. Chem. Acc.*, **117**, 483–489.
101. Eickerling, G. and Reiher, M. (2008) *J. Chem. Theory Comput.*, **4**, 286–296.
102. Frisch, M.J., Trucks, G.W., Schlegel, H.B., Scuseria, G.E., Robb, M.A., Cheeseman, J.R., Scalmani, G., Barone, V., Mennucci, B., Petersson, G.A., Nakatsuji, H., Caricato, M., Li, X., Hratchian, H.P., Izmaylov, A.F., Bloino, J., Zheng, G., Sonnenberg, J.L., Hada, M., Ehara, M., Toyota, K., Fukuda, R., Hasegawa, J., Ishida, M., Nakajima, T., Honda, Y., Kitao, O., Nakai, H., Vreven, T., Montgomery, J., Peralta, J.E., Ogliaro, F., Bearpark, M., Heyd, J.J., Brothers, E., Kudin, K.N., Staroverov, V.N., Kobayashi, R., Normand, J., Raghavachari, K., Rendell, A., Burant, J.C., Iyengar, S.S., Tomasi, J., Cossi, M., Rega, N., Millam, J.M., Klene, M., Knox, J.E., Cross, J.B., Bakken, V., Adamo, C., Jaramillo, J., Gomperts, R., Stratmann, R.E., Yazyev, O., Austin, A.J., Cammi, R., Pomelli, C., Ochterski, J.W., Martin, R.L., Morokuma, K., Zakrzewski, V.G., Voth, G.A., Salvador, P., Dannenberg, J.J., Dapprich, S., Daniels, A.D., Farkas, Ö., Foresman, J.B., Ortiz, J.V., Cioslowski, J., and Fox, D.J. (2009) Gaussian09 Revision B.01, Gaussian, Inc., Wallingford, CT.
103. Eickerling, G., Mastalerz, R., Herz, V., Scherer, W., Himmel, H.-J., and Reiher, M. (2007) *J. Chem. Theor. Comput.*, **3**, 2182–2197.
104. Bader, R.F.W., MacDougall, P.J., and Lau, C.D.H. (1984) *J. Am. Chem. Soc.*, **106**, 1594–1605.
105. Morse, P.M. and Feshbach, H. (1953) *Methods of Theoretical Physics*, McGraw-Hill.
106. Bader, R.F.W. and Beddall, P.M. (1972) *J. Chem. Phys.*, **56**, 3320–3329.
107. Bader, R.F.W. and Essén, H. (1984) *J. Chem. Phys.*, **80**, 1943–1960.
108. Sagar, R.P., Ku, A.C.T., Smith, V.H., and Simas, A.M. (1988) *J. Chem. Phys.*, **88**, 4367–4374.
109. Shi, Z. and Boyd, R.J. (1988) *J. Chem. Phys.*, **88**, 4375–4377.
110. Chan, W.-T. and Hamilton, I.P. (1998) *J. Chem. Phys.*, **108**, 2473–2485.
111. McGrady, G.S., Haaland, A., Verne, H.P., Volden, H.V., Downs, A.J., Shorokhov, D., Eickerling, G., and Scherer, W. (2005) *Chem. Eur. J.*, **11**, 4921–4934.
112. Dewar, M.J.S. (1951) *Bull. Soc. Chim. Fr.*, **18**, C71–79.

113. Chatt, J. and Duncanson, L.A. (1953) *J. Chem. Soc.*, 2939–2947.
114. Holladay, A., Leung, P., and Coppens, P. (1983) *Acta Crystallogr. A*, **39**, 377–387.
115. Popelier, P.L.A. and Logothetis, G. (1998) *J. Organomet. Chem.*, **555**, 101–111.
116. Pantazis, D.A., McGrady, J.E., Besora, M., Maseras, F., and Etienne, M. (2008) *Organometallics*, **27**, 1128–1134.
117. Bader, R.F.W., Matta, C.F., and Cortés-Guzmán, F. (2004) *Organometallics*, **23**, 6253–6263.
118. McGrady, G.S., Sirsch, P., Chatterton, N.P., Ostermann, A., Gatti, C., Altmannshofer, S., Herz, V., Eickerling, G., and Scherer, W. (2009) *Inorg. Chem.*, **48**, 1588–1598.
119. Pidun, U. and Frenking, G. (1995) *Organometallics*, **14**, 5325–5336.
120. Frenking, G. and Fröhlich, N. (2000) *Chem. Rev.*, **100**, 717–774.
121. Macchi, P., Proserpio, D.M., and Sironi, A. (1998) *J. Am. Chem. Soc.*, **120**, 1447–1455.
122. Rohrmoser, B., Eickerling, G., Presnitz, M., Scherer, W., Eyert, V., Hoffmann, R.-D., Rodewald, U.C., Vogt, C., and Pöttgen, R. (2007) *J. Am. Chem. Soc.*, **129**, 9356–9365.
123. Larsen, F.K. and Hansen, N.K. (1984) *Acta Crystallogr. B*, **40**, 169–179.
124. Besnainou, S., Roux, M., and Daudel, R. (1955) *C.R. Acad. Sci. Paris*, **241**, 311.
125. Sakata, M. and Sato, M. (1990) *Acta Crystallogr. A*, **46**, 263–270.
126. Iversen, B.B., Larsen, F.K., Souhassou, M., and Takata, M. (1995) *Acta Crystallogr. B*, **51**, 580–591.
127. Iversen, B.B., Jensen, J.L., and Danielsen, J. (1997) *Acta Crystallogr. A*, **53**, 376–387.
128. de Vries, R.Y., Briels, W.J., and Feil, D. (1996) *Phys. Rev. Lett.*, **77**, 1719–1722.
129. Jayatilaka, D. (1998) *Phys. Rev. Lett.*, **80**, 798–801.
130. Pendás, A.M., Blanco, M.A., Costales, A., Sánchez, P.M., and Luaña, V. (1999) *Phys. Rev. Lett.*, **83**, 1930–1933.
131. Reiher, M. (1998) *Development and Implementation of Numerical Algorithms for the Solution of Multi-Configuration Self-Consistent Field Equations for Relativistic Atomic Structure Calculations*, Bielefeld University.
132. Bersuker, G.I., Peng, C., and Boggs, J.E. (1993) *J. Phys. Chem.*, **97**, 9323–9329.
133. Madsen, G.K.H., Blaha, P., and Schwarz, K. (2002) *J. Chem. Phys.*, **117**, 8030–8035.
134. Stewart, R.F. (1977) *Acta Crystallogr. A*, **33**, 33–38.
135. Dovesi, R., Angonoa, G., and Causà, M. (1982) *Philos. Mag. B*, **45**, 601–606.
136. Dovesi, R., Pisani, C., Ricca, F., and Roetti, C. (1982) *Phys. Rev. B*, **25**, 3731–3739.
137. Burdett, J.K. (1997) *Chemical Bonds: A Dialog*, John Wiley & Sons, Ltd, Chichester, New York.
138. Ruedenberg, K. (1962) *Rev. Mod. Phys.*, **34**, 326–376.
139. McMullan, R.K., Kvick, Å., and Popelier, P. (1992) *Acta Crystallogr. B*, **48**, 726–731.
140. Laidig, K.E. (1994) *Chem. Phys. Lett.*, **225**, 285–292.
141. Matta, C.F. (2010) *J. Comput. Chem.*, **31**, 1297–1311.
142. Kresse, G. and Hafner, J. (1993) *Phys. Rev. B*, **47**, 558–561.
143. Kresse, G. and Furthmüller, J. (1996) *Phys. Rev. B*, **54**, 11169–11186.
144. Kresse, G. and Hafner, J. (1994) *Phys. Rev. B*, **49**, 14251–14269.
145. Kresse, G. and Furthmüller, J. (1996) *Comput. Mater. Sci.*, **6**, 15–50.
146. Togo, A., Oba, F., and Tanaka, I. (2008) *Phys. Rev. B*, **78**, 134106.
147. Kohout, M., Savin, A., and Preuss, H. (1991) *J. Chem. Phys.*, **95**, 1928–1942.
148. Hebben, N., Himmel, H.-J., Eickerling, G., Herrmann, C., Reiher, M., Herz, V., Presnitz, M., and Scherer, W. (2007) *Chem. Eur. J.*, **13**, 10078–10087.
149. Vogt, C., Hoffmann, R.-D., Rodewald, U.C., Eickerling, G., Presnitz, M., Eyert, V., Scherer, W., and Pöttgen, R. (2009) *Inorg. Chem.*, **48**, 6436–6451.
150. Becker, P.J. and Coppens, P. (1974) *Acta Crystallogr. A*, **30**, 129–147.

10
The ELF Perspective of chemical bonding

Yuri Grin, Andreas Savin, and Bernard Silvi

10.1
Introduction

10.1.1
Context

10.1.1.1 Chemical structure given by the electronic structure

In the drawings of molecular or crystal structures, atoms are connected by lines symbolizing bonds. At the beginning of the twentieth century it was assumed that the lines represent pairs of electrons [1]. The modern, quantum mechanical picture of a molecule and – especially – of solids seems to be disconnected from it, as pairs of electrons are not easily recognizable in the molecular Schrödinger equation. In the latter, one only has the information that electrons and nuclei are present, and interact via Coulomb forces. The electron pairs are hidden in the Pauli principle that determines the structure of the wave function. The latter depends on the coordinates of all the electrons, and their spins, and is too complicated to be directly related to the classical, nineteenth century representation of molecular structure. Much effort has been devoted to establish a connection between the classical and the quantum perspectives, and this chapter reviews one of these approaches.

10.1.1.2 Molecules and crystals are objects in three dimensions

Molecules are seen as objects in three-dimensional space: ball-and-stick models are omnipresent in chemistry. For crystals, such models are often more difficult to perceive, but in such situations they are often replaced by other spatial entities, such as polyhedra packings. It is commonly accepted that the three-dimensional view, initiated by Van't Hoff's tetrahedra, relies on the electronic structure. We will thus need a tool that transforms the many-body picture of quantum mechanics in Hilbert space to the three dimensions of real space.

The Chemical Bond: Fundamental Aspects of Chemical Bonding, First Edition.
Edited by Gernot Frenking, Sason Shaik.

10.1.2 Choices

10.1.2.1 Fix the nuclei

We consider fixed nuclear positions.[1] This does not mean that we cannot treat nuclear movement, but that we have to let the electronic structure evolve in time with nuclear displacements.

10.1.2.2 Compute, then analyze

Electrons are moving, and quantum mechanics tells us that deciding about the position gives infinite uncertainty to momentum. A simple electrostatic arrangement of nuclei and electrons is thus doomed to fail. Models of different complexity have been developed, such as Lewis' cubic atom [1], Linnett's double quartet [3], or Gillespie's VESPR [4]. Can such phenomenological chemical models be recovered from quantum mechanics? Should they?

One can defend the viewpoint that with more and more sophisticated quantum mechanical approximations, and with increasing computing power, we can ask the computer to provide us with any observable data we want to know about. Our viewpoint is, however, that simple models are useful for understanding and thinking about molecules and crystals. Here again, a decision has to be taken. One can decide to use simple models that do not need more than a back-of-an-envelope derivation. An alternative view is to use computers to analyze the complicated data produced by quantum mechanical calculations, and provide simple explanations in terms of a model. Although the latter approach seems not to have predictive power, accumulated experience with it does not exclude a later use for simple, computer-independent explanations. We will pursue in this chapter an approach that uses quantum mechanical computations, and extracts data from them in order to provide explanations about the electronic structure of molecules and crystals.

10.1.2.3 Choose regions of space

A picture that would be compatible with the quantum mechanical picture is to search for the most probable spatial arrangement of electrons [5, 6]. It turns out, however, that this is quite sensitive to the quality of the wave function used. The single Slater determinant (the most frequently used approximation for a wave function) yields an arrangement qualitatively different from that provided by better quality (correlated) wave functions (see, e.g., [7, 8]). While for Slater determinants electrons get paired at the same point of space, the two electrons choose different positions in space when better quality wave functions are used. One also notices that electrons may do it in different ways, corresponding to different maxima of the wave function squared [7]. Furthermore, it may happen that the probability of a given electron arrangement is decreasing rapidly around the global maximum, and a lower, local maximum is "broader." In such a situation, one might find the latter situation to represent better the electron arrangement in a molecule. Taking

1) For a recent discussion of extracting the nuclear structure from quantum mechanical calculations, see, for example, Ref. [2].

into account these remarks, it seems preferable to search for domains of space that can be assigned to electron arrangements rather than to search for specific electron locations. A view in real space was already promoted by Daudel [9] and by Bader [10].

10.2 Definitions

10.2.1 Definition of the Electron Localization Function (ELF)

As the electron pair takes a central place in chemistry (bond, lone pair, ...), we will concentrate in this chapter on a method that, starting from a given wave function, defines a region of space that can be associated with an electron pair. The function we will deal with is called the electron localization function (ELF), and was introduced by Becke and Edgecombe in 1990. Its definition, for a closed-shell Slater determinant, is given in the box *What is ELF?*. This function is obtained in each point of space, once the wave function is given.

Box 10.1 What is ELF?

For a system described by a closed-shell single-Slater determinant wave function, the electron localization function (ELF) is calculated starting from the orbitals building this determinant, φ_i that are functions of the position vector **r**. Thus, ELF depends on **r**. The explicit formula for ELF is

$$\chi_\sigma(\mathbf{r}) = \frac{D_\sigma(\mathbf{r})}{D_\sigma^0(\mathbf{r})}, \qquad ELF = \eta(\vec{r}) = \frac{1}{1+\chi(\vec{r})^2},$$

in which

$$D_\sigma(\mathbf{r}) = t_\sigma(\mathbf{r}) - \frac{1}{4}\frac{|\nabla\rho_\sigma(\mathbf{r})|^2}{\rho_\sigma(\mathbf{r})}$$

and

$$\rho_\sigma(\mathbf{r}) = \sum_{i=1}^{N}|\varphi_i|^2; \; t_\sigma(\mathbf{r}) = \sum_{i=1}^{N}|\nabla\varphi_i|^2$$

is the difference of the actual definite positive kinetic energy density $t_\sigma(\mathbf{r})$ and the von Weizsacker kinetic energy density functional [72], whereas

$$D_\sigma^0(\mathbf{r}) = \frac{3}{5}(6\pi^2)^{\frac{2}{3}}\rho_\sigma^{\frac{5}{3}}(\mathbf{r})$$

In practice, the φ_i (and their derivatives) are computed on a grid of points, and the given formula is used to obtain ELF on the same grid. Computer programs drawing isosurfaces in 3D for functions given on a grid are commonly available. Programs for locating the maxima and the basins for functions given on a grid also exist.

For a generalization of ELF to time-dependent processes, see Refs [73, 74].

Maybe the best way to approach ELF is to imagine a hypothetical system, in which space is partitioned in a number of regions (localization regions) equal to the number of pairs of electrons. Furthermore, in each of these regions, there is one, and only one pair of electrons.[2] If we choose such a region of space, we are sure to find one and only one pair of electrons in it. (The probability of finding a pair of electrons in a region of space is maximal, equal to 1, when we choose it to be the localization region defined before.) For this hypothetical system, if the wave function is described by a single Slater determinant, ELF is equal to 1 inside these regions, and equal to 0 at their border [11]. For a realistic system, the probability of finding a pair of electrons in it may not reach its maximal value, and ELF may not reach these limiting values. However, by definition, the function ELF takes in every point of space a value η that lies between 0 and 1. We will identify regions where η is large. These are the regions associated with electron pairs. Furthermore, the regions where η is low will be considered to belong to the border regions.

There are several alternative definitions for ELF, all coinciding for the closed-shell single-determinant wave function.

10.2.2 Definition of auxiliary quantities

10.2.2.1 ELF maxima

Let us now discuss in more detail how to look at ELF that is a function of the spatial coordinates, x, y, z. First, the maxima of the ELF can be identified. We see in Figure 10.1 those present in the ethylene molecule. We find a maximum corresponding to electron pairs: to the two C cores ($\eta = 1.00$), centered on the C nuclei, to the four C–H bonds ($\eta = 1.00$), located close to the positions of the protons, and two maxima of the C–C for the double bond, suggesting "banana bonds" ($\eta = 0.93$).

We can connect the maxima of ELF in the ethylene molecule around a given C nucleus. In this case we obtain tetrahedra, reminding us of those of van't Hoff [12] as can be seen in Figure 10.1.[3]

(a) (b)

Figure 10.1 (a) Maxima of ELF in the ethylene molecule, marked by spheres, corresponding to the C cores, C–H, and C–C bonds, colored magenta, gray and green, respectively. (b) The maxima are connected to form tetrahedra.

2) In order to analyze open-shell systems, one can trivially extend this pictures, including spatial domains where one single unpaired electron is present.

3) In more complex molecules, such a construction does not yield van't Hoff pictures.

10.2.2.2 Spatial regions: f-localization domains

We can use different approaches to define two types of regions around the maxima of ELF [13]:

- f-localization domains,
- basins.

Let us first discuss the f-localization domains. For a given value f, the f-localization domain contains all points where ELF takes values larger than f ($\eta > f$). They can be visualized by isosurfaces with $\eta = f$. As ELF is defined between 0 and 1, so must be f. For example, we see in Figure 10.2a, such an isosurface corresponding to $f = 0.8$.

We start by choosing a value of f slightly lower than the largest value that ELF can reach in the given molecule. The isosurface encloses the points in three-dimensional space where this maximum (or maxima) is reached. As f is lowered, other maxima start to be enclosed by the isosurface. In the case of the ethylene molecule, Figure 10.2b, $f = 0.99$ produces a surface that encloses the maxima corresponding to the C cores and C–H bonds. For $f = 0.93$, a domain also encloses each of the maxima corresponding to the C–C double bond. As long as the value of f is not too low, the isosurface encloses individual maxima. As the chosen value f is lowered, regions enclosing several maxima are formed. In ethylene, this happens for $f = 0.92$: the two domains of the double bond maxima merge. Around $f = 0.66$, domains of the C–H bonds merge pairwise, and at a slightly lower value of f these merge with the domain of the C–C double bond: all valence domains have merged. For very low values of f, all domains have merged, and all maxima lie within the f-localization domain.

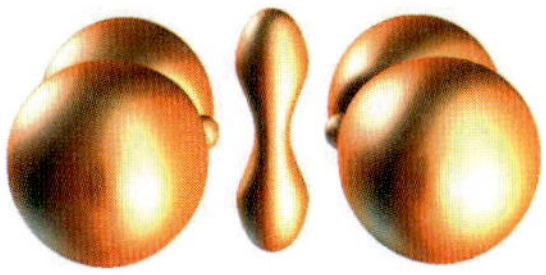

(a)

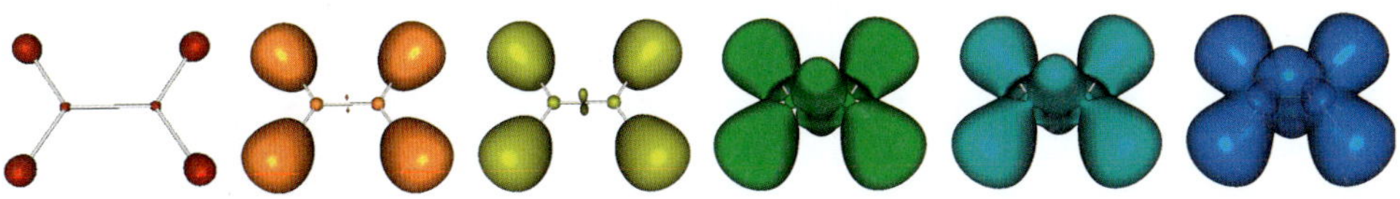

(b)

Figure 10.2 (a) Isosurface of the electron localization function in the ethylene molecule, corresponding to the value $f = 0.8$. (b) ELF f-localization domains in the ethylene molecule, from left to right: (i) $f = 0.99$, red, around maxima of C–H bonds, and the C nuclei (cores); (ii) $f = 0.93$, orange, domains for the double C–C bond show up; (iii) $f = 0.92$, yellow, the two domains corresponding to the double bond have merged; (iv) $f = 0.65$, green, the domains of the adjacent C–H bonds have merged; (v) $f = 0.64$, cyan, all valence domains have merged; (vi) $f = 0.50$, blue.

As the isosurfaces are already in three dimensions, one could imagine visualizing all of them in a movie by a continuous change of the chosen value for f. In such a movie, time would be the fourth dimension corresponding to f. Alternatively, one can take snapshots, as was done in Figure 10.2b.

10.2.2.3 **Bifurcation diagrams**

We have seen that when lowering the value f, the topology of the f-localization domains can change: domains merge. Notice, however, that although the shape of the valence domains looks different for $f = 0.64$ and $f = 0.50$, they are not different in a topological sense, that is, one domain can be deformed into the other without cutting or gluing. (Although the coffee mug and the doughnut look different, a topologist does not distinguish them.) One can try to catch only the values of f where the topology changes (domains merge or separate). This can be done by using a bifurcation diagram that shows a line for each isolated spatial domain versus the value of η. When two domains merge, the corresponding two lines merge into a single one. We can see such a diagram in Figure 10.3, for ethylene. We have chosen in this figure the vertical axis to show the change in f. We see

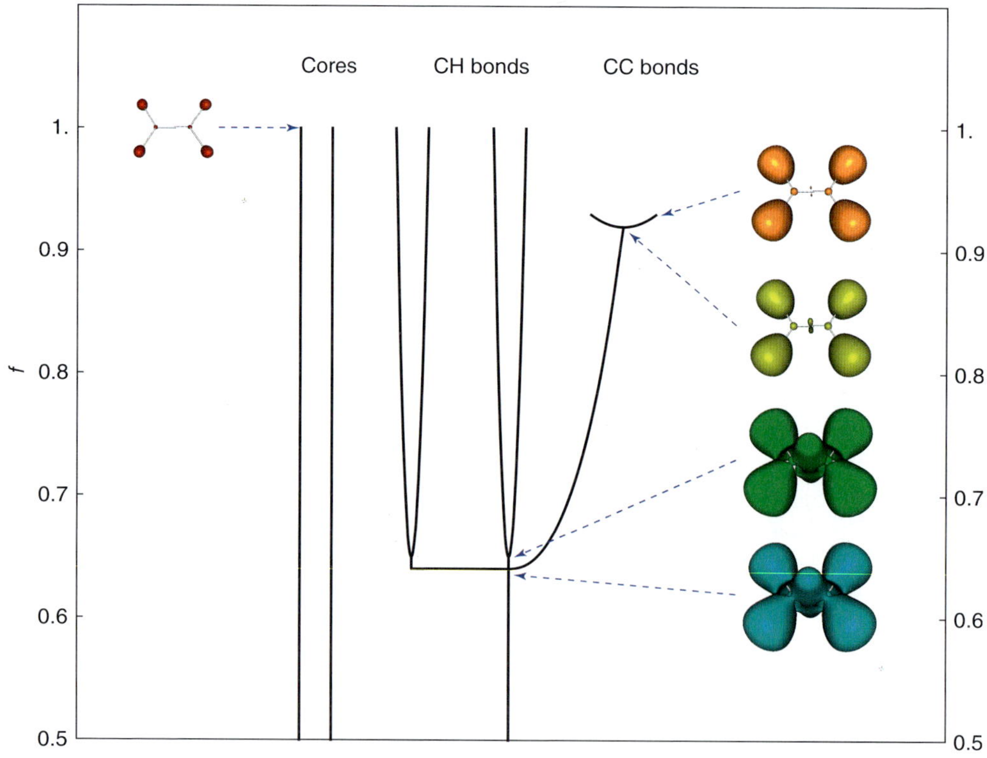

Figure 10.3 Bifurcation diagram for ELF in the ethylene molecule. Plots of f-localization domains, as in Figure 10.2b, are added for guidance; dashed arrows connect the figures to the corresponding bifurcation point in the diagram.

that the maxima for the C cores and the C–H maxima show up first, then those for the C–C double bond. For a slightly lower value of f, the latter merge, and so do the lines. For an even lower value, domains of C–H bonds merge pairwise, as do the corresponding curves. And at $f = 0.64$ all valence domains have merged.

10.2.2.4 **Spatial regions: Basins**

To define basins, we "climb" from each point of space on the steepest path to a maximum. All points connected to the same maximum form a region of space called *basin*.

The ELF basins corresponding to the two C–C bonds and the two C cores, found in ethylene are shown in Figure 10.4. The C–H basins are not shown, as they would cover the basins presented in the figure. Notice that the upper C–C basin appears cut, owing to the finite box used in the calculation.

Basins are partitioning space. Thus, it is possible to visualize a single basin, but when several basins are shown, one basin may hide another. An alternative is to indicate the basins (or the types of basins) by coloring the f-localization domains: each point on the isosurface gets a color according to the basin it belongs to. For example, Figure 10.5 shows the same isosurface as Figure 10.2a. However, the points of the surface are colored differently when they belong to core, C–H, or C–C basins.

10.2.2.5 **ELF terminology**

With ELF, one essentially searches to recover the Lewis picture by associating the basins to chemical concepts. One could thus use the classical terminology of

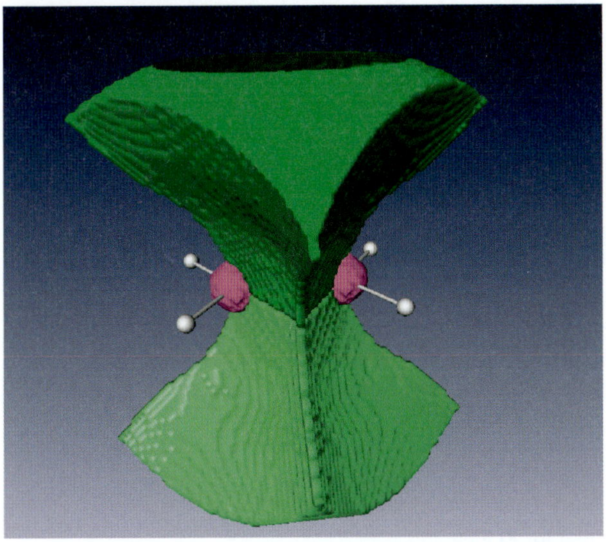

Figure 10.4 ELF basins in the C_2H_4 molecule. The core basins are colored in magenta, the two C–C basins- in dark and light green, C–H basins are not shown for clarity.

chemistry, and speak about bonds, lone pairs, and so on. However, for the sake of rigor, one often uses a specific language that is explained in the box *ELF terminology*.

Box 10.2 ELF terminology

The ELF topological analysis of the bonding in molecules and crystalline solids often uses a vocabulary that is not familiar for a great number of experimentally working chemists. This vocabulary takes its origin, on the one hand, from the mathematical theory, the theory of dynamical systems [75, 76], used to perform the partition of the molecular space and, on the other hand, from our intention to avoid any confusion between the concepts specific to the ELF approach and other chemical concepts.

In the framework of the dynamical system theory, the points corresponding to the local maxima of the investigated function (in our case, ELF) are called attractors. The basin of an attractor is a region of space containing the points by which a steepest ascent leads to the attractor, thus yielding a non-overlapping partition of the space. This topological partition using ELF [13, 47] yields basins of attractors that are considered as corresponding to atomic cores, bonds, and lone pairs recovering the Lewis picture of bonding and the electronic domains of the VSEPR approach. As the ELF is totally symmetrical, the attractors can be single points (general case), circles (off axis attractors of linear molecules), or spheres (off center attractors of atoms) according to their location and to the molecular symmetry.

In a molecule, one can find two types of basins. On the one hand are core basins surrounding nuclei with atomic number $Z > 2$ and labeled C(A), where A is the atomic symbol of the element, and on the other hand are the valence basins. For atoms heavier than neon, there are several core shells and therefore several core basins that are, in practice, unified in the unique superbasin labeled C(A). The valence basins are characterized by the number of atomic valence shells to which they participate. This number is called the *synaptic order*. Thus, there are *monosynaptic, disynaptic, trisynaptic* basins, and so on. Monosynaptic basins, labeled V(A), correspond to the lone pairs of the Lewis model, and polysynaptic basins to the shared pairs of the Lewis model. In particular, disynaptic basins, labeled V(A, X), correspond to two-center bonds, trisynaptic basins, labeled V(A, X, Y), to three-center bonds, and so on V(A, X, Y, ...) for higher *polysynaptic* basins. The valence shell of a molecule is the union of its valence basins. A polysynaptic basin encompassing a proton constitutes the valence shell of a bonded hydrogen atom, and is often said to be protonated polysysnaptic.

10.2.2.6 **Quantities obtained for ELF basins**

Graphically representing ELF gives a qualitative, general overview of atomic interactions. Quantitative characterizations can be further obtained by a common

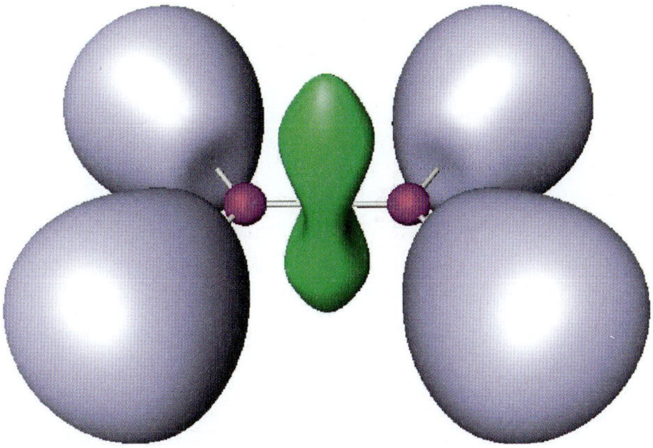

Figure 10.5 f-localization domain in the ethylene molecule. Surfaces are colored according to the basin they belong to: magenta for core basins, light blue for the C–H bonds, and green for the C–C banana bonds.

analysis of ELF and other properties. For example, one can integrate the electron density over a basin, and obtain the average number of electrons in a basin [14]. Or, one can obtain unpaired-electron numbers, by integrating spin densities within the basins [15].

Furthermore, one can give even more information by not only providing averages but also variances and fluctuations [14]. For example, a C–H basin in ethylene and core basin have populations of around 2.1 electrons. However, the C–H basin gives a standard deviation of 0.8, while this is lower for the core basin (0.5): fluctuations are more important between valence electrons than between core and valence electrons.

One can even obtain a more detailed view by looking at the probability of having ν electrons in a given basin ($\nu = 0, 1, 2, \dots$) [16]. For example, one finds that the probability is maximal for finding two electrons both in the C–H and in one of the C–C banana bond basins; see Figure 10.6. However, the probability is larger for the C–H bond than for one of the double bonds, and the probability distribution is broader for the latter.

10.2.2.7 ELF from experimental data

Although mainly obtained from quantum mechanical calculations, ELF can be obtained from experimental data [17, 18].

It is, in principle, possible to extract the exact Kohn–Sham potential corresponding to any density. This technique has been used since the eighties [19] with accurate computed densities. The Kohn–Sham orbitals are generated as solutions of the Kohn–Sham equations with this potential, enabling determinination of ELF [20].

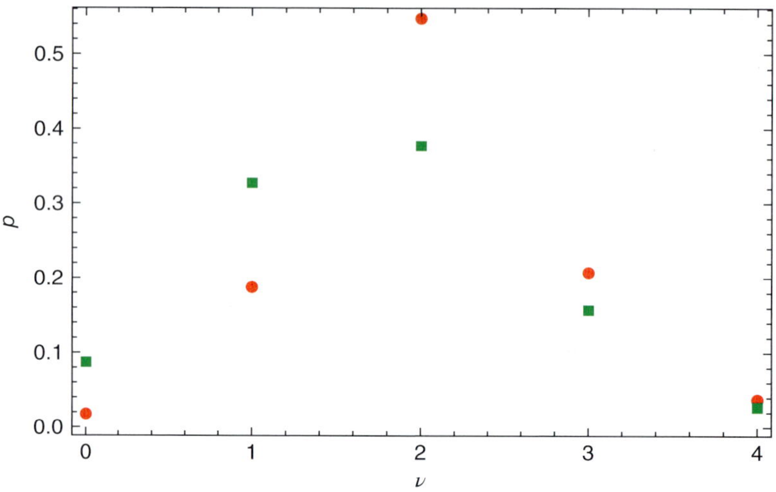

Figure 10.6 Probability of finding ν and only ν electrons in one of the C–H bond basins (red circles) or C–C banana bond basins (green squares) of ethylene molecule.

10.2.2.8 **Simplified forms of ELF**

In order to simplify the access to ELF from experimental data, alternative formulas that use no other input than the electron density (see, e.g. [18, 21, 22],) were presented.

10.2.3
Hints for interpretation

10.2.3.1 **ELF and mesomery**

Often, a single Lewis structure cannot properly describe the electronic structure of a molecule (e.g., benzene) or crystal (e.g., metal). In such cases, ELF presents a picture that corresponds to an averaging of the Lewis structures.

10.2.3.2 **How important is a maximum?**

It can happen that maxima of ELF exist, but that a small lowering of f merges together domains that were surrounding different maxima. In the case of the ethylene molecule, discussed earlier, this happened twice: once, when the domains of the double bond have merged, and another time, when all the domains associated with the valence electrons have merged. Such situations are typical for metals, such as the sodium crystal. There, maxima are observed between the atoms, but a very small lowering of f produces a domain that contains the whole crystal, except the atomic cores. In such situations, we can ask what importance one should assign to such shallow maxima. Is it not preferable to speak about a metal, than about many bonds between sodium atoms? Is it not better to speak about a double bond in C_2H_4, than about two banana bonds? Is it not more natural to speak about a valence shell than about two pairs of C–H bonds and a C–C double bond? At the moment, there is no standardized way to treat such situations. Some suggestions are given in the paragraphs on examples (10.3 and 10.4).

10.2.3.3 **When several maxima merge into a single one**

ELF is constructed to respect symmetry. As a result, when several symmetry-equivalent situations are possible, ELF yields an average. In the Ne atom, the classical picture of the valence would give four electron pairs at the corners of a regular tetrahedron. Owing to spherical symmetry, however, their orientation is arbitrary. Thus, with ELF, one finds only a spherical shell for the valence region: the four valence domains are merged.

10.2.3.4 **Hidden bonding**

In case of complete transfer of the valence electrons from one species to another within a molecule, a maximum between them does not appear (cf NaCl in the Examples). One may ask, however, what happens in charge-shift bonds [23], as in He_2^+; that is best described by the mesomery

$$\mathrm{He}\cdots\mathrm{He}^+ \leftrightarrow \mathrm{He}^+\cdots\mathrm{He}$$

ELF will provide a picture that averages out the two situations, and there will be no ELF maximum between the two nuclei (see Figure 10.7). The calculations indicate, however, that the value of ELF between the two He nuclei in He_2^+ does not go below 0.4, while in He_2, it goes down to ≈ 0.001. Low values are also encountered in ionic pairs. These systems will be discussed later.

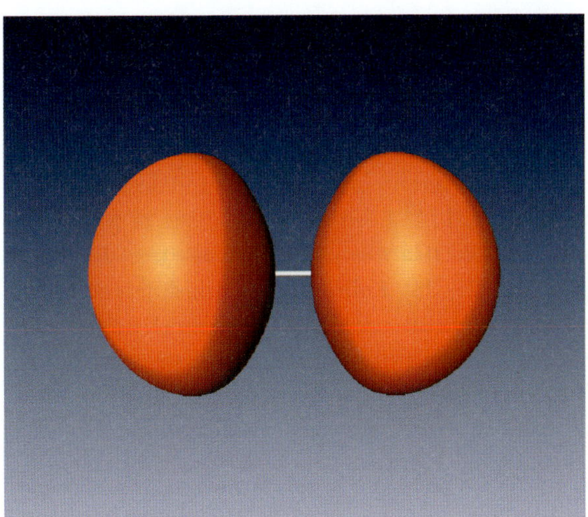

Figure 10.7 ELF isosurface ($\eta = 0.8$) in the He_2^+ molecular ion.

10.2.3.5 Electron counting: oxidation numbers and formal charges

Electron counting (providing formal charges to atoms) is a useful tool in chemistry. ELF provides a simple way to do it: if a valence basin has a common surface to a single core (i.e., for a *monosynaptic* basin), the average number of electrons is attributed to the nucleus to which the core belongs[4)] (for other basin properties, see this box.). If a basin has a common surface with more than one core (i.e., a *multisynaptic* basin), the electron count is attributed to the most electronegative atom. The latter can be estimated either using the suitable electronegativity scale or by analyzing the contributions of different atomic basins (QTAIM atoms) to the given valence basin [24].

Box 10.3 Basin properties

This box provides the definitions of the different integrated basin properties available in the ELF analysis. The following symbols have been used to define the calculated properties:

- Ω_i: basin labeled by i
- $\rho(\mathbf{r})$: spinless electron density
- $\rho^{\alpha}(\mathbf{r})$: α-spin electron density
- $\rho^{\beta}(\mathbf{r})$: β-spin electron density
- $\pi(\mathbf{r}, \mathbf{r}')$: spinless electron pair density
- $\pi^{\alpha\beta}(\mathbf{r}, \mathbf{r}')$: $\alpha\beta$ contribution to electron pair density
- $\pi^{\alpha\alpha}(\mathbf{r}, \mathbf{r}')$: $\alpha\alpha$ contribution to electron pair density
- $\pi^{\beta\beta}(\mathbf{r}, \mathbf{r}')$: $\beta\beta$ contribution to electron pair density.

10.2.4 Sensitivity of ELF to technical details

By experience, the ELF images are not overly sensitive to technical details, such as the quality of the basis set used. Table 10.1 shows the populations of typical ELF basins obtained with different methods (Hartree–Fock, post Hartree–Fock with a different treatment of correlation [25–28], Kohn–Sham with different density functionals), and different basis sets. One can see that the numbers hardly change. A noticeable change is observed when using an extremely poor basis set, STO-3G. However, even in this case, the results are still qualitatively correct, although such a basis set would be considered unreliable for almost any calculation. In cases when the maxima of ELF are not pronounced, their position can be sensitive to technical details, as can be seen for elemental metals (Section 10.4.3).

4) Bonds involving hydrogen provide a special case, as the hydrogen atom does not possess core electrons. In such a case, the H nucleus plays the role of the core.

Table 10.1 Basis set and correlation effects on the ELF basin populations of indexes of CO, H_2O, NH_3, and BF_3.

	CO					H_2O			NH_3			BF_3			
Basis set (HF)	C(C)	C(O)	V(C)	V(O)	V(C,O)	C(O)	V(O)	V(O,H)	C(N)	V(N)	V(N,H)	C(B)	C(F)	V(F)	V(B,F)
STO-3G	2.03	2.02	2.80	4.36	2.80	2.03	2.59	1.39	2.03	2.52	1.77	2.05	2.04	7.32	0.62
3-21G	2.06	2.10	2.48	4.70	2.52	2.17	2.44	1.47	2.15	2.64	1.77	2.09	2.19	6.56	1.21
6-21G	2.06	2.10	2.52	4.74	2.57	2.12	2.46	1.48	2.10	2.56	1.78	2.06	2.12	6.63	1.21
6-31G	2.07	2.11	2.63	4.61	2.57	2.15	2.46	1.46	2.11	2.38	1.83	2.09	2.19	6.57	1.21
6-31G**	2.07	2.12	2.40	4.07	3.34	2.15	2.26	1.67	2.11	2.12	1.92	2.06	2.16	6.49	1.32
6-311G**	2.07	2.11	2.43	4.06	3.32	2.14	2.22	1.71	2.11	2.10	1.93	2.07	2.13	6.37	1.48
6-311++G(2df,2pd)	2.06	2.11	2.37	4.25	3.21	2.14	2.22	1.70	2.10	1.85	2.01	2.07	2.13	6.39	1.45
aug-cc-PVTZ	2.07	2.11	2.38	4.28	3.15	2.14	2.22	1.71	2.11	2.12	1.92	2.06	2.14	6.34	1.49
Correlation method															
MP2	2.08	2.12	2.65	3.94	3.22	2.23	2.21	1.67	2.11	2.12	1.92	2.07	2.14	6.42	1.41
MP3	2.08	2.11	2.56	3.98	3.26	2.23	2.24	1.65	2.11	2.13	1.91	2.07	2.14	6.44	1.40
MP4	2.08	2.11	2.60	3.99	3.21	2.24	2.22	1.66	2.11	2.14	1.91	2.07	2.14	6.43	1.40
CASSCF	2.08	2.11	2.57	3.98	3.25	2.14	2.24	1.69	2.11	2.13	1.92	2.07	2.12	6.35	1.49
CISD	2.08	2.11	2.56	4.00	3.25	2.14	2.26	1.67	2.11	2.13	1.91	2.07	2.13	6.42	1.43
QCISD	2.08	2.11	2.59	4.01	3.21	2.21	2.22	1.67	2.11	2.14	1.91	2.07	2.14	6.43	1.40
C–CSD	2.08	2.11	2.58	4.00	3.23	2.14	2.26	1.67	2.14	2.10	1.92	2.08	2.13	6.43	1.40
Density functional method															
LDA	2.08	2.12	2.64	4.01	3.15	2.15	2.26	1.67	2.11	2.14	1.91	2.08	2.12	6.38	1.45
BP86	2.07	2.12	2.65	4.01	3.15	2.14	2.27	1.66	2.11	2.15	1.91	2.08	2.12	6.39	1.44
BLYP	2.07	2.12	2.65	4.02	3.14	2.14	2.28	1.65	2.11	2.16	1.90	2.08	2.12	6.36	1.47
B3LYP	2.07	2.12	2.60	4.02	3.18	2.14	2.27	1.66	2.11	2.15	1.91	2.08	2.12	6.38	1.46
PBE1PBE	2.07	2.11	2.59	4.01	3.22	2.14	2.26	1.67	2.11	2.14	1.91	2.08	2.14	6.41	1.43
VSXC	2.07	2.12	2.64	3.99	3.19	2.14	2.27	1.66	2.11	2.15	1.91	2.07	2.12	6.36	1.51
HCTH93	2.07	2.12	2.63	4.02	3.16	2.14	2.27	1.65	2.11	2.16	1.91	2.08	2.14	6.41	1.43

10.3
Simple examples

10.3.1
Atoms and ions

10.3.1.1 Atomic shells and cores

It is not easy to find spatial functions that reproduce the atomic shell structure. The textbook formula (the radial density defined by $4\pi r^2 \rho(r)$, where r is the distance of the electron from the nucleus, and ρ is the electron density, fails for heavier atoms). Also, the Laplacian of the density fails for heavier atoms [29–31].

ELF was designed to find electron pairs, not atomic shells. However, ELF was also designed to follow the symmetry properties of the nuclear framework. In particular, it will have spherical symmetry in an atom. For Be, for example, this does not pose a problem. For Ne, the K shell of the atom is well separated, and in the corresponding region, ELF reaches values close to its maximal value, 1. In the L shell, however, there are four electron pairs. Owing to the spherical symmetry of the system, ELF yields an average picture with a single basin for all four pairs together, reaching a value of approximately 0.8.

Figure 10.8 shows ELF for the Xe atom as a function of the distance from the core. The innermost (K) shell is so narrow that it can be hardly noticed on the scale of the plot. Increasing the resolution shows that ELF approaches 1 close to the nucleus. For the other shells, one can see the lowering of the maxima of ELF to values around 0.7–0.8.

The cores can be identified in ELF by nearly spherical regions around the nuclei that mostly remain practically unaffected by the atomic environment.

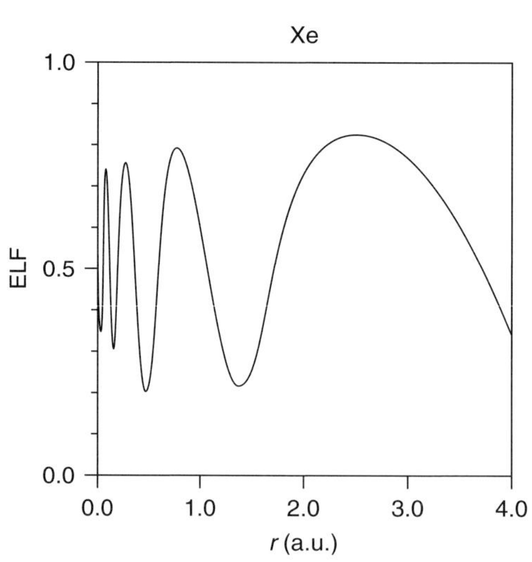

Figure 10.8 ELF in the Xe atom, as a function of the distance from the nucleus.

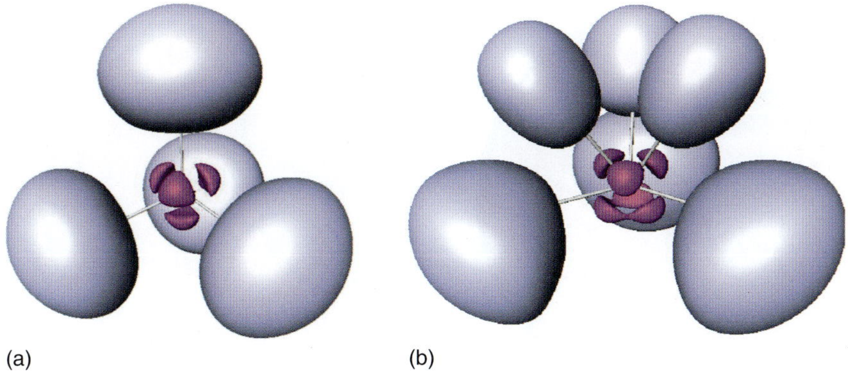

Figure 10.9 ELF isosuface ($\eta = 0.8$) in TiH_4 and CrH_6. Cores are colored in magenta.

10.3.1.2 **Structured cores: TiH_4, CrH_6**

Although the penultimate shells are believed to be inert, one can notice with ELF that they can be deformed according to VESPR rules [32]. Figure 10.9 shows that the cores in TiH_4 and CrH_6 are split by the ligands. In TiH_4 the splitting produces maxima, tetrahedrally oriented, in directions avoiding the tetrahdrally oriented valence basins. A similar effect is seen in CrH_6, where there are six ligands and six basins for the core. It may happen that the number of maxima in the penultimate shell differs from the number of ligands, especially when symmetry of the coordination sphere is lowered. In this case, one still notices that the penultimate shell maxima of ELF avoid the ligands.

10.3.1.3 **Ions**

In ionic systems, for sufficiently large f, one typically obtains isosurfaces that resemble those of a noble gas atom. However, for lower f values, as well as for basins, some deformation of the atomic spheres shows up; it can be explained by the Pauli repulsion between the ions. The fact that the ions get together in spite of the repulsion is due to some attractive force, that is not visible in the picture for LiH and LiF (Figure 10.10). We know that it is the electrostatic force that brings together the ions. Notice that usually the cations are harder than the anions, and that the deformation of the latter is more important.

10.3.1.4 **Squeezing effects**

In a molecule such as He_2, there is no ELF maximum between the He nuclei (see Figure 10.11). This is natural, as van der Waals binding is not a manifestation of electron pairing. The deformation of the atomic shells is a manifestation of the Pauli repulsion. In general, one may observe such deformations when two atoms are held together by forces that are not covalent in nature, such as electrostatic interaction (as had been also noticed in Figure 10.10), van der Waals interaction, mediation of a third atom, or nonequilibrium structures.

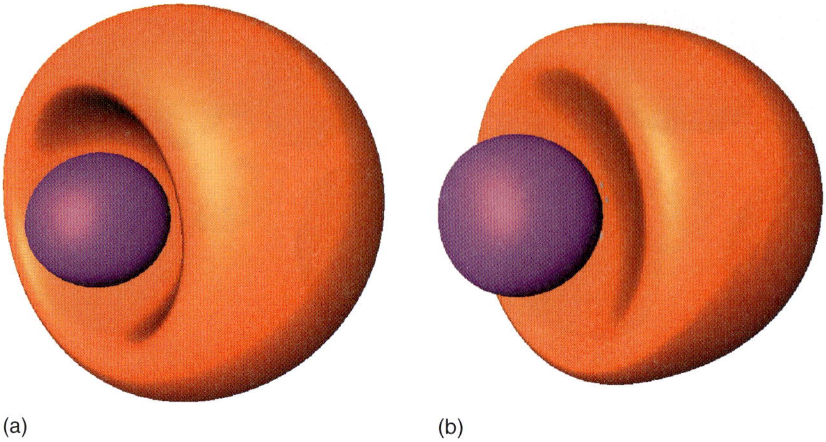

Figure 10.10 ELF, $\eta = 0.5$ for the LiH (a) and LiF (b) molecules.

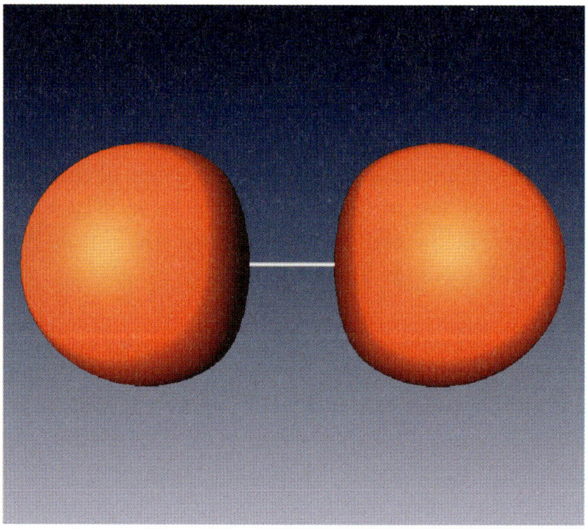

Figure 10.11 ELF isosurface ($\eta = 0.8$) for the He_2 molecule, calculated for the internuclear distance of 5.62 Å, at the minimum of the potential energy curve [33].

10.3.2 Bonds and lone pairs

10.3.2.1 Bonds: C_2H_6, diamond

Typically covalent bonds, such as in C_2H_6 or in diamond crystal, show that ELF achieves to describe regions of space that can be attributed to electron pairs. One can see that the C–C bond in ethane and diamond look quite similar (Figure 10.12). For further examples of solids, see subsequent text.

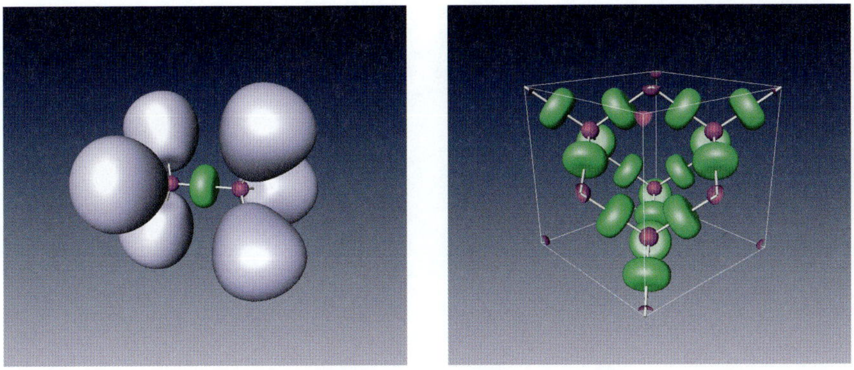

(a) (b)

Figure 10.12 ELF isosuface ($\eta = 0.8$) in C_2H_6 (a), and in diamond (b); C–C bonds in green.

10.3.2.2 **Multiple bonds: Allene**

For double bonds, ELF provides images reminding those of a banana bond, such as in C_2H_4 (see Figure 10.2a). The different orientation of the double bonds in allene can be seen when an f-localization domain is plotted (see Figure 10.13). The formation of banana bonds is usually observed in simpler molecules, where the local symmetry forces such distribution of ELF. For less symmetrical systems and solids, the splitting of such bonding basins and definition of the bond multiplicity may be more complex.

10.3.2.3 **Lone pairs: NH_3, H_2O, ice**

ELF basins for lone pairs are similar to those for bonds. One can see their tetrahedral arrangement in NH_3 (Figure 10.14) is similar to that in TiH_4 (Figure 10.9a).

Note that the lone pairs tend to be less well separated than the bond pairs: in H_2O, the bifurcation of the lone pairs occurs at a higher value of f than for the OH bonds. One could see it as a "memory" of the atomic shell structure.

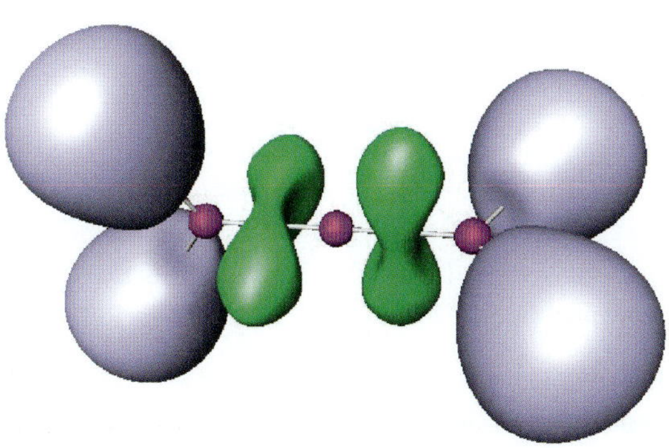

Figure 10.13 ELF isosurface ($\eta = 0.8$) in the allene molecule, C_3H_4.

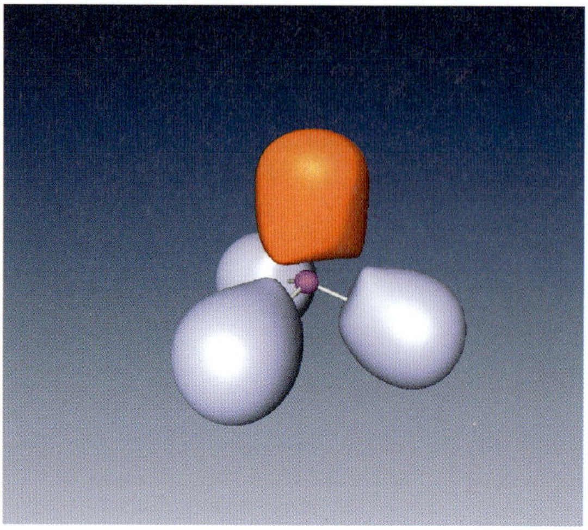

Figure 10.14 ELF isosurface ($\eta = 0.8$) in the ammonia molecule, NH_3.

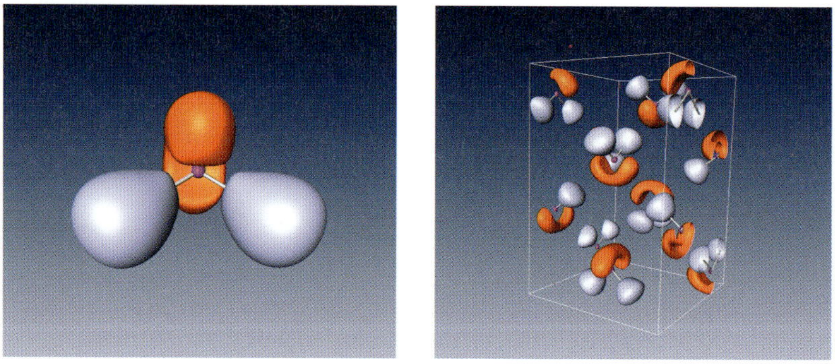

(a) (b)

Figure 10.15 ELF isosuface ($\eta = 0.8$) for an isolated water molecule (a), and in crystalline modification of ice VIII (b). Lone-pair localization domains are shown in orange.

In the crystal structure of ice VIII, Figure 10.15, one can see the orientation of the lone pairs toward the H atoms of the neighboring molecules.

10.3.2.4 Multicenter bonds: B_2H_6

ELF helps recognizing multicenter bonds (see Figure 10.16). A typical example is given by B_2H_6. A less trivial example of a three-center bond is provided by $Al_2H_4(CH_3)_2$, where two CH_3 groups make the bridge between the two AlH_2 fragments. Further examples of complex borane species can be found in Ref. [34].

10.3.2.5 Cylindrical symmetry effects: C_2H_2, HF

In molecules, as for atoms, symmetry is conserved. For example, there is a triple bond between the two C atoms of the C_2H_2, but the molecule has cylindrical symmetry. As a result, the image presented by ELF shows a torus, a cylindrically

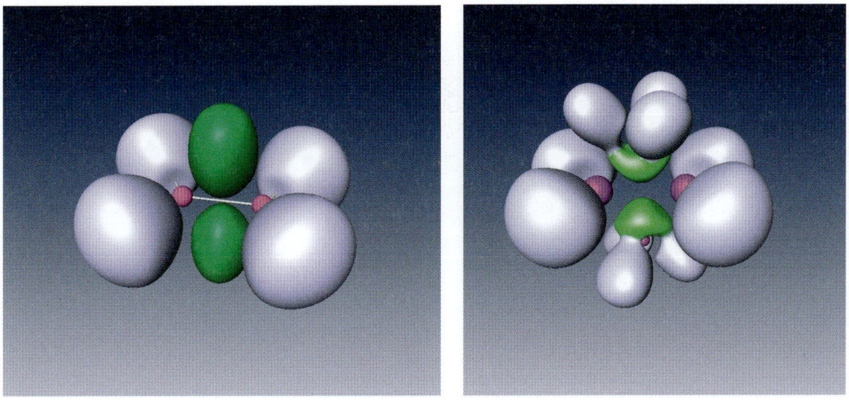

(a) (b)

Figure 10.16 ELF isosurface ($\eta = 0.8$) in B_2H_6 (a), and $Al_2H_4(CH_3)_2$ (b): core basins are colored in magenta, three-center bonds (B–H–B or Al–C–Al) – in green, two-center bonds (B–H, C–H, or Al–H) – in gray.

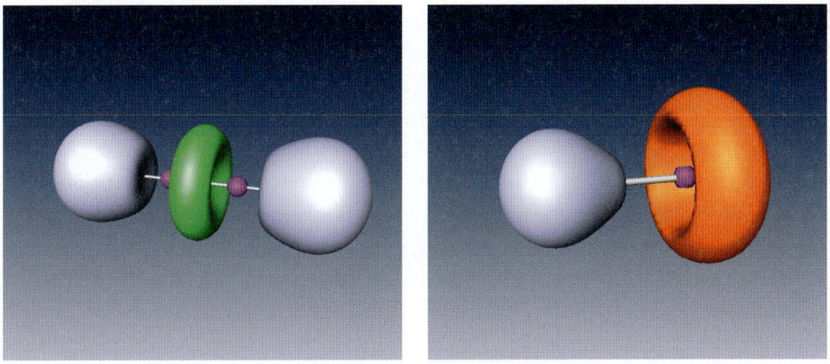

(a) (b)

Figure 10.17 ELF isosuface ($\eta = 0.8$) for C_2H_2 (a), and HF (b). The C–C bond in C_2H_2 is colored green, the lone pair on the F atom in HF – in orange.

symmetric f-localization domain between the two C atoms (see Figure 10.17a). The same type of ELF distribution occurs also for the three pairs of lone pairs in the HF molecule (see Figure 10.17b).

10.3.2.6 **Delocalization: Butadiene, benzene**

The eye has more difficulty in recognizing immediately the delocalization effect. In ethane or diamond, Figure 10.12, the f-localization domain was close to cylindrically symmetric around the line joining the two C nuclei. In trans-butadiene, Figure 10.18, the central C–C bond gets a deformation toward the shape of a double bond. The effect gets more pronounced for benzene, where no distinction can be made between successive C–C bonds.

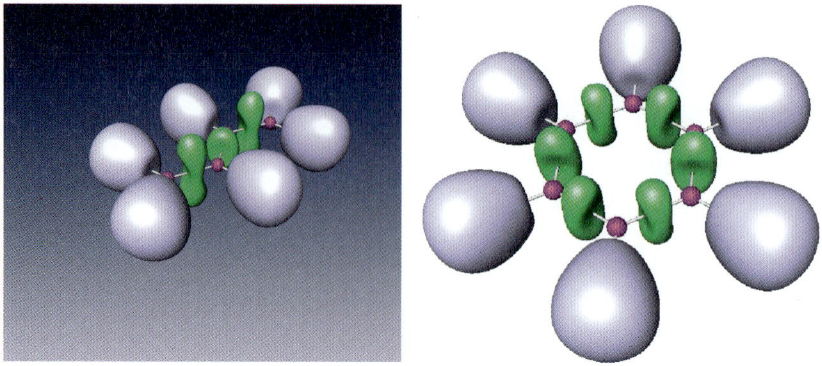

Figure 10.18 ELF isosufaces ($\eta = 0.8$) for trans-butadiene (a), and benzene (b).

10.3.3 Molecular reactions

Taking snapshots of the ELF localization domains along a reaction pathway enables the study of molecular reactions and displacive pressure-induced phase transitions from the "classical" point of view of bond reorganization. This is possible because ELF depends on the electronic state and the nuclear geometry of the investigated system. In practice, once the electronic transition state is located, the reaction path is determined by the intrinsic reaction coordinate (IRC) method [35] and the ELF analysis carried out for a selection of atomic arrangements along the reaction path. The evolution of the basin populations provides an indication of the electron density transfers, which can easily be represented by curly arrows in the mechanism representation. This method has been initially presented within the catastrophe theory framework [36]. In this context, the system visits the different bonding structures represented by the ELF basins linking the reactants to the products. The art is then to locate the turning points at which the number and/or the type of the basins are changed. This method has been extensively applied to reactions belonging to both organic [37, 38] and inorganic chemistry as well as to pressure-induced phase transitions. Shown here are a few examples taken from organic chemistry, which clearly illustrate the method.

10.3.3.1 Proton transfer in malonaldehyde

The intramolecular proton transfer in malonaldehyde is an example of tautomerization, which has been studied by the ELF technique by Krokidis *et al.* [39]. The conventional Lewis representation, represented subsequently, implies the breaking of an OH bond on the first oxygen atom followed by the formation of an OH bond on the second oxygen and also a counterclockwise migration of the double bonds.

The ELF analysis of the reaction summarized by Figure 10.19 shows that the proton transfer implies two transformations of the bonding. The first one

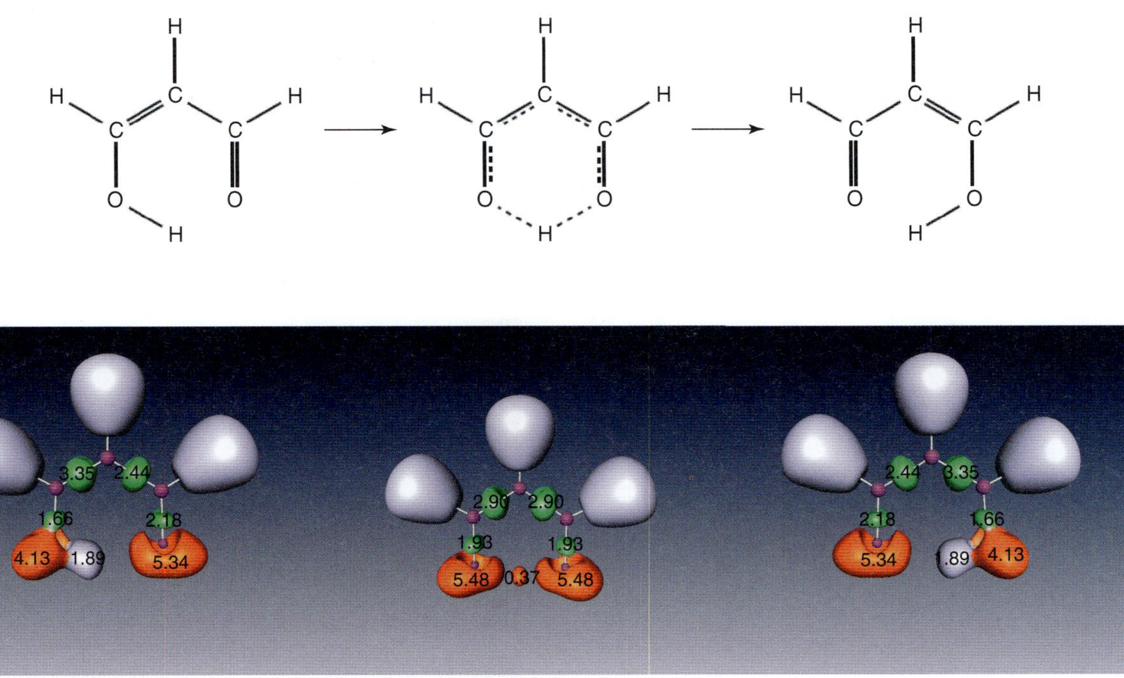

Figure 10.19 ELF localization domains ($f = 0.8$) and basin populations for reactant, transition state and product of the intramolecular proton transfer in malonaldehyde. Color code: magenta for cores, gray for C–H and O–H bonds, green for the C–C bonds, and orange for lone pairs and "dressed proton"-configurations.

corresponds to the covalent dissociation of the OH bond and the second to the formation of another OH bond from a hydrogen atom and an oxygen lone pair. The process involves an intermediate structure in which the hydrogen is detached and within which a significant electron transfer from the hydrogen toward the oxygen lone pairs brings its ionic character to the reaction. The presence of the detached hydrogen, or "dressed proton," in the intermediate state is a consequence of the attractive Coulomb potential of the proton, which has no Pauli repulsion counterpart. There is no bonding rearrangement in the skeleton but rather a concerted charge transfer from one bond to another.

10.3.3.2 **Aliphatic nucleophilic substitution S_N2**

The reaction mechanism of the backside attack in four nucleophilic substitutions, $XCH_3 + X^- \rightarrow X^- + CH_3X$ has been investigated with ELF by Polo *et al.* [40] who found three structural stability domains along the reaction path. In the $Cl^- + H_3CBr \rightarrow ClCH_3 + Br^-$ reaction displayed in Figure 10.20, the three structural stability domains correspond successively to the chemical structures of the reactant, the transition state, and the products. The transition structures are characterized by ionic species without the presence of disynaptic basin connecting the halogens and the carbon atom.

The population of the V(C,Br) basin is 1.25 *e* in the reactant. As bromine dissociates, about 80% of the V(C,Br) population is transferred toward the bromine lone pairs and the remaining toward the V(C,H) basins; on chlorine, the formation of the V(C,Cl) basin implies transfers from both V(C,H) and V(Cl) basins.

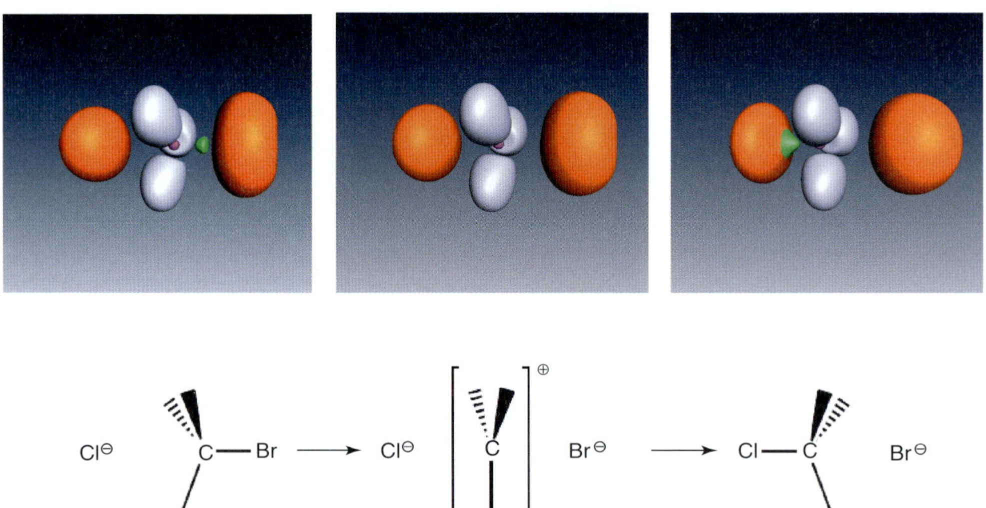

Figure 10.20 Evolution of the ELF localization domains along the reaction path of the $Cl^- + H_3CBr \rightarrow ClCH_3 + Br^-$ nucleophilic substitution. Color code: magenta for cores, orange for lone pairs, light blue for C–H bonds, and green for the C-halogen bonds.

10.3.3.3 Diels Alder addition

This class of reaction has been the topic of several analyses by the ELF technique [41]. The addition of cyclopropene on cyclopentadiene yields two stereoisomers:

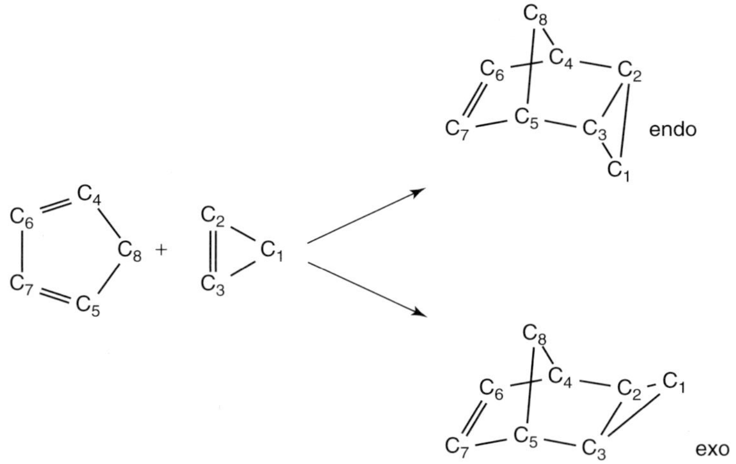

For both routes, six bonding rearrangements take place along the reaction path as shown for the exo reaction in Figure 10.21. The first part of the intrinsic reaction pathway links the reactants to the activated complex: the reactants get close together and adapt their internal geometries to satisfy the steepest ascent energy requirement. In particular, the double bonds of the cyclopropene and of the cyclopentadiene increase by about 0.025 Å which implies a density transfer from the C = C double bonds toward the adjacent C–C single bonds. In the ELF picture, the decrease in the double bond character is accounted for by the merging of the relevant V(C,C) basins. This occurs first for the cyclopropene moiety and a few steps after for the cyclopentadiene. After the transition step, the system relaxes toward the product following a steepest descent path on the Born–Oppenheimer energy hypersurface, in which angular deformations play an important role. The pyramidalization of the bonds around, on the one hand, the C(2), C(3) and, on the other hand, the C(4) and C(5) carbons yields the appearance of monosynaptic basins on top of these atoms. Again, the topological changes take place first on the cyclopropene side. As the C(2)–C(4) and C(3)–C(5) distances shorten, the monosysnaptic basins on these atoms merge yielding the V(C2,C4) and V(C3,C5) disynaptic basins, which materialize the formation of the corresponding single bonds. Finally, the V(C6,C7) valence basin splits to give a double bond.

The graphical properties of the ELF approach are exploited to make animations of chemical reactions such as those presented at the following URL: *http://www.lct.jussieu.fr/pagesperso/silvi/recherche˙en.html*

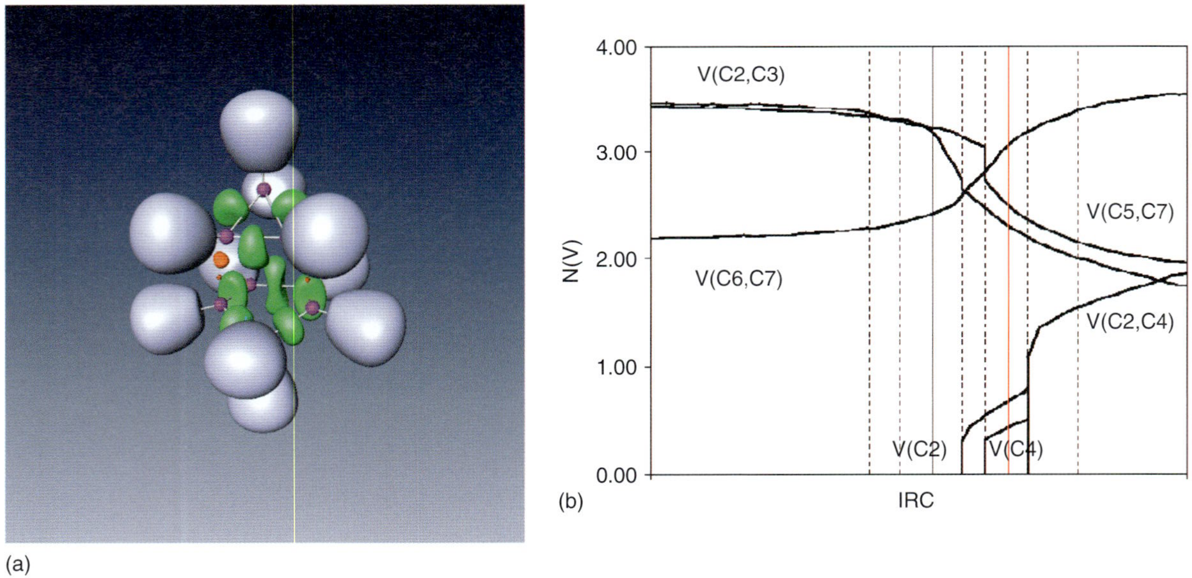

Figure 10.21 $C_5H_6 + C_3H_4$ exo addition reaction. (a) snapshot of the ELF localization domains ($f = 0.8$) taken before the formation of the C(2)–C(4) and C(3)–C(5) bonds and showing the presence of the four valence basins, V(C2), V(C3), V(C4) and V(C5), involved in the cyclization. Color code: magenta for cores, gray for C–H and O–H bonds, green for the C–C bonds, and red for lone pairs. (b) basin population along the reaction path; the vertical dashed lines correspond to the changes in the ELF topology, accounting for the evolution of the bonding structures, the black vertical full line locates the transition state and the red one corresponds to the geometry shown on the left part of the figure.

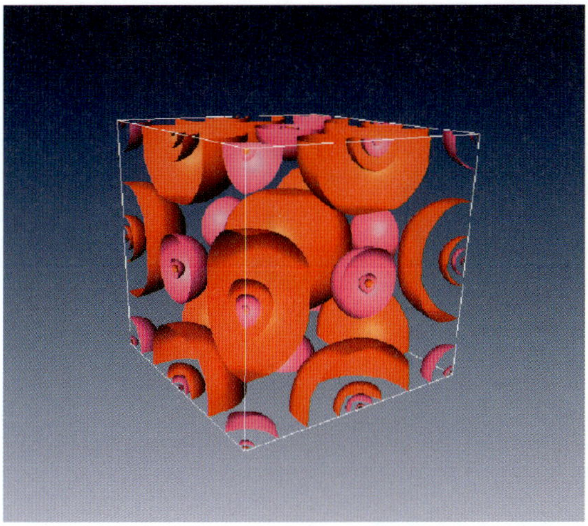

Figure 10.22 ELF distribution in the solid NaCl: first and second shells are shown with magenta isosurfaces, third shells are represented with red surfaces.

10.4 Solids

The importance of the ELF as a suitable tool for analysis of atomic interactions not only in molecules [42] but also in solids was recognized by R. Nesper and H. G. von Schnering, who initialized the first systematic investigations [43, 44]. The definition of ELF within LDA DFT [43] opened the opportunity to perform the first solid state calculations with the TB-LMTO-ASA technique [45].

10.4.1 Ionic compounds

In ionic solids, one typically obtains close to spherical distribution of ELF around the nuclei that resembles that of a noble gas atom. Figure 10.22 shows an ELF isosurface in NaCl. One easily identifies the small Na and large Cl species. While chlorine shows three shells, the last (third) shell of sodium is not present, suggesting the charge transfer from Na to Cl. Indeed, integration of the electron density within the basins yields the effective charge transfer of 0.98 electrons from Na and Cl [46]. The spherical character of the ELF distribution is more pronounced for the harder Na cation than for the softer Cl anion. As previously noticed for molecular systems (see paragraph 10.3.1.3), the fact that the ions get together in spite of the repulsion is due to some attractive force that is not directly visible in the ELF picture and we know that it is the electrostatic force that brings together the ions.

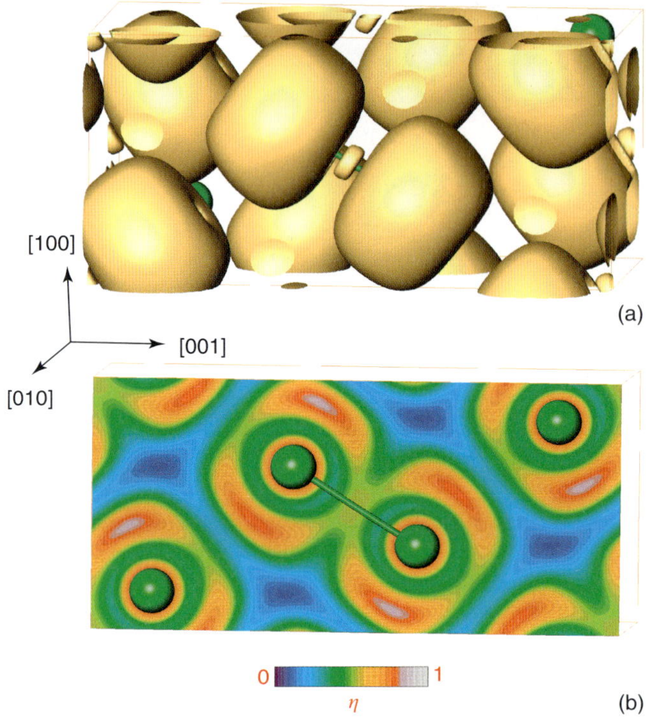

Figure 10.23 ELF in the crystalline iodine formed by I_2 molecules: (a) isosurface $\eta \approx$; (b) ELF distribution in the (010) plane.

10.4.2
Molecular compounds

The I_2 molecules form a molecular crystal. When high values for ELF are chosen, one can identify the electron pair domains making the bond between I atoms, and a torus, for the three lone pairs on each of the I atoms (Figure 10.23). For low values, one can see how molecules are packed in a crystal. Note the distortion of the f-localization domains by the packing in the crystal, reminding us of that observed for single ions in NaCl, Figure 10.22.

10.4.3
Elemental metals

Topology of the ELF obtained by different calculation techniques [47–50][5)] for cubic metals differs in the location of ELF maxima (Figure 10.24), as the maxima themselves are shallow, and not strongly pronounced. So, for the bcc (body-centered

5) For technical details, and differences between Hartree–Fock (HF), and the density functional methods LMTO-ASA and FPLO, see [48, 49] .

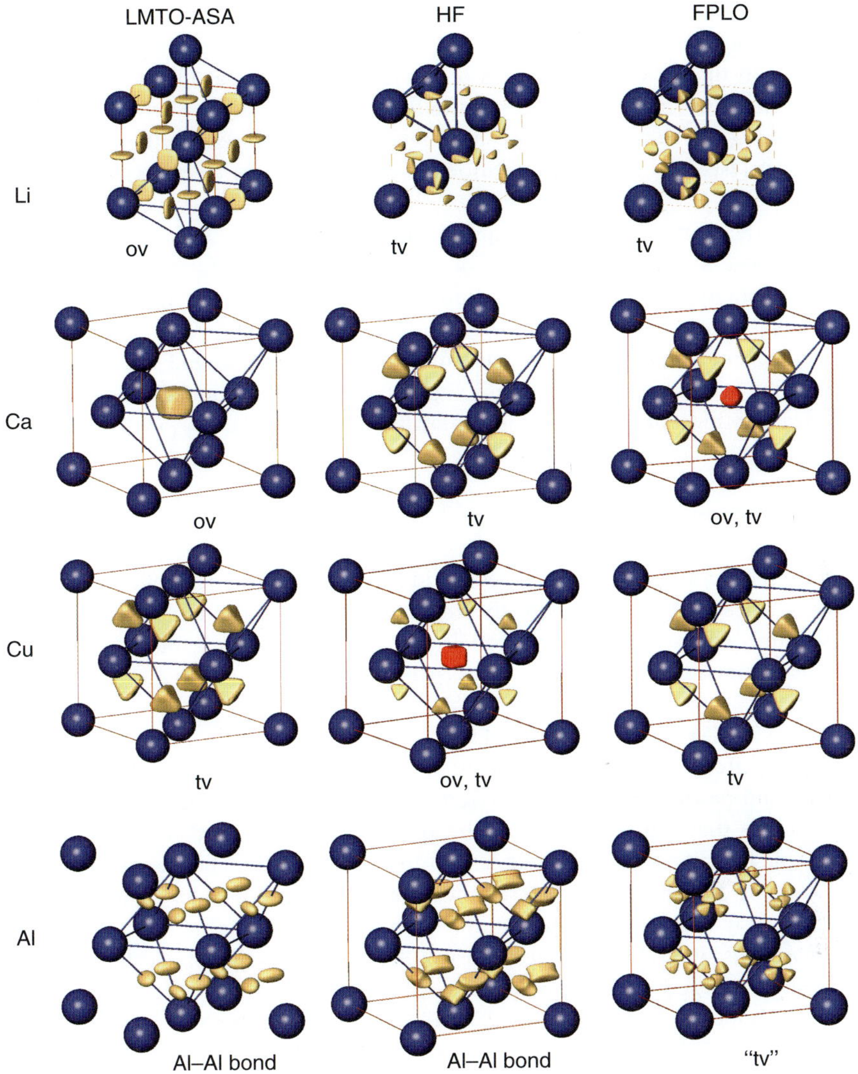

Figure 10.24 Isosurfaces of ELF in the cubic elemental metals, illustrating the position of the ELF maxima (yellow or red) depending on the applied calculation technique. The conventional cells are shown by red lines, the octahedral (ov) and tetrahedral voids (tv) by blue ones.

cubic) modification of Li, the LMTO-ASA calculation yields ELF attractors in the octahedral voids (ov), whereas the HF and FPLO calculations reveal ELF attractors in the tetrahedral voids (tv). Such differences are observed independently of the type of the structure and position of the elements in the periodic table. For the fcc (face-centered cubic) modifications of Ca and Cu, the maxima of ELF are found either in the octahedral or in the tetrahedral, or in both types of holes depending on the method of calculation. The most striking case of such differences represents the fcc structure of aluminum. The LMTO-ASA and HF calculations reveal ELF maxima on the bond lines between the nuclei; the FPLO calculation shows that the positions of the maxima are shifted toward the centers of the tetrahedral holes, being the trend found for other metals. This trend was further confirmed by the analysis of the hcp modifications of metals with the electron localizability indicator [51].

The general trend in the topology of ELF in elemental metals is the location of its attractors in the tetrahedral or octahedral holes within the crystal structure. The synapticity of such attractors is four and six, respectively. In this way, the bonding picture is characterized by the multicenter interactions. So, the four- and six-center bonds in elemental metals can be considered as the ELF representation of metallic bonding, as it was recently suggested for the distribution of the electron localizability indicator [52] in the intermetallic compound Al_5Co_2 [53].

10.4.4 Intermetallic compounds

Intermetallic compounds exist in the solid state only and do not follow the usual valence rules. The complexity of the crystal structures and the bonding patterns hinder the application of the traditional quantum chemical tools for analysis of the atomic interactions in this group of inorganic substances.

10.4.4.1 Zintl–Klemm concept and ELF

A special group of materials within the field of intermetallic compounds is called Zintl phases (Ref. [54], and references therein), that is, those intermetallic compounds that can be understood by application of the Zintl–Klemm concept [55]. This definition of Zintl phases requires – beside the charge transfer from the cationic to the anionic part of the crystal structure – that the number of homonuclear bonds per atom within the anionic part is defined by the general Pearson rule [56, 57][6]. Thus, it is no wonder that ELF was applied for bonding analysis in a large group of Zintl phases [44].

The example of $Ba_2Mg_3Si_4$ (Ref. [58], Figure 10.25) illustrates that the topology of ELF in the Zintl phases has three main features. First, it shows that the last shells of barium (the sixth one) and magnesium (the third one) are not present, revealing the charge transfer and ionic interaction between the Ba and Mg cations and anionic substructure build of two different silicon anions. The formation of the covalent Si–Si bonds within the $[Si_2]$ and $[Si_6]$ anions is shown by the maxima on the Si–Si contacts. And, finally, the 'lone pairs' on the Si atoms, instead of avoiding the Mg cores, point toward them, thus visualizing the strongly polar Mg–Si interactions being in agreement with the charge transfer in the crystal structure. Further examples of the ELF in Zintl phases, especially illustrating the role of the lone pairs may be found in Ref. [59].

6) According to the Pearson rule, the anions form as many bonds as necessary to achieve octet configuration.

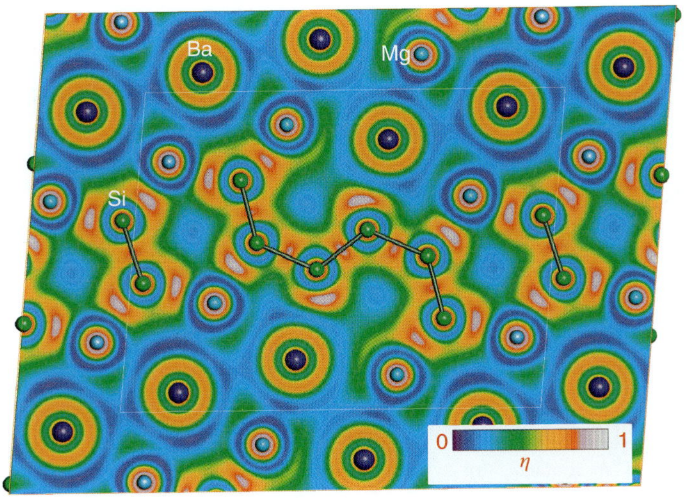

Figure 10.25 ELF distribution in the (010) plane of $Ba_2Mg_3Si_4$.

10.4.4.2 ELF for penultimate shells of transition metals

Additional features appear in the distribution of ELF, if – together with the main group elements – the transition metals are constituents of an intermetallic compound. So, the crystal structure of the filled skutterudite $NaFe_4Sb_{12}$ [60] consists of three-dimensional framework built by vertices-condensed octahedrons $[FeSb_6]$ with the sodium embedded into the framework voids (Figure 10.26). In the ELF

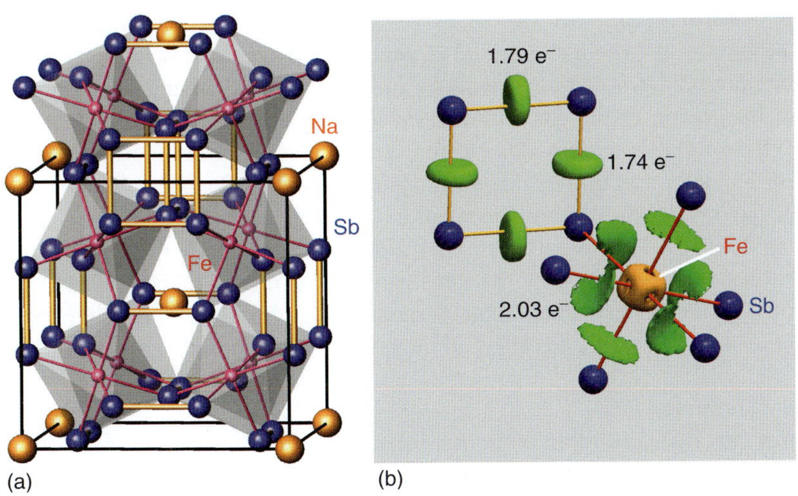

Figure 10.26 (a) Crystal structure of the filled skutterudite $NaFe_4Sb_{12}$ as a framework built of the vertice-condensed octahedrons $[FeSb_6]$ with the sodium atoms embedded into the cavities of the framework; (b) Isosurfaces of ELF ($\eta = 0.56$, green, $\eta = 0.53$, light green and $\eta = 0.72$, orange) in visualizing the Sb–Sb and Fe–Sb bonds as well the structuring of the penultimate shell of Fe, respectively.

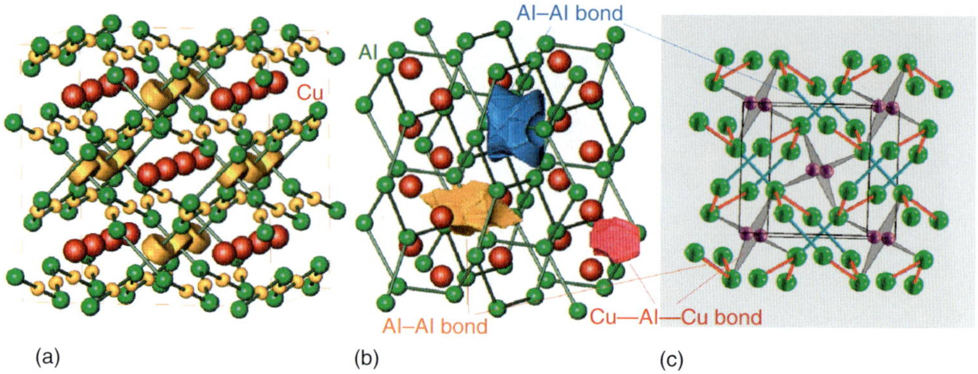

Figure 10.27 (a) ELF in Al_2Cu: Isosurface of ELF ($\eta = 0.55$) reveals the location of ELF maxima on the Al–Al contacts; (b) Basins of three types of ELF maxima; (c) Interpenetrating graphite-like aluminum nets formed by the two-center Al–Al bonds interacting with the copper by three-center Cu–Al–Cu bonds.

distribution, the Sb–Sb bonds interconnecting the octahedrons as well as the Fe–Sb bonds within the octahedrons are visualized by ELF maxima, suggesting covalent interactions within the framework and ionic interaction of the sodium cations with the framework. In addition, the structuring of the penultimate shell of Fe is observed. This topological feature is shown to be the fingerprint of the participation of the electrons of the penultimate shell in the interactions within the valence region [61].

10.4.4.3 Case of Al_2Cu

The distribution of ELF in the crystal structure of Al_2Cu [62] reveals only three types of maxima. Two of them visualize two-center Al–Al interactions (Figure 10.27a). Integration of the electron density within these basins yields counts of ca. 1.7 electrons, that is, very close to the two-electron bonding (Figure 10.27b). By these bonds, a spectacular three-dimensional framework of interpenetrating graphite-like nets of aluminum is formed. In the framework channels, the copper atoms are located. They interact with the framework by means of the three-center Cu–Al–Cu bonds, as shown by the third type of ELF maxima (Figure 10.27b,c). None of the previous crystallographic descriptions of this structural pattern was able to reveal or – at least – to reflect all these three interactions.

10.4.4.4 Surprises

The most surprising example of atomic interactions revealed by ELF was found in $RhBi_4$ [64]. The cubic crystal structure of this intermetallic compound consists of two interpenetrating frameworks built of edge-condensed tetragonal antiprisms [$RhBi_8$] (Figure 10.28c, labyrinth-like part of space within the orange surface). The Rh–Bi interaction within the antiprism is covalent, whereby the structuring of the penultimate shell of Rh reveals participation of these electrons in the interaction in the valence region (Figure 10.28a). On the outer side of the prisms are located

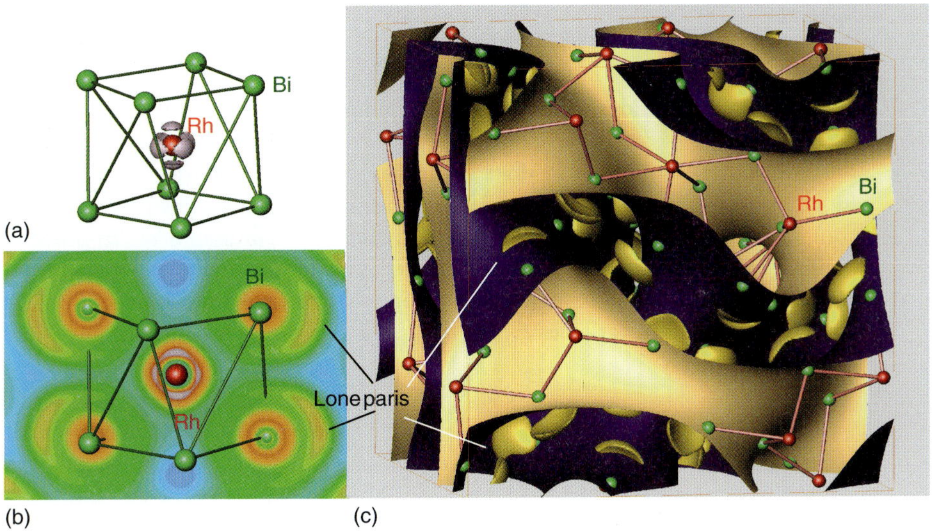

Figure 10.28 ELF in $RhBi_4$: (a) isosurface of ELF illustrating the structuring of the penultimate shell of Rh in relation to its coordination polyhedron–tetragonal antiprism $[RhBi_8]$; (b) distribution of ELF around the antiprism showing – beside the Rh–Bi interaction – formation of the Bi lone pairs on the outer side of the antiprism; (c) a periodic nodal surface (Ref. [63]) outlines two different parts of the space in the crystal structure: one (orange labyrinth, region enclosed within the orange surface) contains the Rh–Bi framework, the isosurface of ELF ($\eta = 0.7$, yellow) visualizes the positions of the lone-pair ELF maxima in the other labyrinth (blue).

ELF maxima (Figure 10.28b); each of them is only monosynaptic to one Bi nucleus. So, the whole space between the interpenetrating frameworks is "filled" by the lone pairs of the bismuth atoms visualizing a van der Waals interaction in this part of the structure (visualized by the blue surface in Figure 10.28c). Note that the ELF does not contain explicitly the information about this kind of chemical bonds! Further examples of practical application of ELF can be found on the website *http://www.cpfs.mpf.de/ELI*.

10.5 Perspectives

ELF is a tool to recover chemical "prejudices" from quantum chemical calculation. At first sight, in particular, because of the strong expression of the graphical representations, it looks like a computer game. In practice, however, it is a powerful quantum chemical tool in real space and has not only served to guide the chemist toward a correct explanation and understanding of atomic interactions in different classes of chemical compounds but also to improve their experimental design.

Appendix

10.A
Mathematical expressions of calculated basin properties

Assuming the Born–Oppenheimer separation, the electronic wave function is a function of $4N$ variables: each electron, labeled by i, is described by the three components of its position vectors $\mathbf{r}_i$ and by its spin coordinates $\boldsymbol{\sigma}_i$. In order to simplify the notation, $\mathbf{r}_i$ and $\boldsymbol{\sigma}_i$ are gathered in a four-component (space+spin) vector ξ_i. The probability of finding k electrons with the space and spin coordinates specified by the list of variables $\xi_1, \xi_2, \ldots, \xi_k$, the remaining $N-k$ being anywhere is given by

$$\Gamma^{(k)}(\xi_1, \xi_2, \ldots, \xi_k) = \frac{N!}{(N-k)!} \int \Psi(\xi_1, \xi_2, \ldots, \xi_k, \xi_{k+1}, \ldots, \xi_N) \Psi^*(\xi_1, \xi_2, \ldots, \xi_k, \xi_{k+1}, \ldots, \xi_N) d\xi_{k+1}, \ldots, d\xi_N$$

Particularly important are the first- and second-order distribution functions $\Gamma^{(1)}(\xi)$ and $\Gamma^{(2)}(\xi, \xi')$. Integration of $\Gamma^{(1)}(\xi)$ over the spin coordinate $\boldsymbol{\sigma}$ yields the electron density $\rho(\mathbf{r})$; as the spin coordinate is discrete, this integration is actually the sum of two contributions $\rho_\alpha(\mathbf{r})$ and $\rho_\beta(\mathbf{r})$, which correspond to the two values $\pm 1/2$ of σ:

$$\rho(\mathbf{r}) = \int \Gamma^{(1)}(\xi) d\sigma = \Gamma^{(1)}(\mathbf{r}, \frac{1}{2}) + \Gamma^{(1)}(\mathbf{r}, -\frac{1}{2}) = \rho_\alpha(\mathbf{r}) + \rho_\beta(\mathbf{r})$$

The spin density is defined as

$$Q_z(\mathbf{r}) = \frac{1}{2}[\rho_\alpha(\mathbf{r}) - \rho_\beta(\mathbf{r})]$$

Integrating the spin density over the entire space yields $\langle S_z \rangle = M_S$. The same procedure is applied to $\Gamma^{(2)}(\xi, \xi')$ in order to obtain the pair function and its four spin components:

$$\begin{aligned}\pi(\mathbf{r}, \mathbf{r}') &= \int \int \Gamma^{(2)}(\xi, \xi') d\sigma d\sigma' \\ &= \pi_{\alpha\alpha}(\mathbf{r}, \mathbf{r}') + \pi_{\alpha\beta}(\mathbf{r}, \mathbf{r}') + \pi_{\beta\alpha}(\mathbf{r}, \mathbf{r}') + \pi_{\beta\beta}(\mathbf{r}, \mathbf{r}')\end{aligned}$$

10.A.1
Basin populations

The basin populations are obtained by integrating the density distribution functions over the basins.

- α basin population: $\overline{N}^\alpha[\Omega_I] = \int_{\Omega_I} \rho^\alpha(\mathbf{r}) d\mathbf{r}$.
- β basin population: $\overline{N}^\beta[\Omega_I] = \int_{\Omega_I} \rho^\beta(\mathbf{r}) d\mathbf{r}$.
- total basin population: $\overline{N}[\Omega_I] = \overline{N}^\alpha[\Omega_I] + \overline{N}^\beta[\Omega_I] = \int_{\Omega_I} \rho(\mathbf{r}) d\mathbf{r}$.
- basin spin density: $\langle S_z \rangle_{\Omega_I} = \frac{1}{2}(\overline{N}^\alpha[\Omega_I] - \overline{N}^\beta[\Omega_I])$

- $\alpha\beta$ pair population $\overline{\Pi}^{\alpha\beta}[\Omega_I,\Omega_I] = \int_{\Omega_I}\int_{\Omega_I}\pi^{\alpha\beta}(\mathbf{r},\mathbf{r}')\mathrm{d}\mathbf{r}\mathrm{d}\mathbf{r}'$.
- $\alpha\alpha$ pair population $\overline{\Pi}^{\alpha\alpha}[\Omega_I,\Omega_I] = \int_{\Omega_I}\int_{\Omega_I}\pi^{\alpha\alpha}(\mathbf{r},\mathbf{r}')\mathrm{d}\mathbf{r}\mathrm{d}\mathbf{r}'$.
- $\beta\beta$ pair population $\overline{\Pi}^{\beta\beta}[\Omega_I,\Omega_I] = \int_{\Omega_I}\int_{\Omega_I}\pi^{\beta\beta}(\mathbf{r},\mathbf{r}')\mathrm{d}\mathbf{r}\mathrm{d}\mathbf{r}'$.
- pair population $\overline{\Pi}[\Omega_I,\Omega_I] = \int_{\Omega_I}\int_{\Omega_I}\pi(\mathbf{r},\mathbf{r}')\mathrm{d}\mathbf{r}\mathrm{d}\mathbf{r}'$.

10.A.2 Variance and covariance of basin populations

The basin populations are averages; they can be expressed as expectation values of basin population operators. The basin population operator enabling the calculation of $\overline{N}[\Omega_I]$ has been introduced by Diner and Claverie [65] in the case of Bader's partition:

$$\hat{N}^{\alpha}[\Omega_I] = \sum_i^N \hat{y}(\mathbf{r}_i,\sigma_i) \text{ with } \hat{y}(\mathbf{r}_i,\sigma_i)\begin{cases}\hat{y}(\mathbf{r}_i i,\sigma_i)=1 & \mathbf{r}_i\in\Omega_I \text{ and } \sigma_i=\frac{1}{2}\\ \hat{y}(\mathbf{r}_i)=0 & \text{otherwise}\end{cases}$$

$$\hat{N}^{\beta}[\Omega_I] = \sum_i^N \hat{y}(\mathbf{r}_i,\sigma_i) \text{ with } \hat{y}(\mathbf{r}_i,\sigma_i)\begin{cases}\hat{y}(\mathbf{r}_i i,\sigma_i)=1 & \mathbf{r}_i\in\Omega_I \text{ and } \sigma_i=-\frac{1}{2}\\ \hat{y}(\mathbf{r}_i)=0 & \text{otherwise}\end{cases}$$

and

$$\hat{N}[\Omega_I] = \hat{N}^{\alpha}[\Omega_I] + \hat{N}^{\beta}[\Omega_I]$$

These expressions can be generalized to pair population spin components in a straightforward manner; for example:

$$\begin{aligned}\hat{\Pi}^{\alpha\alpha}[\Omega_I,\Omega_I] &= \hat{N}^{\alpha}[\Omega_I]\hat{N}^{\alpha}[\Omega_I] - \hat{N}^{\alpha}[\Omega_I]\\ \hat{\Pi}^{\alpha\alpha}[\Omega_I,\Omega_J] &= \hat{N}^{\alpha}[\Omega_I]\hat{N}^{\alpha}[\Omega_J]\\ \hat{\Pi}^{\alpha\beta}[\Omega_I,\Omega_I] &= \hat{N}^{\alpha}[\Omega_I]\hat{N}^{\beta}[\Omega_I]\end{aligned}$$

and

$$\begin{aligned}\hat{\Pi}[\Omega_I,\Omega_I] &= \hat{\Pi}^{\alpha\alpha}[\Omega_I,\Omega_I] + \hat{\Pi}^{\alpha\beta}[\Omega_I,\Omega_I] + \hat{\Pi}^{\beta\alpha}[\Omega_I,\Omega_I]\\ &\quad + \hat{\Pi}^{\beta\beta}[\Omega_I,\Omega_I] = \hat{N}[\Omega_I]\hat{N}[\Omega_I] - \hat{N}[\Omega_I]\end{aligned}$$

They are used to build the variance and covariance operators:

$$\begin{aligned}\widehat{\sigma^2}(\Omega_I,\Omega_I) &= \hat{N}[\Omega_I]\hat{N}[\Omega_I] - \overline{N}[\Omega_I]^2 = \hat{\Pi}[\Omega_I,\Omega_I] + \hat{N}[\Omega_I] - \overline{N}[\Omega_I]^2\\ \widehat{cov}(\Omega_I,\Omega_J) &= \hat{N}[\Omega_I]\hat{N}[\Omega_J] - \overline{N}[\Omega_I]\overline{N}[\Omega_J] = \hat{\Pi}[\Omega_I,\Omega_J] - \overline{N}[\Omega_I]\overline{N}[\Omega_J]\end{aligned}$$

which yield:

- variance $\langle\sigma^2(\Omega_I)\rangle = \int_{\Omega_I}\int_{\Omega_I}\pi(\mathbf{r},\mathbf{r}')\mathrm{d}\mathbf{r}\mathrm{d}\mathbf{r}' + \overline{N}\Omega_I - \overline{N}^2(\Omega_I)$.
- covariance matrices $\langle\widehat{cov}(\Omega_I,\Omega_J)\rangle = \overline{\Pi}(\Omega_I,\Omega_J) - \overline{N}(\Omega_I)\overline{N}(\Omega_J))$.
- standard deviation $\sigma(\Omega_I) = \sqrt{\sigma^2(\Omega_I)}$
- correlation coefficient matrices: $R_{\Omega_I,\Omega_J} = \frac{\langle\widehat{cov}(\Omega_I,\Omega_J)\rangle}{\sqrt{\langle\sigma^2(\Omega_I)\rangle}\sqrt{\langle\sigma^2(\Omega_J)\rangle}}$

10.A.3
Probability distributions

- Probability of finding exactly n electrons within a given basin Ω_i the remaining $N - n$ being located in the other basins [66]:

$$P_n(\Omega_I) = \binom{N}{n} \int_{\Omega_I} d\mathbf{x}_1 \ldots d\mathbf{x}_n \int_{\mathbb{R}^3-\Omega_I} \Psi^* \Psi d\mathbf{x}_{n+1} \ldots d\mathbf{x}_N$$

- Electron number probability distribution functions, that is, probabilities $P(n_1, n_2,...,n_m)$ of having exactly n_1 electrons in Ω_1, n_2 electrons in Ω_2, … , and n_m electrons $(n_1 + n_2 + \cdots + n_m = N)$ in Ω_n [67]:

$$P(n_1, n_2,...,n_m) = \frac{N!}{n_1!n_2! \ldots n_m!} \int_{\Omega_1} d\mathbf{x}_1 \ldots d\mathbf{x}_{n_1} \int_{\Omega_2} d\mathbf{x}_{n_1+1} \ldots d\mathbf{x}_{n_1+n_2} \cdots \times \int_{\Omega_m} \Psi^* \Psi d\mathbf{x}_{N+1-n_m} d\mathbf{x}_N$$

10.A.4
Basin electrostatic moments

The basin electrostatic multipoles are defined as minus the integral over the basin volume of the product of the electron density distribution times a multipole operator [68], that is:

$$\langle Q_l^m \rangle_{\Omega_I} = -\int_{\Omega_I} Q_l^m(\mathbf{r} - \mathbf{r}_c)\rho(\mathbf{r})d\mathbf{r}$$

where $Q_l^m(\mathbf{r} - \mathbf{r}_c)$ is the multipole operator expressed on the basis of the spherical harmonics:

$$Q_l^m(\mathbf{r} - \mathbf{r}_c) = \left(\frac{4\pi}{2l+1}\right)^{\frac{1}{2}} |\mathbf{r} - \mathbf{r}_c|^l Y_l^m(\theta, \varphi)$$

Except for the unipole ($l = 0$), the multipole operators are origin dependent. As a general rule, the origin is taken as the attractor of the considered basin, except for attractors degenerated on a circle in which case the center of the circle defines the origin.

10.A.5
Combining ELF and AIM approaches

Combining the ELF and QTAIM approaches, Raub and Jansen [24] have introduced a bond polarity index defined as

$$pxy = \frac{\overline{N}[\mathrm{V(X,Y)}|\mathrm{X}] - \overline{N}[\mathrm{V(X,Y)}|\mathrm{Y}]}{\overline{N}[\mathrm{V(X,Y)}|\mathrm{X}] + \overline{N}[\mathrm{V(X,Y)}|\mathrm{Y}]}$$

where $\overline{N}[V(X, Y)|X]$ denotes the contribution of the X atomic basin to the population of the V(X,Y) basin.

The orbital contributions to the valence basins is another piece of information which, for example, enables to quantitatively estimate the σ-donation and π-back-donation.

10.A.6
Potential energy contributions

The generalization to ELF basins of the interacting quantum atoms energy Decomposition [69], can be achieved for only potential energy because the ELF basin separatrix are not zero flux surfaces of the electron density gradient field. The electron–nucleus contribution is given by

$$V_{eN}(\Omega_i) = -\sum_A \int_{\Omega_i} \frac{Z_A}{|\mathbf{r} - \mathbf{R}_A|} \rho(\mathbf{r})d\mathbf{r}$$

The decomposition of the interelectron repulsion into two-basin contributions:

$$V_{ee}(\Omega, \Omega') = \int_{\Omega} \int_{\Omega'} \frac{\pi(\mathbf{r}, \mathbf{r}')}{|\mathbf{r}' - \mathbf{r}|} d\mathbf{r}d\mathbf{r}'$$

decomposes in a purely Coulomb energy

$$V_C(\Omega, \Omega') = \int_{\Omega} \int_{\Omega'} \frac{\rho(\mathbf{r})\rho(\mathbf{r}')}{|\mathbf{r}' - \mathbf{r}|} d\mathbf{r}d\mathbf{r}'$$

and an exchange-correlation contribution

$$V_{xc}(\Omega, \Omega') = \int_{\Omega} \int_{\Omega'} \frac{\Pi_{xc}(\mathbf{r}, \mathbf{r}')}{|\mathbf{r}' - \mathbf{r}|} d\mathbf{r}d\mathbf{r}'$$

is very interesting, being directly related to the VSEPR model [70].

10.A.7
Miscellaneous

- Basin volume: $V(\Omega_I) = \int_{\Omega_I} d\mathbf{r}$, for $\rho(\mathbf{r}) \geq 0.002$ (the cutoff value is chosen in accordance with Ref. [10])
- Basin population molecular orbital contributions: $\overline{N}_\nu[\Omega_I] = \int_{\Omega_I} \varphi_\nu(\mathbf{r})d\mathbf{r}$
- Condensed Fukui functions f^+, f^0, f^-:

$$\langle f^- \rangle_{\Omega_I} = \int_{\Omega_I} [\rho_N(\mathbf{r}) - \rho_{N-1}(\mathbf{r})]d\mathbf{r}$$

$$\langle f^+ \rangle_{\Omega_I} = \int_{\Omega_I} [\rho_{N+1}(\mathbf{r}) - \rho_N(\mathbf{r})]d\mathbf{r}$$

$$\langle f^0 \rangle_{\Omega_I} = \frac{1}{2}(\langle f^- \rangle + \langle f^+ \rangle)$$

The Fukui function, $f(\mathbf{r})$, is a local reactivity descriptor defined in the framework of the conceptual DFT as the derivative of the electron density with respect to the number of electrons, the external potential $\nu(\mathbf{r})$ being kept constant [71],

$$f(\mathbf{r}) = \left(\frac{\partial\rho(\mathbf{r})}{\partial N}\right)_{\nu(\mathbf{r})}$$

In fact, this equation yields three descriptors according to the direction of variation of N, the electrophilic attack is governed by

$$f^{-}(\mathbf{r}) = \left(\frac{\partial\rho(\mathbf{r})}{\partial N}\right)^{-}_{\nu(\mathbf{r})} \approx \rho_{N_0}(\mathbf{r}) - \rho_{N_0-1}(\mathbf{r})$$

the nucleophilic attack by

$$f^{+}(\mathbf{r}) = \left(\frac{\partial\rho(\mathbf{r})}{\partial N}\right)^{+}_{\nu(\mathbf{r})} \approx \rho_{N_0+1}(\mathbf{r}) - \rho_{N_0}(\mathbf{r})$$

and the radicalar attack by

$$f^{0}(\mathbf{r}) = \left(\frac{\partial\rho(\mathbf{r})}{\partial N}\right)^{0}_{\nu(\mathbf{r})} = \frac{1}{2}(f^{-}(\mathbf{r}) + f^{+}(\mathbf{r}))$$

where the subscripts $N_0 - 1, N_0, N_0 + 1$ are the numbers of electrons of the system.

References

1. Lewis, G.N. (1916) *J. Am. Chem. Soc.*, **38**, 762–786.
2. Mátyus, E., Hutter, J., Müller-Herold, U. and Reiher, M. (2011) *Phys. Rev. A*, **83**, 052512.
3. Linnett, J.W. (1961) *J. Am. Chem. Soc.*, **83**, 2643–2653.
4. Gillespie, R.J. and Nyholm, R.S. (1957) *Q. Rev. Chem. Soc.*, **11**, 339–380.
5. Artmann, K. (1946) *Z. Naturf.*, **1**, 426–432.
6. Schubert, K. (1964) *Kristallstrukturen zweikomponentiger Phasen*, Springer-Verlag, Berlin, Göttingen, Heidelberg.
7. Scemama, A., Caffarel, M. and Savin, A. (2007) *J. Comput. Chem.*, **28**, 442.
8. Lüchow, A. and Petz, R. (2011) *J. Comput. Chem.*, **32**, 2619–2626.
9. Daudel, R. (1953) *C.R. Acad. Sci.*, **237**, 601–603.
10. Bader, R.F.W. (1990) *Atoms in Molecules: A Quantum Theory*, Oxford University Press, Oxford.
11. Savin, A. (2005) *J. Chem. Sci.*, **117**, 473–475.
12. Ramberg, P. (2000) *Int. J. Philos. Chem.*, **6**, 35–61.
13. Silvi, B. and Savin, A. (1994) *Nature*, **371**, 683–686.
14. Savin, A., Silvi, B. and Colonna, F. (1996) *Can. J. Chem.*, **74**, 1088–1096.
15. Beltrán, A., Andrés, J., Noury, S. and Silvi, B. (1999) *J. Phys. Chem. A*, **103**, 3078–3088.
16. Chamorro, E., Fuentalba, P. and Savin, A. (2003) *J. Comput. Chem.*, **24**, 496–504.
17. Grimwood, D.J., Bytheway, I. and Jayatilaka, D. (2003) *J. Comput. Chem.*, **24**, 470–483.
18. Tsirelson, V. and Stash, A. (2002) *Chem. Phys. Lett.*, **351**, 142–148.
19. Almbladh, C.O. and Pedroza, A.C. (1984) *Phys. Rev. A*, **29**, 2322.
20. Kohout, M. and Savin, A. (1997) *J. Comput. Chem.*, **18**, 1431–1439.
21. Gadre, S.R., Kulkarni, S.A. and Pathak, R.K. (1993) *J. Chem. Phys.*, **98**, 3574–3576.
22. Fuentealba, P. (1998) *Int. J. Quant. Chem.*, **69**, 559–565.
23. Shaik, S.S., Maitre, P., Sini, G. and Hiberty, P.C. (1992) *J. Am. Chem. Soc.*, **114**, 7861–7866.

24. Raub, S. and Jansen, G. (2001) *Theor. Chem. Acc.*, **106**, 223–232.
25. Kohout, M., Pernal, K., Wagner, F.R. and Grin, Y. (2004) *Theor. Chem. Acc.*, **112**, 453–459.
26. Kohout, M., Pernal, K., Wagner, F.R. and Grin, Y. (2005) *Theor. Chem. Acc.*, **113**, 287–293.
27. Matito, E., Silvi, B., Duran, M. and Solà, M. (2006) *J. Chem. Phys.*, **125**, 024301.
28. Feixas, F., Matito, E., Duran, M., Solá, M. and Silvi, B. (2010) *J. Chem. Theory Comput.*, **6**, 2736–2742.
29. Sagar, R.P., Ku, A.C.T., Vedene, H., Smith, J. and Simas, A.M. (1988) *J. Chem. Phys.*, **88**, 4367–4374.
30. Shi, Z. and Boyd, R.J. (1988) *J. Chem. Phys.*, **88**, 4375–4377.
31. Kohout, M., Savin, A. and Preuss, H. (1991) *J. Chem. Phys.*, **95**, 1928–1942.
32. Gillespie, R.J., Noury, S., Pilmé, J. and Silvi, B. (2004) *Inorg. Chem.*, **43**, 3248–3256.
33. Anderson, J.B. (2001) *J. Chem. Phys.*, **115**, 4546.
34. Kellner, R. Berger, N. Stöckle, R. Vaas, K. Söylemez, S. Pfitzner, A. Binder, H. Hönle, W. Grin, Yu. Burkhardt, A. Schultheiss, M. Wedig, U. Nuß, J. von Schnering, H. G. Borrmann, H. and Simon, S. (1997) in *Advances in Boron Chemistry* (ed. W. Siebert), The Royal Society of Chemistry, Cambridge, pp. 38–45.
35. Gonzalez, C. and Schlegel, H.B. (1989) *J. Chem. Phys.*, **90**, 2154–2161.
36. Krokidis, X., Noury, S. and Silvi, B. (1997) *J. Phys. Chem. A*, **101**, 7277–7282.
37. Polo, V., Andres, J., Berski, S., Domingo, L.R. and Silvi, B. (2008) *J. Phys. Chem. A*, **112**, 7128–7136.
38. Andrés, J., Berski, S., Domingo, L.R., Polo, V. and Silvi, B. (2011) *Curr. Org. Chem.*, **15**, 3566–3575.
39. Krokidis, X., Goncalves, V., Savin, A. and Silvi, B. (1998) *J. Phys. Chem. A*, **102**, 5065–5073.
40. Polo, V., Gonzalez-Navarrete, P., Silvi, B. and Andrés, J. (2008) *Theor. Chem. Acc.*, **120**, 341–349.
41. Berski, S., Andrés, J., Silvi, B. and Domingo, L. (2003) *J. Phys. Chem. A*, **107**, 6014–6024.
42. Savin, A., Becke, A.D., Flad, J., Nesper, R., Preuss, H. and von Schnering, H.G. (1991) *Angew. Chem., Int. Ed. Engl.*, **30**, 409–412.
43. Savin, A., Jepsen, O., Flad, J., Andersen, O.K., Preuss, H. and von Schnering, H.G. (1992) *Angew. Chem., Int. Ed. Engl.*, **31**, 187–188.
44. Savin, A., Nesper, R., Wengert, S. and Fässler, T.F. (1997) *Angew. Chem., Int. Ed. Engl.*, **36**, 1808–1832.
45. van Schilfgarde, M., Paxton, T.A., Jepsen, O., Andersen, O.K. and Krier, G. (1994) *Program TB-LMTO*, Max-Planck-Institut für Festkörperforschung, Stuttgart.
46. Kohout, M. (2002) PhD thesis, University Stuttgart.
47. Häussermann, U., Wengert, S. and Nesper, R. (1994) *Angew. Chem., Int. Ed. Engl.*, **33**, 2069–2072.
48. Ormeci, A., Rosner, H., Wagner, F.R., Kohout, M. and Grin, Yu. (2006) *J. Phys. Chem. A*, **110**, 1100–1105.
49. Silvi, B. and Gatti, C. (2000) *J. Phys. Chem.*, **104**, 947–953.
50. Gu, Q.F., Krauss, G., Grin, Yu. and Steurer, W. (2009) *Phys. Rev. B*, **79**, 134121.
51. Baranov, A.V. and Kohout, M. (2008) *J. Comput. Chem.*, **29**, 2161–2171.
52. Kohout, M. (2004) *Int. J. Quant. Chem.*, **97**, 651–658.
53. Ormeci, A. and Grin, Yu. (2011) *Isr. J. Chem.*, **51**, 1304–1313.
54. Kniep, R. (1996) in *Chemistry, Structure and Bonding of Zinlt Phases and Ions* (ed. S. Kauzlarich), VCH Publishers, Weinheim, pp. xvii–xxx.
55. Klemm, W. and Bussmann, E. (1963) *Z. Anorg. Allg. Chem.*, **319**, 297–311.
56. Schäfer, H., Eisenmann, B. and Müller, W. (1973) *Angew. Chem.*, **85**, 742–777.
57. Pearson, W.B. (1964) *Acta Crystallogr.*, **17**, 1–15.
58. Wengert, S. and Nesper, R. (1998) *Z. Anorg. Allg. Chem.*, **624**, 1801–1806.
59. Fässler, T.F. (2003) *Chem. Soc. Rev.*, **32**, 80–86.
60. Leithe-Jasper, A., Schnelle, W., Rosner, H., Baenitz, M., Rabis, A., Gippius, A.A., Morozova, E.N., Borrmann, H., Burkhardt, U., Ramlau, R., Schwarz, U.,

Mydosh, J.A. and Grin, Yu. (2004) *Phys. Rev. B*, **70**, 214418.
61. Kohout, M., Wagner, F.R. and Grin, Yu. (2002) *Theor. Chem. Acc.*, **108**, 150–156.
62. Grin, Yu., Armbrüster, M., Kohout, M., Leithe-Jasper, A., Schwarz, U., Wedig, U. and von Schnering, H.G. (2006) *J. Solid State Chem.*, **179**, 1707–1719.
63. Grin, Yu. and Nesper, R. (2011) *Z. Kristallogr.*, **226**, 692–710.
64. Grin, Yu., Wedig, U. and von Schnering, H.G. (1995) *Angew. Chem., Int. Ed. Engl.*, **34**, 1204–1206.
65. Diner, S. and Claverie, P. (1976) in *Localization and Delocalization in Quantum Chemistry*, Vol. II (eds O. Chalvet, R. Daudel, S. Diner and J.P. Malrieu), Reidel, Dordrecht, pp. 395–448.
66. Cancès, E., Keriven, R., Lodier, F. and Savin, A. (2004) *Theor. Chem. Acc.*, **111**, 373–380.
67. Francisco, E., Martin Pendás, A. and Blanco, M. (2008) *Comput. Phys. Commun.*, **178**, 621–634.
68. Pilme, J. and Piquemal, J.-P. (2008) *J. Comput. Chem.*, **29**, 1440–1449.
69. Martin Pendás, A., Blanco, M.A. and Francisco, E. (2004) *J. Chem. Phys.*, **120**, 4581–4592.
70. Martin Pendás, A., Francisco, E. and Blanco, M. (2008) *Chem. Phys. Lett.*, **454**, 396–403.
71. Parr, R.G. and Yang, W. (1984) *J. Am. Chem. Soc.*, **106**, 4049–4050.
72. von Weizsäcker, C.F. (1935) *Z. Phys.*, **96**, 431–458.
73. Burnus, T., Marques, M.A.L. and Gross, E.K.U. (2005) *Phys. Rev. A*, **71**, 010501(R).
74. Dobson, J.F. (1993) *J. Chem. Phys.*, **98**, 8870.
75. Abraham, R.H. and Shaw, C.D. (1992) *Dynamics: The Geometry of Behavior*, Addison-Wesley, Redwood City, CA.
76. Abraham, R.H. and Marsden, J.E. (1994) *Foundations of Mechanics*, Addison-Wesley, Redwood City, CA.

11
Relativity and Chemical Bonding

Peter Schwerdtfeger

11.1
Introduction

The concept of chemical bonding within a nonrelativistic quantum-chemical framework is well understood [1–6], albeit heavily discussed and debated at times for the many different chemical models and approximations applied. Questions such as how many bonds are shared between atoms [7–10], what the oxidation state is of a specific atom in a molecule or the solid state [11], or how covalent or ionic a chemical bond is [12], can often lead to (very) different opinions depending on the chemical bonding model favored [13]. In other words, although the term chemical bond is, without any doubt, an important concept in chemistry (and not one chemist will debate this), it is *not* a quantum-mechanical observable, and there perhaps lie all its difficulties associated with many misunderstandings and disputes [14–16]. The many bonding models used may only be loosely connected to quantum-mechanical observables (if at all), the latter being strictly represented in quantum theory by self-adjoint operators acting on specific Hilbert (L_2) spaces. Here we quote (for convenience) Charles Alfred Coulson:

> *One is almost tempted to say … at last I can almost see a bond. But that will never be, for a bond does not really exist at all: it is a most convenient fiction which, as we have seen, is convenient both to experimental and theoretical chemists* [17].[1)]

If Einstein's relativity is introduced, the situation would become even more complicated. From the principle of special relativity it follows that the velocity c of light is constant ($c = 299\,792\,458$ m s^{-1}) [18]. From this one can derive that the velocity v of a massive particle is also bound by the velocity of light; that is, $v < c$, which limits the accuracy in the measurement of a particle's momentum according

1) A very similar quote by Coulson is "Sometimes it seems to me that a bond between two atoms has become so real, so tangible, so friendly, that I can almost see it. Then I awake with a little shock, for a chemical bond is not a real thing. It does not exist. No one has ever seen one. No one ever can. It is a figment of our own imagination."

The Chemical Bond: Fundamental Aspects of Chemical Bonding, First Edition.
Edited by Gernot Frenking, Sason Shaik.

to the Heisenberg uncertainty principle ($\Delta p \Delta t \sim \hbar/c$) [19]. A strict relativistic (Lorentz-covariant) formalism implies that space and time have to be treated on equal footing (e.g., same order in space and time derivatives have to be used in a relativistic Hamiltonian), which for an electron of spin ½ in an external field leads to the famous Dirac–Coulomb equation [20]

$$H_D \begin{pmatrix} \phi_L \\ \phi_S \end{pmatrix} = \begin{pmatrix} V & c\vec{\sigma}\left(\vec{p} + c^{-1}\vec{A}\right) \\ c\vec{\sigma}(\vec{p} + c^{-1}\vec{A}) & V - 2mc^2 \end{pmatrix} \begin{pmatrix} \phi_L \\ \phi_S \end{pmatrix} = E \begin{pmatrix} \phi_L \\ \phi_S \end{pmatrix} \quad (11.1)$$

Here $\vec{\sigma}$ are the (2×2) Pauli spin matrices, $\vec{p}$ is the momentum of the electron, $\vec{A}$ is the magnetic vector potential introduced through the minimal coupling scheme, and ϕ_L and ϕ_S are, respectively, the large and small Dirac components of the electronic wave function $\psi = (\phi_L, \phi_S)$. As Eq. (11.1) is a (4×4) matrix differential equation, the Dirac operator H_D acts on a *four-component* wave function containing (in a way) the two spin directions and the particle/antiparticle symmetry of the electron. Hence, the spin of an electron is naturally contained within the Dirac equation. The potential V in Eq. (11.1) may contain corrections beyond the instantaneous Coulomb interaction between the two electrons. This leads to the so-called Dirac–Coulomb–Breit equation [21, 22] that covers most of the physics in electronic structure theory and is sufficient for the discussion of chemical bonding up to the heaviest elements in the periodic table [23].

The Dirac equation is plagued with many (albeit often technical) difficulties [21–24]. While the nonrelativistic Hamiltonian of the Schrödinger equation is bound in energy from below, ensuring that the Ritz–Rayleigh variational principle works for the minimization of the total electronic energy, the Dirac operator is not [25], and one has to deal with a continuum of negative energy states. These negative energy states have been the reason for many misconceptions in the relativistic quantum theory [26]. While a complete Lorentz-covariant picture is given by quantum electrodynamics (QED), or more precisely by the so-called *standard model* in physics, which is not yet fully accessible for the accurate treatment of many-electron systems, we have learned only in the past few decades how to handle the no-virtual-pair (i.e., without including electron–positron pair creation by QED) many-electron Dirac–Coulomb equation for molecules within mean-field theories such as the Hartree–Fock theory or the density functional theory (DFT), or beyond [27].

The discussion of relativistic effects in chemical bonding requires the correct nonrelativistic limit of the Dirac equation. Here we should note that, in principle, relativistic effects in molecular properties – that is, the change from the nonrelativistic (Schrödinger) picture to the relativistic (Dirac) picture (or vice versa) – cannot be observed directly, because we are living in a "relativistic world" with a fixed velocity of light c (if we neglect the fact that the fundamental constants in physics, such as the fine-structure constant, may vary in time and space [28]). The nonrelativistic limit can be obtained by either a unitary transformation to a two-component (spin–orbit-coupled) Schrödinger picture H_S or by elimination of the small Dirac component (ϕ_S) [29], followed by taking the nonrelativistic limit,

$\lim(c \to \infty)H_S$. There are many such transformations one can perform to obtain (regular) *two-component* relativistic Hamiltonians to a certain order in the expansion (in $1/c^2$), and there are a number of different relativistic approximations in use in standard quantum chemical program packages (Breit-Pauli, Douglas-Kroll, Barysz-Sadlej-Snijders, zero-order regular approximation, and X2C to name a few; see Refs [27, 29–34] for more details). Further, if we separate off spin-dependent terms in the two-component formalism, we obtain a spin-averaged (*one-component* or *scalar-relativistic*) formalism most widely used in quantum-chemistry program packages. We mention that the use of different relativistic perturbation operators to the nonrelativistic Schrödinger equation gives rise to different physical interpretations of relativistic effects if not taken to infinite order in the two-component expansion. Moreover, scalar-relativistic (spin-free) schemes differ in the way the spin has been separated off [35]. Further, the nonrelativistic limit, $\lim(c \to \infty)H_S$, has to be considered carefully, as properties connected to the electromagnetic field may vanish. This is clearly seen from the Dirac–Coulomb equation (Eq. (11.1)), as the vector potential $\vec{A}$ vanishes in the nonrelativistic limit, which implies that all magnetic fields vanish as well. Moreover, some properties are intrinsically relativistic. For example, parity violation effects that lift the degeneracy between left- and right-handed chiral molecules are only nonzero if spin–orbit effects are included [36]. Hence, one has to analyze carefully which relativistic perturbation operators are selected and which are omitted for the nonrelativistic limit. In the past, relativistic effects in molecular properties were mostly determined within the pseudopotential approximation that is used in a nonrelativistic, scalar-relativistic, or two-component (spin–orbit-coupled) forms [37, 38]. This valence-only approximation leads to accurate results for valence properties compared to all-electron calculations [39].

In what follows, we do not discuss all the difficulties and technical details associated with the picture as it changes from a relativistic to a quasi- or nonrelativistic framework or vice versa (see Ref. [40] for details), and assume that for molecular properties P (in general connected to observables) obtained correctly within the Dirac picture, P_R, and the Schrödinger picture, P_{NR}, the relativistic effects in P are simply defined as follows:

$$\Delta_R P = P_R - P_{NR} \tag{11.2}$$

As relativistic effects in most (but not all) atomic molecular properties scale approximately as Z^2 (Z being the nuclear charge), it is convenient to introduce the (dimensionless) relativistic enhancement factor f_P for a specific property P as [41, 42]

$$\frac{\Delta_R P}{P_R} = f_P\left(\frac{Z}{c}\right)^2 + \quad \ldots \tag{11.3}$$

where the velocity of light is in a.u. as $c \cong 137.036$. For example, we can analyze relativistic effects in the one-electron density, which is important in the discussion of chemical bonding by

$$\rho_R(\vec{r}) = \rho_{NR}(\vec{r}) + \Delta_R \rho(\vec{r}) \tag{11.4}$$

An interesting consequence of the Dirac equation is that the large (ϕ_L) and small (ϕ_S) component orbitals obtained in Dirac mean-field theories alternate in their radial notes, and therefore the radial density $\rho_i(r)$ for a specific orbital i does not become zero anymore in the range $0 < r < \infty$ [43]. However, as the density of the small Dirac component constitutes only little to the total density ($\leq$1%), and although much less in the important valence region, this fact is perhaps not so important for the discussion of chemical bonding.

The development of relativistic program packages using either the Dirac–Hamiltonian, or its transformed two-component (spin–orbit-coupled) or scalar-relativistic (spin–orbit-averaged) versions, made it possible over the past 30 years to study relativistic effects in molecular properties and chemical bonding in considerable detail. It leads to the correct interpretation of many anomalies observed in the chemical and physical behavior of heavy elements [44, 45], and to a deeper understanding of periodic trends down to the superheavy elements [46], and to very accurate results for molecular properties compared to high precision experiments for light molecules, such as H_2 [47] or He_2 [48]. As a prime example for the discussion of chemical bonding, H_2 is perhaps the most widely investigated molecule [4]. Therefore, in Table 11.1 we list the different contributions to the dissociation energy of H_2 and D_2 beyond a Born–Oppenheimer nonrelativistic treatment [47]. The results clearly demonstrate how much relativistic quantum chemistry advanced in the past few decades toward the accurate prediction of molecular properties including relativistic effects and beyond. This is also the case for molecules containing heavier atoms, albeit to a lesser extent as the main bottleneck here is still the accurate determination of electron correlation in many-electron systems.

Table 11.1 Dissociation energies D_0 and corresponding corrections for H_2 and D_2 (in cm^{-1}) at various levels of theory compared to experimental data.

Term	H_2	D_2
Born–Oppenheimer	36112.5927(1)	36746.1623(1)
Adiabatic	5.7711(1)	2.7725(1)
Nonadiabatic	0.4339(2)	0.1563(2)
Total nonrelativistic	36118.7978(2)	36749.0910(2)
One-electron relativistic	−0.4809(2)	−0.4748(2)
Two-electron relativistic	−0.0510(1)	−0.0528(1)
Total relativistic	−0.5319(3)	−0.5276(3)
Lowest-order QED	−0.1948(2)	−0.1983(2)
Higher-order QED	−0.0016(8)	−0.0016(8)
Total theory	36118.0695(10)	36748.3633(9)
Experimental (Refs [49, 50])	36118.0696(4)	36748.343(10)

The theoretical value for D_2 is probably more accurate compared to the most recent experimental result.
(Data are taken from Ref. [47].)

While relativistic effects do not alter our view of chemical bonding for molecules containing light elements and are only required for high spectroscopy accuracy, the situation changes completely when moving down the periodic table. It has become clear over the past 30 years that relativistic effects can substantially alter our understanding of chemical bonding in molecules (or the solid state) containing heavy elements, and are required to achieve even qualitative correct results [34, 44, 45]. For example, relativistic effects are responsible for AuCs to crystallize in an ionic lattice instead of being an intermetallic compound (Here two metals do not form a metallic bond!) [51]; mercury to be the only metallic liquid at room temperature [44, 52] with a melting temperature of $T_{\mathrm{m}} = -38.83°\mathrm{C}$ (although gallium is close with $T_{\mathrm{m}} = 29.8°\mathrm{C}$); the yellow color of bulk gold due to a relativistic decrease in the electronic $5d$–$6s$ band excitation energy [53, 54]; singlet–triplet transitions being allowed through spin–orbit coupling, for example, the strong 254 nm line in a mercury–vapor lamp belongs to the spin-forbidden ${}^3\mathrm{P}_0 \rightarrow {}^1\mathrm{S}_0$ transition [55]; the high stability of gold compounds in the oxidation state +3 (and +5 for AuF_6^-) [44, 56], with implications to homogeneous Au^{+3} catalysis [57]; catalytic methane activation by Pt^+ catalyst [58]; and finally the lead–acid battery that would not work without relativistic effects [59]. In Section 11.2, we discuss the main relativistic effects responsible for significantly changing our (nonrelativistic) view on chemical bonding in heavy element chemistry; that is, we use the convenient terms of direct and indirect relativistic effects and spin–orbit coupling. Some prime examples of how relativistic effects significantly alter our understanding of chemical bonding in heavy-element-containing molecules are presented in Section 11.3.

11.2 Direct and Indirect Relativistic Effects and Spin–Orbit Coupling

11.2.1 Scalar-Relativistic Effects

Special relativity predicts an increase in mass m of a particle with rest mass $m_0 \neq 0$ traveling at a certain velocity v relative to an outside observer at rest. This implies a relativistic decrease of the Bohr radius in a hydrogenic atom

$$a = \frac{\hbar^2}{me^2} = \frac{\hbar^2}{m_0 e^2}\sqrt{1-\frac{v^2}{c^2}} = a_0\sqrt{1-\frac{v^2}{c^2}} \quad \Rightarrow \quad \frac{\Delta_{\mathrm{R}} a_{1s}}{a_{1s}} \approx -\frac{1}{2}\left(\frac{Z}{c}\right)^2 \tag{11.5}$$

The second part of the formula (in a.u.) follows from the fact that for a hydrogenic atom we have for the average radial velocity of a $1s$ electron $\langle v_r \rangle = Z$. For the heavy element Pb ($Z = 82$), this results in a velocity for a $1s$ electron of 60% of the velocity of light. Equation (11.5) gives a rough estimate for the relativistic enhancement factor $f_a = -1/2$ for a relativistic orbital contraction. The question is, whether this simple picture holds for a many-electron quantum system.

Table 11.2 shows relativistic enhancement factors for the Group 11 series of elements obtained from orbital radial $\langle r \rangle$ expectation values of relativistic

Table 11.2 Relativistic enhancement factors $f_{<r>}$ for the Group 11 elements from Cu (Z = 29) to Rg (Z = 111) in the $nd^{10}(n+1)s^1$ configuration, and for the elements No (Z=102, $5f^{14}7s^2$) and 118 ($7s^27p^6$).

Orbital	Cu	Ag	Au	Rg	No	118
$1s$	−0.333	−0.351	−0.410	−0.551	−0.497	−0.607
$2s$	−0.351	−0.376	−0.438	−0.579	−0.525	−0.633
$2p_{1/2}$	−0.395	−0.437	−0.528	−0.749	−0.660	−0.847
$2p_{3/2}$	−0.040	−0.061	−0.076	−0.085	−0.083	−0.086
$3s$	−0.313	−0.293	−0.335	−0.425	−0.391	−0.458
$3p_{1/2}$	−0.340	−0.303	−0.354	−0.465	−0.422	−0.509
$3p_{3/2}$	−0.037	−0.054	−0.078	−0.089	−0.086	−0.090
$3d_{3/2}$	−0.006	−0.046	−0.085	−0.102	−0.098	−0.105
$3d_{5/2}$	0.243	0.039	0.001	−0.009	−0.007	−0.011
$4s$	−0.477	−0.269	−0.296	−0.346	−0.324	−0.368
$4p_{1/2}$		−0.286	−0.305	−0.362	−0.335	−0.390
$4p_{3/2}$		−0.032	−0.063	−0.069	−0.067	−0.070
$4d_{3/2}$		0.006	−0.058	−0.069	−0.066	−0.070
$4d_{5/2}$		0.159	0.014	−0.002	0.001	−0.004
$5s$		−0.506	−0.310	−0.316	−0.306	−0.328
$4f_{5/2}$			0.050	0.003	0.009	0.000
$4f_{7/2}$			0.089	0.035	0.041	0.031
$5p_{1/2}$			−0.331	−0.327	−0.317	−0.343
$5p_{3/2}$			−0.049	−0.054	−0.056	−0.053
$5d_{3/2}$			−0.014	−0.047	−0.050	−0.047
$5d_{5/2}$			0.140	0.018	0.021	0.015
$6s$			−0.629	−0.379	−0.410	−0.349
$5f_{5/2}$				0.053	0.108	0.037
$5f_{7/2}$				0.087	0.176	0.067
$6p_{1/2}$				−0.410	−0.462	−0.368
$6p_{3/2}$				−0.047	−0.050	−0.028
$6d_{3/2}$				−0.008	0.108	0.000
$6d_{5/2}$				0.133	0.176	0.074
$7s$				−0.832	−0.374	−0.451
$7p_{1/2}$						−0.517
$7p_{3/2}$						−0.012

Dirac–Coulomb and nonrelativistic Hartree–Fock calculations, as well as for two selected heavy elements, the actinide nobelium and the rare gas element with nuclear charge 118. As the action of the relativistic perturbation operator is in the near vicinity of the nucleus (or more precisely in the K- and L-shell region [60]), s- and $p_{1/2}$-orbitals that have considerable density in the inner core region are most susceptible to such *direct relativistic effects*. These orbitals get "pulled-in" by their inner density tails, and as a result, s- and $p_{1/2}$-orbitals contract and energetically stabilize (e.g., by the direct action of the relativistic kinetic mass–velocity operator in the Pauli picture [41]). This is clearly evident from the *s*- and $p_{1/2}$-enhancement factors that are listed in Table 11.2. What is perhaps surprising (or even paradoxical) is that the valence *s*- and $p_{1/2}$-electrons, which from a classical point of view have velocities far smaller than the velocity of light ($v \ll c$), also undergo large relativistic contractions. In fact, for the Group 11 elements the valence *s*-electrons all have $f_{<r>}$ values that are significantly larger (!) compared to their deeper-lying (core) *s*-shells. This rather unexpected result is perhaps the reason why this large valence–shell relativistic contraction has been overlooked for such a long time [61], which led, for example, Glashow in 1988 to conclude that

> *Relativity becomes important only when velocities become comparable to the speed of light. Electrons in atoms move rather slowly, at a mere one percent of light speed. Thus it is that a satisfactory description of the atom can be obtained without Einstein's revolutionary theory* [62].

Owing to the large contraction of all *s*- and $p_{1/2}$-shells, the nucleus becomes more screened and the effective nuclear charge Z_{eff} therefore decreases (*relativistic shielding*). This leads to an energetic destabilization and expansion of all orbitals (*indirect relativistic effect*). This indirect effect is especially important for the outer and more diffuse high angular-momentum functions, that is, for the *d*- and *f*-orbitals. The interplay between the direct and indirect effects can however be quite subtle as the $f_{<r>}$-values listed in Table 11.2 show. For example, for the Group 11 elements these direct and indirect relativistic effects almost cancel out for the outer $d_{3/2}$-orbitals, while the outer $d_{5/2}$-orbitals undergo a sizeable expansion, and therefore become valence-like and accessible to chemical bonding. Further, the results for nobelium show that both $5f_{5/2}$- and $5f_{7/2}$-orbitals expand because of the relativistically screened nucleus, and in contrast to the lanthanides, become more valence-like and therefore contribute to chemical bonding in early actinide compounds. The very large relativistic contraction of the valence $p_{1/2}$-valence shell for superheavy element 118 (but to a lesser extent compared to the deep core $2p_{1/2}$-orbital) is remarkable, indicating rather strong spin–orbit coupling effects, which will be discussed further later in this chapter.

What is the reason behind this rather large and unexpected relativistic valence *s*-shell contraction? In early days of relativistic quantum chemistry it was attributed to the orthogonality constraints among the relativistically contracted atomic *s*-orbitals [63, 64]; that is, if 1*s*-orbital contracts, then the other follows. While this explanation is intuitive, it turned out that the orthogonality constraint of the valence atomic orbitals on the relativistically contracted inner-core orbitals leads to a slight expansion effect for valence *s*-orbitals [40]. The perturbation theory shows a more subtle (and perhaps unexpected) effect that mixing in higher bound and continuum orbitals by relativistic perturbation operators is responsible for the contraction. The relativistic orbital contraction is thus mainly due to the direct action of relativistic perturbation operator, which dominates for *s*-orbitals as discussed earlier. This is the reason why for Eq. (11.5) we use the unscreened nuclear charge Z (and not Z_{eff}).

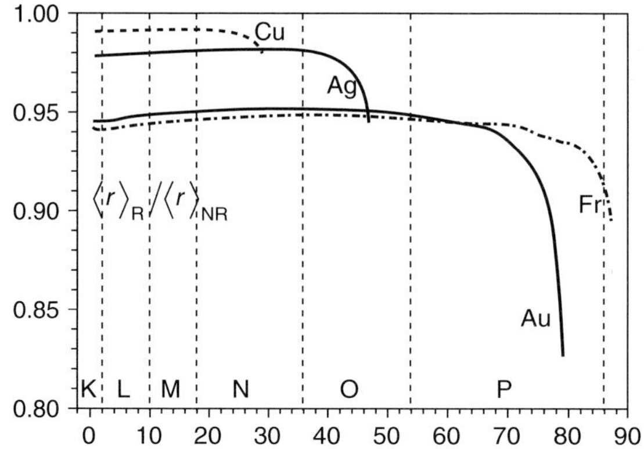

Figure 11.1 Valence relativistic s-shell contraction $<r>_R/<r>_{NR}$ (4s for Cu, 5s for Ag, 6s for Au, and 7s for Fr). The *ns*-shell remains singly occupied and the *x*-axis gives the total number of electrons *N* with electrons filled in successively starting from the innermost shells. For example, $N=5$ for Au describes the occupation $1s^2 2s^2 6s^1$ and $N=11$ the occupation $1s^2 2s^2 2p^6 6s^1$. (The data are taken from Ref. [66].)

Still we need to explain why for gold, for example, the relativistic 6*s*-contraction is significantly larger compared to the 1*s*-contraction.

In 1962, Corbett [65] attributed the high first ionization potential of mercury to the rather poor nuclear shielding by the lower-lying 5*d* and 4*f* electrons. The lower-lying relativistically destabilized (expanded) *d*- and *f*-orbitals cause an *indirect stabilization* by increasing Z_{eff} (*relativistic deshielding*) for the other orbitals. Figure 11.1 shows that much of the relativistic valence *s*-contraction (stabilization) originates from the filling of the lower-lying *d*-shell for the Group 11 elements. This effect is less pronounced for core *p*-shells as shown for francium in Figure 11.1. Removing an electron from the 5*d*-shell in Au substantially reduces the relativistic 6*s*-contraction [66]. Hence, the relativistic valence *s*-shell contraction diminishes for incompletely filled *d*-shells, thus explaining the rather large relativistic effects in the Group 11 and 12 series of elements; the Group 11 showing the maximum of relativistic effects [63, 67] compared to the other elements in the period. This maximum shifts to the Group 12 elements if the *s*-level is doubly occupied [66] (for more details the reader is referred to Refs [41, 42, 66]).

The relativistic enhancement factors listed in Table 11.2 confirm that the valence *s*-shell values *increase* from Cu down to Rg ($Z=111$). This is perhaps not surprising as the nonrelativistic valence *s*-orbitals become more diffuse and more polarizable going down the group for the *s*-block elements, thus it becomes easier for relativistic perturbation operators to pull in these orbitals; in fact, the very strong relativistic 7*s*-contraction for Rg results in a similar atomic radius compared to Cu(!). Table 11.3 lists ionization potentials and electron affinities for the Group 11 elements. Again, we see very large relativistic enhancement factors (for Cu and Ag the coupled cluster values have to be taken with some care as the published ionization potentials and electron affinities [68] are not of sufficient accuracy to produce accurate relativistic

Table 11.3 Nonrelativistic (NR) and relativistic (R) ionization potentials ΔE_{IP} and electron affinities ΔE_{EA} (in eV), relativistic effects Δ_R and relativistic enhancement factors f for the Group 11 elements.

Method	Cu	Ag	Au	Rg
Ionization potential				
Experimental	7.726	7.576	9.226	–
HF ΔE_{IP} (R)	6.564	6.339	7.682	8.93
ΔE_{IP} (NR)	6.408	5.909	5.915	5.395
$\Delta_R \Delta E_{IP}$	0.156	0.431	1.766	3.535
f_{IP}	0.531	0.577	0.692	0.603
CC ΔE_{IP} (R)	7.733	7.495	9.197	10.6
ΔE_{IP} (NR)	7.555	6.934	7.057	5.87
$\Delta_R \Delta E_{IP}$	0.236	0.561	2.14	4.73
f_{IP}	0.514	0.636	0.700	0.680
Electron affinity				
Experimental	1.163	1.303	2.309	–
HF ΔE_{EA} (R)	0.033	0.123	0.666	0.244
ΔE_{EA} (NR)	0.011	0.043	0.099	0.069
$\Delta_R \Delta E_{EA}$	0.022	0.08	0.567	0.175
f_{EA}	14.89	5.529	2.562	1.093
CC ΔE_{EA} (R)	1.253	1.254	2.295	1.565
ΔE_{EA} (NR)	1.165	1.054	1.283	1.054
$\Delta_R \Delta E_{EA}$	0.088	0.212	1.012	0.511
f_{EA}	1.568	1.356	1.327	0.498

For Au and Rg the Fock-space coupled cluster (CC) results are taken from Kaldor and coworkers [70, 71], for Cu and Ag the Douglas–Kroll scalar-relativistic CCSD(T) results are used from Sadlej and coworkers [68]. Experimental values are from Refs [72, 73]. For Rg the relativistic ground states are as a reference, for example, the $^2D_{5/2}$ state for neutral Rg.

enhancement factors). For superheavy element Rg the relativistic 7*s*-contraction is so large that the electronic ground state becomes $^2D_{5/2}$ $(6d^4_{3/2}d^5_{5/2}7s^2_{1/2})$ instead of the expected $^2S_{1/2}$ $(6d^4_{3/2}d^6_{5/2}7s^1_{1/2})$ state [69]. Therefore, the first ionization (or uptake) of an electron is from (into) the $6d_{5/2}$ shell and not from the $7s_{1/2}$ shell, which diminishes relativistic effects for both the ionization potential and electron affinity. For element 118 the relativistic $8s_{1/2}$-contraction becomes so large that it becomes the first rare-gas element with a positive electron affinity (0.056 eV at the coupled cluster level of theory) [71].

11.2.2
Spin–Orbit Effects

Atomic orbitals of angular momentum $l > 0$ separate into $j = l \pm 1/2$ levels due to spin–orbit interaction, which is a relativistic effect as it is correctly derived only from the Dirac equation. This lifting of degeneracy in atomic levels becomes very large in heavy elements as clearly seen from the relativistic enhancement factors that are listed in Table 11.2. This was already known from early days of atomic spectroscopy [72], but its implication to chemical bonding was realized rather late, except perhaps for electronic transitions in atoms and molecules to correctly assign spectra, and where forbidden electronic transitions become allowed due to spin–orbit coupling (such as in the case of the intense $^3P_0 \rightarrow {}^1S_0$ transition in mercury vapor [55]). Spin–orbit coupling was included in early ligand and crystal field theory as well [74], as also in Jahn–Teller or Renner–Teller symmetry-breaking effects that are sensitive to spin–orbit effects [75]. Because spin–orbit effects can break symmetry between degenerate energy levels, it is the most well known and studied of all relativistic effects.

The relativistic enhancement factors listed in Table 11.2 show that $p_{1/2}$-orbitals contract strongly, while $p_{3/2}$-orbitals remain mostly unaffected (or expand slightly for valence shells); valence $d_{5/2}$-, $f_{5/2}$-, and $f_{7/2}$-orbitals expand, while the $d_{3/2}$-orbital can contract or expand depending on subtle shell–structure effects. Because of the strong relativistic stabilization of the $p_{1/2}$-shell, spin–orbit splitting of the $p_{1/2}/p_{3/2}$ levels can reach the size of typical bond dissociation energies for the sixth and seventh row *p*-block elements. For example, the $^2P_{1/2}/{}^2P_{3/2}$ spin–orbit splitting in thallium is 0.966 eV [72], and that for element 113 is 2.793 eV [76]. Spin–orbit splitting becomes even larger moving to the right-side within the *p*-block elements in the periodic table. For astatine, the $^2P_{3/2}/{}^2P_{1/2}$ spin–orbit splitting is 1.969 eV [72], while for element 117 it is 7.733 eV(!) at the Dirac–Fock level (including Breit and the lowest-order QED effects). Moreover, the relativistic (*nlj*) orbitals for $l > 0$ ($p_{1/2}$, $p_{3/2}$, $d_{3/2}$, etc.) are quite different to the nonrelativistic (*nl*) orbitals, the latter used extensively in Cartesian coordinates (e.g., p_z and d_{z^2}) to discuss chemical bonding in standard quantum chemistry [77, 78]. For example, a $p_{1/2}$-orbital consists of a (complex) admixture of all three Cartesian *p*-spin-orbitals. The (positive) combination of two $p_{1/2}$-orbitals along a molecular axis leads to 1/3 σ-bond and 2/3 π-antibonding. Hence, in order to form a pure σ-bond in a molecule one requires an admixture of $p_{3/2}$- into $p_{1/2}$-orbitals, which is not always energetically favored if spin–orbit splitting becomes large. In other words, strong spin–orbit coupling will destabilize a σ-bond and will lead to a closed-shell system if the $p_{1/2}$-orbital becomes doubly occupied. As an example, we mention superheavy element Fl ($Z = 114$) with a $7s^2 7p_{1/2}^2$ closed-shell configuration, which is assumed to be chemically inert. Current one-atom-at-a-time experiments are designed around this chemical property [79]. The relativistic stabilization of the $p_{1/2}$-shell is already seen in the trend down the periodic table for the Group 13 and 14 elements. An important implication of the spin–orbit coupling is that the special stability of a half-filled *p*-shell (p^3) loses its significance for the heavy elements [41].

A further complication arises from the fact that spin-symmetry has to be included allowing for half-integer angular momenta. Bethe [80] therefore introduced an additional rotation *R* by 360° into molecular point-groups (which for single point-groups leads to the identity), resulting in the so-called double groups, as it doubles the number of operations. Here either the Bethe (Γ_j-notation) or the Mulliken–Herzberg notation (labeled according to their dimensionality) is used for labeling irreducible representations in double groups, except for linear molecules where one uses conveniently the ω–ω-coupling notation. For example, a $^3\Pi$ state of a diatomic molecule splits into four states of 0^+, 0^-, 1, and 2, originating from products of spin and spatial symmetry. Double-group symmetry is extensively used in ligand- or crystal-field theory when strong spin–orbit coupling is encountered [81]. For example, a ^{4}F atomic state (e.g., the ground state of Cr^{3+}) splits into four levels because of spin–orbit coupling ($^4F_{3/2}$, $^4F_{5/2}$, $^4F_{7/2}$, $^4F_{9/2}$), and further in a total of nine levels in a cubic ligand field, which are characterized by their double-group irreducible representations. For further information the reader is referred to standard textbooks in group theory [82], and to an excellent review by Leszczynski and coworkers [83]. Double-group symmetry has been implemented into two- and four-component molecular SCF codes to save computer time [84].

The discussion presented in Section 11.2 should be sufficient to understand the subtle interplay of relativistic effects in atomic shells qualitatively, and to discuss its implications to chemical bonding in heavy-element-containing compounds further. We refer to the published work by Schwarz for a more detailed analysis of relativistic effects on atomic shell structure [41, 42].

11.3
Chemical Bonding and Relativistic Effects

It has become clear over the past decades that the influence relativistic effects have on valence orbitals will substantially alter our understanding of chemical bonding in heavy elements. The field has been extensively reviewed [23, 34, 41, 44, 63], and we restrict ourselves to the discussion of a few prime examples to discuss the importance of direct, indirect, and spin–orbit relativistic effects on chemical bonding.

The filling of the *d*-levels in the transition metal series results in an increase in the contraction/stabilization of the valence *s*-orbital as Figure 11.1 shows, and to the well-known maximum of relativistic effects at the Group 11 series of elements in the periodic table. The strong relativistic stabilization of the 6*s*-orbital in Au leads to the notion that bulk gold is chemically more inert compared to their lighter congeners [44]. Table 11.4 shows relativistic and nonrelativistc solid-state calculations for bulk gold by Takeuchi *et al.* [86]. The atomic volume increases by 14%, the cohesive energy by 52%, and the bulk modulus by 80% due to relativistic effects. The relativistic effects lead to many anomalies in the properties of bulk gold, for example, we have for the specific resistivity Cu 1.72, Ag 1.62, Au 2.4 [10^{-8} Wm]; the thermal conductivity Cu 3.85, Ag 4.18, Au 3.1 [J cm^{-1} s^{-1} K^{-1}]; the electronic heat capacity Cu 6.926, Ag 6.411, Au 6.918 10^{-4} J $K^{-1}mol^{-1}$; the melting point Cu 1083, Ag 961, Au 1064°C; and the boiling point Cu 2567, Ag 2212, Au 3080°C to name but a few [66]. Christensen and Seraphin determined the interband edge in the optical spectrum as 2.38 eV in agreement with experiment, which is much smaller compared to that of silver (3.7 eV) [53]. More recently, the onset of the interband transition in gold has been calculated at 1.9 eV using

Table 11.4 Lattice constants a and volume V, cohesive energy E_{coh} and bulk modulus B for *fcc* gold from nonrelativistic and relativistic pseudopotential density functional calculations [86].

Au (*fcc*)	a [Å]	V [Å^3]	E_{coh} [eV]	B [Mbar]
Nonrelativistic	4.314	20.072	2.27	0.996
Relativistic	4.104	17.282	3.46	1.790
Experimental	4.079	16.967	3.81	1.732

Experimental values are from Refs. [87, 88].

scalar-relativistic time-dependent DFT, compared to 3.5 eV at the nonrelativistic level [54]. The low-absorption edge calculated at the relativistic level is seen as the primary reason for the yellow color of gold.

From the relativistic increase in both the ionization potential ΔE_{IP} and the electron affinity ΔE_{EA} we can estimate the change in the electronegativity χ using Mulliken's empirical formula

$$\chi \cong 0.187(\Delta E_{IP} + \Delta E_{EA}) + 0.17 \tag{11.6}$$

where ΔE_{IP} and ΔE_{EA} are in units of eV. This leads to a relativistic change in χ of 0.07 for Cu, 0.14 for Ag, 0.58 for Au, and 1.14 for Rg [66]. For the heavier metals, these changes are significant yielding rather large electronegativities of 2.37 for Au and 2.44 for Rg [89]. Hence, both Au and Rg may be considered as pseudo-halides having the largest electronegativities of all metals in the periodic table (for comparison, iodine has a Pauling electronegativity of 2.2). A direct consequence of this shift to higher electronegativities is that the ionicity increases in gold bonding with electropositive ligands such as Li or Cs, and decreases with more electronegative ligands such as F or Cl. A Mulliken population analysis supports this, for example, we obtain at the Hartree–Fock level a Mulliken charge of −0.68 (nonrelativistic) and −0.74 (relativistic) for gold in AuCs, and +0.73 (nonrelativistic) and +0.59 (relativistic) in AuF. This relativistic change of the ionic bonding character has direct consequences for bulk properties.

Self-consistent linear muffin-tin orbital (LMTO) band-structure calculations by Christensen and Kollar in 1983 showed that nonrelativistic AuCs is a metal (as one expects), whereas relativistic AuCs is a semiconductor in agreement with experimental results [51]. The band gap has been determined to be of 2.6 eV [90]. The electronic band gap at the relativistic level originates from the scalar-relativistic effects and not from the spin–orbit coupling [51]. Hence, two metals do not form a metallic bond in this case.

Turning to electronegative ligands, the unusual zigzag chain-like structures as shown in Figure 11.2, known for the bulk gold halides AuCl, AuBr, and AuI, have puzzled inorganic chemists for a long time, as simple ionic models predict cubic-like structures as found, for example, for NaCl or AgCl. Here the relativistically reduced ionicity (increased covalency) in the gold–halide bonding leads to a

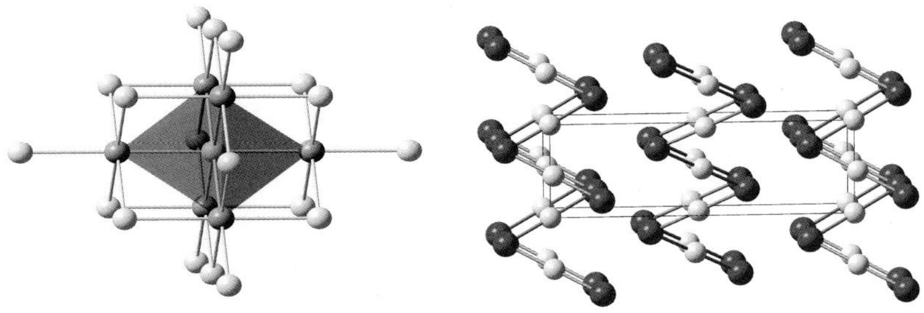

(a) (b)

Figure 11.2 Cubic (rock-salt) solid-state structure of AgCl (a) in comparison to chain-like zigzag structures for AuI (b).

stabilization of more than 100 kJ mol^{-1} for the chain-like structure of AuCl, AuBr, and AuI compared to the cubic arrangement [91]. Similarly, the unusual cinnabar structure of HgS and the montroydite structure of HgO, can both be attributed to the relativistic 6*s*-stabilization and a reduced ionicity in mercury–ligand bonding [92, 93].

Concerning the stability of gold–ligand bonding we can use Pauling's empirical formula for a diatomic gold molecule (which defines the Pauling electronegativity scale), and divide relativistic effects into covalent and ionic contributions

$$\begin{aligned} D_e(\mathrm{AuL}) &= \frac{1}{2}\{D_e(\mathrm{Au}_2) + D_e(\mathrm{L}_2)\} + f\{\chi_{\mathrm{Au}} - \chi_{\mathrm{L}}\}^2 \\ \Rightarrow \Delta_{\mathrm{R}} D_e(\mathrm{AuL}) &= \frac{1}{2}\Delta_{\mathrm{R}} D_e(\mathrm{Au}_2) + f\Delta_{\mathrm{R}}\chi^2_{\mathrm{Au}} - 2f\chi_{\mathrm{L}}\Delta_{\mathrm{R}}\chi_{\mathrm{Au}} \\ &\approx 280 - 80\chi_{\mathrm{L}}\ [\mathrm{kJ}\cdot\mathrm{mol}^{-1}] \end{aligned} \tag{11.7}$$

where D_e(Au-L) is the dissociation energy of diatomic Au-L, χ the electronegativity of a specific atom, and f an empirical factor to be determined. Here we neglect relativistic effects in the ligand L. Using the relativistic coupled-cluster calculations we obtain $\Delta_R D_e(Au_2)$=78 kJ mol^{-1} [94]. The empirical factor can be obtained from calculations of relativistic effects in dissociation energies for series of diatomic gold compounds, or directly through Eq. (11.7), which yields an f-value of about 80–100 kJ mol^{-1}. Equation (11.7) correctly predicts that AuF is relativistically destabilized with respect to dissociation into Au and F. The instability of AuF is the reason behind its late discovery, which has been synthesized only in the gas phase [95]. This relativistic destabilization can also be explained by a depletion of 6*s*-density from Au on bonding with electronegative ligands, leading to a reduced relativistic 6*s*-participation in all occupied bonding orbitals, and therefore to a reduced orbital stabilization in such orbitals compared to the atomic level [44].

Electropositive ligands shift electron density into gold leading to an increase in the relativistic 6*s*-stabilization at the molecular level. As a result, intermetallic gold compounds undergo a rather large relativistic bond stabilization leading to some unusually high dissociation energies [96]. This has implications for gold nanocatalysis where the 6*s*-density in gold clusters can be fine-tuned by doping with appropriate electropositive metals [97]. An interesting consequence of the large relativistic stabilization through bonding with electropositive ligands is that the two closed-shell $6s^2$ atoms Au^- and Ba give the strongest closed-shell interaction predicted so far with a dissociation energy of 1.48 eV [98].

The linear relationship (Eq. (11.7)) also works reasonably well for estimating relativistic bond contractions in diatomic gold compounds [66]

$$\Delta_R r_e(\text{AuL}) \approx -0.26 + 0.02\chi_L[\text{Å}] \tag{11.8}$$

As shown in Eq. (11.8) all Au-L bonds contract because of relativistic effects. The largest relativistic bond contraction is obtained in bonding with electropositive ligands (metals) leading to very short bond distances, which can become as small as those in the equivalent compounds with the lightest Group 11 element, copper. In contrast, for coordination compounds of gold in the oxidation state +3, the 6*s*-density is substantially reduced by the surrounding electronegative ligands, and the relativistic bond contraction is therefore much smaller compared to the corresponding coordination compounds of gold in the oxidation state +1. This leads to an interesting anomaly for the crystal structure of $Cs_2[AuCl_2][AuCl_4]$, where the shorter Au–Cl distance belongs for the $[AuCl_2]^-$ unit, and the longer distance to the $[AuCl_4]^-$ unit, and not vice versa as one expects [99]. Even larger relativistic bond contractions result for the compounds containing the superheavy element Rg [100].

From the chemical hardness η defined according to Parr and Pearson (HSAB model) we get

$$\eta = \frac{1}{2}(\Delta E_{\text{IP}} - \Delta E_{\text{EA}}) \quad \Rightarrow \quad \Delta_R \eta = \frac{1}{2}(\Delta_R \Delta E_{\text{IP}} - \Delta_R \Delta E_{\text{EA}}) \tag{11.9}$$

where ΔE_{IP} and ΔE_{EA} are in eV [101]. Equation (11.9) gives chemical hardness values (in eV) of 6.28 for Cu^+, 6.95 for Ag^+, and 5.64 for Au^+. For comparison, the nonrelativistic hardness is 7.19 for Au^+ [66]. Hancock *et al.* [102] investigated the aqueous chemistry of the positively charged Group 11 metals. They found that both Au^+ and Rg^+ are strong Lewis acids, and the softness ($S \sim 1/\eta$) follows approximately the trend $Rg^+ > Au^+ > Ag^+ > Cu^+$. Au^+ therefore binds preferentially to soft ligands, and it is well known that relativistic effects dominate the coordination chemistry of Au^+ [66].

Concluding this section on gold chemistry we note that the easy availability of the *d*-electrons in chemical bonding due to relativistic effects leads to a stabilization of the higher oxidation states +3 and +5 in gold, and even more so for Rg. At the nonrelativistic level, AuF_6^-, with gold in the oxidation state +5, would not be known [100]. Many other properties, such as gold–ligand stretching force constants or gold–nuclear quadrupole coupling constants, show extremely large relativistic effects [66, 103]. Thus, the chemistry of gold cannot be understood without the inclusion of relativistic effects.

Another consequence of the relativistic valence *s*-stabilization is that the 6*s* (and 7*s*) electrons become more inert and less active in chemical bonding for the 6*p* (and 7*p*) main group elements. While the inert pair effect can be explained by a natural periodic trend down the p-block elements of the periodic table, the relativistic effects substantially alter it for the heaviest group members [104]. It leads to a destabilization of the higher oxidation state, that is, for Tl^{3+} or Pb^{4+}, and even more so for the superheavy elements 113 and 114 (Fl), where, in addition the $7p_{1/2}$-shell becomes chemically inert as well [105]. One interesting consequence of this is that the voltage of the lead–acid battery changes from +0.39 V to +2.13 V; thus, cars only start because of relativity (as Pyykkö phrased it [59]). We also note that relativistic effects contribute to the lanthanide, and even more so to the actinide contraction [106].

The role of indirect relativistic effects to chemical bonding is less understood and studied computationally. The close proximity of the $5d$- and $6s$-orbitals for gold gives rise to higher oxidation states, but this is mostly due to the relativistic $6s$-contraction, which brings the $6s$-orbitals closer to the $5d$, and only partly due to the relativistic $5d$-expansion. The most spectacular example here is HgF_4 (Hg in the oxidation state +4), which has been theoretically predicted by Kaupp and co-workers [107, 108], and experimentally found by Wang *et al.* [109]. This molecule would not exist at the nonrelativistic level. Participation of the $5d$-orbitals [108] in the chemical bonding is evident from the MO pictures shown in Figure 11.3. Going further down the Group 12 elements of the periodic table, CpF_4 is thermodynamically even more stable compared to HgF_4 [110].

The indirect relativistic expansion of the $5f$-shell in the actinides makes f-participation in chemical bonding more accessible, which implies that the $5f$-electrons are not quite as core-like anymore compared to the lanthanides. However, the importance of $5f$-participation in actinide bonding is often a matter of debate [111]. For example, the addition of an extra $5f$ electron to UO_2^{2+} forming UO_2^{+} causes a change in the bond length of 0.07 Å [112], but this is not because of overlap of U $5f$ with oxygen $2p$, but more likely due to the additional screening of the nucleus. Figure 11.4 shows the two highest occupied valence orbitals for UO_2^{2+}, with large $5f$-participation in the U–O bonding. Gagliardi and Roos have recently proposed a quintuple bond for U_2 clearly involving $5f$-participation (beside $7s$ and

(a) (b) (c)

Figure 11.3 Bonding of F $2p$-orbitals with Hg $5d_{xy}$ (a), $5d_{z^2}$ (b) and $6s$ (c) orbitals in HgF_4. Note that the Mulliken population analysis for Hg gives $5d^{9.75}6s^{0.81}6p^{0.56}$ at the DFT level.

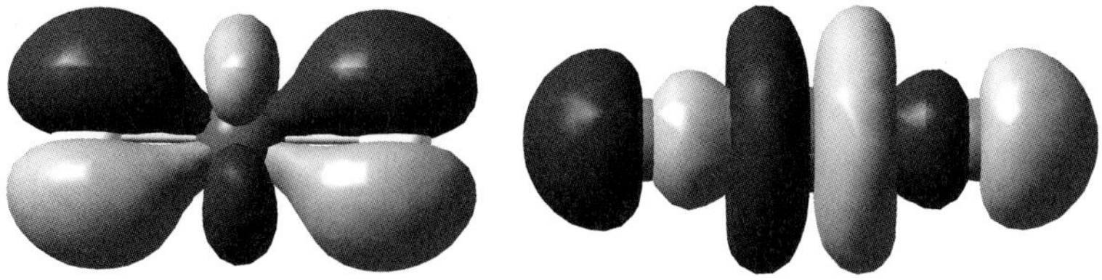

(a) (b)

Figure 11.4 The two highest occupied orbitals of π_u (a) and σ_u (b) symmetry of UO_2^{2+} showing large 5*f*-participation in the U–O bond (see Ref. [113] for details).

6*d*) in chemical bonding [9]. Spin–orbit coupling leads to a destabilization in this dimer; that is, from 40.2 kcal mol^{-1} at the scalar-relativistic down to 30.5 kcal mol^{-1} for the dissociation energy [9]. Liu *et al.* [114] pointed out that the low stability of the lighter actinocene anions $M(C_8H_8)_2^-$ is a relativistic effect as the 5*f*-orbitals are relativistically destabilized, thus lowering the electron affinity in the actinocenes.

In ligand-field theory, spin–orbit coupling is treated as a perturbation in strong ligand fields. This leads to the notation that spin–orbit coupling is quenched in strong ligand fields, as different *j*-states of the same *l* quantum number mix in a molecular field. This may be rationalized by considering the static dipole polarizabilities for the different *j*-levels. It is evident that orbitals in the lower $j=l-1/2$ state are less polarizable than the ones in the corresponding high angular momentum state of $j=l+1/2$. For example, for Tl the $^2P_{1/2}$ ground state has a polarizability of 51.6 a.u., whereas the excited $^2P_{3/2}$ state of Tl has larger polarizabilities of 105.0 a.u. ($M_j=3/2$) and 55.7 a.u ($M_j=1/2$). [115]. In an increasing electrostatic field these atomic levels change in energy according to the quadratic Stark shift. This allows for better mixing of $p_{1/2}$- and $p_{3/2}$-orbitals to form σ- or π-bonds. Moreover, spin–orbit-coupled valence levels can become depleted in density because of electron-withdrawing ligands, as in the case of TlF_3. Large spin–orbit coupling effects are expected where strong ligand fields are absent, for example, for the diatomic compounds TlH, Tl_2, or Pb_2. In molecules of Tl and Pb, the spin–orbit destabilization can be estimated from spin–orbit averaging of known atomic *J*-states [72], which results in an atomic spin–orbit stabilization of 0.64 eV for the $^2P_{1/2}$ state of Tl and 1.06 eV for the $^3P_0$ state of Pb compared to the scalar-relativistic level. As spin–orbit coupling at the molecular level is much reduced in a strong ligand field compared to the atomic limit, these estimates represent upper bounds. As a result, we obtain rather small dissociation energies of 2.06 eV for TlH, 0.43 eV for Tl_2, and 0.86 eV Pb_2. For comparison, without spin–orbit coupling at the scalar-relativistic level one obtains dissociation energies of 2.52 eV for TlH, 1.03 eV for Tl_2, and 1.83 eV Pb_2 [113, 116]. As a result of this net spin–orbit destabilization, the dimers Tl_2 and Pb_2 have the lowest dissociation energies in the main group element series (except for the superheavy elements). This translates into low cohesive energies for the bulk metals, where we see a monotonic decreasing trend in the Group 14 series of elements down to superheavy element Fl. Bulk lead has a calculated cohesive energy of 1.98 eV, which compares to 3.01 eV at the scalar-relativistic level and 3.22 eV at the nonrelativistic level [117]. For comparison, bulk Fl ($Z = 114$) undergoes a very large spin–orbit destabilization resulting in a cohesive energy of only 0.50 eV. Furthermore, because of scalar-relativistic effects, bulk Pb changes from the diamond structure to *fcc*, and because of spin–orbit coupling bulk Fl changes further from *fcc* to *hcp* [117]. Interestingly, low cohesive energies do not necessarily translate into a low melting point T_M as we have T_M(Tl) = 303°C > T_M(In) = 156°C, and T_M(Pb) = 327°C > T_M(Sn) = 232°C, which clearly is a many-body effect and needs to be analyzed further.

Jahn–Teller (JT) geometric distortions can be rather large even for small energy differences between high (undistorted) symmetry state and the lower symmetry distorted state, often resulting in a dynamic rather than a static JT effect [118]. It is clear that for such systems relativistic effects can have substantial influence to JT distortions. Spin–orbit effects can often quench a JT distortion. For example, both the $^1E'(e'^2)$ and the $^3E'(a''^1_2e'^1)$ states in Pb_3 (D_{3h}) undergo first-order $E \otimes e$ JT distortion to the corresponding $^1A_1/^1B_2$ and $^3A_2/^3B_1$ states, respectively, with bonding angles away from the ideal angle of $\alpha = 60^\circ$. If, however, we assume strong spin–orbit coupling and Pb in the $J = 0$ ($6p^2_{1/2}$) closed-shell state, we expect restoration of the high symmetry D_{3h} state. In fact, for the intermediate coupling scheme the situation is far more complicated, with the spin–orbit-coupled energy hypersurface showing multiple minima [119]. In the scalar-relativistic domain, the $^3B_2$ ($\alpha = 60^\circ$) and $^1A_1$ ($\alpha = 86.5^\circ$) states compete for the ground state, while in the spin–orbit-coupled domain these states mix and they become separated with A_1 being the ground state at an angle of $\alpha = 106.9^\circ$ [119]. A similar situation is found for Pb_6, where spin–orbit coupling suppresses completely the expected JT distortion from the high symmetry O_h structure to the D_{4h} symmetry [120]. For a detailed discussion of relativistic effects in Jahn–Teller or Renner–Teller systems the reader is referred to the work by Domcke and coworkers [121–123].

Finally, we mention a few more spectacular examples of spin–orbit effects in superheavy-element-containing compounds. Mixing of $a'_{1\text{HOMO}} \otimes e'_{\text{LUMO}}$ orbitals results in a second-order JT effect for the Group 17 trihalides EX_3 (such as IF_3) yielding potential energy hypersurfaces with a Mexican hat topology. These molecules distort from a high symmetry D_{3h} geometry to a T-shaped C_{2v} structure. Both scalar and spin–orbit effects have considerable influence in predicting accurately such JT hypersurfaces [124, 125]. For $(117)F_3$ the situation is quite different [124, 126]. Here the $7p_{3/2}$ electrons are more loosely bound due to relativistic effects, and the molecule can approximately be described by ionic bonding, that is, 117^{+3} bonding to three negatively charged fluorine atoms. The ion 117^{+3} is spherically symmetrical because the $7p^2_{1/2}$ configuration is of closed shell due to strong spin–orbit coupling. This predicts a D_{3h} high symmetry structure rather than a JT distortion to a T-shaped structure [124, 125]. Another example is $(118)F_4$, which changes from a planar D_{4h} structure predicted at the scalar-relativistic level to a tetrahedral T_d structure on inclusion of spin–orbit effects [127].

11.4 Conclusions

Relativistic quantum chemistry has advanced to such an extent that accurate predictions of molecular properties can now be made for heavy-element-containing compounds. This allows for the discussion of relativistic effects comparing properties obtained from the Dirac–Coulomb–(Breit+QED) equation with those from a nonrelativistic Schrödinger treatment. For a qualitative discussion it is convenient to categorize relativistic effects into direct, indirect, and spin–orbit effects. Valence orbitals important for chemical bonding are directly affected by the action of the relativistic perturbation operator leading to a contraction/stabilization of all orbitals (*direct-relativistic effect*). This leads to additional screening of the nucleus, and a possible relativistic expansion/destabilization (*indirect-relativistic effect*) of diffuse higher angular momentum orbitals. *Spin–orbit coupling* lifts the degeneracy in the shells with $l > 0$, and the corresponding level splitting can reach the size of typical bond dissociation energies in heavy-element-containing compounds. A delicate interplay between relativistic effects in different shells and the resulting nuclear screening is responsible for the large relativistic *s*-shell contraction observed for the late transition elements such as the Group 11 or 12 elements of the periodic table. The study of the chemistry and bonding of the sixth row elements requires the inclusion of relativistic effects, and some unusual anomalies down the periodic table can be attributed to such effects. Going even further down the periodic table, the chemistry of superheavy elements cannot be understood without inclusion of special relativity. In fact, the correct placement of the elements in the periodic table beyond element 120 requires the inclusion of relativistic effects [23, 128]. Dirac had laid the foundation of the relativistic quantum theory as early as in 1926, but only in recent history has it made it into mainstream chemistry. Over the past 30 years or so we have realized that relativistic effects in molecular properties and chemical bonding of heavy-element-containing compounds play a much greater role than anticipated originally [44].

Acknowledgments

This work was supported through a Humboldt Research Prize of the Alexander von Humboldt Foundation (Bonn). The author wishes to thank Prof. Gernot Frenking (Marburg) for being such a wonderful host and Daniel Götz (Darmstadt) for critically reading this chapter.

References

1. Pauling, L. (1960) *The Nature of the Chemical Bond*, Cornell University Press.
2. Weinhold, F. and Landis, C. (2005) *Valency and Bonding*, Cambridge University Press, Cambridge.
3. Ruedenberg, K. (1962) *Rev. Mod. Phys.*, **34**, 326–376.
4. Kutzelnigg, W. (1990) in *Theoretical Models of Chemical Bonding*, vol. **2** (ed. Z.B. Maksić), Springer, Berlin, pp. 1–44.

5. Ruedenberg, K. and Schmidt, M.W. (2009) *J. Phys. Chem. A*, **113**, 1954–1968.
6. Bader, R.F.W. (1990) *Atoms in Molecules: A Quantum Theory*, Oxford University Press, Oxford.
7. Shaik, S., Danovich, D., Wu, W., Su, P., Rzepa, H.S., and Hiberty, P.C. (2012) *Nat. Chem.*, **4**, 195–200; see also Grunenberg, J. (2012) *Nat. Chem.*, **4**, 154-155.
8. Shaik, S., Rzepa, H.S., and Hoffmann, R. (2013) *Angew. Chem. Int. Ed.*, **52**, 3020-3033.
9. Gagliardi, L. and Roos, B.O. (2005) *Nature*, **433**, 848–851.
10. Ritter, S.K. (2007) *C&N News*, **85** (5), 37–40.
11. Kerridge, A., Coates, R., and Kaltsoyannis, N. (2009) *J. Phys. Chem. A*, **113**, 2896–2905.
12. Kaltsoyannis, N., *Inorg. Chem.* 2013, **52**, 3407–3413.
13. Weinhold, F. (2012) *J. Comput. Chem.*, **33**, 2363–2379.
14. Primas, H. (1981) *Chemistry, Quantum Mechanics and Reductionism – Perspectives in Theoretical Chemistry*, Lecture Notes in Chemistry, vol. **24**, Springer, Berlin.
15. Scerri, E.R. (2000) *J. Chem. Educ.*, **77**, 1492–1494.
16. Bader, R.F.W. (2010) *J. Phys. Chem. A*, **114**, 7431–7444.
17. Coulson, C.A. (1993) What is a chemical bond?, in *From Chemical Philosophy to Theoretical Chemistry*, vol. **261** (ed. M. Jo Nye).
18. Franck, P. and Rothe, H. (1911) *Ann. Phys.*, **34**, 825–855.
19. Landau, L.D. and Peierls, R. (1930) *Z. Phys.*, **62**, 188–200.
20. Feshbach, H. and Villars, F. (1958) *Rev. Mod. Phys.*, **30**, 24–45.
21. Dyall, K.G. and Fægri, K. (2007) *Introduction to Relativistic Quantum Chemistry*, Oxford University Press, Oxford.
22. Lindgren, I. (2011) *Relativistic Many-Body Theory*, Springer, New York.
23. Pyykkö, P. (2012) *Chem. Rev.*, **112**, 371–384.
24. Plesset, M.S. (1932) *Phys. Rev.*, **41**, 278–290.
25. Thaller, B. (1992) *The Dirac Equation*, Springer, Berlin.
26. Fanchi, J.R. and Wilson, W.J. (1983) *Found. Phys.*, **13**, 571–605.
27. Liu, W. (2012) *Phys. Chem. Chem. Phys.*, **14**, 35–48.
28. Uzan, J.-P. (2003) *Rev. Mod. Phys.*, **75**, 403–455.
29. Peng, D. and Reiher, M. (2012) *Theor. Chem. Acc.*, **131**, 1081–1101.
30. Autschbach, J. (2012) *J. Chem. Phys.*, **136**, 150902-1–150902-15.
31. Liu, W. (2010) *Mol. Phys.*, **108**, 1679–1706.
32. Nakajima, T. and Hirao, K. (2012) *Chem. Rev.*, **112**, 385–402.
33. Saue, T. (2011) *ChemPhysChem*, **12**, 3077–3094.
34. Iliaš, M., Kellö, V., and Urban, M. (2010) *Acta Phys. Slovaca*, **60**, 259–391.
35. Reiher, M. and Wolf, A. (2009) *Relativistic Quantum Chemistry*, Wiley-VCH Verlag GmbH, Weinheim.
36. Schwerdtfeger, P. (2010) in *Computational Spectroscopy* (ed. J. Grunenberg), Wiley-VCH Verlag GmbH, Weinheim, pp. 201–221.
37. Schwerdtfeger, P. (2011) *Chem. Phys. Chem.*, **12**, 3143–3155.
38. Dolg, M. and Cao, X. (2012) *Chem. Rev.*, **112**, 403–480.
39. Schwerdtfeger, P., Brown, J.R., Laerdahl, J.K., and Stoll, H. (2000) *J. Chem. Phys.*, **113**, 7110–7118.
40. Baerends, E.J., Schwarz, W.H.E., Schwerdtfeger, P., and Snijders, J.G. (1990) *J. Phys. B: At. Mol. Opt. Phys.*, **23**, 3225–3240.
41. Schwarz, W.H.E. (1990) in *The Concept of the Chemical Bond* (ed. Z.B. Maksic), Springer, Berlin, pp. 593–643.
42. Wang, S.-G., Liu, W., and Schwarz, W.H.E. (2002) *J. Phys. Chem. A*, **106**, 795–803.
43. Rosen, M.E. (1961) *Relativistic Electron Theory*, John Wiley & Sons, Inc, New York.
44. Pyykkö, P. (1988) *Chem. Rev.*, **88**, 563–594.
45. Pyykkö, P. (2012) *Annu. Rev. Phys. Chem.*, **63**, 45–64.
46. Pershina, V. (2011) *Radiochim. Acta*, **99**, 459–476.

47. Piszczatowski, K., Łach, G., Przybytek, M., Komasa, J., Pachucki, K., and Jeziorski, B. (2009) *J. Chem. Theory Comput.*, **5**, 3039–3048.

48. Cencek, W., Przybytek, M., Komasa, J., Mehl, J.B., Jeziorski, B., and Szalewicz, K. (2012) *J. Chem. Phys.*, **136**, 224303-1–224303-31.

49. Liu, J., Salumbides, E.J., Hollenstein, U., Koelemeji, J.C.J., Eikema, K.S.E., Ubachs, W., and Merkt, F. (2009) *J. Chem. Phys.*, **130**, 174306-1–174306-8.

50. Zhang, Y.P., Cheng, C.H., Kim, J.T., Stanojevic, J., and Eyler, E.E. (2004) *Phys. Rev. Lett.*, **92**, 203003-1–203003-4.

51. Christensen, N.E. and Kollar, J. (1983) *Solid State Commun.*, **46**, 727–730.

52. Calvo, F., Pahl, E., Schwerdtfeger, P., and Spiegelman, F. (2012) *J. Chem. Theory Comput.*, **8**, 639–648.

53. Christensen, N.E. and Seraphin, B.O. (1971) *Phys. Rev. B*, **4**, 3321–3344.

54. Romaniello, P. and de Boeij, P.L. (2005) *J. Chem. Phys.*, **122**, 164303-1–164303-6.

55. Sansonetti, J.E. and Martin, W.C. (2005) *J. Phys. Chem. Ref. Data*, **34**, 1559–2259.

56. Schwerdtfeger, P., Boyd, P.D.W., Brienne, S., and Burrell, A.K. (1992) *Inorg. Chem.*, **31**, 3411–3422.

57. Lein, M., Hashmi, S.K., and Schwerdtfeger, P. (2010) *Organometallic*, **29**, 2206–2210.

58. Schwarz, H. (2003) *Angew. Chem. Int. Ed.*, **42**, 4442–4454.

59. Ahuja, R., Blomqvist, A., Larsson, P., Pyykkö, P., and Zaleski-Ejgierd, P. (2011) *Phys. Rev. Lett.*, **106**, 018301-1–018301-4.

60. Schwarz, W.H.E., van Wezenbeck, E.M., Baerends, E.J., and Snijders, J.G. (1989) *J. Phys. B: At. Mol. Opt. Phys.*, **22**, 1515–1530.

61. Note that in Hartree's 1958 paper on the "Variation of atomic wave functions with atomic number" there is only a small hint that the nonrelativistic Schrödinger equation is not physically significant for large values of the atomic number (a) Hartree, D.R. (1958) *Rev. Mod. Phys.*, **30**, 63–68; However, A. O. Williams carrying out Hartree calculations on the closed-shell atom Cu^+ noted as early as in that *The charge density of each single electron turns out to resemble that for the nonrelativistic case, but with the maxima "pulled in" and raised .... The size of the relativistic corrections appears to be just too small to produce important corrections in atomic form factors or other secondary characteristics of the whole atom. ... However, it must be noticed that copper is a relatively light ion, and the corrections for such an ion as mercury would be enormously greater* (b) Williams, A.O. (1940) *Phys. Rev.*, **58**, 723–726. Later S. Cohen in 1955 (Ph.D. Thesis, Cornell, 1955); and Mayers D. F. in 1957 pointed out that relativistic effects are very important even in the valence region of heavy elements such as mercury: *The values of the energy parameter ε ... show a substantial increase over the nonrelativistic values, the relation depending more on the l than on the n value of the wave function. The increase is particularly marked for the inner shells, as was expected. For the outer groups, indirect effects have become appreciable; the 5d group, which is known to be very sensitive to slight changes in the potential, has been so affected by the contraction of the inner groups, that its energy parameter has in fact decreased* (c) Mayers, D.F. (1957) *Proc. R. Soc. London, Ser. A*, **241**, 93–109; Later in 1967 V. M. Burke and I. P. Grant stated that *Little attention appears to have been paid to the effect of relativity on atomic wave functions...* (d) Burke, V.M. and Grant, I.P. (1967) *Proc. Phys. Soc.*, **90**, 297–314; Of course, relativistic quantum chemistry started with P. Pyykkö and J. P. Desclaux's seminal work on relativistic quantum chemistry in the early seventies (for a review see) (e) Pyykkö, P. and Desclaux, J.P. (1979) *Acc. Chem. Res.*, **12**, 276–281.

62. Glashow, S.L. (1988) *Interactions*, Warner Books, New York.

63. Pyykkö, P. (1978) *Adv. Quantum Chem.*, **11**, 353–409.

64. Balasubramanian, K. and Pitzer, K.S. (1987) *Adv. Chem. Phys.*, **57**, 287–319.

65. Corbett, J.D. (1962) *Inorg. Chem.*, **1**, 700–703.

66. Schwerdtfeger, P. and Lein, M. (2009) in *Gold Chemistry. Current Trends and Future Directions* (ed. F. Mohr), Wiley-VCH Verlag GmbH, Weinheim, pp. 183–247.
67. Autschbach, J., Siekierski, S., Seth, M., Schwerdtfeger, P., and Schwarz, W.H.E. (2002) *J. Comput. Chem.*, **23**, 804–813.
68. Neogrády, P., Kellö, V., Urban, M., and Sadlej, A.J. (1997) *Int. J. Quantum Chem.*, **63**, 557–565.
69. Eliav, E., Kaldor, U., Schwerdtfeger, P., Hess, B., and Ishikawa, Y. (1994) *Phys. Rev. Lett.*, **73**, 3203–3206.
70. Eliav, E., Kaldor, U., and Ishikawa, Y. (1994) *Phys. Rev. Lett.*, **49**, 1724–1729.
71. Eliav, E., Kaldor, U., Ishikawa, Y., and Pyykkö, P. (1996) *Phys. Rev. Lett.*, **77**, 5350–5352.
72. Moore, C.E. (1958) National Bureau of Standard (U.S.) Circ. No. 467. *Atomic Energy Levels*, U.S. GPO, Washington, DC.
73. Hotop, H. and Lineberger, W.C. (1975) *J. Phys. Chem. Ref. Data*, **14**, 731–750.
74. Haberditzl, W. (1979) *Quantenchemie – Komplexverbindungen*, Hüthig Verlag, Heidelberg.
75. Pople, J.A. (1960) *Mol. Phys.*, **3**, 16–22.
76. Eliav, E., Kaldor, U., Ishikawa, Y., Seth, M., and Pyykkö, P. (1994) *Phys. Rev. A*, **53**, 3926–3933.
77. White, H.E. (1931) *Phys. Rev.*, **38**, 513–520.
78. Szabo, A. (1969) *J. Chem. Educ.*, **46**, 678–680.
79. Gäggeler, H.W. (2011) *Radiochim. Acta*, **99**, 503–513.
80. Bethe, H.A. (1929) *Ann. Phys.*, **3**, 133–206.
81. Watanabe, H. (1966) *Operator Methods in Ligand Field Theory*, Prentice Hall, New Jersey.
82. Ludwig, W. and Falter, C. (1996) *Symmetries in Physics*, Springer, Berlin.
83. Majumdar, D., Roszak, S., and Leszczynski, J. (2010) in *Relativistic Methods for Chemists* (eds M. Barysz and Y. Ishikawa), Springer, Dordrecht, pp. 373–405.
84. Visscher, L. (1996) *Chem. Phys. Lett.*, **253**, 20–26.
85. Barysz, M. and Ishikawa, Y. (2010) *Relativistic Methods for Chemists*, Springer, Dordrecht.
86. Takeuchi, N., Chan, C.T., and Ho, K.M. (1989) *Phys. Rev. B*, **40**, 1565–1570.
87. (2013) *CRC Handbook of Chemistry and Physics*, Haynes W. M. (ed.), 94th edition, CRC Press, Taylor and Francis, Boca Radon.
88. Moruzzi, V.L., Janak, J.F., and Williams, A.R. (1978) *Calculated Electronic Properties of Metals*, Pergamon Press, New York.
89. Schwerdtfeger, P. (1991) *Chem. Phys. Lett.*, **183**, 457–463.
90. Skottke-Klein, M., Böttcher, A., Imbeck, R., Kennou, S., Morgante, A., and Ertl, G. (1991) *Thin Solid Films*, **203**, 131–145.
91. Söhnel, T., Hermann, H.L., and Schwerdtfeger, P. (2001) *Angew. Chem. Int. Ed.*, **40**, 4381–4385.
92. Biering, S., Hermann, A., Furthmüller, J., and Schwerdtfeger, P. (2009) *J. Phys. Chem. A*, **113**, 12427–12432; Erratum 2010, **114**, 8020.
93. Biering, S. and Schwerdtfeger, P. (2012) *J. Chem. Phys.*, **136**, 034504-1–034504-15.
94. Wesendrup, R., Laerdahl, J.K., and Schwerdtfeger, P. (1999) *J. Chem. Phys.*, **110**, 9457–9462.
95. Schröder, D., Hrusak, J., Tornieporth-Oetting, I.C., Klapötke, M., and Schwarz, H. (1994) *Angew. Chem.*, **33**, 212–214.
96. Schwerdtfeger, P. and Dolg, M. (1991) *Phys. Rev. A*, **43**, R1644–R1646.
97. Schwerdtfeger, P. (2003) *Angew. Chem. Int. Ed.*, **42**, 1892–1895.
98. Wesendrup, R. and Schwerdtfeger, P. (2000) *Angew. Chem. Int. Ed.*, **39**, 907–910.
99. Eijnhoven, J.C.M.T. and Verschoor, G.C. (1987) *Mater. Res. Bull.*, **9**, 1667–1670.
100. Seth, M., Cooke, F., Pelissier, M., Heully, J.-L., and Schwerdtfeger, P. (1998) *J. Chem. Phys.*, **109**, 3935–3943.
101. Parr, R.G. and Pearson, R.G. (1983) *J. Am. Chem. Soc.*, **105**, 7512–7516.

102. Hancock, R.D., Bartolotti, L.J., and Kaltsoyannis, N. (2006) *Inorg. Chem.*, **45**, 10780–10785.
103. Schwerdtfeger, P., Dolg, M., Schwarz, W.H.E., Bowmaker, G.A., and Boyd, P.D.W. (1989) *J. Chem. Phys.*, **91**, 1762–1774.
104. Schwerdtfeger, P., Heath, G.A., Dolg, M., and Bennett, M.A. (1992) *J. Am. Chem. Soc.*, **114**, 7518–7527.
105. Seth, M., Faegri, K., and Schwerdtfeger, P. (1998) *Angew. Chem., Int. Ed. Engl.*, **37**, 2493–2496.
106. Seth, M., Dolg, M., Fulde, P., and Schwerdtfeger, P. (1995) *J. Am. Chem. Soc.*, **117**, 6597–6598.
107. Kaupp, M. and von Schnering, H.G. (1993) *Angew. Chem. Int. Ed.*, **32**, 861–863.
108. Kaupp, M., Dolg, M., Stoll, H., and von Schnering, H.G. (1994) *Inorg. Chem.*, **33**, 2122–2131.
109. Wang, X., Andrews, L., Riedel, S., and Kaupp, M. (2007) *Angew. Chem. Int. Ed.*, **46**, 8371–8375.
110. Seth, M., Schwerdtfeger, P., and Dolg, M. (1997) *J. Chem. Phys.*, **106**, 3623–3632.
111. Kaltsoyannis, N., Hay, P.J., Li, J., Blaudeau, J.P., and Bursten, B.E. (2005) in *The Chemistry of the Actinide and Transactinide Elements*, 3rd edn, vol. **3** (eds L.R. Morss, N.M. Edelstein, and J. Fuger), Springer, New York, pp. 1893–2012.
112. Gagliardi, L., Roos, B.O., Malmqvist, P.-Å., and Dyke, J.M. (2001) *J. Phys. Chem. A*, **105**, 10602–10606.
113. Roos, B.O. and Malmqvist, P.-Å. (2004) *Phys. Chem. Chem. Phys.*, **6**, 2919–2927.
114. Liu, W., Dolg, M., and Fulde, P. (1997) *J. Chem. Phys.*, **107**, 3584–3591.
115. Borschevsky, A., Zelovich, T., Eliav, E., and Kaldor, U. (2012) *Chem. Phys.*, **395**, 104–107.
116. Lee, H.-S., Han, Y.-K., Kim, M.C., Bae, C., and Lee, Y.S. (1998) *Chem. Phys. Lett.*, **293**, 97–102.
117. Hermann, A., Furthmüller, J., Gäggeler, H., and Schwerdtfeger, P. (2010) *Phys. Rev. B*, **82**, 155116-1–155116-8.
118. Bersuker, I.B. (2001) *Chem. Rev.*, **101**, 1067–1114.
119. Balasubramanian, K. and Majumdar, D. (2001) *J. Chem. Phys.*, **115**, 8795–8809.
120. Armbruster, M.K., Weigend, F., van Wüllen, C., and Klopper, W. (2008) *Phys. Chem. Chem. Phys.*, **10**, 1748–1756.
121. Poluyanov, L.V. and Domcke, W. (2004) *Chem. Phys.*, **301**, 111–127.
122. Mishra, S., Vallet, V., Poluyanov, L.V., and Domcke, W. (2005) *J. Chem. Phys.*, **123**, 124104-1–124104-8.
123. Poluyanov, L.V. and Domcke, W. (2010) *Chem. Phys.*, **374**, 86–93.
124. Schwerdtfeger, P. (1996) *J. Phys. Chem.*, **100**, 2968–2973.
125. Bae, C.B., Han, Y.K., and Lee, Y.S. (2003) *J. Phys. Chem. A*, **107**, 852–858.
126. Yang, D.-D. and Wang, F. (2012) *Phys. Chem. Chem. Phys.*, **14**, 15816–15825.
127. Nash, C.S. and Bursten, B.E. (1999) *Angew. Chem. Int. Ed.*, **38**, 151–153.
128. Pyykkö, P. (2011) *Phys. Chem. Chem. Phys.*, **13**, 161–168.

Index

The Chemical Bond: Fundamental Aspects of Chemical Bonding, First Edition.
Edited by Gernot Frenking, Sason Shaik.

d